AF676285

Small Molecules in Neurodegeneration

Neurodegenerative diseases, like Alzheimer's, Parkinson's, and Huntington's, cast a long shadow over millions of lives, stealing memories, independence, and futures. The rising tide of these age-related disorders threatens to overwhelm healthcare systems worldwide, prompting an urgent search for effective treatments. While current options offer temporary relief, they fail to halt the relentless march of these devastating conditions. However, a beacon of hope shines at the intersection of cutting-edge neuroscience and the world of small molecules. These tiny chemical warriors, with their inherent advantages of cost-effectiveness, scalability, and diverse functionalities, are emerging as powerful weapons in the fight against neurodegeneration. This book stands as a testament to this increasing revolution. It investigates deep into the intricacies of small molecules, exploring their potential to unravel the mysteries of neurodegenerative diseases and pave the way for innovative therapeutic interventions.

Key Features:

- Provides a comprehensive and accessible overview of the emerging field of small molecules in neurodegenerative diseases.
- Bridges the gap in current literature by addressing the specific impact of small molecules on these devastating conditions.
- Offers an in-depth exploration of the various mechanisms by which small molecules can combat neurodegeneration.
- Serves as a valuable resource for a wide audience, from students to professionals in the field.
- Lays the foundation for a deeper understanding of neurodegenerative diseases and unlocks new avenues for the development of effective therapies.

Small Molecules in Neurodegeneration

Edited by
Nawab John Dar and Shahnawaz Ali Bhat

CRC Press
Taylor & Francis Group
Boca Raton London New York

CRC Press is an imprint of the
Taylor & Francis Group, an **informa** business

Designed cover image: Shutterstock Illustration Contributor Boris Nikolaev Ilchev

First edition published 2025
by CRC Press
2385 NW Executive Center Drive, Suite 320, Boca Raton FL 33431

and by CRC Press
4 Park Square, Milton Park, Abingdon, Oxon, OX14 4RN

CRC Press is an imprint of Taylor & Francis Group, LLC

© 2025 selection and editorial matter, Nawab John Dar and Shahnawaz Ali Bhat; individual chapters, the contributors

Reasonable efforts have been made to publish reliable data and information, but the author and publisher cannot assume responsibility for the validity of all materials or the consequences of their use. The authors and publishers have attempted to trace the copyright holders of all material reproduced in this publication and apologize to copyright holders if permission to publish in this form has not been obtained. If any copyright material has not been acknowledged please write and let us know so we may rectify in any future reprint.

Except as permitted under U.S. Copyright Law, no part of this book may be reprinted, reproduced, transmitted, or utilized in any form by any electronic, mechanical, or other means, now known or hereafter invented, including photocopying, microfilming, and recording, or in any information storage or retrieval system, without written permission from the publishers.

For permission to photocopy or use material electronically from this work, access www.copyright.com or contact the Copyright Clearance Center, Inc. (CCC), 222 Rosewood Drive, Danvers, MA 01923, 978-750-8400. For works that are not available on CCC please contact mpkbookspermissions@tandf.co.uk

Trademark notice: Product or corporate names may be trademarks or registered trademarks and are used only for identification and explanation without intent to infringe.

ISBN: 978-1-032-85728-2 (hbk)
ISBN: 978-1-032-85944-6 (pbk)
ISBN: 978-1-003-52061-0 (ebk)

DOI: 10.1201/9781003520610

Typeset in Times
by codeMantra

Contents

Preface

Neurodegenerative diseases, such as Alzheimer's, Parkinson's, and Huntington's, present an escalating threat to global health, especially as populations worldwide continue to age. These disorders are marked by a gradual deterioration of the nervous system, leading to significant cognitive, motor, and functional impairments that profoundly impact patients, families, and healthcare systems. Despite the growing prevalence of these diseases, current treatment options are limited, primarily focusing on symptom relief rather than halting or reversing the underlying disease processes. The need for more effective, accessible, and affordable therapies is more urgent than ever. *Small Molecules in Neurodegeneration* addresses this pressing need by presenting a comprehensive examination of the role of small molecules in the treatment of neurodegenerative diseases. Small molecules, due to their unique properties such as their ability to cross the blood–brain barrier, cost-effectiveness, ease of synthesis, and potential for oral administration have emerged as promising candidates for novel therapeutic approaches. Unlike larger, more complex biological therapies, small molecules offer the potential for broader accessibility, greater patient compliance, and fewer logistical challenges. This book explores the multifaceted roles small molecules play in neurodegeneration, from inhibiting protein aggregation, a key pathological hallmark in diseases like Alzheimer's and Parkinson's, to providing neuroprotection by regulating cell death pathways and mimicking neurotrophic factors. With a focus on the latest research, including natural products and repurposed drugs, this book investigates the therapeutic promise of small molecules and their future potential in clinical applications. Each chapter in this book has been carefully curated to offer a detailed exploration of the different aspects of small molecule therapy in neurodegenerative diseases. From understanding the basic pathology of these diseases to discussing the challenges and limitations of small molecule therapies, this book aims to bridge the gap between current knowledge and future therapeutic possibilities. Topics range from introduction of the diseases, pathophysiology, medicinal chemistry, and drug design to advances in drug delivery systems and the ethical and economic considerations of deploying these therapies.

This book is designed to be a valuable resource for a diverse audience, including undergraduate and graduate students in fields such as neuroscience, biochemistry, molecular biology, neuropharmacology, and medicinal chemistry. It is also aimed at researchers, clinicians, and professionals engaged in the study and treatment of neurodegenerative diseases. By compiling the latest knowledge and insights from experts in the field, this book offers a timely and significant contribution to understanding the role of small molecules in the fight against neurodegeneration. As research methodologies and technologies advance, it is crucial to consolidate and disseminate this emerging knowledge. We hope this book will inspire further research and innovation, paving the way for new therapeutic strategies and, ultimately, improved outcomes for patients suffering from neurodegenerative diseases.

Nawab John Dar and Shahnawaz Ali Bhat
Editors

The Editors

Dr. Nawab John Dar is at the Salk Institute for Biological Studies in La Jolla, California, where he is part of a research group led by Professor Pamela Maher. His research journey began after graduating from the CSIR – Indian Institute of Integrative Medicine in India, and he has since been dedicated to advancing the understanding of neurodegenerative disorders, particularly Alzheimer's disease. Dr. Dar's work focuses on exploring the neuroprotective potential of bioactive compounds and investigating the role of oxytosis/ferroptosis in Alzheimer's disease. His goal is to identify targets for effective therapies against Alzheimer's. Beyond his scientific endeavors, Dr. Dar is a committed educator and advocate for raising awareness about neurodegenerative diseases, emphasizing the importance of small molecules in targeting various aspects of neurodegeneration. His work not only sheds light on the potential of these molecules but also offers hope for effective treatments for millions affected by these conditions.

Dr. Shahnawaz Ali Bhat is an Assistant Professor in the Department of Zoology at Aligarh Muslim University (AMU), Aligarh, India. Currently, his research program focuses on deciphering the molecular and cellular processes/pathways involved in the pathogenesis of neurodegenerative diseases and metabolic disorders. Further, he is interested in neuropharmacological and neurotoxicological studies to screen or identify various bioactive molecules for effective prevention of neurodegenerative diseases and hypertension-induced neurodegeneration.

Contributors

Aida Adlimoghaddam
Department of Neurology, Dale and Deborah Smith Center for Alzheimer's Research and Treatment
Neuroscience Institute, Southern Illinois University School of Medicine
Springfield, Illinois, USA

Shakir Ahamad
Department of Chemistry
Aligarh Muslim University
Aligarh, Uttar Pradesh, India

Mushtaq Ahmad
Department of Radiation Oncology
Sher-i-Kashmir Institute of Medical Sciences
Soura, Jammu and Kashmir, India

Muzamil Ahmad
Pharmacology Division
CSIR-Indian Institute of Integrative Medicine
Srinagar, Jammu and Kashmir, India

Saeed Ahmed
Department of Medicine and Surgery
Rawalpindi Medical University
Rawalpindi, Punjab, Pakistan

Zabeer Ahmed
Pharmacology Division
CSIR-Indian Institute of Integrative Medicine
Jammu and Kashmir, India

Nargis Bano
Department of Zoology
Aligarh Muslim University
Aligarh, Uttar Pradesh, India

Gh Nabi Bhat
Aazam Healthcare
Tral.i.payeen, Jammu and Kashmir, India

Gh Rasool Bhat
Aazam Healthcare
Tral.i.payeen, Jammu and Kashmir, India

Javeed Ahmad Bhat
Department of Biochemistry and Biophysics
University of Rochester
Rochester, New York, USA

Sajad Ahmad Bhat
Department of Internal Medicine
Government Sub District Hospital
Tral-i-Bala, Jammu and Kashmir, India

Shahnawaz Ali Bhat
Department of Zoology
Aligarh Muslim University
Aligarh, Uttar Pradesh, India

Duilio Cervantes
Department of Neurology, Dale and Deborah Smith Center for Alzheimer's Research and Treatment
Neuroscience Institute, Southern Illinois University School of Medicine
Springfield, Illinois, USA

Nawab John Dar
CNB
SALK Institute of Biological Sciences
La Jolla, California, USA

Showkat Dar
CSIR – IMTech
Chandigarh, India

Mir Mohd Faheem
Pharmacology Division
CSIR-Indian Institute of Integrative Medicine
Srinagar, Jammu and Kashmir, India

Faaizah Fazal
Department of Zoology
Aligarh Muslim University
Aligarh, Uttar Pradesh, India

Bisma Gul
Pharmacology Division
CSIR-Indian Institute of Integrative Medicine
Srinagar, Jammu and Kashmir, India

Rukhsana Hassan
Department of Human Genetics
Advanced Centre for Human Genetics
Sher-i-Kashmir Institute of Medical Sciences
Soura, Jammu and Kashmir, India

Mir Mohammad Hussain
Department of Medical Oncology
Sher-i-Kashmir Institute of Medical Sciences
Soura, Jammu and Kashmir, India

Ishtiaq Jeelani
Division of Endocrinology and Metabolism, Department of Medicine
University of California
San Diego, California, USA

Urmilla John
School of Studies in Neuroscience
Jiwaji University
Gwalior, Madhya Pradesh, India

Umeair Kazime
Department of Pediatrics
Government Sub District Hospital
Tral-i-Bala, Jammu and Kashmir, India

Sameera Khan
Department of Zoology
Aligarh Muslim University
Aligarh, Uttar Pradesh, India

Abhishek Kumar Mishra
Government Shaheed Gendsingh College, Charama
Uttar Bastar Kanker, Chhattisgarh, India

Dipak Kumar
Department of Zoology
Munger University
Munger, Bihar, India

Rahul Kumar
Department of Pharmacy
School of Medical & Allied Sciences, Galgotias University
Greater Noida, Uttar Pradesh, India

Mudassir Jan Makhdoomi
Department of Pediatrics
All India Institute of Medical Sciences (AIIMS) New Delhi
New Delhi, India

Feroze Ahmad Mir
Department of Neurology
Sher-i-Kashmir Institute of Medical Sciences
Soura, Jammu and Kashmir, India

Sameer A. Mir
National Cancer Institute (NCI)
National Institutes of Health (NIH)
Bethesda, Maryland, USA

Prem Shankar Mishra
Department of Pharmacy
School of Medical & Allied Sciences, Galgotias University
Greater Noida, Uttar Pradesh, India

Allah Nawaz
Joslin Diabetes Center
Harvard Medical School, Harvard University
Cambridge, Boston, Massachusetts, USA

Sakshi Patel
Department of Pharmacy
School of Medical & Allied Sciences, Galgotias University
Greater Noida, Uttar Pradesh, India

S. K. Abdul Rahaman
Department of Pharmacy
School of Medical & Allied Sciences, Galgotias University
Greater Noida, Uttar Pradesh, India

Faheem Rasool
Department of Higher Education
Government of J&K
Jammu & Kashmir, India

Aamir Sharif
Department of Neurophysiology
Faculty of Medicine, University of Yamanashi
Kofu, Yamanashi, Japan

Shikha Sharma
CSIR – IMTech
Chandigarh, India

Abid Ali Sheikh
National Institute of Biologicals
Noida (Uttar Pradesh), India

Alisha Sheikh
Department of Ayurvedic Medicine and Surgery
Jammu Institute of Ayurveda and Research
University of Jammu
Jammu & Kashmir, Jammu, India

Aarizo Tabasum
Department of Obstetrics and Gynecology
Government Sub District Hospital
Tral-i-Bala, Jammu and Kashmir, India

Insha Zahoor
Department of Neurology
Henry Ford Health
Detroit, Michigan, USA

1 Introduction to Neurodegenerative Diseases

Urmilla John, Javeed Ahmad Bhat, Nawab John Dar, and Shahnawaz Ali Bhat

ABBREVIATIONS

AD Alzheimer's Disease
ALS Amyotrophic Lateral Sclerosis
AVs Autophagic Vacuoles
Aβ Amyloid-Beta
BBB Blood–Brain Barrier
CNS Central Nervous System
CSF Cerebrospinal Fluid
ER Endoplasmic Reticulum
FTD Frontotemporal Dementia
GBA Glucocerebrosidase
HD Huntington's Disease
MRI Magnetic Resonance Imaging
PD Parkinson's Disease
PET Positron Emission Tomography
ROS Reactive Oxygen Species

1.1 INTRODUCTION

Neurodegenerative diseases (NDDs) are a group of disorders characterized by the progressive degeneration and death of neurons. This leads to a decline in cognitive and motor functions, severely impacting the quality of life. The hallmark of these diseases is the accumulation of misfolded proteins and the formation of pathological aggregates in the brain and spinal cord (Kovacs, 2019). These aggregates disrupt normal cellular functions, leading to neuronal death and subsequent clinical symptoms. NDDs can be broadly classified based on the type of protein aggregates, the affected brain regions, and the clinical symptoms. For instance, Alzheimer's disease (AD) is primarily associated with amyloid-beta (Aβ) plaques and tau tangles in the hippocampus and cortex, leading to memory loss and cognitive decline (Dar and Glazner, 2020). Parkinson's disease (PD), on the contrary, is characterized by alpha-synuclein aggregates in the substantia nigra, resulting in motor symptoms such as tremors and rigidity (GBD 2017 US Neurological Disorders Collaborators, 2020). Other notable NDDs include Huntington's disease (HD), which involves the accumulation of mutant huntingtin (HTT) protein, and amyotrophic lateral sclerosis (ALS), characterized by the degeneration of motor neurons (Thorpe and Miller, 2021).

The classification of NDDs is multifaceted, involving clinical presentation, anatomical regions, and the types of cells affected. For example, frontotemporal dementia (FTD) primarily affects the frontal and temporal lobes, leading to changes in personality and behavior (Dugger & Dickson, 2017). In contrast, spinocerebellar ataxias affect the cerebellum, causing coordination and balance issues (GBD 2021 Collaborators, 2024). The molecular classification of these diseases is equally

DOI: 10.1201/9781003520610-1

important. Proteinopathies, such as tauopathies and synucleinopathies, are classified based on the specific proteins involved. Tauopathies include diseases like AD and progressive supranuclear palsy, where tau protein aggregates are the primary pathological feature (Alzheimer's Association, 2021). Synucleinopathies, including PD and multiple system atrophy, are characterized by alpha-synuclein aggregates (Parkinson's Foundation, 2021).

1.2 EPIDEMIOLOGY AND IMPACT

1.2.1 Global Prevalence and Incidence Rates

NDDs are a significant global health concern, affecting millions of people worldwide. Population aging is a key factor contributing to an increased global prevalence of neurodegenerative disorders and the associated social and economic impacts on health systems, caregivers, and society as a whole. AD alone affects approximately 50 million people globally, with the number expected to triple by 2050 (World Health Organization, 2024). The prevalence of these diseases increases with age, making them a major concern for aging populations. AD mainly affects individuals aged 65 and older, with prevalence rates doubling every half decade after this age (Tahami Monfared et al., 2022). The annual worldwide incidence of AD is estimated to be approximately 10 million new cases, and its diagnosis is more common in women owing to their longer life expectancy. The incidence rates of NDDs also vary by region and demographic factors. For instance, the incidence of AD is higher in developed countries, likely due to longer life expectancies (Wolters et al., 2019). PD is the second most common NDD and affects over 10 million people worldwide (Petersen and Walter, 2023) (Petersen and Walter, 2023) . Similar to AD, the prevalence and incidence of PD increase substantially with age, seeing a dramatic surge post 60 years old. There are approximately 15–20 new cases of PD per 100,000 people each year with a male predominance (Pringsheim et al., 2014; Lee and Gilbert, 2016). The prevalence and incidence of PD also vary between regions due to certain factors such as genetic vulnerability and environmental and lifestyle factors. It is often more prevalent in high-income countries than in lower- and middle-income countries (LMICs) (Dorsey et al., 2018; Ben-Shlomo et al., 2024).

HD is rare in occurrence but has a high genetic burden. The global prevalence of HD is estimated to be five to ten cases per 100,000 people, with the highest prevalence found in populations of European descent (Pringsheim et al., 2012). HD is unique among all of the common neurodegenerative disorders because it is entirely genetic, overwhelmingly caused by mutation in a single gene, and individuals typically develop symptoms between the ages of 30 and 50 years (Bhat et al., 2023) The worldwide incidence remains stable at approximately 0.5–1 new case per 100,000 people each year (Harper, 2002). However, HD has a significant impact on families and communities because of its hereditary pattern. Furthermore, ALS, also known as Lou Gehrig's disease, is another type of NDD with an estimated worldwide prevalence of two to three cases per 100,000 people (Chiò et al., 2013; Sacks et al., 2022). ALS shows little variation in incidence rates across different regions of the world, even though Western countries tend to have slightly higher rates. The incidence of ALS is consistently reported at one to two new cases per 100,000 people annually and has been linked to environmental and occupational exposures such as pesticides and heavy metals recognized as contributing factors to ALS etiology (Newell et al., 2022). In addition to ALS, other NDDs such as multiple sclerosis (MS) and FTD also contribute to the global disease burden. MS has been reported to affect around 2.8 million people across the world, with the highest incidences found in North America and Europe (Wallin et al., 2019; Walton et al., 2020). Although less common, FTD is prominently recognized as a significant cause of dementia in people below the age of 65, with the global prevalence reported at 15–22 per 100,000 people aged 45–64 years (von Gunten et al., 2019; Luukkainen, 2023)

Thus, the global incidence and prevalence rates of these neurodegenerative conditions are affected by a multitude of factors such as genetic susceptibility, aging populations, environmental exposures, and diagnostic advancements. The prevalence of these disorders is generally greater in higher-income countries, possibly as a result of better access to healthcare systems and longer life expectancies (Rocca et al., 2011). However, rates in LMICs are increasing faster. However, there are rising numbers of these diseases in LMICs due to advancements in healthcare and longevity that have likely increased the recognition of these conditions (Dorsey and Bloem, 2018; Nichols et al., 2022).

1.2.2 Economic and Social Impact

1.2.2.1 Healthcare Costs

The economic burden of NDDs is substantial, with costs associated with medical care, long-term care, and lost productivity. In the United States, the annual cost of AD and related dementias is estimated to be over $290 billion. This includes direct medical costs, such as hospitalizations and medications, as well as indirect costs, such as lost productivity and informal caregiving (Alzheimer's Association, 2021). Furthermore, disease management can also be expensive for PD, ALS, and other forms of NDDs, which require specialized care, medications, and assistive devices (Maresova et al., 2020).

1.2.2.2 Loss of Productivity

Considering that these diseases frequently impact people in the most productive part of life, there is the possibility for people to leave working life early and face early retirement, disability, and loss of wages. The financial burdens related to these diseases are not limited to direct medical care costs but also include indirect costs associated with the loss of productivity such as informal caregiving (Maresova et al., 2020; Tay et al., 2024). In cases of younger-onset conditions such as HD, the economic impact can be even greater because of the longer years of disability (To et al., 2023). From a macroeconomic perspective, sectors and industries are affected by the loss of experienced workers especially in countries with declining populations, where workforce shortages are a major concern (Johnson et al., 2011).

1.2.2.3 Long-Term Care and Support Services

In line with the progression of the disease, patients generally have to obtain specialized care in nursing homes/long-term care facilities. The costs of long-term care can be enormous and pose a significant financial burden on the affected families and the healthcare system (Rapp et al., 2015; Culberson et al., 2023). Full-time care in the home or institutional settings is needed by many individuals as their condition deteriorates. The informal caregiving, often provided by family members, represents a hidden cost and is often not included in economic analyses. A common source of strain for informal caregivers is the loss of income from work and the need to reduce work hours or resign from work to provide care (Armstrong et al., 2022; Hellmers, 2024). The economic value of the informal caregiving associated with dementia alone was estimated to be in the hundreds of billions of dollars globally (Farré et al., 2016; Cantarero-Prieto et al., 2020).

1.2.2.4 Societal Impact

The social impact of NDDs is also profound. These diseases not only affect patients but also their families, caregivers, communities, and entire societies. Caregivers often experience significant emotional and physical stress, leading to a higher risk of health problems (Roland and Chappell, 2019). NDDs, such as AD, PD, and ALS, cause progressive physical and cognitive deficits, which ultimately result in a degree of dependence and a requirement for long-term care. This places a burden on healthcare systems, as they provide continuous medical care, specialized assistance, and care to the patients and their caretakers. The primary caregivers of those diagnosed with NDDs face significant stress and challenges as they watch the deteriorated condition of their loved ones. It is common for caregivers to experience emotional, physical, and financial hardship, and the responsibility of caregiving can be overwhelming. These stressors can result in caregivers experiencing a downgrade in their quality of life and health issues. On the community level, the social consequences of NDDs include social isolation for both the individual and their caregivers (Lwi et al., 2017). There is a fear of being stigmatized because of cognitive and physical limitations that are associated with neurodegeneration and the fear of exclusion from social activities and support networks (National Alliance for Caregiving, 2020a). This context of social isolation can cause damage to the person's well-being and can deteriorate community solidarity and support structures.

1.3 PATHOPHYSIOLOGY

1.3.1 Molecular Mechanisms

1.3.1.1 Protein Misfolding and Aggregation

Protein misfolding and aggregation are central to the pathogenesis of many NDDs. Misfolded proteins can form toxic aggregates that disrupt cellular functions. For example, in AD, Aβ peptides misfold and aggregate into plaques, while tau proteins form neurofibrillary tangles (Hardy & Selkoe, 2002). Similarly, in PD, alpha-synuclein aggregates into Lewy bodies (Spillantini et al., 1997). The process of protein misfolding begins when proteins fail to achieve their native conformation. This can be due to genetic mutations, environmental factors, or cellular stress. Misfolded proteins expose hydrophobic regions that are normally buried inside the protein structure, leading to aggregation (Chiti & Dobson, 2006). These aggregates can be intracellular, such as tau tangles in AD, or extracellular, like amyloid plaques (Selkoe, 2003). The cellular response to misfolded proteins involves molecular chaperones and protein degradation pathways. Chaperones assist in refolding misfolded proteins, while the ubiquitin-proteasome system and autophagy pathways degrade irreversibly misfolded proteins (Ciechanover & Kwon, 2015). However, in NDDs, these protective mechanisms are often overwhelmed, leading to the accumulation of toxic aggregates (Ross & Poirier, 2004).

1.3.1.2 Oxidative Stress and Mitochondrial Dysfunction

Oxidative stress and mitochondrial dysfunction are closely linked to neurodegeneration. Mitochondria are the primary source of reactive oxygen species (ROS) in cells. Under normal conditions, ROS are neutralized by antioxidant defenses. However, in NDDs, excessive ROS production or impaired antioxidant defenses lead to oxidative stress (Lin & Beal, 2006 and Dar et al., 2017). Oxidative stress damages cellular components, including lipids, proteins, and DNA. In neurons, this can disrupt synaptic function and lead to cell death. Mitochondrial dysfunction further exacerbates oxidative stress by impairing ATP production and increasing ROS generation (Dar et al., 2018). In AD, for example, Aβ aggregates can interact with mitochondrial membranes, disrupting their function and increasing ROS production (Reddy & Beal, 2008). In PD, mutations in genes like PINK1 and Parkin, which are involved in maintaining mitochondrial quality, result in mitophagy defects and an accumulation of damaged mitochondria (Malpartida et al., 2021). In ALS, mitochondrial defects in motor neurons are related to disruptions in energy metabolism and a heightened vulnerability to oxidative stresses (Bozzo et al., 2017). These energy deficits are particularly harmful to motor neurons because they have a high metabolic rate and almost entirely rely on the ATP produced by the mitochondria for their function (Vandoorne et al., 2018).

1.3.2 Cellular Mechanisms

1.3.2.1 Neuronal Loss and Synaptic Dysfunction

Neuronal loss and synaptic dysfunction are key features of NDDs (Shah et al., 2016). Neuronal death can occur through various mechanisms, including apoptosis, necrosis, and autophagy. In AD, for instance, Aβ and tau aggregates trigger apoptotic pathways, leading to neuronal death

(Mattson, 2000). The degeneration of cholinergic neurons, which release the neurotransmitter acetylcholine, induces cognitive declines in AD (Chen et al., 2022). In PD, the death of dopaminergic neurons in the substantia nigra pars compacta induces typical motion symptoms like tremors, rigidity, and bradykinesia (Kumar et al., 2022). Furthermore, the loss of medium spiny neurons in the striatum in HD is associated with motor dysfunction, psychiatric disturbances, and cognitive deficits (Bergonzoni, et al., 2021). Additionally, the progressive degeneration of motor neurons in the brain stem and spinal cord in ALS pertains to the loss of muscle strength and paralysis (Ragagnin et al., 2019).

Synaptic dysfunction precedes neuronal loss and is a critical early event in neurodegeneration. Synapses are the sites of communication between neurons, and their dysfunction leads to impaired neuronal signaling. In AD, Aβ plaques and tau tangles accumulate and alter synaptic function, leading to long-term potentiation deficits and synaptic weakening (Rajmohan et al., 2017; Zhang et al., 2022). ALS likewise features synaptic dysfunction, most notably at the neuromuscular junction, where motor neurons interact with muscle fibers (Verma et al., 2022). In PD, alpha-synuclein aggregates disrupt synaptic vesicle trafficking, leading to synaptic dysfunction (Chung et al., 2009; Sharma et al., 2023). Similarly, in HD, mutant HTT protein disrupts synaptic function by interfering with synaptic proteins and signaling pathways (Zuccato et al., 2010; Barron et al., 2021).

1.3.2.2 Inflammation and Immune Responses

Neuroinflammation is a common feature of NDDs. It involves the activation of glial cells, including microglia and astrocytes, which release pro-inflammatory cytokines and chemokines. Chronic neuroinflammation can exacerbate neuronal damage and contribute to disease progression (Glass et al., 2010). Microglia, the sentinels of the brain, play a central role in this process. In response to injury, infection, or disease-related pathology, microglia are activated and adopt a pro-inflammatory phenotype. Activated microglia release various cytokines, chemokines, and ROS that can further exacerbate neuronal damage (Thameem Dheen et al., 2007). Chronic microglial activation, as observed in neuropathologies such as AD, PD, and ALS, leads to a maladaptive inflammatory response that is unfavorable for neuronal survival (Mizuno, 2015). In AD, for example, Aβ aggregates activate microglia, leading to the release of inflammatory mediators that further damage neurons (Heneka et al., 2015). Astrocytes, the most abundant glial cell in the central nervous system (CNS), also participate in neuroinflammation. In normal physiology, astrocytes play a supporting role in maintaining the neuronal function, regulating neurotransmitter levels, maintaining the integrity of the blood–brain barrier (BBB), and providing metabolic support (Lee et al., 2022). However, in the context of NDDs, astrocytes may become activated or reactive, and this is typically characterized by alterations in morphology and gene expression (Li et al., 2019). Reactive astrocytes can contribute to the inflammatory milieu, release neurotoxic factors, and potentially do not provide adequate support to neurons, all of which together could drive disease progression. For example, in PD, alpha-synuclein aggregates can activate astrocytes, leading to a pro-inflammatory response that contributes to neuronal death (Ransohoff, 2016).

1.3.2.3 Autophagy and Lysosomal Dysfunction

Autophagy is a cellular degradation pathway that is responsible for the removal of damaged organelles and misfolded proteins. The failure of autophagy is the pathological hallmark of many NDDs and eventually leads to the accumulation of toxic proteins and organelles. In addition, lysosomal dysfunction has also been observed in NDDs, which further exacerbate protein aggregate accumulation. In AD, dysfunctional autophagy causes the accumulation of autophagic vacuoles (AVs) in neurons. These AVs are frequently seen within dystrophic neurites, which contain undegraded proteins including Aβ and hyperphosphorylated tau, which are both critical in AD pathogenesis (Liu and Li, 2019). Lysosomal dysfunction in AD is characterized by the accumulation of lysosomal enzymes and other lysosomal constituents in neurons, indicating that the degradation process is impeded (Nixon, 2017). The accumulation of Aβ and tau protein levels disrupts cellular homeostasis, which in turn causes synaptic dysfunction, neuronal death, and cognitive deficits (Rajmohan and Reddy, 2017). Furthermore, the process of autophagy responsible for the elimination of alpha-synuclein is also seen to be impaired in PD patients and leads to the presence of these toxic protein aggregates. Certain mutations in genes associated with autophagy including LRRK2 and PINK1 have been linked to familial forms of PD, further emphasizing the importance of the autophagic process in the maintenance of neuronal health (Senkevich and Gan-Or, 2022). Lysosomal dysfunction has also been reported in PD, with mutations in the gene encoding glucocerebrosidase (GBA) being a significant risk factor in the disease. The mutations in GBA cause the accumulation of glucosylceramide, which can eventually lead to lysosomal malfunction and exacerbate alpha-synuclein aggregation (Beavan et al., 2013).

Autophagic failure in HD is linked to defects in cargo recognition, leading to the accumulation of mutant HTT and other hazardous proteins (Martinez-Vicente et al., 2010). Moreover, lysosomal malfunction is also evident in HD as indicated by reduced activity of lysosomal enzymes and defective clearance of autophagic cargo, which further contributes to neurotoxicity and degeneration (Fraldi et al., 2016). In ALS, autophagy has been shown to be hindered by mutations in several genes such as superoxide dismutase 1 (SOD1), transactive response DNA-binding protein (TDP-43), and fused-in sarcoma (FUS). In particular, the aggregation of the mutant SOD1 protein is impervious to the autophagy-lysosomal process and thus resistant to degradation (Ramesh and Pandey, 2017). Lysosomal dysfunction has also been observed in ALS with impaired acidification and a decrease in enzyme activity, contributing to the accumulation of incompletely digested cellular waste, which in turn would lead to the death of motor neurons (Lee et al., 2015).

1.3.2.4 Calcium Homeostasis and Signaling

Calcium ions have an important role in a variety of fundamental cellular processes, such as neurotransmitter release, gene expression, synaptic plasticity, and apoptosis. The accurate regulation of intracellular calcium levels is vital for the maintenance of neuronal function and survival. However, in the context of NDDs, there are often disturbances in calcium homeostasis that can have deleterious effects on neuronal health. Excessive influx of calcium normally results from excitotoxicity, oxidative stress, or dysfunction of calcium channels and pumps (Mehta et al., 2013 and Khan et al., 2024). The high levels of intracellular calcium activate a myriad of signaling pathways and enzymes like calpains and caspases, which result in the degradation of cytoskeletal proteins, mitochondrial dysfunction, and ultimately cell death via necrosis or apoptosis. For example, calpains are calcium-activated proteases and have been demonstrated to cleave major neuronal proteins and cause cytoskeletal disruption and loss of synaptic integrity (Metwally et al., 2023). Similarly, caspases are central to the execution of apoptosis and can become activated by increased levels of calcium, leading to the initiation of programmed cell death and progressive neuronal loss (Friedlander, 2023).

In AD, the pathology of both Aβ and tau are closely associated with the imbalance in calcium homeostasis, leading to synaptic dysfunction and neurodegeneration. Aβ peptides can create calcium-permeable pores in the plasma membrane or alternatively engage with N-methyl-D-aspartate receptors, resulting in an influx of calcium into neurons (Demuro et al., 2010). This abnormal calcium signaling results in synaptic dysfunction, oxidative stress, and the initiation of proteolytic enzymes, which lead to neurodegenerative changes in the brain (Plascencia-Villa and Perry, 2023). Hyperphosphorylated tau, a key feature of AD, disrupts microtubule functioning and impairs the ability of the endoplasmic reticulum (ER) to maintain calcium homeostasis, which together contributes to the dysregulation of calcium levels and increased vulnerability of neurons to neuropathological insults (Stutzmann, 2007; Guan et al., 2021). Similarly, PD is characterized by progressive degeneration of dopaminergic neurons in the substantia nigra. These neurons are especially sensitive to calcium dysregulation because they possess a unique and critical reliance on calcium channels, specifically L-type Voltage-gated calcium channels to maintain pacemaker activity (Leandrou et al., 2019). The continuous influx of calcium through these channels promotes metabolic stress on mitochondria, leading to mitochondrial dysregulation, increased ROS production, and impaired calcium buffering (Surmeier et al., 2010; Calì et al., 2011). Mitochondrial stress is believed to contribute to the selective vulnerability of dopaminergic neurons. Additionally, the abnormally folded alpha-synuclein (an aggregated protein that forms Lewy bodies) protein significantly impairs calcium homeostasis by disrupting both ER and mitochondria function, contributing to the ongoing oxidative stress and ultimately the death of neurons (Dar et al., 2022 and Erustes et al., 2022).

Furthermore, the dysregulation of calcium homeostasis is integral to the pathology of HD. In this disorder, the mutant HTT protein composed of pathologically expanded polyglutamine (polyQ) tracts interacts abnormally with inositol 1,4,5-triphosphate receptors on the ER and results in excessive calcium efflux from the ER to the cytoplasm (Tang et al., 2003; Czeredys, 2020). The excessive levels of calcium overwhelm mitochondrial calcium buffering, inducing mitochondrial calcium overload, ROS production, and cell death pathways (Duchen, 2000; Baev et al., 2022). Striatal neurons, which are highly vulnerable to perturbations of calcium signaling, are among the first to preferentially degenerate in HD, leading to the motor symptoms associated with this disease (Raymond, 2017). In ALS, pathological alterations in calcium signaling and defective buffering of calcium govern the selective vulnerability of motor neurons. Mutations in genes including SOD1, TDP-43, FUS, and C9orf72 associated with familial forms of ALS cause disruptions in mitochondrial functioning and impaired calcium buffering capacity (Dafinca et al., 2020). Motor neurons, which have a high metabolic demand and use calcium signaling heavily for neurotransmitter release and muscle contraction, are particularly vulnerable to changes in cellular calcium homeostasis (Grosskreutz et al., 2010). The resulting calcium overload results in downstream activation of calcium-dependent enzymes, including calpains and caspases that contribute to neuronal injury and apoptosis. Furthermore, abnormal accumulation of toxic protein aggregates and aberrant RNA processing further contribute to the dysregulation of calcium homeostasis, driving the selective degeneration of motor neurons that underlies the muscle weakness and atrophy seen in ALS disease (Droppelmann et al., 2014).

1.3.3 Disease-Specific Pathological Mechanisms

1.3.3.1 Amyloid Cascade Hypothesis (AD)

According to the amyloid cascade hypothesis, deposits of abnormal Aβ accumulation, in addition to its aggregation, initiate a pathological cascade of events that leads to further propagation of neurodegeneration in AD patients (Hardy and Selkoe, 2002). The initiation of Aβ deposits results from a disturbance in the balance of Aβ production and elimination within the CNS, which leads to the deposition of Aβ in the form of filaments and plaques (Kim and Priefer, 2020). Filaments and plaques of Aβ disturb cellular homeostasis, lead to a shortage of functional synapses, and result in a restorative response by astrocytes and recruitment of specialized cell types (microglia) as an inflammatory response. The overall dysfunction of normal neuronal or other cell activity in AD patients is due to the blocking of functional synapses from Aβ oligomers rather than the build-up of Aβ plaques (Tu et al., 2014). A result of this process is the hyper-phosphorylation of tau, which is believed to support or facilitate the progression of the neurodegenerative pathology that sweeps through the brain of an AD patient (Ittner and Götz, 2011).

1.3.3.2 Loss of Dopaminergic Neurons (PD)

PD is characterized by a selective loss or degradation of dopaminergic neurons residing in the substantia nigra pars compacta, causing a pervasive lack of dopamine in the striatum and leading to a change in motility and additional motor symptoms of PD. The loss of dopaminergic neurons in the substantia nigra is the result of multiple disease mechanisms, some of which include oxidative stress, mitochondrial dysfunction, impaired autophagy, proteasomal degradation, and inflammatory responses (Morris et al., 2024). Familial PD is primarily induced by genetic causes, including mutations in the synuclein alpha (SNCA)gene, LRRK2, and other genes (Klein and Westenberger, 2012). Lewy bodies are a pathological signature that is primarily composed of aggregated alpha-synuclein.

A possible factor in the neurodegenerative process of PD is the transmission of misfolded alpha-synuclein from one cell to another (Emin et al., 2022).

1.3.3.3 Polyglutamine Expansion (HD)

HD occurs due to the elongation of CAG trinucleotide repeats within the HTT gene, leading to the formation of an elongated polyglutamine stretch in the HTT protein (Podvin et al., 2019). The mutant HTT protein builds up as aggregates within neurons, disturbing various cellular processes that include transcription, proteolysis, and mitochondrial functioning (McColgan and Tabrizi, 2018). In HD, the primary degeneration occurs in the striatum, particularly in the medium spiny neurons of the caudate and putamen, leading to motor symptoms such as chorea and dystonia (Pandey and Rajamma, 2018). The mechanisms underlying the neurodegenerative aspects of HD remain largely elusive; however, it is reported that the mutant HTT protein has both toxic gain-of-function and normal HTT loss-of-function contributions (Mees et al., 2022).

1.3.3.4 Excitotoxicity and Glutamate Toxicity (ALS)

In ALS patients, there is an increased amount of glutamate excitotoxicity, thus increasing the specificity of damage experienced by motor neurons. In the CNS, glutamate acts as the principal excitatory neurotransmitter and is tightly regulated by various transporters, including the excitatory amino acid transporter 2 (EAAT2) expressed by astrocytes (Malik and Willnow, 2019). In ALS, defects in glutamate uptake or clearance within the synaptic cleft lead to the prolonged activation of the glutamate receptors (AMPA and NMDA receptors) and result in the influx of calcium ions into these neurons (Heath and Shaw, 2002). This influx of calcium triggers a cascade of enzymes such as proteases, lipases, and endonucleases, which leads to the degradation of important cellular components and consequently cell death (Dar et al., 2024 and Grosskreutz et al., 2010). The familial ALS is often associated with mutations in the superoxide dismutase 1 (SOD1) gene. Several genes, including SOD1, TARDBP (encodes TDP-43), FUS, and C9orf72, cause motor neuron degeneration in ALS by misfolding or aggregation, resulting in motor neuron toxicity (Lattante et al., 2013).

1.4 MAJOR NDDS

1.4.1 Alzheimer's Disease

1.4.1.1 Pathophysiology and Symptoms

AD is the most common cause of dementia, characterized by progressive cognitive decline, memory loss, and changes in behavior and personality. Early symptoms often include difficulty remembering recent events, while later stages involve severe cognitive impairment and loss of independence (Wolk & Dickerson, 2021). The pathophysiology of AD involves the accumulation of Aβ plaques and tau tangles in the brain, leading to neuronal death and brain atrophy (Jack et al., 2018). Neuroinflammation and oxidative stress also play significant roles in disease progression (Heneka et al., 2015).

1.4.1.2 Advances in Treatment and Management

Current treatments for AD primarily aim to alleviate symptoms. These include cholinesterase inhibitors (e.g., donepezil, rivastigmine) and NMDA receptor antagonists (e.g., memantine) (Cummings et al., 2019). Recent advances include the development of monoclonal antibodies targeting Aβ, such as aducanumab, which has shown promise in reducing amyloid plaques in the brain (Sevigny et al., 2016). Other investigational therapies target tau pathology, neuroinflammation, and synaptic dysfunction (Hampel et al., 2021). Non-pharmacological interventions, such as cognitive training, physical exercise, and dietary modifications, are also important for managing AD and improving quality of life (Ahlskog et al., 2011).

1.4.2 Parkinson's Disease

1.4.2.1 Pathophysiology and Symptoms

PD is a neurodegenerative disorder characterized by motor symptoms such as tremors, rigidity, bradykinesia, and postural instability. These symptoms result from the loss of dopaminergic neurons in the substantia nigra and the presence of Lewy bodies composed of alpha-synuclein (Poewe et al., 2017). Non-motor symptoms, including cognitive impairment, mood disorders, and autonomic dysfunction, are also common (Chaudhuri et al., 2006).

1.4.2.2 Advances in Treatment and Management

Treatment for PD includes pharmacological therapies such as levodopa, dopamine agonists, and MAO-B inhibitors, which aim to replenish or mimic dopamine (Connolly & Lang, 2014). Deep brain stimulation is a surgical option for patients with advanced PD who do not respond adequately to medication (Weaver et al., 2009). Ongoing research is exploring neuroprotective strategies and gene therapies to slow disease progression (Kalia & Lang, 2015). Recent studies have also investigated the role of gut microbiota in PD pathogenesis and potential therapeutic interventions (Sampson et al., 2016).

1.4.3 Huntington's Disease

1.4.3.1 Genetic Basis and Clinical Manifestations

HD is an autosomal dominant disorder caused by a CAG repeat expansion in the HTT gene, leading to the production of mutant HTT protein. Clinical manifestations include motor dysfunction, cognitive decline, and psychiatric symptoms (Walker, 2007). The disease typically presents in mid-adulthood and progresses over 15–20 years.

1.4.3.2 Ongoing Research and Therapeutic Strategies

Research in HD focuses on understanding the pathogenesis of the disease and developing targeted therapies. Current therapeutic strategies include antisense oligonucleotides that aim to reduce the production of mutant HTT protein (Tabrizi et al., 2019). Other approaches include gene editing technologies such as CRISPR/Cas9 and small molecule inhibitors targeting specific pathways involved in HD (Ross & Tabrizi, 2011). Clinical trials are ongoing to evaluate the safety and efficacy of these novel therapies (Wild & Tabrizi, 2017).

1.4.4 Amyotrophic Lateral Sclerosis

1.4.4.1 Disease Progression and Clinical Management

ALS is a progressive neurodegenerative disorder characterized by the degeneration of motor neurons, leading to muscle weakness, atrophy, and eventually respiratory failure. The disease can be sporadic or familial, with mutations in genes such as SOD1, C9orf72, and TARDBP being implicated in familial cases (Taylor et al., 2016).

1.4.4.2 Emerging Therapies and Research Insights

Current treatments for ALS are limited and primarily focus on symptom management and improving quality of life. Riluzole and edaravone are the only FDA-approved drugs that modestly extend survival (Miller et al., 2012). Emerging therapies include gene therapies targeting specific genetic mutations and stem cell-based approaches aimed at replacing damaged neurons (Van Damme et al., 2017). Recent research has also explored the role of neuroinflammation and the potential of anti-inflammatory treatments in ALS (Philips & Rothstein, 2014).

1.5 DIAGNOSTIC APPROACHES

1.5.1 Neuroimaging Techniques

1.5.1.1 MRI, PET, and Other Imaging Modalities

Neuroimaging techniques are essential for diagnosing and monitoring NDDs. Magnetic resonance imaging (MRI) provides detailed images of brain structures and is used to detect atrophy, white matter changes, and other structural abnormalities. Functional MRI measures brain activity by detecting changes in blood flow, revealing functional impairments in specific brain regions (Márquez & Yassa, 2019). Diffusion tensor imaging, an advanced MRI technique, assesses white matter integrity and connectivity, which can be disrupted in NDDs (Márquez & Yassa, 2019). Positron emission tomography (PET) uses radioactive tracers to visualize metabolic processes and pathological aggregates in the brain. Amyloid PET imaging detects Aβ plaques in AD, while tau PET imaging visualizes tau tangles (Young et al., 2020). Single photon emission computed tomography is similar to PET but uses different tracers and is often used to assess cerebral blood flow and dopamine transporter levels in PD (Brooks, 2010). Emerging imaging techniques include the development of novel PET ligands to target specific pathological proteins and the integration of machine learning algorithms with neuroimaging data to improve diagnostic accuracy and predict disease progression (Young et al., 2020). These advancements offer tremendous potential for early diagnosis and personalized treatment strategies.

1.5.2 Biomarkers

1.5.2.1 Cerebrospinal Fluid Biomarkers and Blood Tests

Cerebrospinal fluid (CSF) biomarkers are widely used in diagnosing NDDs. In AD, decreased levels of Aβ 42 and increased levels of total tau and phosphorylated tau in CSF are indicative of the disease (Blennow & Zetterberg, 2018). These biomarkers reflect the underlying pathology and can aid in early diagnosis and monitoring of disease progression. Blood-based biomarkers are an emerging area of research, offering a less invasive alternative to CSF biomarkers. Recent studies have identified plasma Aβ, tau, and neurofilament light chain as potential biomarkers for AD and other neurodegenerative conditions (Hampel et al., 2018). These biomarkers are being validated in large-scale studies and hold promise for widespread clinical use.

1.5.3 Genetic Testing

1.5.3.1 Role of Genetic Screening in Diagnosis and Prognosis

Genetic testing plays a significant role in diagnosing and prognosing NDDs, particularly those with a known genetic basis. In HD, genetic testing for the CAG repeat expansion in the HTT gene confirms the diagnosis and provides information on disease onset and progression (Walker, 2007). Similarly, genetic testing for mutations in the SOD1, C9orf72, and TARDBP genes is used in familial cases of ALS (Taylor et al., 2016). In AD, genetic testing for the apolipoprotein E ε4 allele (associated with increased risk of AD) can provide information on an individual's risk of developing the disease, although it is not used as a definitive diagnostic tool (Liu et al., 2013). Advances in next-generation sequencing and genome-wide association studies are identifying new genetic risk factors and potential therapeutic targets for NDDs (Naj et al., 2011).

1.6 TREATMENT AND MANAGEMENT

1.6.1 Current Therapies

1.6.1.1 Pharmacological Treatments and Their Limitations

Pharmacological treatments for NDDs aim to alleviate symptoms and improve quality of life, although they do not cure the diseases. In AD, cholinesterase inhibitors (e.g., donepezil, rivastigmine) and NMDA receptor antagonists (e.g., memantine) are commonly prescribed to manage cognitive symptoms (Cummings et al., 2019). However, these drugs have limited efficacy and can cause side effects such as gastrointestinal disturbances and dizziness (Birks and Harvey, 2018) . In PD, levodopa remains the gold standard treatment for motor symptoms. It is often combined with carbidopa to enhance its effectiveness and reduce side effects (Connolly & Lang, 2014). Dopamine agonists (e.g., pramipexole, ropinirole) and MAO-B inhibitors (e.g., selegiline, rasagiline) are also used, but they can cause side effects such as hallucinations and impulse control disorders (Poewe et al., 2017). For ALS, riluzole and edaravone are the only FDA-approved drugs. Riluzole modestly extends survival by reducing glutamate toxicity, while edaravone is thought to reduce oxidative stress (Miller et al., 2012). However, both drugs have limited efficacy, and there is a critical need for more effective treatments.

1.6.1.2 Non-Pharmacological Interventions

Non-pharmacological interventions play a crucial role in managing NDDs. Physical therapy, occupational therapy, and speech therapy are essential components of care for patients with PD and ALS, helping to maintain mobility, function, and communication abilities (Ortega-Hombrados et al., 2021; Ransmayr, 2011). Cognitive training and rehabilitation are also beneficial for patients with AD, aiming to enhance cognitive function and delay progression (Cavedo et al., 2015). Other non-pharmacological approaches include lifestyle modifications such as regular exercise, a healthy diet, and social engagement, which have been shown to improve overall well-being and potentially slow disease progression (Ahlskog et al., 2011). Mindfulness-based interventions and psychological support can help manage the emotional and psychological impact of NDDs (Zhang et al., 2021) .

1.7 CHALLENGES AND FUTURE DIRECTIONS

1.7.1 Research Gaps

1.7.1.1 Unresolved Questions and Areas Needing Further Exploration

Despite significant advances in understanding and treating NDDs, many research gaps remain. One major area needing further exploration is the precise molecular mechanisms underlying these diseases. While protein misfolding and aggregation are well-documented, the exact triggers and progression pathways are not fully understood. Recent studies suggest that small molecules may play a crucial role in modulating disease progression by targeting the formation, stability, or degradation of amyloid fibrils, which could alleviate the associated neurotoxic effects (Wareham et al., 2022). Additionally, the role of genetic and environmental factors in disease onset and progression requires more investigation. Genomic analyses may provide valuable insights into the pathological processes of neurodegeneration, emphasizing the need to complement classic case–control genetic association studies with wider population-based studies to better capture the full range of susceptibility variants (De Jager et al., 2018). Another critical area is the development of reliable biomarkers for early diagnosis and disease monitoring. Although CSF and blood-based biomarkers show promise, efforts are being made to improve their sensitivity and specificity (Blennow & Zetterberg, 2018). Furthermore, understanding the interplay between neuroinflammation, oxidative stress, and mitochondrial dysfunction in neurodegeneration could reveal new therapeutic targets. The roles of stress granule pathways and liquid-phase transitions in pathological protein aggregation are emerging mechanisms that need further exploration (Gan et al., 2018). Emerging research is also focusing on the role of the gut–brain axis in NDDs. Studies suggest that gut microbiota may influence neuroinflammation and neurodegeneration, presenting a novel area for therapeutic intervention (Sampson et al., 2016). Additionally, the impact of lifestyle factors such as diet, exercise, and sleep on NDD progression is an area of growing interest. Future studies are needed to understand the role of neurovascular disruptions and whether this is a cause or a consequence of NDD (Zlokovic et al., 2018).

1.7.2 Policy and Public Health

1.7.2.1 Strategies for Improving Patient Care and Support

Effective policy and public health strategies are essential for improving patient care and support for individuals with NDDs. One key strategy is enhancing access to diagnostic and therapeutic services. This includes increasing the availability of specialized care centres and ensuring that healthcare providers are trained in the latest diagnostic and treatment protocols (Phillips et al., 2023). Public health initiatives should also focus on raising awareness about NDDs and reducing stigma. Education campaigns can help the public understand these conditions and encourage early diagnosis and intervention (WHO, 2021). Additionally, policies that support caregivers, such as providing financial assistance and respite care, are crucial for improving the quality of life for both patients and their families (National Alliance for Caregiving, 2020). Investment in research funding is also critical. Governments and private organizations must continue to support research into the causes, diagnosis, and treatment of NDDs. Collaborative efforts between academia, industry, and healthcare providers can accelerate the development of new therapies and improve patient outcomes (Phillips et al., 2023).

1.7.3 Innovative Approaches

1.7.3.1 Advances in Technology and Personalized Medicine

Advances in technology and personalized medicine hold great promise for the future of NDD treatment. One exciting area is the use of artificial intelligence (AI) and machine learning to analyze large datasets and identify patterns that could lead to new diagnostic and therapeutic approaches (Young et al., 2020). AI can also be used to develop predictive models for disease progression and treatment response, enabling more personalized care (Márquez & Yassa, 2019). Personalized medicine, which tailors treatment to an individual's genetic makeup, lifestyle, and environment, is another promising approach. Advances in genomics and proteomics are enabling the identification of specific biomarkers and therapeutic targets for individual patients (Liu et al., 2013). This approach has the potential to improve treatment efficacy and reduce side effects by targeting therapies to those most likely to benefit. Additionally, the development of novel drug delivery systems, such as nanoparticles and liposomes, can enhance the delivery of therapeutic agents to the brain, overcoming the challenges posed by the BBB (Pardridge, 2012). These technologies can improve the bioavailability and efficacy of small molecule drugs and other therapeutic agents.

1.8 CONCLUSION AND FUTURE DIRECTIONS

NDDs present a formidable challenge due to their progressive nature and the complexity of underlying pathophysiological mechanisms. These diseases, including AD, HD, and ALS, are marked by specific protein misfolding and aggregation, oxidative stress, and mitochondrial dysfunction, leading to neuronal loss and synaptic dysfunction. Despite significant strides in understanding these mechanisms, effective treatments remain limited, primarily focusing on symptomatic relief rather than disease modification. Advancements in diagnostic approaches, such as neuroimaging techniques, biomarker identification, and genetic testing, have improved early detection and monitoring of NDDs. Pharmacological treatments and non-pharmacological interventions, including physical therapy and lifestyle modifications, play essential roles in managing symptoms and improving patient quality of life. Notably, small molecules have emerged as promising therapeutic agents due to their ability to cross the BBB and target specific pathological processes. Despite these advancements, challenges persist, including the need for a deeper understanding of molecular mechanisms, the development of reliable biomarkers, and addressing the impact of lifestyle and environmental factors. Moreover, enhancing access to diagnostic and therapeutic services, raising public awareness, and supporting caregivers are critical for comprehensive patient care.

Future research and treatment strategies for NDDs should focus on several key areas. Understanding the precise molecular and cellular mechanisms driving these diseases is crucial, including exploring the gut–brain axis, neuroinflammation, and the role of non-neuronal cells in disease progression. Continued efforts to identify and validate biomarkers with high sensitivity and specificity will enhance early diagnosis and monitoring of disease progression, facilitating timely intervention and improving patient outcomes. Additionally, advances in gene therapy, stem cell research, and personalized medicine hold promise for developing disease-modifying treatments, with small molecules targeting specific pathological processes and offering the potential for more effective therapies. Moreover, integrating AI and machine learning in diagnostic tools and treatment planning can lead to more accurate and personalized approaches. Addressing NDDs from a public health perspective involves improving access to diagnostic and therapeutic services, especially in underserved regions, and raising awareness and providing caregiver support. Encouraging interdisciplinary and international collaboration is vital for advancing our understanding and treatment of these diseases.

In conclusion, while NDDs continue to pose significant challenges, ongoing research and technological advancements offer hope for the future. Small molecules, with their ability to target specific pathological processes, represent a promising approach in the fight against these debilitating conditions. By investing in research, public health strategies, and innovative technologies, we can make meaningful progress in understanding, diagnosing, and treating NDDs, ultimately improving the lives of those affected.

REFERENCES

Ahlskog, J.E., Geda, Y.E., Graff-Radford, N.R. and Petersen, R.C. 2011. Physical exercise as a preventive or disease-modifying treatment of dementia and brain aging. *Mayo Clinic Proceedings*, *86*(9), pp. 876–8846.

Alzheimer's Association. 2021. Alzheimer's Disease Facts and Figures. Alzheimer's Association.

Armstrong, M.J., Alliance, S., Corsentino, P., Lunde, A. and Taylor, A. 2022. Informal caregiver experiences at the end-of-life of individuals living with dementia with Lewy bodies: an interview study. *Dementia*, *21*(1), pp. 287–303.

Baev, A.Y., Vinokurov, A.Y., Novikova, I.N., Dremin, V.V., Potapova, E.V. and Abramov, A.Y. 2022. Interaction of mitochondrial calcium and ROS in neurodegeneration. *Cells*, *11*(4), p. 706.

Barron, J.C., Hurley, E.P. and Parsons, M.P. 2021. Huntingtin and the Synapse. *Frontiers in Cellular Neuroscience*, *15*, p. 689332.

Beavan, M.S. and Schapira, A.H. 2013. Glucocerebrosidase mutations and the pathogenesis of Parkinson disease. *Annals of Medicine*, *45*(8), pp. 511–521.

Ben-Shlomo, Y., Darweesh, S., Llibre-Guerra, J., Marras, C., San Luciano, M. and Tanner, C. 2024. The epidemiology of Parkinson's disease. *The Lancet*, *403*(10423), pp. 283–292.

Bergonzoni, G., Döring, J. and Biagioli, M. 2021. D1R-and D2R-medium-sized spiny neurons diversity: insights into striatal vulnerability to Huntington's disease mutation. *Frontiers in Cellular Neuroscience*, *15*, p. 628010.

Bhat, S.A., Ahamad, S., Dar, N.J., Siddique, Y.H. and Nazir, A. 2023. The emerging landscape of natural small-molecule therapeutics for Huntington's disease. *Current Neuropharmacology*, *21*(4), p. 867.

Birks, J.S., Harvey, R.J., 2018. Donepezil for dementia due to Alzheimer's disease. Cochrane Database Syst. Rev. 2018. https://doi.org/10.1002/14651858.CD001190.pub3

Blennow, K. and Zetterberg, H. 2018. Biomarkers for Alzheimer's disease: current status and prospects for the future. *Journal of Internal Medicine*, *284*(6), pp. 643–663.

Bozzo, F., Mirra, A. and Carrì, M.T. 2017. Oxidative stress and mitochondrial damage in the pathogenesis of ALS: new perspectives. *Neuroscience Letters*, *636*, pp. 3–8.

Brooks, D.J. 2010. Imaging approaches to Parkinson disease. *Journal of Nuclear Medicine*, *51*(4), pp. 596–6093.

Calì, T., Ottolini, D. and Brini, M. 2011. Mitochondria, calcium, and endoplasmic reticulum stress in Parkinson's disease. *Biofactors*, *37*(3), pp. 228–240.

Cantarero-Prieto, D., Leon, P.L., Blazquez-Fernandez, C., Juan, P.S. and Cobo, C.S. 2020. The economic cost of dementia: a systematic review. *Dementia*, *19*(8), pp. 2637–2657.

Cavedo, E., Kasper, E., Ochmann, S., Hoffmann, W., Schneider, W., Hampel, H., S., T., 2015. COGNITIVE REHABILITATION IN ALZHEIMER'S DISEASE – A CONCEPTUAL AND METHODOLOGICAL REVIEW. J. Prev. Alzheimer's Dis. 1–11. https://doi.org/10.14283/jpad.2015.58

Chaudhuri, K.R., Healy, D.G. and Schapira, A.H.V. 2006. Non-motor symptoms of Parkinson's disease: diagnosis and management. *The Lancet Neurology*, *5*(3), pp. 235–2454.

Chen, Z.R., Huang, J.B., Yang, S.L. and Hong, F.F. 2022. Role of cholinergic signaling in Alzheimer's disease. *Molecules*, *27*(6), p. 1816.

Chiò, A., Logroscino, G., Traynor, B.J., Collins, J., Simeone, J.C., Goldstein, L.A. and White, L.A. 2013. Global epidemiology of amyotrophic lateral sclerosis: a systematic review of the published literature. *Neuroepidemiology*, *41*(2), pp. 118–130.

Chiti, F. and Dobson, C.M. 2006. Protein misfolding, functional amyloid, and human disease. *Annual Review of Biochemistry*, *75*, pp. 333–36614.

Chung, C.Y., Koprich, J.B., Siddiqi, H. and Isacson, O. 2009. Dynamic changes in presynaptic and axonal transport proteins combined with striatal neuroinflammation precede dopamine depletion in a mouse model of Parkinson's disease. *Journal of Neuroscience*, *29*(11), pp. 3365–33736.

Ciechanover, A. and Kwon, Y.T. 2015. Degradation of misfolded proteins in neurodegenerative diseases: therapeutic targets and strategies. *Experimental & Molecular Medicine*, *47*(3), p. e1471.

Connolly, B.S. and Lang, A.E. 2014. Pharmacological treatment of Parkinson disease: a review. *JAMA*, *311*(16), pp. 1670–16835.

Culberson, J.W., Kopel, J., Sehar, U. and Reddy, P.H. 2023. Urgent needs of caregiving in ageing populations with Alzheimer's disease and other chronic conditions: support our loved ones. *Ageing Research Reviews*, *90*, p. 102001.

Cummings, J., Lee, G., Ritter, A., Sabbagh, M., and Zhong, K. 2019. Alzheimer's disease drug development pipeline: 2019. *Alzheimer's & Dementia: Translational Research & Clinical Interventions*, *5*, pp. 272–2932.

Czeredys, M. 2020. Dysregulation of neuronal calcium signaling via store-operated channels in Huntington's disease. *Frontiers in Cell and Developmental Biology*, *8*, p. 611735.

Dafinca, R., Barbagallo, P., Farrimond, L., Candalija, A., Scaber, J., Nida'a, A.A., Sathyaprakash, C., Vowles, J., Cowley, S.A. and Talbot, K. 2020. Impairment of mitochondrial calcium buffering links mutations in C9ORF72 and TARDBP in iPS-derived motor neurons from patients with ALS/FTD. *Stem Cell Reports*, *14*(5), pp. 892–908.

Dar, N.J. and Glazner, G.W. 2020. Deciphering the neuroprotective and neurogenic potential of soluble amyloid precursor protein alpha (sAPPα). *Cellular and Molecular Life Sciences*, *77*, pp. 2315–2330.

Dar, N.J., Bhat, J.A., Satti, N.K., Sharma, P.R., Hamid, A. and Ahmad, M. 2017. Withanone, an active constituent from Withania somnifera, affords protection against NMDA-induced excitotoxicity in neuron-like cells. *Molecular Neurobiology*, *54*(7), pp. 5061–5073.

Dar, N.J., John, U., Bano, N., Khan, S. and Bhat, S.A. 2024. Oxytosis/ferroptosis in neurodegeneration: the underlying role of master regulator glutathione peroxidase 4 (GPX4). *Molecular Neurobiology*, *61*(3), pp. 1507–1526.

Dar, N.J., Na, R. and Ran, Q. 2022. Functional deficits of 5× fad neural stem cells are ameliorated by glutathione peroxidase 4. *Cells*, *11*(11), p. 1770.

Dar, N.J., Satti, N.K., Dutt, P., Hamid, A. and Ahmad, M. 2018. Attenuation of glutamate-induced excitotoxicity by withanolide-A in neuron-like cells: role for PI3K/Akt/MAPK signaling pathway. *Molecular Neurobiology*, *55*(4), pp. 2725–2739.

De Jager, P.L., Ma, Y., McCabe, C., et al. 2018. A multi-omic atlas of the human frontal cortex for aging and Alzheimer's disease research. *Scientific Data*, *5*, Article number: 180142.

Demuro, A., Parker, I. and Stutzmann, G.E. 2010. Calcium signaling and amyloid toxicity in Alzheimer disease. *Journal of Biological Chemistry*, *285*(17), pp. 12463–12468.

Dorsey, E.R. and Bloem, B.R. 2018. The Parkinson pandemic—a call to action. *JAMA Neurology*, *75*(1), pp. 9–10.

Dorsey, E.R., Elbaz, A., Nichols, E., Abbasi, N., Abd-Allah, F., Abdelalim, A., Adsuar, J.C., Ansha, M.G., Brayne, C., Choi, J.Y.J. and Collado-Mateo, D. 2018. Global, regional, and national burden of Parkinson's disease, 1990–2016: a systematic analysis for the Global Burden of Disease Study 2016. *The Lancet Neurology*, *17*(11), pp. 939–953.

Droppelmann, C.A., Campos-Melo, D., Ishtiaq, M., Volkening, K. and Strong, M.J. 2014. RNA metabolism in ALS: when normal processes become pathological. *Amyotrophic Lateral Sclerosis and Frontotemporal Degeneration*, *15*(5–6), pp. 321–336.

Duchen, M.R. 2000. Mitochondria and calcium: from cell signalling to cell death. *The Journal of Physiology*, *529*(1), pp. 57–68.

Dugger, B.N. and Dickson, D.W. 2017. Pathology of Neurodegenerative Diseases. Cold Spring Harbor Perspectives in Biology.

Emin, D., Zhang, Y.P., Lobanova, E., Miller, A., Li, X., Xia, Z., Dakin, H., Sideris, D.I., Lam, J.Y., Ranasinghe, R.T. and Kouli, A. 2022. Small soluble α-synuclein aggregates are the toxic species in Parkinson's disease. *Nature Communications*, *13*(1), p. 5512.

Erustes, A.G., Guarache, G.C., Guedes, E.D.C., Leão, A.H.F.F., Pereira, G.J.D.S. and Smaili, S.S. 2022. α-Synuclein interactions in mitochondria-ER contacts: a possible role in Parkinson's disease. *Contact*, *5*, p. 25152564221119347.

Farré, M., Haro, J.M., Kostov, B., Alvira, C., Risco, E., Miguel, S., Cabrera, E. and Zabalegui, A. 2016. Direct and indirect costs and resource use in dementia care: a cross-sectional study in patients living at home. *International Journal of Nursing Studies*, *55*, pp. 39–49.

Fraldi, A., Klein, A.D., Medina, D.L. and Settembre, C. 2016. Brain disorders due to lysosomal dysfunction. *Annual Review of Neuroscience*, *39*(1), pp. 277–295.

Friedlander, R.M. 2003. Apoptosis and caspases in neurodegenerative diseases. *New England Journal of Medicine*, *348*(14), pp. 1365–1375.

Gan, L., Cookson, M.R., Petrucelli, L., et al. 2018. Converging pathways in neurodegeneration, from genetics to mechanisms. *Nature Neuroscience*, *21*(10), pp. 1300–1309.

GBD 2017 US Neurological Disorders Collaborators. 2020. Burden of neurological disorders across the US from 1990–2017: a global burden of disease study. *JAMA Neurology*, *78*(2), pp. 165–176

GBD 2021 Collaborators. 2024. Global burden of disease, injuries, and risk factor study. *The Lancet Neurology*, 403, pp. 1958–1959.

Glass, C.K., Saijo, K., Winner, B., Marchetto, M.C. and Gage, F.H. 2010. Mechanisms underlying inflammation in neurodegeneration. *Cell*, *140*(6), pp, 918–9348.

Grosskreutz, J., Van Den Bosch, L. and Keller, B.U. 2010. Calcium dysregulation in amyotrophic lateral sclerosis. *Cell Calcium*, *47*(2), pp. 165–174.

Guan, P.P., Cao, L.L. and Wang, P. 2021. Elevating the levels of calcium ions exacerbate Alzheimer's disease via inducing the production and aggregation of β-amyloid protein and phosphorylated tau. *International Journal of Molecular Sciences*, *22*(11), p. 5900.

Hampel, H., Hardy, J., Blennow, K., Chen, C., Perry, G., Kim, S.H., Villemagne, V.L., Aisen, P., Vendruscolo, M., Iwatsubo, T., Masters, C.L., Cho, M., Lannfelt, L., Cummings, J. and Vergallo, A. 2021. The future of Alzheimer's disease: the next 10 years. *Progress in Neurobiology*, *194*, p. 1018882.

Hampel, H., O'Bryant, S.E., Molinuevo, J.L., Zetterberg, H., Masters, C.L., Lista, S., Kiddle, S.J., Batrla, R. and Blennow, K. 2018. Blood-based biomarkers for Alzheimer disease: mapping the road to the clinic. *Nature Reviews Neurology*, *14*(11), pp. 639–6525.

Hardy, J. and Selkoe, D.J. 2002. The amyloid hypothesis of Alzheimer's disease: progress and problems on the road to therapeutics. *Science*, *297*(5580), pp. 353–356.

Harper, P.S. 2002. The epidemiology of Huntington's disease. *Oxford Monographs on Medical Genetics*, *45*, pp. 159–197.

Heath, P.R. and Shaw, P.J. 2002. Update on the glutamatergic neurotransmitter system and the role of excitotoxicity in amyotrophic lateral sclerosis. *Muscle & Nerve: Official Journal of the American Association of Electrodiagnostic Medicine*, *26*(4), pp. 438–458.

Hellmers, N. 2024. *Work-Life Balance among Family Caregivers of Adults with Parkinson's Disease* (Doctoral dissertation). New York University, New York.

Heneka, M.T., Golenbock, D.T. and Latz, E. 2015. Innate immunity in Alzheimer's disease. *Nature Immunology*, *16*(3), pp. 229–2369.

Ittner, L.M. and Götz, J. 2011. Amyloid-β and tau—a toxic pas de deux in Alzheimer's disease. *Nature Reviews Neuroscience*, *12*(2), pp. 67–72.

Jack, C.R., Bennett, D.A., Blennow, K., Carrillo, M.C., Dunn, B., Haeberlein, S.B., Holtzman, D.M., Jagust, W., Jessen, F., Karlawish, J. and Liu, E. 2018. NIA-AA research framework: toward a biological definition of Alzheimer's disease. *Alzheimer's & Dementia*, *14*(4), pp. 535–5622.

Johnson, S., Davis, M., Kaltenboeck, A., Birnbaum, H., Grubb, E., Tarrants, M. and Siderowf, A. 2011. Early retirement and income loss in patients with early and advanced Parkinson's disease. *Applied Health Economics and Health Policy*, *9*, pp. 367–376.

Kalia, L.V. and Lang, A.E. 2015. Parkinson's disease. *The Lancet*, *386*(9996), pp. 896–9127.

Khan, S., Bano, N., Ahamad, S., John, U., Dar, N.J. and Bhat, S.A. 2024. Excitotoxicity, oxytosis/ferroptosis, and neurodegeneration: emerging insights into mitochondrial mechanisms. *Aging and Disease*, *15*(1), pp. 125–144.

Kim, D.E. and Priefer, R. 2020. Therapeutic potential of direct clearance of the amyloid-β in Alzheimer's disease. *Brain Sciences*, *10*(2), p. 93.

Klein, C. and Westenberger, A. 2012. Genetics of Parkinson's disease. *Cold Spring Harbor Perspectives in Medicine*, *2*(1), p. a008888.

Kovacs, G.G. 2019. Molecular pathology of neurodegenerative diseases: principles and practice. *Journal of Clinical* Pathology, *72*(11), pp. 725–735.

Kumar, S., Goyal, L. and Singh, S. 2022. Tremor and rigidity in patients with Parkinson's disease: emphasis on epidemiology, pathophysiology and contributing factors. *CNS & Neurological Disorders-Drug Targets (Formerly Current Drug Targets-CNS & Neurological Disorders)*, *21*(7), pp. 596–609.

Lattante, S., Rouleau, G.A. and Kabashi, E. 2013. TARDBP and FUS mutations associated with amyotrophic lateral sclerosis: summary and update. *Human Mutation*, *34*(6), pp. 812–826.

Leandrou, E., Emmanouilidou, E. and Vekrellis, K. 2019. Voltage-gated calcium channels and α-synuclein: implications in Parkinson's disease. *Frontiers in Molecular Neuroscience*, *12*, p. 237.

Lee, A. and Gilbert, R.M. 2016. Epidemiology of Parkinson disease. *Neurologic Clinics*, *34*(4), pp. 955–965.

Lee, H.G., Wheeler, M.A. and Quintana, F.J. 2022. Function and therapeutic value of astrocytes in neurological diseases. *Nature Reviews Drug Discovery*, *21*(5), pp. 339–358.

Lee, J.K., Shin, J.H., Lee, J.E. and Choi, E.J. 2015. Role of autophagy in the pathogenesis of amyotrophic lateral sclerosis. *Biochimica et Biophysica Acta (BBA)-Molecular Basis of Disease*, 1852(11), pp. 2517–2524.

Li, K., Li, J., Zheng, J. and Qin, S. 2019. Reactive astrocytes in neurodegenerative diseases. *Aging and Disease*, *10*(3), p. 664.

Lin, M.T. and Beal, M.F. 2006. Mitochondrial dysfunction and oxidative stress in neurodegenerative diseases. *Nature*, *443*(7113), pp. 787–7953.

Liu, C.C., Kanekiyo, T., Xu, H. and Bu, G. 2013. Apolipoprotein E and Alzheimer disease: risk, mechanisms and therapy. *Nature Reviews Neurology*, *9*(2), pp. 106–1186.

Liu, J. and Li, L. 2019. Targeting autophagy for the treatment of Alzheimer's disease: challenges and opportunities. *Frontiers in Molecular Neuroscience*, *12*, p. 203.

Luukkainen, L. 2023. Epidemiology of early-onset frontotemporal dementia and molecular genetics of early-onset neurodegenerative dementia (Doctoral dissertation). University of Oulu, Finland.

Lwi, S.J., Ford, B.Q., Casey, J.J., Miller, B.L. and Levenson, R.W. 2017. Poor caregiver mental health predicts mortality of patients with neurodegenerative disease. *Proceedings of the National Academy of Sciences*, *114*(28), pp. 7319–7324.

Malik, A.R. and Willnow, T.E. 2019. Excitatory amino acid transporters in physiology and disorders of the central nervous system. *International Journal of Molecular Sciences*, *20*(22), p. 5671.

Malpartida, A.B., Williamson, M., Narendra, D.P., Wade-Martins, R. and Ryan, B.J. 2021. Mitochondrial dysfunction and mitophagy in Parkinson's disease: from mechanism to therapy. *Trends in Biochemical Sciences*, *46*(4), pp. 329–343.

Maresova, P., Hruska, J., Klimova, B., Barakovic, S. and Krejcar, O. 2020. Activities of daily living and associated costs in the most widespread neurodegenerative diseases: a systematic review. *Clinical Interventions in Aging*, *15*, pp. 1841–1862.

Márquez, F. and Yassa, M.A. 2019. Neuroimaging biomarkers for Alzheimer's disease. *Molecular Neurodegeneration*, *14*, p. 211.

Martinez-Vicente, M., Talloczy, Z., Wong, E., Tang, G., Koga, H., Kaushik, S., De Vries, R., Arias, E., Harris, S., Sulzer, D. and Cuervo, A.M. 2010. Cargo recognition failure is responsible for inefficient autophagy in Huntington's disease. *Nature Neuroscience*, *13*(5), pp. 567–576.

Mattson, M.P. 2000. Apoptosis in neurodegenerative disorders. *Nature Reviews Molecular Cell Biology*, *1*(2), pp. 120–1295.

McColgan, P. and Tabrizi, S.J. 2018. Huntington's disease: a clinical review. *European Journal of Neurology*, *25*(1), pp. 24–34.

Mees, I., Li, S., Beauchamp, L.C., Barnham, K.J., Dutschmann, M., Hannan, A.J. and Renoir, T. 2022. Loss-of-function and gain-of-function studies refute the hypothesis that tau protein is causally involved in the pathogenesis of Huntington's disease. *Human Molecular Genetics*, *31*(12), pp. 1997–2009.

Mehta, A., Prabhakar, M., Kumar, P., Deshmukh, R. and Sharma, P.L. 2013. Excitotoxicity: bridge to various triggers in neurodegenerative disorders. *European Journal of Pharmacology*, *698*(1–3), pp. 6–18.

Metwally, E., Al-Abbadi, H.A., Hussain, T., Murtaza, G., Abdellatif, A.M. and Ahmed, M.F. 2023. Calpain signaling: from biology to therapeutic opportunities in neurodegenerative disorders. *Frontiers in Veterinary Science*, *10*, p. 1235163.

Miller, R.G., Mitchell, J.D., Lyon, M. and Moore, D.H. 2012. Riluzole for amyotrophic lateral sclerosis (ALS)/ motor neuron disease (MND). *Cochrane Database of Systematic Reviews*, (2), CD001447.

Mizuno, T. 2015. Neuron–microglia interactions in neuroinflammation. *Clinical and Experimental Neuroimmunology*, *6*(3), pp. 225–231.

Morris, H.R., Spillantini, M.G., Sue, C.M. and Williams-Gray, C.H. 2024. The pathogenesis of Parkinson's disease. *The Lancet, 403*(10423), pp. 293–304.

Naj, A.C., Jun, G., Beecham, G.W., Wang, L.S., Vardarajan, B.N., Buros, J., Gallins, P.J., Buxbaum, J.D., Jarvik, G.P., Crane, P.K., Larson, E.B., Bird, T.D., Boeve, B.F., Graff-Radford, N.R., De Jager, P.L., Evans, D., Schneider, J.A., Carrasquillo, M.M., Ertekin-Taner, N., Younkin, S.G., Cruchaga, C., Kauwe, J.S.K., Nowotny, P., Kramer, P., Hardy, J., Haines, J.L., Farrer, L.A., Pericak-Vance, M.A., Schellenberg, G.D. and Goate, A.M. 2011. Common variants at MS4A4/MS4A6E, CD2AP, CD33 and EPHA1 are associated with late-onset Alzheimer's disease. *Nature Genetics, 43*(5), pp. 436–4416.

National Alliance for Caregiving. 2020. Caregiving in the U.S.

Newell, M.E., Adhikari, S. and Halden, R.U. 2022. Systematic and state-of the science review of the role of environmental factors in Amyotrophic Lateral Sclerosis (ALS) or Lou Gehrig's Disease. *Science of the Total Environment, 817*, p. 152504.

Nichols, E., Steinmetz, J.D., Vollset, S.E., Fukutaki, K., Chalek, J., Abd-Allah, F., Abdoli, A., Abualhasan, A., Abu-Gharbieh, E., Akram, T.T. and Al Hamad, H. 2022. Estimation of the global prevalence of dementia in 2019 and forecasted prevalence in 2050: an analysis for the Global Burden of Disease Study 2019. *The Lancet Public Health, 7*(2), pp. e105–e125.

Nixon, R.A. 2017. Amyloid precursor protein and endosomal–lysosomal dysfunction in Alzheimer's disease: inseparable partners in a multifactorial disease. *The FASEB Journal, 31*(7), p. 2729.

Ortega-Hombrados, L., Molina-Torres, G., Galán-Mercant, A., Sánchez-Guerrero, E., González-Sánchez, M., Ruiz-Muñoz, M., 2021. Systematic Review of Therapeutic Physical Exercise in Patients with Amyotrophic Lateral Sclerosis over Time. Int. J. Environ. Res. Public Health 18, 1074. https://doi.org/10.3390/ijerph18031074

Pandey, M. and Rajamma, U. 2018. Huntington's disease: the coming of age. *Journal of genetics, 97*(3), pp. 649–664.

Pardridge, W.M. 2012. Drug transport across the blood-brain barrier. *Journal of Cerebral Blood Flow & Metabolism, 32*(11), pp. 1959–19726.

Parkinson's Foundation. 2021. Understanding Parkinson's Disease. Parkinson's Foundation.

Petersen, R.B., Walter, B., 2023. Editorial: Insights into Parkinson's disease and aging related movement disorders. Front. Aging Neurosci. 15. https://doi.org/10.3389/fnagi.2023.1193197

Philips, T. and Rothstein, J.D. 2014. Glial cells in amyotrophic lateral sclerosis. *Experimental Neurology, 262*, pp. 111–1202.

Phillips, K.A. and Yassa, M.A. 2023. Ten health policy challenges for the next 10 years. *Health Affairs Scholar, 1*(1), p. qxad0106.

Plascencia-Villa, G. and Perry, G. 2023. Roles of oxidative stress in synaptic dysfunction and neuronal cell death in Alzheimer's disease. *Antioxidants, 12*(8), p. 1628.

Podvin, S., Reardon, H.T., Yin, K., Mosier, C. and Hook, V. 2019. Multiple clinical features of Huntington's disease correlate with mutant HTT gene CAG repeat lengths and neurodegeneration. *Journal of Neurology, 266*, pp. 551–564.

Poewe, W., Seppi, K., Tanner, C.M., Halliday, G.M., Brundin, P., Volkmann, J., Schrag, A.E. and Lang, A.E. 2017. Parkinson disease. *Nature Reviews Disease Primers, 3*, p. 170133.

Pringsheim, T., Jette, N., Frolkis, A. and Steeves, T.D. 2014. The prevalence of Parkinson's disease: a systematic review and meta-analysis. *Movement Disorders, 29*(13), pp. 1583–1590.

Pringsheim, T., Wiltshire, K., Day, L., Dykeman, J., Steeves, T. and Jette, N. 2012. The incidence and prevalence of Huntington's disease: a systematic review and meta-analysis. *Movement Disorders, 27*(9), pp. 1083–1091.

Ragagnin, A.M., Shadfar, S., Vidal, M., Jamali, M.S. and Atkin, J.D. 2019. Motor neuron susceptibility in ALS/FTD. *Frontiers in Neuroscience, 13*, p. 532.

Rajmohan, R. and Reddy, P.H. 2017. Amyloid-beta and phosphorylated tau accumulations cause abnormalities at synapses of Alzheimer's disease neurons. *Journal of Alzheimer's Disease, 57*(4), pp. 975–999.

Ramesh, N. and Pandey, U.B. 2017. Autophagy dysregulation in ALS: when protein aggregates get out of hand. *Frontiers in Molecular Neuroscience, 10*, p. 263.

Ransmayr, G., 2011. Physical, occupational, speech and swallowing therapies and physical exercise in Parkinson's disease. J. Neural Transm. 118, 773–781.

Ransohoff, R.M. 2016. How neuroinflammation contributes to neurodegeneration. *Science, 353*(6301), pp. 777–78310.

Rapp, T., Chauvin, P., Costa, N. and Molinier, L. 2015. Health economic considerations in neurodegenerative disorders. *Imaging in Neurodegenerative Disorders, 42*, pp. 42–53.

Raymond, L.A. 2017. Striatal synaptic dysfunction and altered calcium regulation in Huntington disease. *Biochemical and Biophysical Research Communications*, *483*(4), pp. 1051–1062.
Reddy, P.H. and Beal, M.F. 2008. Amyloid beta, mitochondrial dysfunction and synaptic damage: implications for cognitive decline in aging and Alzheimer's disease. *Trends in Molecular Medicine*, *14*(2), pp. 45–534.
Rocca, W.A., Petersen, R.C., Knopman, D.S., Hebert, L.E., Evans, D.A., Hall, K.S., Gao, S., Unverzagt, F.W., Langa, K.M., Larson, E.B. and White, L.R. 2011. Trends in the incidence and prevalence of Alzheimer's disease, dementia, and cognitive impairment in the United States. *Alzheimer's & Dementia*, *7*(1), pp. 80–93.
Roland, K.P. and Chappell, N.L. 2019. Caregiver experiences across three neurodegenerative diseases: Alzheimer's, Parkinson's, and Parkinson's with dementia. *Journal of Aging and Health*, *31*(2), pp. 256–279.
Ross, C.A. and Poirier, M.A. 2004. Protein aggregation and neurodegenerative disease. *Nature Medicine*, *10*(S7), pp. S10–S172.
Ross, C.A. and Tabrizi, S.J. 2011. Huntington's disease: from molecular pathogenesis to clinical treatment. *The Lancet Neurology*, *10*(1), pp. 83–982.
Sacks, B., Bashford, J., Wijesekera, L., Leigh, P.N. and Sreedharan, J. 2022. Motor neuron disease: amyotrophic lateral sclerosis. In *Neuroscience in the 21st Century: From Basic to Clinical* (pp. 4221–4271). Cham: Springer International Publishing. Editor name: Donald W. Pfaff, Nora D. Volkow, John L. Rubenstein
Sampson, T.R., Debelius, J.W., Thron, T., et al. 2016. Gut microbiota regulate motor deficits and neuroinflammation in a model of Parkinson's disease. *Cell*, *167*(6), pp. 1469–1480.e12.
Sampson, T.R., Debelius, J.W., Thron, T., Janssen, S., Shastri, G.G., Ilhan, Z.E., Challis, C., Schretter, C.E., Rocha, S., Gradinaru, V., Chesselet, M.F., Keshavarzian, A., Shannon, K.M., Krajmalnik-Brown, R., Wittung-Stafshede, P., Knight, R. and Mazmanian, S.K. 2016. Gut microbiota regulate motor deficits and neuroinflammation in a model of Parkinson's disease. *Cell*, *167*(6), pp. 1469–14808.
Selkoe, D.J. 2003. Folding proteins in fatal ways. *Nature*, *426*(6968), pp. 900–90415.
Senkevich, K. and Gan-Or, Z. 2020. Autophagy lysosomal pathway dysfunction in Parkinson's disease; evidence from human genetics. *Parkinsonism & Related Disorders*, *73*, pp. 60–71.
Sevigny, J., Chiao, P., Bussière, T., Weinreb, P.H., Williams, L., Maier, M., Dunstan, R., Salloway, S., Chen, T., Ling, Y., O'Gorman, J., Qian, F., Arastu, M., Li, M., Chollate, S., Brennan, M.S., Quintero-Monzon, O., Scannevin, R.H., Arnold, H.M., Engber, T., Rhodes, K., Ferrero, J., Hang, Y., Mikulskis, A., Grimm, J., Hock, C., Nitsch, R.M. and Sandrock, A. 2016. The antibody aducanumab reduces Aβ plaques in Alzheimer's disease. *Nature*, *537*(7618), pp. 50–561.
Shah, A., John Dar, N., Parvaiz Hassan, Q. and Ahmad, M. 2016. Endophytes and neurodegenerative diseases: a hope in desperation. *CNS & Neurological Disorders-Drug Targets (Formerly Current Drug Targets-CNS & Neurological Disorders)*, *15*(10), pp. 1231–1239.
Sharma, M. and Burré, J. 2023. α-Synuclein in synaptic function and dysfunction. *Trends in Neurosciences*, *46*(2), pp. 153–166.
Spillantini, M.G., Schmidt, M.L., Lee, V.M., Trojanowski, J.Q., Jakes, R. and Goedert, M. 1997. Alpha-synuclein in Lewy bodies. *Nature*, *388*(6645), pp. 839–84013.
Stutzmann, G.E. 2007. The pathogenesis of Alzheimers disease—is it a lifelong "Calciumopathy"? *The Neuroscientist*, *13*(5), pp. 546–559.
Surmeier, D.J., Guzman, J.N. and Sanchez-Padilla, J. 2010. Calcium, cellular aging, and selective neuronal vulnerability in Parkinson's disease. *Cell Calcium*, *47*(2), pp. 175–182.
Tabrizi, S.J., Leavitt, B.R., Landwehrmeyer, G.B., Wild, E.J., Saft, C., Barker, R.A., Blair, N.F., Craufurd, D., Priller, J., Rickards, H., Rosser, A., Kordasiewicz, H.B., Czech, C., Swayze, E.E., Norris, D.A., Baumann, T., Gerlach, I., Schobel, S.A., Paz, E., Smith, A.V., Bennett, C.F. and Lane, R.M. 2019. Targeting Huntingtin expression in patients with Huntington's disease. *New England Journal of Medicine*, *380*(24), pp. 2307–231610.
Tahami Monfared, A.A., Byrnes, M.J., White, L.A. and Zhang, Q. 2022. Alzheimer's disease: epidemiology and clinical progression. *Neurology and Therapy*, *11*(2), pp. 553–569.
Tang, T.S., Tu, H., Chan, E.Y., Maximov, A., Wang, Z., Wellington, C.L., Hayden, M.R. and Bezprozvanny, I. 2003. Huntingtin and huntingtin-associated protein 1 influence neuronal calcium signaling mediated by inositol-(1, 4, 5) triphosphate receptor type 1. *Neuron*, *39*(2), pp. 227–239.
Tay, L.X., Ong, S.C., Tay, L.J., Ng, T. and Parumasivam, T. 2024. Economic burden of Alzheimer 's disease: a systematic review. *Value in Health Regional Issues*, *40*, pp. 1–12.
Taylor, J.P., Brown, R.H. and Cleveland, D.W. 2016. Decoding ALS: from genes to mechanism. *Nature*, *539*(7628), pp. 197–20611.

Thameem Dheen, S., Kaur, C. and Ling, E.A. 2007. Microglial activation and its implications in the brain diseases. *Current Medicinal Chemistry*, *14*(11), pp. 1189–1197.

Thorpe, K.E., Levey, A.I. and Thomas, J. 2021. U.S. Burden of Neurodegenerative Disease. Fight Chronic Disease.

To, T.M., Ta, J.T., Patel, A.M., Arndorfer, S., Abbass, I.M. and Gandhy, R. 2023. Healthcare resource utilization and cost among individuals with late-onset versus adult-onset Huntington's disease: a claims-based retrospective cohort study. *Journal of Medical Economics*, *26*(1), pp. 862–870.

Tu, S., Okamoto, S.I., Lipton, S.A. and Xu, H. 2014. Oligomeric Aβ-induced synaptic dysfunction in Alzheimer's disease. *Molecular Neurodegeneration*, *9*, pp. 1–12.

Van Damme, P., Robberecht, W. and Van Den Bosch, L. 2017. Advances in the genetic and molecular understanding of amyotrophic lateral sclerosis. *Nature Reviews Neurology*, *13*(11), pp. 661–6782.

Vandoorne, T., De Bock, K. and Van Den Bosch, L. 2018. Energy metabolism in ALS: an underappreciated opportunity? *Acta Neuropathologica*, *135*, pp. 489–509.

Verma, S., Khurana, S., Vats, A., Sahu, B., Ganguly, N.K., Chakraborti, P., Gourie-Devi, M. and Taneja, V. 2022. Neuromuscular junction dysfunction in amyotrophic lateral sclerosis. *Molecular Neurobiology*, *59*(3), pp. 1502–1527.

von Gunten, A., Nogueira, E., Parmentier, H. and Gomes, I. 2019. Neurocognitive disorders in old age: Alzheimer's disease, frontotemporal dementia, dementia with Lewy bodies, and prion and infectious diseases. In *Primary Care Mental Health in Older People: A Global Perspective*, Carlos Augusto de Mendonça Lima, Gabriel Ivbijaro (eds.). Springer Cham, Switzerland, pp. 251–298.

Walker, F.O. 2007. Huntington's disease. *The Lancet*, *369*(9557), pp. 218–2289.

Wallin, M.T., Culpepper, W.J., Campbell, J.D., Nelson, L.M., Langer-Gould, A., Marrie, R.A., Cutter, G.R., Kaye, W.E., Wagner, L., Tremlett, H. and Buka, S.L. 2019. The prevalence of MS in the United States: a population-based estimate using health claims data. *Neurology*, *92*(10), pp. e1029–e1040.

Walton, C., King, R., Rechtman, L., Kaye, W., Leray, E., Marrie, R.A., Robertson, N., La Rocca, N., Uitdehaag, B., van Der Mei, I. and Wallin, M. 2020. Rising prevalence of multiple sclerosis worldwide: insights from the Atlas of MS. *Multiple Sclerosis Journal*, *26*(14), pp. 1816–1821.

Wareham, L.K., Liddelow, S.A., Temple, S., et al. 2022. Solving neurodegeneration: common mechanisms and strategies for new treatments. *Molecular Neurodegeneration*, *17*, Article number: 23.

Weaver, F.M., Follett, K., Stern, M., Hur, K., Harris, C., Marks, W.J. Jr., Rothlind, J., Sagher, O., Reda, D., Moy, C.S., Pahwa, R., Burchiel, K., Hogarth, P., Lai, E.C., Duda, J.E., Holloway, K., Samii, A., Horn, S. and Bronstein, J. 2009. Bilateral deep brain stimulation vs best medical therapy for patients with advanced Parkinson disease: a randomized controlled trial. *JAMA*, *301*(1), pp. 63–736.

Wild, E.J. and Tabrizi, S.J. 2017. Therapies targeting DNA and RNA in Huntington's disease. *The Lancet Neurology*, *16*(10), pp. 837–8472.

Wolk, D.A. and Dickerson, B.C. 2021. Clinical features and diagnosis of Alzheimer disease.

Wolters, F.J., Tinga, L.M., Dhana, K., Koudstaal, P.J., Hofman, A., Bos, D., Franco, O.H. and Ikram, M.A. 2019. Life expectancy with and without dementia: a population-based study of dementia burden and preventive potential. *American Journal of Epidemiology*, *188*(2), pp. 372–381.

World Health Organization (WHO). 2021. Global action plan on the public health response to dementia 2017–2025.

World Health Organization. 2024. Over 1 in 3 people affected by neurological conditions, the leading cause of illness and disability worldwide.

Young, P.N.E., Estarellas, M., Coomans, E., Srikrishna, M., Beaumont, H., Maass, A., Venkataraman, A.V., Lissaman, R., Jiménez, D., Betts, M.J., McGlinchey, E., Berron, D., O'Connor, A., Fox, N.C., Pereira, J.B., Jagust, W., Carter, S.F., Paterson, R.W. andSchöll, M. 2020. Imaging biomarkers in neurodegeneration: current and future practices. *Alzheimer's Research & Therapy*, *12*, p. 492.

Zhang, D., Lee, E.K.P., Mak, E.C.W., Ho, C.Y., Wong, S.Y.S., 2021. Mindfulness-based interventions: an overall review. Br. Med. Bull. 138, 41–57. https://doi.org/10.1093/bmb/ldab005

Zhang, H., Jiang, X., Ma, L., Wei, W., Li, Z., Chang, S., Wen, J., Sun, J. and Li, H. 2022. Role of Aβ in Alzheimer's-related synaptic dysfunction. *Frontiers in Cell and Developmental Biology*, *10*, p. 964075.

Zlokovic, B.V., Gottesman, R.F., Bernstein, K.E., et al. 2018. Vascular contributions to cognitive impairment and dementia (VCID): a report from the National Institute of Neurological Disorders and Stroke and National Institute on Aging VCID Biomarkers Consortium. *Alzheimer's & Dementia*, *14*(11), p. 1510–1523.

Zuccato, C., Valenza, M. and Cattaneo, E. 2010. Molecular mechanisms and potential therapeutical targets in Huntington's disease. *Physiological Reviews*, *90*(3), pp. 905–9817.

2 Small Molecules
An Emerging Therapeutic Frontier

Urmilla John, Nawab John Dar, and Shahnawaz Ali Bhat

ABBREVIATIONS

AChE	Acetylcholinesterase
AD	Alzheimer's Disease
ADME	Absorption, Distribution, Metabolism and Excretion
ALS	Amyotrophic Lateral Sclerosis
BBB	Blood–Brain Barrier
CNS	Central Nervous System
CRISPR	Clustered Regularly Interspaced Short Palindromic Repeats
CSF	Cerebrospinal Fluid
DAT	Dopamine Transporter
HD	Huntington's Disease
HDAC	Histone Deacetylase
HTS	High-Throughput Screening
log *P*	Partition Coefficient
MAO-B	Monoamine Oxidase-B
miRNAs	microRNAs
NMDA	N-Methyl-D-Aspartate
PD	Parkinson's Disease
PET	Positron Emission Tomography
PK	Pharmacokinetics
SBDD	Structure-Based Drug Design
V_d	volume of distribution

2.1 INTRODUCTION

Neurodegenerative disorders are debilitating conditions characterized by progressive neuronal loss and associated functional decline (Lamptey et al., 2022). Despite a significant amount of studies being carried out to comprehend the cellular mechanisms underpinning the emergence of these disorders, there are currently no identified treatments to appreciably cure, prevent or inhibit the progression of these disorders (Troncoso-Escudero et al., 2020). Furthermore, the pharmaceutical approaches used currently are palliative, and they neither halt nor reverse the disease progression (Durães et al., 2018). Thus, the absence of reliable and clinically significant outcomes for gene and cellular therapy in addition to their high cost and delivery challenges emphasizes the pressing necessity to explore alternative therapeutic options. Regarding the crucial circumstances and various efforts that have contributed to defining and/or revising the pathophysiological processes underlying neurodegenerative disorders, there is an intriguing trajectory for the rational design and development of novel small molecules that target those mechanisms to prevent and/or reverse the advancement of these disorders. The development of small molecule therapeutics has gained traction because they possess desirable characteristics such as low molecular weight, ease of synthesis and affinity to reach intracellular targets (Scott et al., 2016). This chapter explores the crucial attributes of small molecules that render them viable options

DOI: 10.1201/9781003520610-2

for the treatment of these neurodegenerative disorders including their ability to penetrate the blood–brain barrier (BBB), modulating specific biological pathways and their generally favourable physicochemical, pharmacokinetic and pharmacodynamic profiles.

2.2 PROPERTIES OF SMALL MOLECULES

2.2.1 Physicochemical Properties

The physicochemical properties of small molecules significantly impact their potential as therapeutic agents. These properties are important when considering how well these molecules interact with their biological targets, move to their target location or preferable tissues and thus define their overall therapeutic potential.

2.2.1.1 Molecular Size and Weight

The pharmacokinetic characteristics of small molecules are largely determined by molecular size and weight. The 'Rule of Five,' proposed by Lipinski et al., provides guidelines for the drug-like properties of small molecules based on important molecular characteristics, including molecular weight, lipophilicity and hydrogen bonding capabilities (Lipinski et al., 2001). The molecular weight of small molecules is typically less than 900 Da. Specifically, their low molecular size and weight render it easier for them to diffuse across cell membranes, particularly the BBB, which is a significant barrier to the central nervous system (CNS)-targeted therapies (Mikitsh & Chacko, 2014; Pardridge, 2012). The ability to penetrate the BBB is crucial for treating neurodegenerative disorders, as it enables the therapeutic agents to reach the brain parenchyma, the site of the pathological processes. For example, donepezil, a small molecule with a molecular weight of 379.5 Da, is used as one of the therapeutics for Alzheimer's disease (AD) treatment. It is able to cross the BBB effectively and enable the inhibition of acetylcholinesterase (AChE), raising the level of acetylcholine in the brain and enhancing cognitive functions (Zambrano et al., 2019).

2.2.1.2 Lipophilicity and Hydrophilicity

Lipophilicity refers to the property that allows a molecule to dissolve in fats, oils and lipids and is crucial for small molecules targeting CNS disorders. It is assessed by the partition coefficient (log P) and impacts the absorption, distribution, metabolism and excretion (ADME) properties of the small molecules (Mikitsh & Chacko, 2014). Lipophilic molecules possess the ability to access brain tissues more readily by passing through the BBB. It has been reported to be a key determinant of the ability of small molecules to cross the BBB. However, molecules that are excessively lipophilic may have poor solubility and higher toxicity (Chowdhury et al., 2021). Finding the most appropriate balance between lipophilicity and hydrophilicity is critical to achieving the ideal pharmacokinetic profile for the efficacy of a CNS drug (Rankovic, 2015). For instance, memantine is one of the *N*-methyl-D-aspartate (NMDA) receptor antagonists used in the treatment of AD. It holds a log P value of 3.28 and could penetrate into the CNS with appropriate solubility to distribute around the body (Kaur et al., 2020).

2.2.1.3 Solubility and Permeability

The solubility in biological fluids and permeability through biological membranes are the primary factors in determining a drug's bioavailability (Avdeef, 2012). Solubility in both aqueous and lipid environments is crucial for effective drug absorption and distribution of small molecules. However, effective drug absorption and distribution need an appropriate balance between hydrophilicity and lipophilicity (Milanetti et al., 2016). Adequate solubility in aqueous environments ensures that the drug can be transported through the bloodstream and reach its target site. The therapeutic efficiency of poorly soluble drugs may be compromised due to their limited bioavailability (Espinar et al., 2024). Drug development critically depends on techniques to increase permeability and solubility, such as prodrug synthesis and the use of permeability enhancers (Rodriguez-Aller et al., 2015).

For example, the solubility of riluzole, a drug used in amyotrophic lateral sclerosis (ALS), has been optimized to ensure that it reaches therapeutic concentrations in the CNS while maintaining sufficient bioavailability for clinical efficacy (Dash et al., 2018).

2.2.1.4 Chemical Stability and Reactivity

Chemical stability against metabolic degradation is crucial to ensure that the small molecule remains intact and active in the body for sufficient time duration. Thus, compounds that are too labile will degrade rapidly and result in the loss of their inbuilt therapeutic efficiency (Pattee et al., 2023). Therefore, factors that may influence the stability of small molecules within the biological system like vulnerability to enzymatic degradation and oxidative processes must be taken into consideration throughout the design process. An example of a chemically stable small molecule is edaravone, used in ALS, which has been designed to resist oxidative degradation, thus maintaining its efficacy over extended periods (Pattee et al., 2023). Reactivity is concerned with the ability of a molecule to undergo chemical transformations, which may impact its metabolic fate and interaction with biological targets. The optimization of both stability and reactivity is crucial for the development of effective and safe therapeutic agents (Thompson, 2001).

2.2.1.5 Binding Affinity and Selectivity

Finding small molecules possessing good binding affinity and selectivity for their target is essential for their efficacy and safety. High binding affinity ensures that the small molecules interact effectively with the target, while the selectivity decreases the probability of off-target interactions and potential side effects (Hughes et al., 2011). The development of small molecules with improved binding properties could be facilitated by advances in structure-based drug design (SBDD), therefore enhancing the therapeutic potential of these small molecules (Wu et al., 2023).

2.2.2 Pharmacokinetic Properties

The pharmacokinetics (PK) of small molecules is critically important in dictating their efficacy as therapeutic agents, especially as they pertain to neurodegenerative diseases, such as AD, Parkinson's disease (PD) and ALS. Key pharmacokinetic considerations include ADME properties of small molecules and how well they can cross the BBB. A comprehensive understanding of these aspects is critical for enhancing the delivery efficiency to the CNS, thereby maximizing therapeutic efficacy while minimizing undesirable side effects.

2.2.2.1 Absorption

The bioavailability of orally administered drugs refers to the fraction of the administered dose that reaches the systemic circulation in an un-metabolized and active form. For neurodegenerative diseases, oral administration is particularly important because of the ease associated with oral intake, which enhances patient compliance particularly, given the chronicity of these disorders (Bawa et al., 2016). Optimized small molecules for better absorption after oral administration reach therapeutic effective concentrations without any invasive routes of drug administration (Akhtar et al., 2021). Thus, solubility, permeability and stability in the gastrointestinal tract are factors affecting oral bioavailability (Song et al., 2004).

2.2.2.2 Distribution

Effective distribution in the body, specifically across the BBB, is very essential for small molecules targeting the CNS. The potential to cross the BBB enables these molecules to reach and exert their impact on brain tissues. The important parameters regarding the distribution of small molecules include volume of distribution (V_d), plasma protein binding and brain distribution (Chowdhury et al., 2021). The V_d of a drug is a measure of the extent to which it is distributed throughout the body compartments. For drugs that act on the CNS, a high volume of distribution may reflect extensive tissue binding. This type of drug binding might show extensive binding to tissue proteins in the brain. The lower the V_d, the more likely the drug will be confined to the bloodstream (Neumaier et al., 2021). Many small molecules have a higher binding to plasma proteins like albumin. The pharmacologically active fraction is only the unbound fraction of the drug, which is free to cross the BBB. High plasma protein binding has the potential to act as a reservoir that may prolong the drug's duration of action, but it may also lead to potential drug–drug interactions (Tesseromatis and Alevizou, 2008). Thus, the major factors affecting the distribution of small molecules include molecular weight, lipophilicity and the presence of active transport mechanisms.

2.2.2.3 Metabolism

Drug metabolism or the chemical alteration of the orally administered drug occurs predominately in the liver through enzymatic processes. Ideally, the metabolic pathways should produce non-toxic metabolites, maintaining the integrity and prolonged efficacy of the active components without causing harm. An understanding of the metabolism of small molecules is crucial in predicting their PK and potential adverse side effects. The major enzyme system involved in this process is the cytochrome P450 enzyme system. This system is responsible for converting a drug into a more hydrophilic compound that can be easily excreted (Zhao et al., 2021). For small molecules used for neurodegenerative disorders, it is important to balance metabolic stability with the formation of potentially active metabolites. A drug with high metabolic stability will have a longer half-life and less frequent dosing may be sufficient, which could be beneficial for chronic diseases. On the other hand, significant metabolism can result in the production of toxic metabolites, necessitating careful monitoring and optimization during drug development (Thompson, 2001; Słoczyńska et al., 2019). Metabolism of small molecules can be influenced by other medications and this is especially important for elderly patients who frequently require multiple drugs, which may either inhibit or induce metabolic enzymes. Metabolic stability can be enhanced by structural modification that lessens their susceptibility to enzymatic degradation (Nassar et al., 2004).

2.2.2.4 Excretion

Effective removal of small molecules from the body, primarily through renal or hepatic routes, must be assured to prevent accumulation and potential toxicity (Ajavon and Taft, 2016.). The excretory pathways of small molecules should be considered in their design so that they do not build up to toxic levels. Then, the removal mechanisms for small molecules in the body are also a critical consideration to ensure safe and effective elimination (Zhou et al., 2016). Thus, excretion is the final process in the elimination process of small molecules and occurs mainly through the kidneys or biliary system. The rate of excretion is dependent on the drug's chemical and physical characteristics, including molecular size, polarity and the presence of ionized groups that facilitate conjugation and renal clearance (Franklin, 2012.). Renal excretion is the most common route of excretion for small molecules, in particular for metabolites that are hydrophilic. Most drugs are eliminated from the body by metabolism and renal excretion, which generally occurs at a constant rate (i.e., linear). The half-life of a drug is influenced by the rate of metabolism and the rate of excretion and is used to determine dosing schedules (Jang et al., 2001). For the treatment of neurodegenerative disorders, a drug with a long half-life could reduce the dose frequency and improve patient adherence, which is particularly useful for long-term treatment. However, drugs that have prolonged half-lives require monitoring to prevent accumulation and potential toxicity, particularly in patients with a compromised kidney or liver function (Verbeec and Musuamba, 2009).

2.2.3 Pharmacodynamic Properties

The study of pharmacodynamics pertains to the effects of drugs on biochemical and physiological processes and their molecular and cellular modes of action.

2.2.3.1 Dose–Response Relationship

The concept of the dose–response relationship is how the effect of a drug changes in magnitude as its concentration changes. The maximum effect that a drug can produce is known as efficacy, and potency is the concentration required to produce a particular effect regardless of the dose (Tallarida and Jacob, 2012). For example, in AD, small molecules like cholinesterase inhibitors (e.g. donepezil) exhibit a clear dose–response relationship, where higher doses of the drug clearly result in greater efficacy (Imbimbo, 2001). These drugs inhibit the degradation of acetylcholine, resulting in increased acetylcholine levels in the brain and enhanced cognitive function (Tsuno, 2009). The dose–response curve therefore serves as an important tool in determining the optimal dose that provides the most therapeutic benefit while causing minimal side effects.

2.2.3.2 Therapeutic Window and Index

The 'therapeutic window' of a drug is the concentration range where it can be effective without causing serious side effects. When a drug has a narrow therapeutic window, it must be dosed and monitored accordingly to maintain the therapeutic effect, while minimizing the possibility for toxicity. For neurodegenerative diseases, there is a need for a wide therapeutic window due to the long-term drug treatment necessary for these chronic diseases. For instance, memantine which acts as an NMDA receptor antagonist in AD is associated with a favourable therapeutic index, thereby allowing for effective symptom management with reduced side effects (Rogawski and Wenk, 2003).

2.2.3.3 Biomarkers and Surrogate Endpoints

Biomarkers are measurable biological parameters that are used to assess normal processes, pathogenic processes and therapeutic responses. In neurodegenerative diseases, biomarkers are essential for diagnostic and prognostic purposes and for measuring response to therapies. Surrogate endpoints, a group of biomarkers are used as substitutes for clinical endpoints in clinical trials and give indications of effectiveness of a drug earlier than clinical endpoints. Amyloid-beta and tau are important biomarkers in AD, which can be measured in cerebrospinal fluid (CSF) or visualized by positron emission tomography (PET) imaging. Small molecules like BACE1 inhibitors work to decrease the production of Aβ, and the efficacy of these drugs can be measured by measuring changes in Aβ levels in CSF or PET imaging (Vassar, 2014). Similarly, the effect of tau aggregation inhibitors on disease progression can be determined by tau PET imaging data. In PD, biomarkers including α-synuclein, dopamine transporter (DAT) imaging and urate levels are being investigated. Of these, DAT imaging is particularly relevant as a surrogate endpoint and evaluates the integrity of dopaminergic neurons following exposure to drugs like levodopa or dopamine agonists (Algarni and Stoessl, 2016). Monitoring these markers helps to determine the drug effect and genomic effects within the context of therapies aimed at altering the disease course or alleviating symptoms in neurodegenerative disorders.

2.2.3.4 Drug–Receptor Interactions

The pharmacodynamic characteristics of small molecules are fundamentally regulated by drug–receptor interactions. These properties include the potency, efficacy and specificity of a drug. An understanding of drug–receptor interactions is critical for developing therapeutic modalities and reducing the likelihood of side effects. For example, the binding affinity of the drug to its target receptor determines its potency. A drug with a greater affinity requires a lower concentration to produce the desired effect. Drug–receptor selectivity, on the contrary, refers to the ability of a drug to recognize one type of receptor over others. Selectivity reduces the probability of off-target pharmacological effects and drug toxicity. For example, in PD, dopamine receptor agonists like pramipexole and ropinirole are highly selective for dopamine (D1) receptors, which are associated with regulating motor control (Ferraiolo and Hermans, 2023.). This high degree of selectivity reduces side effects compared to non-selective agonists. Furthermore, small molecules can operate as agonists, antagonists and allosteric modulators. Agonists activate receptors to create a biological response, while antagonists prevent receptor activation. Allosteric modulators work in other sites (not the active site) and modulate the response of the receptor to its endogenous ligand. In AD, memantine acts as an NMDA receptor antagonist to prevent excitotoxicity by antagonizing the excessive glutamate activity (Kabir et al., 2019). This mechanism helps to manage symptoms without completely inhibiting the physiological functions of the NMDA receptor.

2.3 MECHANISMS OF ACTION

2.3.1 Enzyme Inhibition or Activation

The disease-regulating pathways can be altered by small molecules through the inhibition or activation of specific enzymes. For example, AChE inhibitors like donepezil are used to treat AD by preventing the breakdown of acetylcholine, thus improving cholinergic transmission (Marucci et al., 2021). Enzyme inhibition can be used to restore normal function by correcting neurotransmitter alterations or imbalances in metabolic pathways.

2.3.2 Receptor Modulation

Neurotransmitter receptors can be targeted by small molecules that bind to and modulate their activity, leading to a restoration of their normal function. For instance, in the case of PD, the loss of dopaminergic neurons can be compensated by dopamine receptor agonists, i.e., pramipexole (Albrecht and Buerger, 2009). Similarly, to compensate for the loss of dopaminergic neurons, the brain converts the small molecule levodopa to dopamine in PD. Dopamine then binds to its corresponding receptors, and motor function is then restored (You et al., 2018). The efficacy of levodopa illustrates the critical importance of the binding of receptors and their downstream signalling in the pharmacodynamics of small molecules. Thus, symptomatic relief may be provided by altering signalling pathways in the brain as a result of receptor modulation.

2.3.3 Inhibition of Protein Aggregation

Protein aggregation is a common feature in many neurodegenerative diseases. For example, in AD, amyloid-beta (Aβ) peptides aggregate to form plaques, while in PD α-synuclein forms Lewy bodies. In Huntington's disease (HD), mutant huntingtin (mHTT) aggregates and ALS, proteins such as TDP-43 misfold and aggregate. These aggregates can be toxic to neurons and contribute to disease progression. Prevention or interference with the assembly of misfolded proteins is a key strategy for therapeutics in neurodegenerative disorders. An example of this approach in the context of AD is the development of small molecule inhibitors to target amyloid-beta (Aβ) aggregation. By targeting the aggregation of Aβ, small molecules can prevent the formation of toxic oligomers and fibrils that lead to neuronal damage. For example, molecular chaperones or pharmacological chaperones potentially could stabilize the normal or desired folding state of a protein, thereby preventing aggregation (Hartl et al., 2011). Some small molecules have been identified that inhibit the nucleation phase of aggregation, preventing the initial formation of oligomers, while other molecules inhibit the elongation phase, where existing oligomers grow into larger aggregates (Habchi, et al., 2017). Many small molecules can induce the depolymerization of preformed aggregates, which potentially reverses the aggregation process. An example of one of these molecules is epigallocatechin gallate, a polyphenol found in green tea, which has shown promise in depolymerizing Aβ aggregates (Lin et al., 2023). Thus, small molecules provide a therapeutic approach to slow down the progression of neurodegenerative disorders by inhibiting protein aggregation in a variety of ways.

2.3.4 Neuroprotection

The development of neuroprotective strategies aims to prevent neuronal loss and promote neuronal survival, growth and functioning. Small molecules with neuroprotective effects are aimed at protecting neurons from apoptosis, oxidative stress and other forms of cellular damage through various mechanisms. Oxidative stress plays a decisive role in neuronal damage in neurodegenerative diseases. Small molecules with antioxidant properties have the ability to scavenge free radicals, mitigate oxidative stress and increase the endogenous antioxidant defence of neurons. For instance, edaravone is a small molecule that has been approved for ALS due to its antioxidant effect (Cho and Shukla, 2020.). Similarly, riluzole has also been used in the treatment of ALS and exerts a neuroprotective effect by regulating glutamate release, thereby reducing excitotoxicity and inhibiting neuronal death (Golmohammadi et al., 2024). Mitochondrial dysfunction presents another critical point for the development of neurodegenerative conditions. Coenzyme Q10 and creatine are small molecules that have been shown to stabilize the mitochondrial membrane, reduce oxidative stress and maintain adenosine triphosphate (ATP) levels (Beal, 2011; Salama et al., 2013). Furthermore, excessive release of glutamate and consequent overactivation of NMDA receptors can cause neurodegeneration and excitotoxicity. Thus, memantine, an NMDA receptor antagonist is used in AD to prevent excitotoxicity (Volbracht et al., 2006). Small molecules can also serve to prevent apoptosis, a form of programmed cell death. This can be done through targeting pathways like caspase cascade and members of the Bcl-2 family of proteins (Dailah, 2022). Thus, neuroprotection by small molecules can be provided to preserve neuronal function and integrity and may mitigate damage-induced neurodegeneration to slow disease progression.

2.3.5 Anti-inflammatory Effects

Neuroinflammation is a common occurrence in neurodegenerative diseases. So reducing neuroinflammation can effectively protect neural tissue and thus alleviate symptoms. The small molecules are able to act as neuroprotective agents by modulating inflammatory pathways to reduce the destructive effects of chronic inflammation on neurons (Ivanenkov et al., 2008). Small molecules have the potential to inhibit the activation of microglia in the brain, which are the resident immune cells, thus reducing the production of proinflammatory cytokines such as TNF-α, IL-1β and IL-6. The antibiotic minocycline has been shown to exhibit anti-inflammatory properties and has been studied for its effect on microglial activation in HD and ALS (Leite, 2021). Small compounds are also able to directly modulate cytokine production. For instance, nonsteroidal anti-inflammatory drugs such as ibuprofen have been studied for their possible application in the attenuation of inflammation in AD by reducing prostaglandin production (Lim et al., 2000). Furthermore, cytokines and chemokines released by astrocytes contribute to neuroinflammation. It has been shown that a number of small molecules such as riluzole used in the treatment of ALS can reduce astrocytic activation and glutamate release, thereby providing anti-inflammatory and neuroprotective effects (Cheah et al., 2010). In addition to suppressing proinflammatory signals, some small molecules can activate endogenous anti-inflammatory pathways. As an example, nuclear factor erythroid 2–related factor 2 (Nrf2) activators, such as dimethyl fumarate (DMF), used in the treatment of multiple sclerosis can induce the transcription of antioxidant and anti-inflammatory genes (Brück et al., 2013).

2.3.6 Gene Expression Regulation

The regulation of gene expression is essential for normal neuronal function and protection against stress conditions. Dysregulation of gene expression is involved in the pathophysiology of several neurodegenerative diseases. At the level of gene regulation, small molecules can regulate gene expression by targeting transcription factors and non-coding RNAs. Some of the small molecules appear to modulate gene expression through their interaction with DNA or RNA or are known to impact epigenetic mechanisms, such as methylation or histone acetylation. Small molecules such as histone deacetylase (HDAC) inhibitors can trigger changes in histone acetylation level, resulting in the modification of the chromatin structure and gene expression (Kumar et al., 2022). For example, HDAC inhibitors have been observed to increase the expression of neuroprotective genes and decrease the expression of pro-apoptotic genes (Dietz and Casaccia, 2010). Some small molecules can also have an effect on the activity of certain transcription factors that regulate neurodegenerative pathways. For example, NRF2 is a transcription factor responsible for the expression of many antioxidant and cytoprotective genes. Small molecules like DMF are able to activate NRF2, which is a very interesting therapeutic approach and offers neuroprotection in multiple sclerosis and other neurodegenerative diseases (Scuderi et al., 2020). Furthermore, non-coding RNAs, such as microRNAs (miRNAs), play a pivotal role in post-transcriptional gene regulation. Small molecules that target miRNAs or miRNA processing can modulate genes that are implicated in neurodegenerative processes. For example, antagomirs are chemically modified oligonucleotides that have been designed to inhibit certain miRNAs and potentially alter disease-related gene expression patterns (Gentile et al., 2022). Moreover, the advent of gene-editing technologies, such as clustered regularly interspaced short palindromic repeats (CRISPR/Cas9) has opened new possibilities for therapeutic gene expression modulation. Small molecules can be employed to modulate CRISPR/Cas9 activity, presenting potential treatments for genetically associated neurodegenerative conditions (Karimian et al., 2020).

2.4 EXAMPLES OF THERAPEUTIC SMALL MOLECULES

2.4.1 Alzheimer's Disease

2.4.1.1 Donepezil

Donepezil is an AChE inhibitor that increases the level of acetylcholine in the brain to enhance cognitive function. It holds good pharmacokinetic properties such as its moderate lipophilicity and excellent oral bioavailability, which enable it to effectively penetrate the BBB (Pooladgar et al., 2022). It has also been found to offer modest benefits regarding cognitive and functional performance for AD patients (Black et al., 2007).

2.4.1.2 Memantine

Memantine is classified as an NMDA receptor antagonist. It is an agent that modulates the activity of glutamate to prevent excitotoxicity (Kutzing et al., 2012). Memantine has been shown to improve both cognitive and functional outcomes in individuals with moderate to severe AD (Winblad et al., 2010).

2.4.2 Parkinson's Disease

2.4.2.1 Levodopa

Levodopa is a dopamine precursor that is used to maintain brain dopamine levels and subsequently alleviates motor symptoms (Clarke, 2004). It is currently the most effective symptomatic treatment for PD, although long-term use has the potential to trigger motor complications, such as dyskinesia (Mueller and Russ, 2006).

2.4.2.2 Rasagiline

Rasagiline is a type of monoamine oxidase-B (MAO-B) inhibitor that works by enhancing dopamine levels by blocking its degradation. With its extended half-life and excellent oral bioavailability, it has a good pharmacokinetic profile that permits once-daily dosage (Thébault et al., 2004). Studies have indicated that it can enhance motor symptoms in PD patients by increasing dopamine neurotransmission and potentially offering neuroprotective functions (Pistacchi et al., 2014).

2.4.3 Amyotrophic Lateral Sclerosis

2.4.3.1 Riluzole

Riluzole is a glutamate-release inhibitor that helps to reduce glutamate-induced excitotoxicity in the brain. In ALS, riluzole is thought to have neuroprotective effects via modifying glutamate signalling. It needs to be taken twice a day due to its short half-life and moderate bioavailability (Cheah et al., 2010). By lowering excitotoxicity and neuroinflammation, it is the only medication approved so far to prolong the life expectancy of individuals with ALS (Bellingham, 2011).

2.4.3.2 Edaravone

Edaravone is an antioxidant that is known for its ability to decrease oxidative stress in neurons, leading to a slowdown in the progression of neurodegenerative diseases (Liu et al., 2014; Nagase et al., 2016). Studies have demonstrated that edaravone can help mitigate the decline in functional ability among individuals with ALS (Cho and Shukla, 2020).

2.5 COST-EFFECTIVENESS OF SMALL MOLECULE THERAPEUTICS

2.5.1 Drug Development Costs

Contrary to the more expensive biologics, small molecule therapeutics are more affordable and cost-effective. The chemical synthesis of small molecules eliminates the need for complex biological production systems required by biologics (Samanen, 2013). In addition, technological advancements in high-throughput screening (HTS), computational drug development and synthetic chemistry have improved the drug discovery process, reducing the amount of time and total costs (Aldewachi et al., 2021).

2.5.2 Manufacturing Expenses

Manufacturing small molecules can be relatively simple and is typically scalable. Chemical synthesis offers several distinct advantages in generating large quantities of a drug with high purity and homogeneity. Some of the key reasons indicating that small molecule therapeutics can be cost-effective include the availability of different synthetic routes and the potential for optimization of production processes (Simić et al., 2021).

2.5.3 Market Pricing and Accessibility

Small molecules frequently result in more affordable medications because they have lower production costs and simpler distribution channels (Makurvet, 2021). These advantages are particularly relevant for neurodegenerative diseases, where patients may need long-term treatment. Affordable treatments have the potential to enhance patient compliance and outcomes, emphasizing the significance of small molecules in mitigating the global prevalence of neurodegenerative disorders (Robert et al., 2019).

2.6 EASE OF SYNTHESIS

2.6.1 Synthetic Pathways and Techniques

Designing synthetic pathways for small molecules with precision is crucial in efficiently executing the synthesis process. Among the most significant aspects in the development of a synthetic pathway is the selection of easily accessible, cost-effective and chemically compatible starting materials and intermediates to facilitate the subsequent chemical transformations (Akhtar et al., 2021).

The target molecule's functional groups and core structure typically dictate the design of the pathway, with emphasis placed on the ability to reduce the number of steps while maximizing overall yield and purity (Wilson & Danishefsky, 2006). Small molecule synthesis encompasses several methods involving organic synthesis, combinatorial chemistry and biocatalysis. Organic synthesis can be a means of constructing complex molecules in a step-wise manner (Wilson & Danishefsky, 2006). Combinatorial chemistry provides access to potent, diverse libraries of molecular structures from which potential drugs can be more readily identified (Kennedy et al., 2008). Biocatalysis entails the use of enzymes to catalyse specific chemical transformations for efficient, sustainable and environmentally friendly synthesis (Wohlgemuth, 2010).

2.6.2 Scalability of Production

One major advantage of small molecule drug development is the scalability of small molecule production. The ability to easily scale a synthetic route is contingent upon its robustness and efficiency, as well as the ease of achieving high product purity and yield on a larger scale. The possibility of chemically synthesizing great quantities of small molecules under the same quality criteria is essential to meet clinical and commercial needs (Wang et al., 2024). One of the first considerations is the availability and cost of starting materials and reagents in scaling up a process. For a process to be feasible economically at scale, the availability and cost of the starting materials and reagents must be readily available on a vast scale, acceptable and inexpensive. In addition to cost and availability, synthetic routes should be designed to use reagents and solvents that are not only effective but also environmentally benign in accordance with the principles of green chemistry (Pinto et al., 2021). Using 'green' reagents and solvents minimize the pollution and environmental impact of the whole production process (Pinto et al., 2021). As a result, small molecules can be synthesized and produced on a much larger scale very successfully, and they can be used to secure a reliable and sustainable supply.

2.6.3 Challenges in Synthesis

While there are a number of advantages to the synthesis of small molecules, it can be quite challenging with respect to complex structures. Challenges such as low yield, difficult purification and expensive reagents or catalysts can affect the cost and feasibility of synthesis (Simić et al., 2022). These obstacles can be confronted under innovative synthetic strategies and process optimization to develop successful small molecule therapeutics.

2.7 CURRENT RESEARCH AND DEVELOPMENT TRENDS

2.7.1 High-Throughput Screening

HTS is one of the most powerful tools in the pharmaceutical industry for identifying therapeutic leads from millions of library molecules (Mishra et al., 2008). It makes it quite possible for the researchers to quickly identify potential therapeutic candidates from large libraries of small compounds. This accelerates the discovery process and produces novel compounds of therapeutic interest. Improvements in automation and analysis have improved the throughput and efficiency of HTS (Sarnaik et al., 2020).

2.7.2 Structure-Based Drug Design

SBDD relies on the three-dimensional structures of biological targets to devise small molecules with high specificity and potency. Advances in computational chemistry and new developments in the field of structural biology have hugely refined SBDD in terms of accuracy and speed (Kim et al., 2020). A variety of techniques, such as X-ray crystallography and cryo-electron microscopy, can gather detailed structural information that may be used for the rational design of new compounds (Brito & Archer, 2020).

2.7.3 Combination Therapies

To maximize the therapeutic benefit and target multiple disease pathways simultaneously, the combination of small molecules with additional therapeutic strategies, including monoclonal antibodies or gene therapy, is currently being researched for their potential to improve efficacy (Vafaei et al., 2022; Zhu et al., 2021). This multi-target strategy can be used to enhance the effectiveness of treatment and address the complex nature of neurodegenerative disorders.

2.7.4 Personalized Medicine

The evolution of personalized medicine allows the development of unique interventions tailored to individual genetic profiles and disease phenotypes. Personalized approaches in medicine are an attempt to find the most responsive populations to optimize therapeutic interventions and personalize disease management strategies. In this regard, recent improvements in genomics and the discovery of biomarkers hold great promise in developing personalized therapies (Lee et al., 2021).

2.7.5 Advanced Delivery Systems

The use of nanoparticles and other advanced delivery systems enhances the ability to cross the BBB and achieve target specificity (Pinheiro et al., 2021). Advanced delivery technologies can be applied to improve the bioavailability and therapeutic effectiveness of small molecules. Such delivery systems being investigated for CNS-targeted therapeutics include lipid-based nanoparticles, polymeric nanoparticles and liposomes (Goyal et al., 2014).

2.8 CHALLENGES IN SMALL MOLECULE DEVELOPMENT

2.8.1 BBB Penetration

It is a significant challenge to achieve a sufficient concentration of small molecules in the brain with minimal systemic toxicity. The key strategies that have been proposed to improve BBB penetration by small molecules of therapeutic interest include optimization of lipophilicity, using carrier-mediated transport and newly developed delivery systems such as nanoparticles (Xiong et al., 2021).

2.8.2 Target Specificity

To decrease side effects and improve therapeutic efficacy, it is required to design small molecule drugs in such a way that they target pathological processes specifically without compromising normal physiological functions. High specificity can be attained using techniques such as SBDD or HTS of compound libraries (Brown & Boström, 2018).

2.8.3 Drug Resistance

Prolonged employment of small molecules can produce neuroadaptation that potentially diminishes their therapeutic effectiveness. Thus, the ability to characterize and intervene against the mechanisms of drug resistance will be vital in preserving therapeutic efficacy and it may require a combination therapy method with a rotation of therapeutic agents.

2.8.4 Metabolic and Toxicity Issues

To ensure patient safety, it is necessary that the metabolites generated during the metabolism of small molecules must be non-toxic. Comprehensive preclinical studies are needed to assess the metabolic profiles and toxicity potential of novel compounds. Structural modification can be employed to improve in vivo metabolism and reduce the amount of toxic metabolites formed (Boethling et al., 2007).

2.9 FUTURE DIRECTIONS

2.9.1 Neurogenesis and Synaptic Plasticity

Targeting pathways that promote neurogenesis and synaptic plasticity is a promising approach to promoting brain repair and functional recovery (Wolf & Abi-Dargham, 2023). Small molecules that enhance these processes may offer novel therapeutic opportunities for neurodegenerative diseases, and as such research efforts are currently aimed at identifying molecules that are potent inducers of neural stem cell proliferation and differentiation and enhance synaptic connectivity and plasticity (Silpa et al., 2022).

2.9.2 Biomarker Development

Biomarker identification that can foresee the response to therapy and monitor the progression of disease is integral to the development of effective treatments. Biomarkers help in identifying suitable candidates for clinical trials and in evaluating therapeutic efficiency. The discovery of novel biomarkers is driven by advances in proteomics, genomics and metabolomics.

2.9.3 Regenerative Medicine

The possible ability of small molecules to stimulate regenerative processes within the CNS represents an exciting opportunity for restoring lost function. Such regenerative medicine strategies would aim for the restoration of tissues that are damaged to recover normal brain function. Strategies include the promotion of endogenous repair mechanisms and the use of small molecules alone or in combination with stem cell therapies.

2.10 CONCLUSION

Small molecules have emerged as a versatile and promising strategy for the treatment of neurodegenerative disorders. Small molecules have the unique ability to finely modulate specific biological targets with high precision. This, in combination with advances in drug delivery and design, significantly improves the potential for patient treatment outcomes. Continued research and technological innovation improve the efficacy, specificity and safety of small molecule therapies. The ongoing evolution and development of this drug discovery and technology offer hope for effective treatments for complex and difficult-to-treat neurodegenerative diseases.

REFERENCES

Agarwal, P., Huckle, J., Newman, J. and Reid, D.L. 2022. Trends in small molecule drug properties: a developability molecule assessment perspective. *Drug Discovery Today*, *27*(12), p. 103366.

Ajavon, A.D. and Taft, D.R., 2016. Pharmacokinetics/ADME of small molecules. In *Preclinical Drug Development*. Mark Rogge and David R. Taft (eds.). CRC Press: Boca Raton, pp. 83–128.

Akhtar, A., Andleeb, A., Waris, T.S., Bazzar, M., Moradi, A.-R., Awan, N.R. and Yar, M. 2021. Neurodegenerative diseases and effective drug delivery: a review of challenges and novel therapeutics. *Journal of Controlled Release*, *330*, pp. 1152–1167. https://doi.org/10.1016/j.jconrel.2020.11.021

Albrecht, S. and Buerger, E., 2009. Potential neuroprotection mechanisms in PD: focus on dopamine agonist pramipexole. *Current Medical Research and Opinion*, *25*(12), pp. 2977–2987.

Aldewachi, H., Al-Zidan, R.N., Conner, M.T. and Salman, M.M., 2021. High-throughput screening platforms in the discovery of novel drugs for neurodegenerative diseases. *Bioengineering*, *8*(2), p. 30.

Algarni, M.A. and Stoessl, A.J., 2016. The role of biomarkers and imaging in Parkinson's disease. *Expert Review of Neurotherapeutics*, *16*(2), pp. 187–203.

Avdeef, A., 2012. *Absorption and Drug Development: Solubility, Permeability, and Charge State*. John Wiley & Sons: New York. https://doi.org/10.1002/9781118286067

Bawa, P., Pradeep, P., Kumar, P., Choonara, Y.E., Modi, G. and Pillay, V. 2016. Multi-target therapeutics for neuropsychiatric and neurodegenerative disorders. *Drug Discovery Today*, *21*(12), pp. 1886–1914. https://doi.org/10.1016/j.drudis.2016.08.001

Beal, M.F., 2011. Neuroprotective effects of creatine. *Amino Acids*, *40*(5), pp. 1305–1313.

Bellingham, M.C., 2011. A review of the neural mechanisms of action and clinical efficiency of riluzole in treating amyotrophic lateral sclerosis: what have we learned in the last decade? *CNS Neuroscience & Therapeutics*, *17*(1), pp. 4–31.

Black, S.E., Doody, R., Li, H., McRae, T., Jambor, K.M., Xu, Y., Sun, Y., Perdomo, C.A. and Richardson, S., 2007. Donepezil preserves cognition and global function in patients with severe Alzheimer disease. *Neurology*, *69*(5), pp. 459–469.

Boethling, R.S., Sommer, E. and DiFiore, D. 2007. Designing small molecules for biodegradability. *Chemical Reviews*, *107*(6), pp. 2207–2227. https://doi.org/10.1021/cr050952t

Brito, J.A. and Archer, M. 2020. Structural biology techniques: X-ray crystallography, cryo-electron microscopy, and small-angle X-ray scattering. In *Practical Approaches to Biological Inorganic Chemistry* (pp. 375–416). Elsevier. https://doi.org/10.1016/B978-0-444-64225-7.00010-9

Brown, D.G. and Boström, J. 2018. Where do recent small molecule clinical development candidates come from? *Journal of Medicinal Chemistry*, *61*(21), pp. 9442–9468. https://doi.org/10.1021/acs.jmedchem.8b00675

Brück, W., Gold, R., Lund, B.T., Oreja-Guevara, C., Prat, A., Spencer, C.M., Steinman, L., Tintoré, M., Vollmer, T.L., Weber, M.S. and Weiner, L.P., 2013. Therapeutic decisions in multiple sclerosis: moving beyond efficacy. *JAMA Neurology*, *70*(10), pp. 1315–1324.

Chaudhary, A., Nagaich, U., Gulati, N., Sharma, V.K. and Khosa, R.L., 2012. Enhancement of solubilization and bioavailability of poorly soluble drugs by physical and chemical modifications: a recent review. *Journal of Advanced Pharmacy Education and Research*, 2(1–2012), pp. 32–67.

Cheah, B.C., Vucic, S., Krishnan, A.V. and Kiernan, M.C., 2010. Riluzole, neuroprotection and amyotrophic lateral sclerosis. *Current Medicinal Chemistry*, *17*(18), pp. 1942–1959.

Chico, L.K., 2009. Small *Molecule Properties* in the *Design* and *Development* of *New* CNS *Therapeutic Candidates* (Doctoral dissertation). Northwestern University.

Cho, H. and Shukla, S., 2020. Role of edaravone as a treatment option for patients with amyotrophic lateral sclerosis. *Pharmaceuticals*, *14*(1), p. 29.

Chowdhury, E.A., Noorani, B., Alqahtani, F., Bhalerao, A., Raut, S., Sivandzade, F. and Cucullo, L. 2021. Understanding the brain uptake and permeability of small molecules through the BBB: a technical overview. *Journal of Cerebral Blood Flow & Metabolism*, *41*(8), pp. 1797–1820. https://doi.org/10.1177/0271678X20985946

Clarke, C.E., 2004. Neuroprotection and pharmacotherapy for motor symptoms in Parkinson's disease. *The Lancet Neurology*, *3*(8), pp. 466–474.

Dahl, S.G., Aarons, L., Gundert-Remy, U., Karlsson, M.O., Schneider, Y.J., Steimer, J.L. and Trocóniz, I.F., 2010. Incorporating physiological and biochemical mechanisms into pharmacokinetic–pharmacodynamic models: a conceptual framework. *Basic & Clinical Pharmacology & Toxicology*, *106*(1), pp. 2–12.

Dailah, H.G., 2022. Potential of therapeutic small molecules in apoptosis regulation in the treatment of neurodegenerative diseases: an updated review. *Molecules*, *27*(21), p. 7207.

Dandapani, S., Rosse, G., Southall, N., Salvino, J.M. and Thomas, C.J., 2012. Selecting, acquiring, and using small molecule libraries for high-throughput screening. *Current Protocols in Chemical Biology*, *4*(3), pp. 177–191.

Dash, R.P., Babu, R.J. and Srinivas, N.R. 2018. Two decades-long journey from riluzole to edaravone: Revisiting the clinical pharmacokinetics of the only two amyotrophic lateral sclerosis therapeutics. *Clinical Pharmacokinetics*, *57*(11), pp. 1385–1398. https://doi.org/10.1007/s40262-018-0655-4

Dietz, K.C. and Casaccia, P., 2010. HDAC inhibitors and neurodegeneration: at the edge between protection and damage. *Pharmacological Research*, *62*(1), pp. 11–17.

Durães, F., Pinto, M. and Sousa, E. 2018. Old drugs as new treatments for neurodegenerative diseases. *Pharmaceuticals*, *11*(2), p. 44. https://doi.org/10.3390/ph11020044

Ferraiolo, M. and Hermans, E., 2023. The complex molecular pharmacology of the dopamine D2 receptor: implications for pramipexole, ropinirole, and rotigotine. *Pharmacology & Therapeutics*, *245*, p. 108392.

Franklin, M.R., 2012. *Drug Absorption, Distribution, Metabolism, and Excretion*. Remington, p. 739.

Galloway, W.R., Isidro-Llobet, A. and Spring, D.R., 2010. Diversity-oriented synthesis as a tool for the discovery of novel biologically active small molecules. *Nature Communications*, *1*(1), p. 80.

García-Nafría, J. and Tate, C.G., 2020. Cryo-electron microscopy: moving beyond X-ray crystal structures for drug receptors and drug development. *Annual Review of Pharmacology and Toxicology*, *60*(1), pp. 51–71.

Gentile, G., Morello, G., La Cognata, V., Guarnaccia, M., Conforti, F.L. and Cavallaro, S., 2022. Dysregulated miRNAs as biomarkers and therapeutical targets in neurodegenerative diseases. *Journal of Personalized Medicine*, *12*(5), p. 770.

Golmohammadi, M., Mahmoudian, M., Hasan, E.K., Alshahrani, S.H., Romero-Parra, R.M., Malviya, J., Hjazi, A., Najm, M.A., Almulla, A.F., Zamanian, M.Y. and Kadkhodaei, M., 2024. Neuroprotective effects of riluzole in Alzheimer's disease: a comprehensive review. *Fundamental & Clinical Pharmacology*, *38*(2), pp. 225–237.

Goyal, K., Koul, V., Singh, Y. and Anand, A. 2014. Targeted drug delivery to central nervous system (CNS) for the treatment of neurodegenerative disorders: trends and advances. *Central Nervous System Agents in Medicinal Chemistry*, *14*(1), pp. 43–59. https://doi.org/10.2174/1871524914666141030145948

Habchi, J., Chia, S., Limbocker, R., Mannini, B., Ahn, M., Perni, M., Hansson, O., Arosio, P., Kumita, J.R., Challa, P.K. and Cohen, S.I., 2017. Systematic development of small molecules to inhibit specific microscopic steps of Aβ42 aggregation in Alzheimer's disease. *Proceedings of the National Academy of Sciences*, *114*(2), pp. E200–E208.

Hartl, F.U., Bracher, A. and Hayer-Hartl, M., 2011. Molecular chaperones in protein folding and proteostasis. *Nature*, *475*(7356), pp. 324–332.

Hughes, J., Rees, S., Kalindjian, S. and Philpott, K. 2011. Principles of early drug discovery. *British Journal of Pharmacology*, *162*(6), pp. 1239–1249. https://doi.org/10.1111/j.1476-5381.2010.01127.x

Imbimbo, B.P., 2001. Pharmacodynamic-tolerability relationships of cholinesterase inhibitors for Alzheimer's disease. *CNS Drugs*, *15*, pp. 375–390.

Ivanenkov, Y.A., Balakin, K.V. and Tkachenko, S.E., 2008. New approaches to the treatment of inflammatory disease: focus on small-molecule inhibitors of signal transduction pathways. *Drugs in R & D*, *9*, pp. 397–434.

Jang, G.R., Harris, R.Z. and Lau, D.T., 2001. Pharmacokinetics and its role in small molecule drug discovery research. *Medicinal Research Reviews*, *21*(5), pp. 382–396.

Kabir, M.T., Sufian, M.A., Uddin, M.S., Begum, M., Akhter, S., Islam, A., Mathew, B., Islam, M.S., Amran, M.S. and Md Ashraf, G., 2019. NMDA receptor antagonists: repositioning of memantine as a multitargeting agent for Alzheimer's therapy. *Current Pharmaceutical Design*, *25*(33), pp. 3506–3518.

Karimian, A., Gorjizadeh, N., Alemi, F., Asemi, Z., Azizian, K., Soleimanpour, J., Malakouti, F., Targhazeh, N., Majidinia, M. and Yousefi, B., 2020. CRISPR/Cas9 novel therapeutic road for the treatment of neurodegenerative diseases. *Life Sciences*, *259*, p. 118165.

Kaur, A., Nigam, K., Srivastava, S., Tyagi, A. and Dang, S. 2020. Memantine nanoemulsion: a new approach to treat Alzheimer's disease. *Journal of Microencapsulation*, *37*(5), pp. 355–365. https://doi.org/10.1080/02652048.2020.1756971

Kennedy, J.P., Williams, L., Bridges, T.M., Daniels, R.N., Weaver, D. and Lindsley, C.W. 2008. Application of combinatorial chemistry science on modern drug discovery. *Journal of Combinatorial Chemistry*, *10*(3), pp. 345–354. https://doi.org/10.1021/cc700187t

Kim, C.S., Oh, J. and Lee, T.H. 2020. Structure elucidation of small organic molecules by contemporary computational chemistry methods. *Archives of Pharmacal Research*, *43*(11), pp. 1114–1127. https://doi.org/10.1007/s12272-020-01277-4

Kumar, V., Kundu, S., Singh, A. and Singh, S., 2022. Understanding the role of histone deacetylase and their inhibitors in neurodegenerative disorders: current targets and future perspective. *Current Neuropharmacology*, *20*(1), p. 158.

Kutzing, M.K., Luo, V. and Firestein, B.L., 2012. Protection from glutamate-induced excitotoxicity by memantine. *Annals of Biomedical Engineering*, *40*, pp. 1170–1181.

Lamptey, R.N.L., Chaulagain, B., Trivedi, R., Gothwal, A., Layek, B. and Singh, J. 2022. A Review of the common neurodegenerative disorders: current therapeutic approaches and the potential role of nanotherapeutics. *International Journal of Molecular Sciences*, *23*(3), p. 1851. https://doi.org/10.3390/ijms23031851

Lee, D., Seo, J., Jeong, H. chan, Lee, H. and Lee, S.B. 2021. The perspectives of early diagnosis of schizophrenia through the detection of epigenomics-based biomarkers in iPSC-derived neurons. *Frontiers in Molecular Neuroscience*, *14*. https://doi.org/10.3389/fnmol.2021.756613

Leite, C.Q.D.C., 2021. *Small molecule properties in the design and development of new CNS therapeutic candidates* (Doctoral dissertation).

Lim, G.P., Yang, F., Chu, T., Chen, P., Beech, W., Teter, B., Tran, T., Ubeda, O., Ashe, K.H., Frautschy, S.A. and Cole, G.M., 2000. Ibuprofen suppresses plaque pathology and inflammation in a mouse model for Alzheimer's disease. *Journal of Neuroscience*, *20*(15), pp. 5709–5714.

Lin, X., Liu, W., Dong, X. and Sun, Y., 2023. Epigallocatechin gallate-derived carbonized polymer dots: a multifunctional scavenger targeting Alzheimer's β-amyloid plaques. *Acta Biomaterialia*, *157*, pp. 524–537.

Lipinski, C.A., Lombardo, F., Dominy, B.W. and Feeney, P.J. 2001. Experimental and computational approaches to estimate solubility and permeability in drug discovery and development settings 1PII of original article: S0169-409X(96)00423-1. The article was originally published in Advanced Drug Delivery Reviews 23 (1997). *Advanced Drug Delivery Reviews*, *46*(1–3), pp. 3–26. https://doi.org/10.1016/S0169-409X(00)00129-0

Lipinski, C.A., Lombardo, F., Dominy, B.W. and Feeney, P.J., 2012. Experimental and computational approaches to estimate solubility and permeability in drug discovery and development settings. *Advanced Drug Delivery Reviews*, *64*, pp. 4–17.

Liu, X., Shao, R., Li, M. and Yang, G., 2014. Edaravone protects neurons in the rat substantia nigra against 6-hydroxydopamine-induced oxidative stress damage. *Cell Biochemistry and Biophysics*, *70*, pp. 1247–1254.

Makurvet, F.D., 2021. Biologics vs. small molecules: drug costs and patient access. *Medicine in Drug Discovery*, *9*, p. 100075.

Marucci, G., Buccioni, M., Dal Ben, D., Lambertucci, C., Volpini, R. and Amenta, F., 2021. Efficacy of acetylcholinesterase inhibitors in Alzheimer's disease. *Neuropharmacology*, *190*, p. 108352.

Mayr, L.M. and Bojanic, D., 2009. Novel trends in high-throughput screening. *Current Opinion in Pharmacology*, *9*(5), pp. 580–588.

Mikitsh, J.L. and Chacko, A.-M. 2014. Pathways for small molecule delivery to the central nervous system across the blood-brain barrier. *Perspectives in Medicinal Chemistry*, *6*, p. PMC.S13384. https://doi.org/10.4137/PMC.S13384

Milanetti, E., Raimondo, D. and Tramontano, A. 2016. Prediction of the permeability of neutral drugs inferred from their solvation properties. *Bioinformatics*, *32*(8), pp. 1163–1169. https://doi.org/10.1093/bioinformatics/btv725

Mishra, K.P., Ganju, L., Sairam, M., Banerjee, P.K. and Sawhney, R.C. 2008. A review of high throughput technology for the screening of natural products. *Biomedicine & Pharmacotherapy*, *62*(2), pp. 94–98. https://doi.org/10.1016/j.biopha.2007.06.012

Mueller, T. and Russ, H., 2006. Levodopa, motor fluctuations and dyskinesia in Parkinson's disease. *Expert Opinion on Pharmacotherapy*, *7*(13), pp. 1715–1730.

Nagase, M., Yamamoto, Y., Miyazaki, Y. and Yoshino, H., 2016. Increased oxidative stress in patients with amyotrophic lateral sclerosis and the effect of edaravone administration. *Redox Report*, *21*(3), pp. 104–112.

Nassar, A.E.F., Kamel, A.M. and Clarimont, C., 2004. Improving the decision-making process in the structural modification of drug candidates: enhancing metabolic stability. *Drug Discovery Today*, *9*(23), pp. 1020–1028.

Neumaier, F., Zlatopolskiy, B.D. and Neumaier, B., 2021. Drug penetration into the central nervous system: pharmacokinetic concepts and in vitro model systems. *Pharmaceutics*, *13*(10), p. 1542.

Pardridge, W.M. 2012. Drug transport across the blood–brain barrier. *Journal of Cerebral Blood Flow & Metabolism*, *32*(11), pp. 1959–1972. https://doi.org/10.1038/jcbfm.2012.126

Pattee, G.L., Genge, A., Couratier, P., Lunetta, C., Sobue, G., Aoki, M., Yoshino, H., Jackson, C.E., Wymer, J., Salah, A. and Nelson, S. 2023. Oral edaravone – introducing a flexible treatment option for amyotrophic lateral sclerosis. *Expert Review of Neurotherapeutics*, *23*(10), pp. 859–866. https://doi.org/10.1080/14737175.2023.2251687

Pinheiro, R.G.R., Coutinho, A.J., Pinheiro, M. and Neves, A.R. 2021. Nanoparticles for targeted brain drug delivery: what do we know? *International Journal of Molecular Sciences*, *22*(21), p. 11654. https://doi.org/10.3390/ijms222111654

Pinto, M.M.M., Palmeira, A., Fernandes, C., Resende, D.I.S.P., Sousa, E., Cidade, H., Tiritan, M.E., Correia-da-Silva, M. and Cravo, S. 2021. From natural products to new synthetic small molecules: a journey through the world of xanthones. *Molecules*, *26*(2), p. 431. https://doi.org/10.3390/molecules26020431

Pistacchi, M., Martinello, F., Gioulis, M. and Zambito Marsala, S., 2014. Rasagiline and rapid symptomatic motor effect in Parkinson's disease: review of literature. *Neurology and Therapy*, *3*, pp. 41–66.

Pooladgar, P., Sakhabakhsh, M., Taghva, A. and Soleiman-Meigooni, S., 2022. Donepezil beyond Alzheimer's disease? A narrative review of therapeutic potentials of donepezil in different diseases. *Iranian Journal of Pharmaceutical Research: IJPR*, *21*(1), e128408.

Rankovic, Z. 2015. CNS drug design: balancing physicochemical properties for optimal brain exposure. *Journal of Medicinal Chemistry*, *58*(6), pp. 2584–2608. https://doi.org/10.1021/jm501535r

Robert, A., Benoit-Vical, F., Liu, Y. and Meunier, B., 2019. Small molecules: the past or the future in drug innovation. *Metal Ions in Life Sciences*, *19*, pp. 9783110527872-008.

Roberts, D.J. and Hall, R.I. 2013. Drug absorption, distribution, metabolism and excretion considerations in critically ill adults. *Expert Opinion on Drug Metabolism & Toxicology*, *9*(9), pp. 1067–1984. doi: 10.1517/17425255.2013.799137.

Rodriguez-Aller, M., Guillarme, D., Veuthey, J.-L. and Gurny, R. 2015. Strategies for formulating and delivering poorly water-soluble drugs. *Journal of Drug Delivery Science and Technology*, *30*, pp. 342–351. https://doi.org/10.1016/j.jddst.2015.05.009

Rogawski, M.A. and Wenk, G.L., 2003. The neuropharmacological basis for the use of memantine in the treatment of Alzheimer's disease. *CNS Drug Reviews*, *9*(3), pp. 275–308.

Salama, M., Yuan, T.F., Machado, S., Murillo-Rodríguez, E., A Vega, J., Menéndez-González, M., E Nardi, A. and Arias-Carrión, O., 2013. Co-enzyme Q10 to treat neurological disorders: basic mechanisms, clinical outcomes, and future research direction. *CNS & Neurological Disorders-Drug Targets (Formerly Current Drug Targets-CNS & Neurological Disorders)*, *12*(5), pp. 641–664.

Samanen, J., 2013. Similarities and differences in the discovery and use of biopharmaceuticals and small-molecule chemotherapeutics. In: *Introduction to Biological and Small Molecule Drug Research and Development.* Robin Ganellin, Roy Jefferis and Stanley M. Roberts (eds) pp. 161–203. Elsevier: Amsterdam.

Sarnaik, A., Liu, A., Nielsen, D. and Varman, A.M. 2020. High-throughput screening for efficient microbial biotechnology. *Current Opinion in Biotechnology*, *64*, pp. 141–150. https://doi.org/10.1016/j.copbio.2020.02.019

Scott, D.E., Bayly, A.R., Abell, C. and Skidmore, J. 2016. Small molecules, big targets: drug discovery faces the protein–protein interaction challenge. *Nature Reviews Drug Discovery*, *15*(8), pp. 533–550. https://doi.org/10.1038/nrd.2016.29

Scuderi, S.A., Ardizzone, A., Paterniti, I., Esposito, E. and Campolo, M., 2020. Antioxidant and anti-inflammatory effect of Nrf2 inducer dimethyl fumarate in neurodegenerative diseases. *Antioxidants*, *9*(7), p. 630.

Silpa, L., Sim, R. and Russell, A.J. 2022. Recent advances in small molecule stimulation of regeneration and repair. *Bioorganic & Medicinal Chemistry Letters*, *61*, p. 128601. https://doi.org/10.1016/j.bmcl.2022.128601

Simić, S., Zukić, E., Schmermund, L., Faber, K., Winkler, C.K. and Kroutil, W. 2022. Shortening synthetic routes to small molecule active pharmaceutical ingredients employing biocatalytic methods. *Chemical Reviews*, *122*(1), pp. 1052–1126. https://doi.org/10.1021/acs.chemrev.1c00574

Słoczyńska, K., Gunia-Krzyżak, A., Koczurkiewicz, P., Wójcik-Pszczoła, K., Żelaszczyk, D., Popiół, J. and Pękala, E., 2019. Metabolic stability and its role in the discovery of new chemical entities. *Acta Pharmaceutica*, *69*(3), pp. 345–361.

Song, N.N., Zhang, S.Y. and Liu, C.X., 2004. Overview of factors affecting oral drug absorption. *Asian Journal of Drug Metabolism and Pharmacokinetics*, *4*(3), pp. 167–176.

Sun, J. and Roy, S., 2021. Gene-based therapies for neurodegenerative diseases. *Nature Neuroscience*, *24*(3), pp. 297–311.

Tallarida, R.J. and Jacob, L.S., 2012. *The Dose—Response Relation in Pharmacology*. Springer Science & Business Media: New York, NY. https://doi.org/10.1007/978-1-4684-6265-4

Tesseromatis, C. and Alevizou, A., 2008. The role of the protein-binding on the mode of drug action as well the interactions with other drugs. *European Journal of Drug Metabolism and Pharmacokinetics*, *33*, pp. 225–230.

Thébault, J.J., Guillaume, M. and Levy, R., 2004. Tolerability, safety, pharmacodynamics, and pharmacokinetics of rasagiline: a potent, selective, and irreversible monoamine oxidase type B inhibitor. *Pharmacotherapy: The Journal of Human Pharmacology and Drug Therapy*, *24*(10), pp. 1295–1305.

Thompson, T.N. 2001. Optimization of metabolic stability as a goal of modern drug design. *Medicinal Research Reviews*, *21*(5), pp. 412–449. https://doi.org/10.1002/med.1017

Troncoso-Escudero, P., Sepulveda, D., Pérez-Arancibia, R., Parra, A.V., Arcos, J., Grunenwald, F. and Vidal, R.L. 2020. On the right track to treat movement disorders: promising therapeutic approaches for Parkinson's and Huntington's disease. *Frontiers in Aging Neuroscience*, *12*. https://doi.org/10.3389/fnagi.2020.571185

Tsuno, N., 2009. Donepezil in the treatment of patients with Alzheimer's disease. *Expert Review of Neurotherapeutics*, *9*(5), pp. 591–598.

Vafaei, S., Zekiy, A.O., Khanamir, R.A., Zaman, B.A., Ghayourvahdat, A., Azimizonuzi, H. and Zamani, M. 2022. Combination therapy with immune checkpoint inhibitors (ICIs); a new frontier. *Cancer Cell International*, *22*(1), p. 2. https://doi.org/10.1186/s12935-021-02407-8

Vassar, R., 2014. BACE1 inhibitor drugs in clinical trials for Alzheimer's disease. *Alzheimer's Research & Therapy*, *6*(9), p. 89.

Verbeeck, R.K. and Musuamba, F.T., 2009. Pharmacokinetics and dosage adjustment in patients with renal dysfunction. *European Journal of Clinical Pharmacology*, *65*, pp. 757–773.

Volbracht, C., Van Beek, J., Zhu, C., Blomgren, K. and Leist, M., 2006. Neuroprotective properties of memantine in different in vitro and in vivo models of excitotoxicity. *European Journal of Neuroscience*, *23*(10), pp. 2611–2622.

Wang, Y.-T., Yang, P.-C., Zhang, Y.-F. and Sun, J.-F. 2024. Synthesis and clinical application of new drugs approved by FDA in 2023. *European Journal of Medicinal Chemistry*, *265*, p. 116124. https://doi.org/10.1016/j.ejmech.2024.116124

Wilson, R.M. and Danishefsky, S.J. 2006. Small molecule natural products in the discovery of therapeutic agents: the synthesis connection. *The Journal of Organic Chemistry*, *71*(22), pp. 8329–8351. https://doi.org/10.1021/jo0610053

Winblad, B., Gauthier, S., Åström, D. and Stender, K., 2010. Memantine benefits functional abilities in moderate to severe Alzheimer's disease. *The Journal of Nutrition, Health and Aging*, *14*(9), pp. 770–774.

Wohlgemuth, R. 2010. Biocatalysis—key to sustainable industrial chemistry. *Current Opinion in Biotechnology*, *21*(6), pp. 713–724. https://doi.org/10.1016/j.copbio.2010.09.016

Wolf, M.E. and Abi-Dargham, A. 2023. Synaptic plasticity as a therapeutic target to modulate circuits in psychiatric disorders. *Neuropsychopharmacology*, *48*(1), pp. 1–2. https://doi.org/10.1038/s41386-022-01458-3

Wu, K., Karapetyan, E., Schloss, J., Vadgama, J. and Wu, Y. 2023. Advancements in small molecule drug design: a structural perspective. *Drug Discovery Today*, *28*(10), p. 103730. https://doi.org/10.1016/j.drudis.2023.103730

Xiong, B., Wang, Y., Chen, Y., Xing, S., Liao, Q., Chen, Y., Li, Q., Li, W. and Sun, H. 2021. Strategies for structural modification of small molecules to improve blood–brain barrier penetration: a recent perspective. *Journal of Medicinal Chemistry*, *64*(18), pp. 13152–13173. https://doi.org/10.1021/acs.jmedchem.1c00910

You, H., Mariani, L.L., Mangone, G., Le Febvre de Nailly, D., Charbonnier-Beaupel, F. and Corvol, J.C., 2018. Molecular basis of dopamine replacement therapy and its side effects in Parkinson's disease. *Cell and Tissue Research*, *373*, pp. 111–135.

Zambrano, P., Suwalsky, M., Jemiola-Rzeminska, M., Strzalka, K., Sepúlveda, B., Gallardo, M.J. and Aguilar, L.F. 2019. The acetylcholinesterase (AChE) inhibitor and anti-Alzheimer drug donepezil interacts with human erythrocytes. *Biochimica et Biophysica Acta (BBA) - Biomembranes*, *1861*(6), pp. 1078–1085. https://doi.org/10.1016/j.bbamem.2019.03.014

Zhao, M., Ma, J., Li, M., Zhang, Y., Jiang, B., Zhao, X., Huai, C., Shen, L., Zhang, N., He, L. and Qin, S., 2021. Cytochrome P450 enzymes and drug metabolism in humans. *International Journal of Molecular Sciences*, *22*(23), p. 12808.

Zhou, W., Wang, Y., Lu, A. and Zhang, G., 2016. Systems pharmacology in small molecular drug discovery. *International Journal of Molecular Sciences*, *17*(2), p. 246.

Zhu, S., Zhang, T., Zheng, L., Liu, H., Song, W., Liu, D., Li, Z. and Pan, C. 2021. Combination strategies to maximize the benefits of cancer immunotherapy. *Journal of Hematology & Oncology*, *14*(1), p. 156. https://doi.org/10.1186/s13045-021-01164-5

3 Small Molecules in Multiple Sclerosis
An Update

Insha Zahoor, Urmilla John, and Nawab John Dar

ABBREVIATIONS

BBB	Blood–Brain Barrier
BTK	Burton's Tyrosine Kinase
CNS	Central Nervous System
DHOD	Dihydroorotate Dehydrogenase
DMF	Dimethyl Fumarate
ESI1	Epigenetic-Silencing-Inhibitor-1;
JAK	Janus Kinase
LPA1	Lysophosphatidic acid receptor 1
MAPK	Mitogen-Activated Protein Kinase
MS	Multiple Sclerosis
OPC	Oligodendrocyte Precursor Cells
P-gp	P-glycoprotein
S1P	Sphingosine-1-phosphate
SPMS	Secondary progressive MS

3.1 INTRODUCTION

Multiple sclerosis (MS) is a chronic inflammatory disease of the central nervous system (CNS), characterized by demyelination, gliosis, axonal loss, and neurodegeneration that builds up neurological defects with irreversible disability, eventually making the patient wheelchair-bound (Noseworthy et al., 2000; Wang et al., 2023). MS is a serious health problem primarily affecting the young adult population and causing devastating effects on the quality of life. Reports suggest an increase in the global occurrence of the disease, which has strongly impacted the socio-economic indicators of disease burden (Zahoor et al., 2024b). Nearly more than 1 million people are living with MS in the US, which is more than twice the previous estimate (Charabati et al., 2023; Wang et al., 2023). These figures call for curative studies to repair the damage caused by the disease vis-a-vis rescuing the young population from becoming crippled as they represent the future of the world. MS patients commonly suffer from relapses accompanied by damage-inducing inflammation, and current treatment methods suppress the whole immune system, resulting in unnecessary and ill-tolerated side effects (Brück, 2005; Van der Walt et al., 2010). While the relapsing phenotype of MS is treatable, the disease remains incurable and results in progressive disability over time. In MS, inflammatory attacks progressively intensify and lead to continued neuronal damage and demyelination, primarily due to the impaired function of oligodendroglial progenitor cells (OPCs) (Kuhlmann et al., 2008). Each inflammatory event in the body is followed by an endogenous mechanism that tries to subside this inflammation. However, in MS patients, this mechanism is inadequate, leading to unresolved inflammation that perpetuates neuronal damage and contributes to progressive disability

DOI: 10.1201/9781003520610-3

(Zahoor and Giri, 2021). Therefore, there exists an urgent need to develop novel therapeutic agents with the potential for myelin regeneration and repair by promoting oligodendrocyte formation and differentiation to rescue neuronal damage in MS. The recently published reports suggest that small molecules could serve as good candidates for the treatment of MS and other demyelinating disorders due to their potential for myelin repair (Manousi and Küry, 2021). Investigating the role of small molecules in neurodegeneration and repair therapy holds significant promise for advancing MS treatment. By targeting multiple biological pathways, these molecules could provide both preventive and therapeutic benefits. Their ability to modulate various mechanisms may not only offer a safer alternative to traditional treatments but also support a combinatorial approach, enhancing overall efficacy. This multifaceted action could contribute to accelerated functional recovery, making small molecules a compelling focus for future research and clinical applications in MS management. In this chapter, we have provided a summary of some recent studies highlighting the role of small molecules in curing demyelinating disorders, including MS. This would give new direction on the therapeutic potential of small molecules and associated receptors as novel treatment targets for inflammatory diseases. Therefore, new treatment strategies can be proposed to enhance repair and remyelination in CNS that would highlight their clinical application in the field of MS.

3.2 SMALL MOLECULES IN MS: UPDATE FROM RECENT STUDIES

Numerous studies have highlighted that omega-fatty acid-derived metabolites can target neuroinflammation, promote remyelination, prevent disease progression, and accelerate neurological recovery in experimental models of MS. However, clinical validation is needed before they can be translated into clinical practice (Poisson et al., 2015; Zahoor et al., 2024a, 2022; Zahoor and Giri, 2021). These small molecules can have pleiotropic effects by acting in a cell-specific manner through different receptors, primarily G-protein-coupled receptors. Their protective role has been reported in immune response, neuroinflammation, neurodegeneration, and inflammation-induced pain (Gregson et al., 2019; Najm et al., 2015) (Figure 3.1). The report by Allimuthu et al. (2019) discovered a group of novel compounds with remyelination potential from a library of 10,000 molecules (Allimuthu et al., 2019). Out of such a huge library, three key hits (CW3388, CW9956, and CW9009) emerged as oligodendrocyte enhancers functionally characterized through OPC assay. However, they need further in vivo and clinical validation (Han and Zhou, 2019). A recent study by Zhai et al. (2023) utilized a machine learning approach to identify ZCAN262 as a small molecule that specifically targets the GluA2 subunit of the α-amino-3-hydroxy-5-methyl-4-isoxazolepropio nic acid receptor (Zhai et al., 2023). This led to the recovery of neurological function and remyelination while suppressing inflammatory response in two preclinical animal models of MS. Similarly, PIPE-791 was found to be the antagonist for lysophosphatidic acid receptor 1 (LPA1). LPA acts as

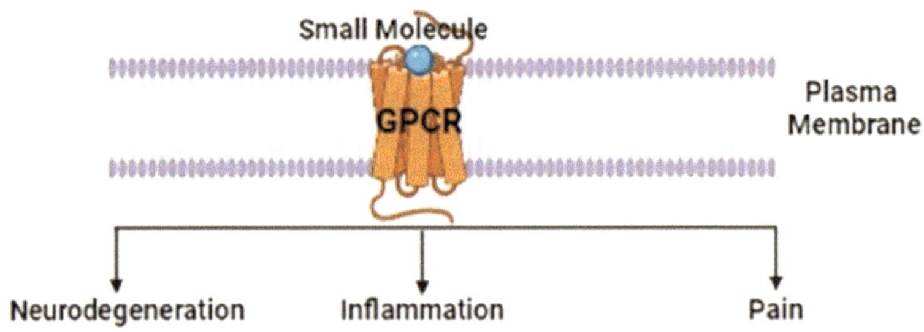

FIGURE 3.1 Overview of the role of small molecules in MS and related disorders: The interplay between small molecules and their receptors activates downstream signaling mechanisms and makes them target multiple points in MS-related pathways. This makes them alternative or complementary options with drug-like properties, targeting multiple disease components in MS such as neurodegeneration, inflammation, and pain-accelerating neurological recovery (Created with BioRender.com).

a pro-inflammatory lipid and mediates its actions through its receptor LPA1, thereby promoting neuroinflammation and demyelination. Oral administration of PIPE-791 was found to cross the blood–brain barrier (BBB) and targeted LPA1 for remyelination (Poon et al., 2024).

Similarly, RA03546849 has been found to cross the BBB and inhibit colony-stimulating factor 1 receptor without inhibiting other kinases, thereby dampening the inflammatory response in the animal model of MS (Nissen et al., 2018). Another study has identified the compound epigenetic-silencing-inhibitor-1 (ESI1) as a potential therapeutic agent for MS. ESI1 targets epigenetic mechanisms by modifying gene expression and influencing the epigenetic barrier in silenced oligodendrocytes within MS lesions. Specifically, it increases the levels of the activating histone marker H3K27ac while reducing repressive histones that contribute to gene silencing. This alteration in the epigenetic landscape promotes the regeneration of myelin, offering a promising approach to repair the damaged myelin sheaths in MS (Liu et al., 2024). Yet another study identified two small molecules, clobetasol and miconazole, with remyelination properties due to their effects on glucocorticoid receptor signaling or mitogen-activated protein kinases (Bierhansl et al., 2022). While clinical validation of small molecules is a gradual process, their discovery and possible receptors and target pathways reflect the progress made in the field of myelin regeneration for MS. Additional validation and functional characterization of therapeutic effects of these small molecules in experimental models of MS would provide further support for the viability of these compounds as potential drug treatments for people with MS. These studies provide a strong rationale and convincing evidence to enable the translation of small molecules into clinical practice to improve the neurological health and well-being of MS patients. Future research around small molecules will identify the mechanisms by which they resolve inflammation, prevent disease progression, improve neurological deficits, and possibly promote neurodegeneration with the potential as an alternative yet safe approach for people with MS.

3.3 TARGETING IMMUNE MODULATION WITH SMALL MOLECULES

Small molecules play a crucial role in modulating the immune response in MS by targeting various components of the immune system through different mechanisms. These compounds offer unique advantages in MS treatment due to their ability to cross the BBB, target intracellular pathways, and potentially provide more selective immunomodulation compared to broader-acting biologic therapies (Gadina et al., 2017). Here, we explore in detail how small molecules affect different aspects of the immune response in MS.

3.3.1 Targeting B Cells

B cells have emerged as important players in MS pathogenesis, and small molecules targeting B cell functions have shown promise in MS treatment.

3.3.1.1 Bruton's Tyrosine Kinase Inhibitors

Bruton's tyrosine kinase (BTK) inhibitors, such as evobrutinib and tolebrutinib, are small molecules that interfere with B cell receptor signaling and activation. These compounds work by:

i. **Inhibiting B cell proliferation and survival**: By blocking BTK, these inhibitors disrupt the signaling cascade necessary for B cell activation and proliferation, potentially reducing the number of autoreactive B cells (Gregson et al., 2019).
ii. **Reducing pro-inflammatory cytokine production**: BTK inhibitors can decrease the production of inflammatory mediators by B cells, which contribute to the inflammatory environment in MS lesions.
iii. **Modulating myeloid cell function**: BTK is also expressed in myeloid cells, and its inhibition may affect the function of microglia and macrophages in the CNS, potentially reducing neuroinflammation.

Recent studies have shown promising results for BTK inhibitors in MS treatment. For example, a phase 2 trial of evobrutinib in relapsing MS demonstrated a significant reduction in gadolinium-enhancing lesions compared to a placebo (Montalban et al., 2019). Similarly, tolebrutinib has shown encouraging results in early clinical trials, with a favorable safety profile and evidence of CNS penetration (Zurmati and Khan, 2023).

3.3.1.2 Sphingosine-1-Phosphate Receptor Modulators

Sphingosine-1-phosphate (S1P) receptor modulators, including fingolimod, siponimod, and ozanimod, are small molecules that affect lymphocyte trafficking. Their mechanisms of action include:

i. **Lymphocyte sequestration**: These compounds cause the retention of lymphocytes, including B cells in lymph nodes, and reduce the number of potentially pathogenic cells that can enter the CNS (Gregson et al., 2019).
ii. **Direct effects on CNS cells**: S1P receptors are known to be expressed on various types of cells within the CNS, including neurons and glial cells, and the modulation of these receptors may provide neuroprotective benefits beyond their role in immune cell trafficking (Mao-Draayer et al., 2017). Recent research has expanded our understanding of S1P receptor modulators. A long-term study of fingolimod in relapsing-remitting MS showed sustained efficacy and safety over 10 years of treatment (Kappos et al., 2015). Siponimod, which was approved for the treatment of secondary progressive MS, has demonstrated efficacy in slowing disability progression and reducing brain atrophy (Kappos, 1998). Ozanimod, the newest-approved S1P modulator, has shown comparable efficacy to interferon beta-1a with a more favorable safety profile (Cohen et al., 2019).

3.3.2 Modulating T Cell Responses

T cells have traditionally been considered central to MS pathology, and several small molecules target various aspects of T cell function.

3.3.2.1 Dihydroorotate Dehydrogenase Inhibitors

Teriflunomide, a dihydroorotate dehydrogenase inhibitor, modulates T cell responses in MS by:

i. **Interfering with pyrimidine synthesis**: This selectively affects rapidly dividing cells such as activated T cells, reducing their proliferation and potentially limiting the expansion of autoreactive T cell populations.
ii. **Altering T cell differentiation:** By affecting metabolic pathways, teriflunomide may influence T cell differentiation, potentially shifting the balance away from pro-inflammatory T cell subsets. Recent studies have provided new insights into the mechanism of action of teriflunomide. A study by Klotz et al. (2019) demonstrated that teriflunomide not only reduces T cell proliferation but also modulates T cell receptor signaling and cytokine production, leading to a more anti-inflammatory T cell phenotype (Klotz et al., 2019).

3.3.2.2 Janus Kinase Inhibitors

Janus kinase (JAK) inhibitors, while not yet approved for MS, are being explored for their potential to modulate T cell responses. These compounds:

i. **Interfere with cytokine signaling**: By inhibiting JAK-STAT pathways, these molecules can disrupt the signaling of multiple cytokines involved in T cell activation and differentiation.

ii. **Modulate T helper cell subsets:** JAK inhibition may affect the balance between different T helper cell subsets, potentially favoring regulatory T cells over pro-inflammatory subsets.

Recent clinical trials have shown promising results for JAK inhibitors in MS. For example, a phase 2 trial of filgotinib, a selective JAK1 inhibitor, demonstrated a significant reduction in gadolinium-enhancing lesions in relapsing MS patients (Jamilloux et al., 2019; Tavakoli et al., 2024). Another JAK inhibitor, baricitinib, is currently being evaluated in a phase 3 trial for relapsing MS (Jamilloux et al., 2019).

3.3.3 Targeting Innate Immune Cells

Small molecules can also modulate the innate immune response, which is increasingly recognized as important in MS pathogenesis.

3.3.3.1 Fumarates

Dimethyl fumarate and diroximel fumarate are approved MS therapies that modulate the innate immune response through:

i. **Activation of the Nrf2 antioxidant pathway**: This leads to reduced oxidative stress and may have neuroprotective effects.
ii. **Modulation of microglial activation**: Fumarates can alter the activation state of microglia, potentially reducing neuroinflammation.
iii. **Effects on T cell responses**: These compounds may also affect T cell responses through metabolic reprogramming, leading to a shift toward anti-inflammatory T cell subsets (Gregson et al., 2019).

Recent research has provided new insights into the mechanism of action of fumarates. A study demonstrated that dimethyl fumarate treatment leads to a shift in B cell subsets, favoring regulatory B cells over pro-inflammatory B cells (Schulze-Topphoff et al., 2016). This suggests that fumarates may have broader effects on the immune system than previously thought.

3.4 EMERGING SMALL MOLECULES FOR NEUROPROTECTION AND REMYELINATION IN MS

Some small molecules aim to address the neurodegenerative aspects of MS, by indirectly modulating the immune response, thereby promoting a more favorable environment for neural repair.

3.4.1 Laquinimod

While not approved, laquinimod has shown potential in clinical trials by:

i. **Modulating both innate and adaptive immune responses**: It modulates the activation of various immune cell types, thereby potentially reducing overall inflammation.
ii. **Promoting neuroprotection**: Laquinimod may have direct neuroprotective effects, reducing axonal damage and potentially slowing disease progression.

Recent studies have provided new insights into laquinimod's mechanism of action. A study demonstrated that laquinimod modulates the phenotype of myeloid cells, promoting an anti-inflammatory state. This effect on innate immune cells may contribute to its neuroprotective properties (Brandstadter and Katz Sand, 2017).

3.4.2 Clemastine

This antihistamine has shown promise in promoting remyelination in MS by:

i. **Enhancing oligodendrocyte differentiation:** This may stimulate myelin repair in demyelinated lesions, potentially reducing the antigenic load that drives the autoimmune response. A clinical trial by Green et al. (2017) demonstrated that clemastine treatment led to improved visual evoked potentials in patients with MS-related optic neuropathy, providing evidence for its remyelinating potential in humans (Green et al., 2017). Small molecules offer diverse approaches to modulate the immune response in MS. Their ability to target specific intracellular pathways and their potential to cross the BBB make them promising candidates for MS therapy. As our understanding of MS pathophysiology evolves, new targets for small molecule intervention continue to emerge, offering hope for more effective and targeted treatments in the future.

   The development of these small molecule therapies represents a shift toward more precise and potentially less toxic treatments for MS, addressing both the inflammatory and neurodegenerative aspects of the disease. However, it's important to note that while many of these approaches show promise, ongoing research and clinical trials are necessary to fully establish their efficacy and safety profiles in MS treatment. Recent advances in mass spectrometry-based techniques have also contributed to our understanding of small molecule effects in MS. For example, a study used high-resolution mass spectrometry to identify novel metabolic pathways affected by dimethyl fumarate treatment in MS patients, providing new insights into its mechanism of action (Bhargava et al., 2017). Similarly, Tabana et al. (2023) reviewed various techniques for target identification of small molecules, highlighting the potential of affinity-based pull-down approaches, coupled with mass spectrometry for identifying the molecular targets of novel MS therapies (Tabana et al., 2023). As research in this field continues to advance, it is likely that we will see the development of even more targeted and effective small molecule therapies for MS. The integration of advanced analytical techniques, such as high-throughput mass spectrometry and artificial intelligence-driven data analysis, will likely play a crucial role in accelerating this process and improving our understanding of MS pathophysiology and treatment.

3.5 NEUROPROTECTION AND REMYELINATION STRATEGIES

Small molecules have emerged as promising therapeutic agents for neuroprotection and remyelination in MS. These compounds offer several advantages, including the ability to cross the BBB, target-specific intracellular pathways, and potentially providing more precise outcomes compared to broader-acting biologic therapies. Recent research has focused on three main areas, i.e., targeting oxidative stress, addressing mitochondrial dysfunction, and enhancing remyelination.

3.5.1 Targeting Oxidative Stress

Oxidative stress plays a significant role in neurodegeneration and demyelination in MS. Several small molecules have shown potential in combating oxidative damage:

3.5.1.1 Dimethyl Fumarate

Dimethyl fumarate (DMF), approved for relapsing-remitting MS under the brand name Tecfidera, activates the Nrf2 antioxidant pathway. Recent studies have provided new insights into its mechanism.

DMF treatment leads to a shift in B cell subsets, favoring regulatory B cells over pro-inflammatory B cells (Schulze-Topphoff et al., 2016). It modulates T cell mitochondrial respiration, affecting T cell activation and differentiation (Klotz et al., 2019). A recent study by Bhargava et al. (2017) used high-resolution mass spectrometry to identify novel metabolic pathways affected by DMF treatment in MS patients, providing new insights into its mechanism of action (Bhargava et al., 2017).

3.5.1.2 MitoQ

MitoQ is a mitochondria-targeted antioxidant that has shown promise in preclinical studies:

- It reduced axonal degeneration and improved neurological deficits in experimental autoimmune encephalomyelitis (EAE) models (Mao et al., 2013).
- Recent work has demonstrated its ability to preserve mitochondrial function and reduce oxidative damage in oligodendrocytes (Fetisova et al., 2017).

3.5.1.3 Lipoic Acid

Lipoic acid is a naturally occurring antioxidant that has shown potential in MS treatment:

- A recent clinical trial demonstrated that lipoic acid supplementation reduced brain atrophy rates in secondary progressive MS patients (Bittner et al., 2017).
- Mechanistic studies have shown that lipoic acid can modulate T cell responses and reduce oxidative stress in the CNS (Salinthone et al., 2008).

3.5.2 Mitochondrial Dysfunction

Mitochondrial dysfunction is increasingly recognized as a key contributor to neurodegeneration in MS. Several small molecules target this aspect.

3.5.2.1 Idebenone

Idebenone, a synthetic analog of coenzyme Q10, has shown potential in protecting neurons from mitochondrial dysfunction:

- It improved neurological outcomes in a mouse model of MS by preserving axonal integrity and reducing inflammation (Mao et al., 2013).
- Recent clinical trials have explored its potential in progressive MS, with preliminary results showing promise in slowing disability progression (Tourbah et al., 2016).

3.5.2.2 Olesoxime

Olesoxime is a cholesterol-like compound that targets the mitochondrial permeability transition pore:

- It has demonstrated neuroprotective effects in preclinical models of MS and other neurodegenerative diseases (Bordet et al., 2007).
- Recent studies have shown its ability to promote oligodendrocyte survival and differentiation, suggesting a potential dual role in neuroprotection and remyelination (Magalon et al., 2012).

3.5.2.3 Bezafibrate

Bezafibrate, a pan-PPAR agonist, has shown promise in addressing mitochondrial dysfunction:

- Recent studies have demonstrated its ability to improve mitochondrial function and reduce neuroinflammation in EAE models (Buonvicino et al., 2019).
- Clinical trials are currently underway to evaluate its potential in progressive MS.

3.5.3 Enhancing Remyelination

Promoting remyelination is a key strategy for repairing damage in MS. Several small molecules have shown promise in this area:

3.5.3.1 Clemastine

Clemastine, an antihistamine, has emerged as a potential remyelinating agent:

- A clinical trial demonstrated its ability to improve visual evoked potentials in patients with MS-related optic neuropathy, providing evidence for its remyelinating potential in humans (Green et al., 2017).
- Recent mechanistic studies have shown that clemastine promotes oligodendrocyte differentiation by modulating the M1 muscarinic receptor (Mei et al., 2014).

3.5.3.2 Bexarotene

Bexarotene, a retinoid X receptor agonist, has shown promise in promoting remyelination:

- It enhanced remyelination in animal models of MS by promoting oligodendrocyte differentiation (Huang et al., 2011).
- A recent phase 2 clinical trial in relapsing-remitting MS patients showed that bexarotene treatment led to an improved magnetization transfer ratio, suggesting enhanced remyelination (Schwartzbach et al., 2017).

3.5.3.3 ESI1

ESI1 is a novel protein function inhibitor that has shown early promise in promoting remyelination:

- Recent studies have demonstrated its ability to regenerate myelin coatings in mouse models of MS and lab-prepared human brain cells (Liu et al., 2024).
- It appears to work by unsilencing repair processes in oligodendrocytes, potentially overcoming barriers to remyelination in MS.

3.5.3.4 Metformin

Metformin, a widely used diabetes drug, has shown potential in promoting remyelination:

- Recent studies have demonstrated its ability to enhance oligodendrocyte differentiation and remyelination in animal models of MS (Correale and Ysrraelit, 2022).
- Clinical trials are currently underway to evaluate its potential as a remyelinating therapy in MS.

Small molecules targeting oxidative stress, mitochondrial dysfunction, and remyelination represent a promising avenue for neuroprotection and repair in MS. While many of these compounds are still in preclinical or early clinical stages, they offer hope for developing more effective therapies that can not only manage symptoms but also promote repair and regeneration in MS and other demyelinating disorders (Villoslada and Steinman, 2020). The development of these small molecule therapies represents a shift toward more precise and potentially less toxic treatments for MS. As research in this field continues to advance, it is likely that we will see the emergence of even more targeted and effective small molecule therapies for MS in the coming years. Future research efforts should focus on optimizing drug delivery methods, identifying biomarkers to predict treatment response, and developing combination therapies that can simultaneously address multiple aspects of MS pathology.

3.6 OVERCOMING BBB BARRIERS IN MS DRUG DEVELOPMENT

The development of small molecules capable of effectively crossing the BBB remains one of the most significant challenges in CNS drug discovery and development. The BBB, a highly specialized interface between the blood and the brain, serves as a protective barrier that regulates the entry of substances into the CNS. While this barrier is crucial for maintaining brain homeostasis and protecting against potentially harmful substances, it also presents a formidable obstacle to the delivery of therapeutic agents to treat neurological disorders (Li et al., 2021). The complexity of designing BBB-penetrating small molecules lies in the need to balance multiple, often conflicting physicochemical properties. Lipophilicity, for instance, is generally favorable for BBB penetration, but excessive lipophilicity can lead to increased plasma protein binding and reduced solubility, ultimately limiting brain exposure (Pardridge, 2023). While a lower molecular weight can enhance BBB penetration, it may also reduce target affinity and specificity (Helms et al., 2020). The presence of efflux transporters, such as P-glycoprotein (P-gp) and breast cancer resistance protein, further complicate drug delivery by actively pumping out many drug molecules out of the brain even if they manage to cross the BBB (Bors and Erdő, 2019). Recent innovations in this field have focused on leveraging advanced computational methods and high-throughput screening techniques to optimize molecular structures for improved BBB penetration while maintaining target activity. Structure–activity relationship studies have become increasingly sophisticated, incorporating machine learning algorithms to predict BBB permeability based on molecular descriptors (Saxena et al., 2019). Prodrug strategies have become increasingly relevant in drug development. These approaches involve designing lipophilic compounds that can effectively cross the BBB. Once inside the brain, these prodrugs are metabolized into their active forms, allowing for targeted therapeutic effects within the CNS. This strategy has shown promise in improving the brain penetration of several compounds, although challenges remain in ensuring efficient and site-specific conversion (Upadhyay, 2014). Another innovative approach involves targeting endogenous transport systems at the BBB. By designing molecules that can utilize carrier-mediated transporters or receptor-mediated transcytosis pathways, researchers aim to hijack the natural uptake mechanisms of the brain. The glucose transporter GLUT1, for example, has been exploited to enhance the brain delivery of glycosylated analogs of various drugs (Patching, 2016). However, the success of this approach often comes at the cost of reduced affinity for the intended CNS target, necessitating careful optimization (Pardridge, 2023). Nanotechnology has opened up new avenues for small molecule delivery across the BBB. While not strictly a small molecule approach, nanocarrier systems such as liposomes, polymeric nanoparticles, and solid lipid nanoparticles have shown potential in encapsulating and delivering small molecule drugs to the brain (Naqvi et al., 2020). These systems can be engineered to target specific BBB transport mechanisms or to respond to external stimuli for controlled release. However, challenges remain in terms of scalability, reproducibility, and long-term safety of these nanocarrier systems (Pinheiro et al., 2021). The way forward in this field likely lies in a multidisciplinary approach that combines advances in medicinal chemistry, computational modeling, and drug delivery technologies. Artificial intelligence and machine learning algorithms are expected to play an increasingly important role in predicting BBB permeability and optimizing molecular design (Vatansever et al., 2021). Additionally, the development of more sophisticated in vitro BBB models, including those derived from human induced pluripotent stem cells, will provide more accurate platforms for screening and optimizing BBB-penetrating compounds (Singh et al., 2023). Furthermore, there is growing interest in strategies that temporarily and locally modulate BBB permeability, such as focused ultrasound techniques. These approaches, when combined with optimized small molecule design, could potentially enhance drug delivery to specific brain regions while minimizing systemic exposure (Chen et al., 2019). While significant challenges remain in designing small molecules that effectively cross the BBB, the field is rapidly evolving with promising innovations. The future of CNS drug discovery will likely involve integrated approaches that

TABLE 3.1
Small Molecules Approved for Multiple Sclerosis (MS) Treatment, Focusing on Their Efficacy, Safety Profiles, and Mechanisms of Action

Small Molecule	Mechanism of Action	Efficacy	Safety Profile	Approval
Siponimod (Mayzent)	Selective sphingosine 1-phosphate receptor modulator (S1P1 and S1P5)	Demonstrated reduction in confirmed disability progression at 3 and 6 months in secondary progressive MS (SPMS)	Common side effects include headache, hypertension, and transient bradycardia upon treatment initiation.	FDA-approved in 2019 for relapsing forms of MS, including SPMS
Diroximel fumarate (Vumerity)	Prodrug of monomethyl fumarate; activates the Nrf2 antioxidant response pathway	Comparable efficacy of dimethyl fumarate in reducing relapse rates and MRI lesions in relapsing MS	Fewer gastrointestinal-related adverse events compared to dimethyl fumarate, improving tolerability.	FDA-approved in 2019 for relapsing forms of MS
Ozanimod (Zeposia)	Sphingosine 1-phosphate receptor modulator (S1P1 and S1P5)	Efficacy in reducing annualized relapse rates and MRI lesions in relapsing MS patients	More favorable cardiac safety profile; no first-dose observation required for most patients.	FDA-approved in 2020 for relapsing forms of MS
Ponesimod (Ponvory)	Selective S1P1 receptor modulator	Demonstrated superior efficacy in reducing annualized relapse rates compared to teriflunomide in relapsing MS	Short half-life allows for faster lymphocyte recovery upon treatment discontinuation; common side effects include infections.	FDA-approved in 2021 for relapsing forms of MS
Monomethyl fumarate (Bafiertam)	Active metabolite of dimethyl fumarate; activates the Nrf2 pathway	Comparable efficacy to dimethyl fumarate in reducing relapse rates and MRI lesions in relapsing MS	May have improved gastrointestinal tolerability compared to dimethyl fumarate.	FDA-approved in 2020 for relapsing forms of MS

combine rational drug design, advanced delivery technologies, and a deeper understanding of BBB physiology. As our knowledge of the molecular mechanisms governing BBB permeability continues to grow, so too will our ability to develop more effective therapies for a wide range of neurological disorders. The key to success will lie in balancing BBB penetration with target engagement, pharmacokinetics, and safety profiles, ultimately leading to the development of more effective and safer CNS therapeutics.

Table 3.1 summarizes the latest small molecules approved for the treatment of MS, highlighting their mechanisms of action, efficacy, safety profiles, and approval details. Each of these therapies represents a significant advancement in the management of MS, offering various mechanisms to target the underlying disease processes while aiming to improve patient outcomes and quality of life.

3.7 CHALLENGES AND LIMITATIONS

The development and application of small molecules for MS treatment face several significant challenges and limitations.

3.7.1 BBB Penetration

Despite recent advancements, designing small molecules that effectively cross the BBB while maintaining their therapeutic efficacy remains a major challenge. Many promising compounds fail due to inadequate CNS penetration.

3.7.2 Target Specificity

Achieving high specificity for intended targets within the CNS while minimizing off-target effects is crucial but often difficult. This challenge is particularly evident in molecules targeting broad pathways like S1P signaling.

3.7.3 Long-Term Safety

The chronic nature of MS necessitates long-term treatment, raising concerns about the long-term safety profiles of these small molecules. Issues such as opportunistic infections, malignancies, and organ toxicity need continuous monitoring.

3.7.4 Heterogeneity of MS

The diverse pathophysiology of MS, including its various subtypes and stages, makes it challenging to develop a one-size-fits-all small molecule approach. Therapies effective in relapsing-remitting MS may not be as effective in progressive forms of the disease.

3.7.5 Balancing Efficacy and Side Effects

Many effective small molecules for MS come with significant side effects. Striking the right balance between therapeutic benefit and tolerability remains an ongoing challenge.

3.7.6 Drug Resistance

There is a potential for the development of drug resistance or reduced efficacy over time, necessitating the continuous development of new therapies.

3.7.7 Combination Therapy Complexities

While combination therapies might offer enhanced efficacy, they also introduce challenges in terms of drug interactions, cumulative toxicities, and complex treatment regimens.

3.7.8 Cost and Accessibility

The high cost of developing and producing these small molecules can limit their accessibility to patients, particularly in resource-limited settings.

3.7.9 Limited Efficacy in Progressive MS

Most approved small molecules show significant efficacy in relapsing forms of MS, but their impact on progressive MS remains limited.

3.7.10 Remyelination and Neuroprotection

While current small molecules are effective in reducing inflammation, their ability to promote remyelination and provide neuroprotection is limited, leaving a significant unmet need in MS treatment.

3.8 CONCLUSION AND FUTURE DIRECTIONS

The field of small molecule therapeutics for MS has witnessed remarkable progress in recent years, with several novel compounds gaining regulatory approval and demonstrating significant efficacy in managing the disease. These advancements have expanded the treatment landscape, offering new options for patients with various forms of MS, particularly those with relapsing-remitting disease. The development of more selective S1P receptor modulators, such as siponimod and ozanimod, has improved the safety profiles of earlier generations while maintaining efficacy. Similarly, the introduction of fumarate-based therapies like diroximel fumarate and monomethyl fumarate has provided alternatives with potentially improved tolerability. However, despite these advancements, significant challenges remain. The complex pathophysiology of MS, particularly in its progressive forms, continues to elude comprehensive treatment. The BBB remains a formidable obstacle in drug delivery, necessitating innovative approaches to enhance CNS penetration. Moreover, the goal of developing therapies that not only suppress inflammation but also promote remyelination and neuroprotection has not yet been fully achieved.

3.9 PROMISING FUTURE DIRECTIONS

3.9.1 Targeted Drug Delivery

Advanced drug delivery systems, including nanoparticles and exosomes, hold promise for enhancing BBB penetration and targeting specific CNS regions. Future research should focus on optimizing these delivery methods for small molecules.

3.9.2 Precision Medicine Approaches

Genomics, proteomics, and advanced imaging techniques are leveraged to develop personalized treatment strategies that account for individual patient characteristics and disease subtypes.

3.9.3 Combination Therapies

Synergistic combinations of small molecules that target different aspects of MS pathophysiology are explored, potentially offering more comprehensive disease control.

3.9.4 Novel Targets

New molecular targets, such as those involved in remyelination and neuroprotection, are investigated. Recent preclinical studies on molecules targeting the glutamate system represent an exciting avenue for future research.

3.9.5 Artificial Intelligence and Machine Learning

These technologies are utilized to accelerate drug discovery, predict BBB permeability, and optimize molecular design for improved efficacy and safety profiles.

3.9.6 Biomarkers for Treatment Response

Reliable biomarkers are developed to predict and monitor treatment response, allowing for more personalized and effective therapeutic strategies.

3.9.7 Long-Term Safety Studies

Comprehensive long-term safety studies are conducted to better understand the risk-benefit profiles of these therapies over extended periods.

3.9.8 Progressive MS Focus

Efforts are intensified to develop small molecules specifically targeting the mechanisms underlying progressive forms of MS, an area of significant unmet need.

3.9.9 Remyelination and Neuroprotection

Research on small molecules that can promote remyelination and provide neuroprotection is advanced, potentially reversing disease progression.

3.9.10 Improved BBB Models

More sophisticated in vitro and in silico models of the BBB are developed to better predict CNS penetration and efficacy of small molecules.

3.9.11 Repurposing Existing Drugs

The potential of repurposing approved drugs from other therapeutic areas is explored for MS treatment, potentially accelerating the development process.

3.9.12 Addressing Immune Senescence

Small molecules that can target age-related changes in the immune system are investigated, which may contribute to MS progression.

In conclusion, while small molecule therapies have significantly advanced MS treatment, the field stands at a crucial juncture. The integration of cutting-edge technologies, a deeper understanding of MS pathophysiology, and innovative approaches to drug design and delivery hold the key to overcoming current limitations. The future of MS treatment likely lies in a more personalized, multifaceted approach that combines highly targeted small molecules with other therapeutic modalities to address the complex and heterogeneous nature of the disease. As research progresses, the hope is to not only better manage MS symptoms and relapses but to develop therapies that can halt or even reverse disease progression, ultimately improving the quality of life for millions of people affected by this challenging condition.

REFERENCES

Allimuthu, D., Hubler, Z., Najm, F.J., Tang, H., Bederman, I., Seibel, W., Tesar, P.J., Adams, D.J., 2019. Diverse chemical scaffolds enhance oligodendrocyte formation by inhibiting CYP51, TM7SF2, or EBP. *Cell Chem. Biol.* 26, pp. 593–599.e4. https://doi.org/10.1016/j.chembiol.2019.01.004

Bhargava, P., Fitzgerald, K.C., Calabresi, P.A., Mowry, E.M., 2017. Metabolic alterations in multiple sclerosis and the impact of vitamin D supplementation. *JCI Insight* 2, p. e95302. https://doi.org/10.1172/jci.insight.95302

Bierhansl, L., Hartung, H.-P., Aktas, O., Ruck, T., Roden, M., Meuth, S.G., 2022. Thinking outside the box: non-canonical targets in multiple sclerosis. *Nat. Rev. Drug Discov.* 21, pp. 578–600. https://doi.org/10.1038/s41573-022-00477-5

Bittner, F., Murchison, C., Koop, D., Bourdette, D., Spain, R., 2017. Lipoic acid pharmacokinetics at baseline and 1 year in secondary progressive MS. *Neurol. Neuroimmunol. Neuroinflam.* 4, p. e380. https://doi.org/10.1212/NXI.0000000000000380

Bordet, T., Buisson, B., Michaud, M., Drouot, C., Galéa, P., Delaage, P., Akentieva, N.P., Evers, A.S., Covey, D.F., Ostuni, M.A., Lacapère, J.-J., Massaad, C., Schumacher, M., Steidl, E.-M., Maux, D., Delaage, M., Henderson, C.E., Pruss, R.M., 2007. Identification and characterization of cholest-4-en-3-one, oxime (TRO19622), a novel drug candidate for amyotrophic lateral sclerosis. *J. Pharmacol. Exp. Ther.* 322, pp. 709–720. https://doi.org/10.1124/jpet.107.123000

Bors, L.A., Erdő, F., 2019. Overcoming the blood–brain barrier. challenges and tricks for CNS drug delivery. *Sci. Pharm.* 87, p. 6. https://doi.org/10.3390/scipharm87010006

Brandstadter, R., Katz Sand, I., 2017. The use of natalizumab for multiple sclerosis. *Neuropsychiatr. Dis. Treat.* 13, pp. 1691–1702. https://doi.org/10.2147/NDT.S114636

Brück, W., 2005. The pathology of multiple sclerosis is the result of focal inflammatory demyelination with axonal damage. *J. Neurol.* 252, pp. v3–v9. https://doi.org/10.1007/s00415-005-5002-7

Buonvicino, D., Ranieri, G., Pratesi, S., Guasti, D., Chiarugi, A., 2019. Neuroimmunological characterization of a mouse model of primary progressive experimental autoimmune encephalomyelitis and effects of immunosuppressive or neuroprotective strategies on disease evolution. *Exp. Neurol.* 322, p. 113065. https://doi.org/10.1016/j.expneurol.2019.113065

Charabati, M., Wheeler, M.A., Weiner, H.L., Quintana, F.J., 2023. Multiple sclerosis: neuroimmune crosstalk and therapeutic targeting. *Cell* 186, pp. 1309–1327. https://doi.org/10.1016/j.cell.2023.03.008

Chen, K.-T., Wei, K.-C., Liu, H.-L., 2019. Theranostic strategy of focused ultrasound induced blood-brain barrier opening for CNS disease treatment. *Front. Pharmacol.* 10, p. 86. https://doi.org/10.3389/fphar.2019.00086

Cohen, J.A., Comi, G., Selmaj, K.W., Bar-Or, A., Arnold, D.L., Steinman, L., Hartung, H.-P., Montalban, X., Kubala Havrdová, E., Cree, B.A.C., Sheffield, J.K., Minton, N., Raghupathi, K., Huang, V., Kappos, L., 2019. Safety and efficacy of ozanimod versus interferon beta-1a in relapsing multiple sclerosis (RADIANCE): a multicentre, randomised, 24-month, phase 3 trial. *Lancet Neurol.* 18, pp. 1021–1033. https://doi.org/10.1016/s1474-4422(19)30238-8

Correale, J., Ysrraelit, M.C., 2022. Multiple sclerosis and aging: the dynamics of demyelination and remyelination. *ASN Neuro.* 14, p. 17590914221118502. https://doi.org/10.1177/17590914221118502

Fetisova, E., Chernyak, B., Korshunova, G., Muntyan, M., Skulachev, V., 2017. Mitochondria-targeted antioxidants as a prospective therapeutic strategy for multiple sclerosis. *Curr. Med. Chem.* 24. https://doi.org/10.2174/0929867324666170316114452

Gadina, M., Gazaniga, N., Vian, L., Furumoto, Y., 2017. Small molecules to the rescue: inhibition of cytokine signaling in immune-mediated diseases. *J. Autoimmun.* 85, pp. 20–31. https://doi.org/10.1016/j.jaut.2017.06.006

Green, A.J., Gelfand, J.M., Cree, B.A., Bevan, C., Boscardin, W.J., Mei, F., Inman, J., Arnow, S., Devereux, M., Abounasr, A., Nobuta, H., Zhu, A., Friessen, M., Gerona, R., von Büdingen, H.C., Henry, R.G., Hauser, S.L., Chan, J.R., 2017. Clemastine fumarate as a remyelinating therapy for multiple sclerosis (ReBUILD): a randomised, controlled, double-blind, crossover trial. *Lancet* 390, pp. 2481–2489. https://doi.org/10.1016/s0140-6736(17)32346-2

Gregson, A., Thompson, K., Tsirka, S.E., Selwood, D.L., 2019. Emerging small-molecule treatments for multiple sclerosis: focus on B cells. *F1000Research 8, F1000 Faculty Rev-245.* https://doi.org/10.12688/f1000research.16495.1

Han, F., Zhou, M.-M., 2019. Small molecules with big promises for curing demyelinating diseases. *Cell Chem. Biol.* 26, pp. 468–470. https://doi.org/10.1016/j.chembiol.2019.04.006

Helms, H.C.C., Kristensen, M., Saaby, L., Fricker, G., Brodin, B., 2020. Drug delivery strategies to overcome the blood–brain barrier (BBB). *Handb. Exp. Pharmacol.* https://doi.org/10.1007/164_2020_403

Huang, J.K., Jarjour, A.A., Nait Oumesmar, B., Kerninon, C., Williams, A., Krezel, W., Kagechika, H., Bauer, J., Zhao, C., Baron-Van Evercooren, A., Chambon, P., Ffrench-Constant, C., Franklin, R.J.M., 2011. Retinoid X receptor gamma signaling accelerates CNS remyelination. *Nat. Neurosci.* 14, pp. 45–53. https://doi.org/10.1038/nn.2702

Jamilloux, Y., El Jammal, T., Vuitton, L., Gerfaud-Valentin, M., Kerever, S., Sève, P., 2019. JAK inhibitors for the treatment of autoimmune and inflammatory diseases. *Autoimmun. Rev.* 18, p. 102390. https://doi.org/10.1016/j.autrev.2019.102390

Kappos, L., 1998. Placebo-controlled multicentre randomised trial of interferon β-1b in treatment of secondary progressive multiple sclerosis. *Lancet* 352, pp. 1491–1497. https://doi.org/10.1016/s0140-6736(98)10039-9

Kappos, L., O'Connor, P., Radue, E.-W., Polman, C., Hohlfeld, R., Selmaj, K., Ritter, S., Schlosshauer, R., von Rosenstiel, P., Zhang-Auberson, L., Francis, G., 2015. Long-term effects of fingolimod in multiple sclerosis: the randomized FREEDOMS extension trial. *Neurology* 84, pp. 1582–1591. https://doi.org/10.1212/WNL.0000000000001462

Klotz, L., Eschborn, M., Lindner, M., Liebmann, M., Herold, M., Janoschka, C., Torres Garrido, B., Schulte-Mecklenbeck, A., Gross, C.C., Breuer, J., Hundehege, P., Posevitz, V., Pignolet, B., Nebel, G., Glander, S., Freise, N., Austermann, J., Wirth, T., Campbell, G.R., Schneider-Hohendorf, T., Eveslage,

M., Brassat, D., Schwab, N., Loser, K., Roth, J., Busch, K.B., Stoll, M., Mahad, D.J., Meuth, S.G., Turner, T., Bar-Or, A., Wiendl, H., 2019. Teriflunomide treatment for multiple sclerosis modulates T cell mitochondrial respiration with affinity-dependent effects. *Sci. Transl. Med.* 11. https://doi.org/10.1126/scitranslmed.aao5563

Kuhlmann, T., Miron, V., Cuo, Q., Wegner, C., Antel, J., Bruck, W., 2008. Differentiation block of oligodendroglial progenitor cells as a cause for remyelination failure in chronic multiple sclerosis. *Brain* 131, pp. 1749–1758. https://doi.org/10.1093/brain/awn096

Li, J., Zheng, M., Shimoni, O., Banks, W.A., Bush, A.I., Gamble, J.R., Shi, B., 2021. Development of novel therapeutics targeting the blood-brain barrier: from barrier to carrier. *Adv. Sci. (Weinheim, Baden-Wurttemberg, Ger).* 8, p. e2101090. https://doi.org/10.1002/advs.202101090

Liu, X., Xin, D.E., Zhong, X., Zhao, C., Li, Z., Zhang, L., Dourson, A.J., Lee, L., Mishra, S., Bayat, A.E., Nicholson, E., Seibel, W.L., Yan, B., Mason, J., Turner, B.J., Gonsalvez, D.G., Ong, W., Chew, S.Y., Ghosh, B., Yoon, S.O., Xin, M., He, Z., Tchieu, J., Wegner, M., Nave, K.-A., Franklin, R.J.M., Dutta, R., Trapp, B.D., Hu, M., Smith, M.A., Jankowski, M.P., Barton, S.K., He, X., Lu, Q.R., 2024. Small-molecule-induced epigenetic rejuvenation promotes SREBP condensation and overcomes barriers to CNS myelin regeneration. *Cell* 187, pp. 2465–2484.e22. https://doi.org/10.1016/j.cell.2024.04.005

Magalon, K., Zimmer, C., Cayre, M., Khaldi, J., Bourbon, C., Robles, I., Tardif, G., Viola, A., Pruss, R.M., Bordet, T., Durbec, P., 2012. Olesoxime accelerates myelination and promotes repair in models of demyelination. *Ann. Neurol.* 71, pp. 213–226. https://doi.org/10.1002/ana.22593

Manousi, A., Küry, P., 2021. Small molecule screening as an approach to encounter inefficient myelin repair. *Curr. Opin. Pharmacol.* 61, pp. 127–135. https://doi.org/10.1016/j.coph.2021.09.008

Mao-Draayer, Y., Sarazin, J., Fox, D., Schiopu, E., 2017. The sphingosine-1-phosphate receptor: a novel therapeutic target for multiple sclerosis and other autoimmune diseases. *Clin. Immunol.* 175, pp. 10–15. https://doi.org/10.1016/j.clim.2016.11.008

Mao, P., Manczak, M., Shirendeb, U.P., Reddy, P.H., 2013. MitoQ, a mitochondria-targeted antioxidant, delays disease progression and alleviates pathogenesis in an experimental autoimmune encephalomyelitis mouse model of multiple sclerosis. *Biochim. Biophys. Acta* 1832, pp. 2322–2331. https://doi.org/10.1016/j.bbadis.2013.09.005

Mei, F., Fancy, S.P.J., Shen, Y.-A.A., Niu, J., Zhao, C., Presley, B., Miao, E., Lee, S., Mayoral, S.R., Redmond, S.A., Etxeberria, A., Xiao, L., Franklin, R.J.M., Green, A., Hauser, S.L., Chan, J.R., 2014. Micropillar arrays as a high-throughput screening platform for therapeutics in multiple sclerosis. *Nat. Med.* 20, pp. 954–960. https://doi.org/10.1038/nm.3618

Montalban, X., Arnold, D.L., Weber, M.S., Staikov, I., Piasecka-Stryczynska, K., Willmer, J., Martin, E.C., Dangond, F., Syed, S., Wolinsky, J.S., 2019. Placebo-controlled trial of an oral BTK inhibitor in multiple sclerosis. *N. Engl. J. Med.* 380, pp. 2406–2417. https://doi.org/10.1056/nejmoa1901981

Najm, F.J., Madhavan, M., Zaremba, A., Shick, E., Karl, R.T., Factor, D.C., Miller, T.E., Nevin, Z.S., Kantor, C., Sargent, A., Quick, K.L., Schlatzer, D.M., Tang, H., Papoian, R., Brimacombe, K.R., Shen, M., Boxer, M.B., Jadhav, A., Robinson, A.P., Podojil, J.R., Miller, S.D., Miller, R.H., Tesar, P.J., 2015. Drug-based modulation of endogenous stem cells promotes functional remyelination in vivo. *Nature* 522, pp. 216–220. https://doi.org/10.1038/nature14335

Naqvi, S., Panghal, A., Flora, S.J.S., 2020. Nanotechnology: a promising approach for delivery of neuroprotective drugs. *Front. Neurosci.* 14. https://doi.org/10.3389/fnins.2020.00494

Nissen, J.C., Thompson, K.K., West, B.L., Tsirka, S.E., 2018. Csf1R inhibition attenuates experimental autoimmune encephalomyelitis and promotes recovery. *Exp. Neurol.* 307, pp. 24–36. https://doi.org/10.1016/j.expneurol.2018.05.021

Noseworthy, J.H., Lucchinetti, C., Rodriguez, M., Weinshenker, B.G., 2000. Multiple sclerosis. *N. Engl. J. Med.* 343, pp. 938–952. https://doi.org/10.1056/nejm200009283431307

Pardridge, W.M., 2023. Physiologically based pharmacokinetic model of brain delivery of plasma protein bound drugs. *Pharm. Res.* 40, pp. 661–674. https://doi.org/10.1007/s11095-023-03484-2

Patching, S.G., 2016. Glucose transporters at the blood-brain barrier: function, regulation and gateways for drug delivery. *Mol. Neurobiol.* 54, pp. 1046–1077. https://doi.org/10.1007/s12035-015-9672-6

Pinheiro, R.G.R., Coutinho, A.J., Pinheiro, M., Neves, A.R., 2021. Nanoparticles for targeted brain drug delivery: what do we know? *Int. J. Mol. Sci.* 22, p. 11654. https://doi.org/10.3390/ijms222111654

Poisson, L.M., Suhail, H., Singh, J., Datta, I., Denic, A., Labuzek, K., Hoda, M.N., Shankar, A., Kumar, A., Cerghet, M., Elias, S., Mohney, R.P., Rodriguez, M., Rattan, R., Mangalam, A.K., Giri, S., 2015. Untargeted plasma metabolomics identifies endogenous metabolite with drug-like properties in chronic animal model of multiple sclerosis. *J. Biol. Chem.* 290, pp. 30697–30712. https://doi.org/10.1074/jbc.M115.679068

Poon, M.M., Lorrain, K.I., Stebbins, K.J., Edu, G.C., Broadhead, A.R., Lorenzana, A.O., Paulson, B.E., Baccei, C.S., Roppe, J.R., Schrader, T.O., Valdez, L.J., Xiong, Y., Chen, A.C., Lorrain, D.S., 2024. Discovery of a brain penetrant small molecule antagonist targeting LPA1 receptors to reduce neuroinflammation and promote remyelination in multiple sclerosis. *Sci. Rep.* 14, p. 10573. https://doi.org/10.1038/s41598-024-61369-9

Salinthone, S., Yadav, V., Bourdette, D., Carr, D., 2008. Lipoic acid: a novel therapeutic approach for multiple sclerosis and other chronic inflammatory diseases of the CNS. *Endocrine, Metab. & Immune Disord. Targets* 8, pp. 132–142. https://doi.org/10.2174/187153008784534303

Saxena, D., Sharma, A., Siddiqui, M.H., Kumar, R., 2019. Blood brain barrier permeability prediction using machine learning techniques: an update. *Curr. Pharm. Biotechnol.* 20, pp. 1163–1171. https://doi.org/10.2174/1389201020666190821145346

Schulze-Topphoff, U., Varrin-Doyer, M., Pekarek, K., Spencer, C.M., Shetty, A., Sagan, S.A., Cree, B.A.C., Sobel, R.A., Wipke, B.T., Steinman, L., Scannevin, R.H., Zamvil, S.S., 2016. Dimethyl fumarate treatment induces adaptive and innate immune modulation independent of Nrf2. *Proc. Natl. Acad. Sci. U. S. A.* 113, pp. 4777–4782. https://doi.org/10.1073/pnas.1603907113

Schwartzbach, C.J., Grove, R.A., Brown, R., Tompson, D., Then Bergh, F., Arnold, D.L., 2017. Lesion remyelinating activity of GSK239512 versus placebo in patients with relapsing-remitting multiple sclerosis: a randomised, single-blind, phase II study. *J. Neurol.* 264, pp. 304–315. https://doi.org/10.1007/s00415-016-8341-7

Singh, S., Agrawal, M., Vashist, R., Patel, R.K., Sangave, S.D., Alexander, A., 2023. Recent advancements on *in vitro* blood-brain barrier model: a reliable and efficient screening approach for preclinical and clinical investigation. *Expert Opin. Drug Deliv.* 20, pp. 1839–1857. https://doi.org/10.1080/17425247.2023.2295940

Tabana, Y., Babu, D., Fahlman, R., Siraki, A.G., Barakat, K., 2023. Target identification of small molecules: an overview of the current applications in drug discovery. *BMC Biotechnol.* 23, p. 44. https://doi.org/10.1186/s12896-023-00815-4

Tavakoli, G.M., Yazdanpanah, N., Rezaei, N., 2024. Targeting Bruton's tyrosine kinase (BTK) as a signaling pathway in immune-mediated diseases: from molecular mechanisms to leading treatments. *Adv. Rheumatol.* 64. https://doi.org/10.1186/s42358-024-00401-y

Tourbah, A., Lebrun-Frenay, C., Edan, G., Clanet, M., Papeix, C., Vukusic, S., De Sèze, J., Debouverie, M., Gout, O., Clavelou, P., Defer, G., Laplaud, D.-A., Moreau, T., Labauge, P., Brochet, B., Sedel, F., Pelletier, J., and on behalf of the MS-SPI study group, 2016. MD1003 (high-dose biotin) for the treatment of progressive multiple sclerosis: a randomised, double-blind, placebo-controlled study. *Mult. Scler.* 22, pp. 1719–1731. https://doi.org/10.1177/1352458516667568

Upadhyay, R.K., 2014. Drug delivery systems, CNS protection, and the blood brain barrier. *Biomed Res. Int.* 2014, pp. 1–37. https://doi.org/10.1155/2014/869269

Van der Walt, A., Butzkueven, H., Kolbe, S., Marriott, M., Alexandrou, E., Gresle, M., Egan, G., Kilpatrick, T., 2010. Neuroprotection in multiple sclerosis: a therapeutic challenge for the next decade. *Pharmacol. & Ther.* 126, pp. 82–93. https://doi.org/10.1016/j.pharmthera.2010.01.006

Vatansever, S., Schlessinger, A., Wacker, D., Kaniskan, H.Ü., Jin, J., Zhou, M.-M., Zhang, B., 2021. Artificial intelligence and machine learning-aided drug discovery in central nervous system diseases: state-of-the-arts and future directions. *Med. Res. Rev.* 41, pp. 1427–1473. https://doi.org/10.1002/med.21764

Villoslada, P., Steinman, L., 2020. New targets and therapeutics for neuroprotection, remyelination and repair in multiple sclerosis. *Expert Opin. Investig. Drugs* 29, pp. 443–459. https://doi.org/10.1080/13543784.2020.1757647

Wang, H., Zhang, X., Li, H., Sun, Z., Zhong, Y., 2023. Gender differences in the burden of multiple sclerosis in China from 1990 to 2019 and its 25-year projection: an analysis of the global burden of diseases study. *Heal. Sci. Rep.* 6, p. e1738. https://doi.org/10.1002/hsr2.1738

Zahoor, I., Giri, S., 2021. Specialized pro-resolving lipid mediators: emerging therapeutic candidates for multiple sclerosis. *Clin. Rev. Allergy Immunol.* 60, pp. 147–163. https://doi.org/10.1007/s12016-020-08796-4

Zahoor, I., Nematullah, M., Ahmed, M.E., Fatma, M., Mir, S., Ayasolla, K., Cerghet, M., Palaniyandi, S., Ceci, V., Carrera, G., Buttari, F., Centonze, D., Mao-Draayer, Y., Rattan, R., Chiurchiù, V., Giri, S., 2024a. Maresin-1 promotes neuroprotection and prevents disease progression in experimental models of multiple sclerosis through metabolic reprogramming and shaping innate and adaptive disease-associated cell types. *bioRxiv Prepr. Serv. Biol.* 2023.09.25.559216. https://doi.org/10.1101/2023.09.25.559216

Zahoor, I., Pan, G., Cerghet, M., Elbayoumi, T., Mao-Draayer, Y., Giri, S., Palaniyandi, S.S., 2024b. Current understanding of cardiovascular autonomic dysfunction in multiple sclerosis. *Heliyon* 10, p. e35753. https://doi.org/10.1016/j.heliyon.2024.e35753

Zahoor, I., Suhail, H., Datta, I., Ahmed, M.E., Poisson, L.M., Waters, J., Rashid, F., Bin, R., Singh, J., Cerghet, M., Kumar, A., Hoda, M.N., Rattan, R., Mangalam, A.K., Giri, S., 2022. Blood-based untargeted metabolomics in relapsing-remitting multiple sclerosis revealed the testable therapeutic target. *Proc. Natl. Acad. Sci. U. S. A.* 119, p. e2123265119. https://doi.org/10.1073/pnas.2123265119

Zhai, D., Yan, S., Samsom, J., Wang, L., Su, P., Jiang, A., Zhang, H., Jia, Z., Wallach, I., Heifets, A., Zanato, C., Tseng, C.-C., Wong, A.H.C., Greig, I.R., Liu, F., 2023. Small-molecule targeting AMPA-mediated excitotoxicity has therapeutic effects in mouse models for multiple sclerosis. *Sci. Adv.* 9, p. eadj6187. https://doi.org/10.1126/sciadv.adj6187

Zurmati, B.M., Khan, J., 2023. Safety and efficacy of tolebrutinib, an oral brain-penetrant BTK inhibitor, in relapsing multiple sclerosis: a phase 2b, randomized, double-blind, placebo-controlled trial by Daniel S Reich et al. *Mult. Scler. Relat. Disord.* 77, p. 104850. https://doi.org/10.1016/j.msard.2023.104850

4 Andrographolide in Neurological Disorders
A Detailed Overview

Bisma Gul, Mir Mohd Faheem, Zabeer Ahmed, and Muzamil Ahmad

ABBREVIATIONS

AD	Alzheimer's Disease
APP	Amyloid Precursor Protein
ARE	Antioxidant Response Element
Aβ	Amyloid Beta
BDNF	Brain-Derived Neurotrophic Factor
CDK	Cyclin-Dependent Kinase
COX-2	Cyclooxygenase 2
CUMS	Chronic Unpredictable Mild Stress
DA	Dopamine
GABA	Gamma Amino Butyric Acid.
HO-1	Heme Oxygenase 1
ICH	Intracerebral Hemorrhage
iNOS	Inducible Nitric Oxide Synthase
LPS	Lipopolysaccharide
MPP	1-Methyl-4-phenyl pyridinium
MPTP	1-Methyl-4-phenyl-1,2,3,6 tetrahydropyridine
NOX2	NADPH Oxidase 2
NQO1	NAD(P)H quinone dehydrogenase
PCP	Phencyclidine
PD	Parkinson's Disease
SOD	Superoxide Dismutase
TBI	Traumatic Brain Injury
TMC	Traditional Chinese Medicine
TRPV1	Transient Receptor Protein Vanilloid 1

4.1 INTRODUCTION

4.1.1 The King of Bitters

Andrographis paniculata (A. paniculata), often referred to as "king of bitters," is a herbaceous plant belonging to the family Acanthaceae. It is characterized by its erect, branched growth and can reach heights of 30–110 cm. The plant has simple, lanceolate leaves that are arranged in opposite pairs. The leaves are dark green, with a pointed tip, and have a characteristic bitter taste due to the presence of andrographolide and other diterpenoids. The flowers of *A. paniculata* are small, tubular, and typically greenish-white to pale lavender. They are arranged in terminal spikes and are bilaterally symmetrical. The plant produces small, ovate fruits that contain numerous seeds

DOI: 10.1201/9781003520610-4

(Hossain et al. 2014, Hossain et al. 2021). *A. paniculata* is native to the tropical and subtropical regions of South and Southeast Asia, such as India, Sri Lanka, and parts of Thailand, Malaysia, and Indonesia. It thrives in well-drained soils and can often be found in a variety of habitats including grasslands, forest edges, and cultivated areas (Bharati et al. 2011). The plant prefers warm climatic conditions (20°C to 35°C) and can tolerate both full sun and partial shade conditions (Akbar 2011). Conventional medical systems, like traditional Chinese medicine and Ayurveda, have been extensively employing the *Andrographis* plant to treat a wide array of diseases. Owing to this, apart from its native regions, *A. paniculata* is cultivated in various other countries for its medicinal properties. It is commonly grown as a garden herb or in agricultural plots. It is generally hardy and can adapt to different soil types, although it prefers loamy or sandy soils (Okhuarobo et al. 2014). Due to its medicinal value, farmers often grow it in controlled environments to ensure high yields of bioactive compounds (Hossain et al. 2021). The major classes of phytochemicals found in *A. paniculata* include diterpenoids, flavonoids, and polysaccharides (Brahmachari 2016). The pharmaceutically relevant compounds include andrographolide, deoxyandrographolide, neoandrographolide, andrographiside, β-sitosterol and various flavonoids. Flavonoids in *A. paniculata* are glycosides and aglycones. Notably, the diverse range of phytochemicals in *A. paniculata* underpins its wide array of medicinal uses contributing to the plant's therapeutic value (Naomi et al. 2022).

4.1.2 Andrographolide, the Primary Phytochemical Present in *A. paniculata*

The principal bioactive compound in *A. paniculata* is andrographolide, which is particularly concentrated in the leaves and stems of the plant (Kumar et al. 2020). Its chemical formula is $C_{20}H_{30}O_5$. Andrographolide is a labdane consisting of a furan ring fused to a labdane skeleton, with hydroxyl and ester groups at specific. Structurally, it exhibits a complex 3D architecture with numerous sp3-hybridized carbons, rich oxygen content, and a polycyclic structure featuring aliphatic side chains but no aromatic rings (Jain et al. 2023). Andrographolide's structure is represented in Figure 4.1.

The presence of diverse functional groups is critical for the pharmacological potential of andrographolide. Importantly, the hydroxyl groups support hydrophilicity and hydrogen-bonding capacity, while the γ-lactone ring plays a crucial role in its biological activity by affecting its contact with biological targets (Brahmachari 2016). Similarly, the presence of double bonds in its structure affects its reactivity and interaction with enzymes and receptors. Since andrographolide is an ester, it is relatively stable under neutral and slightly acidic conditions but can hydrolyze in highly alkaline

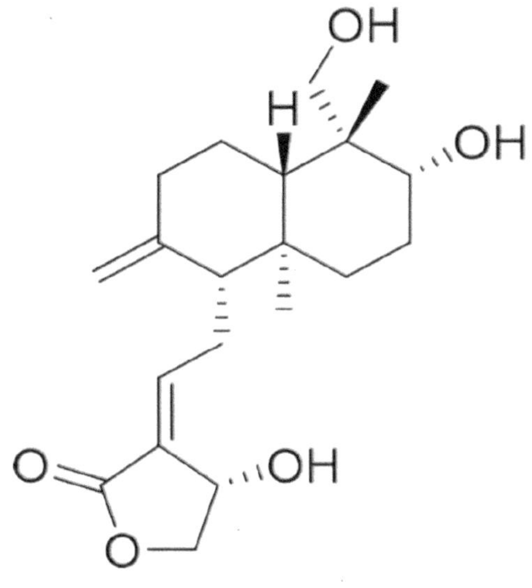

FIGURE 4.1 Structure of andrographolide.

environments, leading to the opening of the lactone ring and loss of biological activity (Yu et al. 1992). Because of its limited aqueous solubility, new delivery methods such as nanoformulations, liposomes, and solid dispersions must be developed to enhance its bioavailability and therapeutic efficacy. Alternatively, chemical modifications, such as acetylation and methylation, are used to enhance its solubility and pharmacokinetic properties. Derivatives of andrographolide often exhibit modified biological activities and improved drug-like properties (Liu and Xu 2008).

4.1.3 Biological Activities of Andrographolide

Preclinical studies on andrographolide have extensively investigated its therapeutic potential across various disease models, highlighting its efficacy and underlying mechanisms of action (Dai et al. 2019). Andrographolide has promise for clinical development because of its antioxidant and anti-inflammatory effects as well as its neuroprotective, anticancer, hepatoprotective, and cardiovascular advantages (Chao and Lin 2010). The key biological activities of andrographolide are summarized in Table 4.1.

TABLE 4.1
Pharmacological Activities of Andrographolide

Pharmacological Activity	Mechanism	Therapeutic Applications	References
Anti-inflammatory	Inhibits NF-κB Reduces inflammatory mediators Limits the production of NO and PGE2 by downregulating iNOS and Cox-2	Rheumatoid arthritis Inflammatory bowel diseases Neurodegenerative diseases Chronic inflammatory conditions	Li et al. (2017), Burgos et al. (2020)
Antioxidant and cytoprotective	Scavenges free radicals and ROS Enhances antioxidant defenses (e.g., SOD, catalase)	Protection against oxidative stress-induced injury	Mussard et al. (2019)
Immunomodulatory	Enhances phagocytosis by macrophages Boosts antibody production by B cells	Immune support Infectious diseases	Wang et al. (2010), Gupta et al. (2018)
Anticancer	Inhibits proliferation of cancer cell lines (breast, prostate, lung, colon) Induces apoptosis through caspase activation and modulation of Bcl-2 proteins	Various cancers (e.g., breast, prostate, lung, colon)	Rajagopal et al. (2003), Peng et al. (2018), Wu et al. (2018)
Hepatoprotective	Protects against liver toxins and oxidative stress Enhances liver antioxidant enzymes (SOD, GPx) Reduces liver injury markers	Liver diseases Hepatotoxicity prevention	Maiti et al. (2010)
Cardiovascular	Reduces inflammation and oxidative stress in cardiovascular tissues Modulates lipid metabolism Improves endothelial function	Cardiovascular health Prevention of heart diseases	Wong et al. (2021)
Neuroprotective	Reduces neuroinflammation and oxidative stress Preserves neuronal function and cognition	Alzheimer's disease, Parkinson's disease, and stroke	Lu et al. (2019), Hossain et al. (2022)
Antiviral	Inhibits viral replication Modulates immune responses	Influenza, herpes simplex virus, human immunodeficiency virus	Gupta et al. (2017), Gupta et al. (2018)
Antimicrobial	Disrupts bacterial cell membranes Inhibits biofilm formation	Bacterial infections Fungal infections	Sule et al. (2012), Banerjee et al. (2017)

In the next section of this chapter, we have discussed how andrographolide modulates different signaling networks to achieve neuroprotection.

4.2 ANDROGRAPHOLIDE AFFORDS NEUROPROTECTION BY MODULATING MULTIPLE SIGNALING CASCADES

Neuroprotection entails the mechanisms and strategies used to shield the nervous system from injury and degeneration. Considerable efforts have been put into studying neuroprotective approaches due to their potential to mitigate the deleterious effects of neurological disorders including traumatic brain injury, stroke, Parkinson's disease (PD), and Alzheimer's disease (AD). Effective neuroprotective strategies aim to preserve neuronal function and structure, consequently maintaining cognitive and motor functions. A plethora of evidence suggests that andrographolide affords robust neuroprotection (Bayazid et al. 2024). The detailed mechanisms by which andrographolide renders neuroprotective effects are discussed in the following subsections and have been summarized in Figure 4.2.

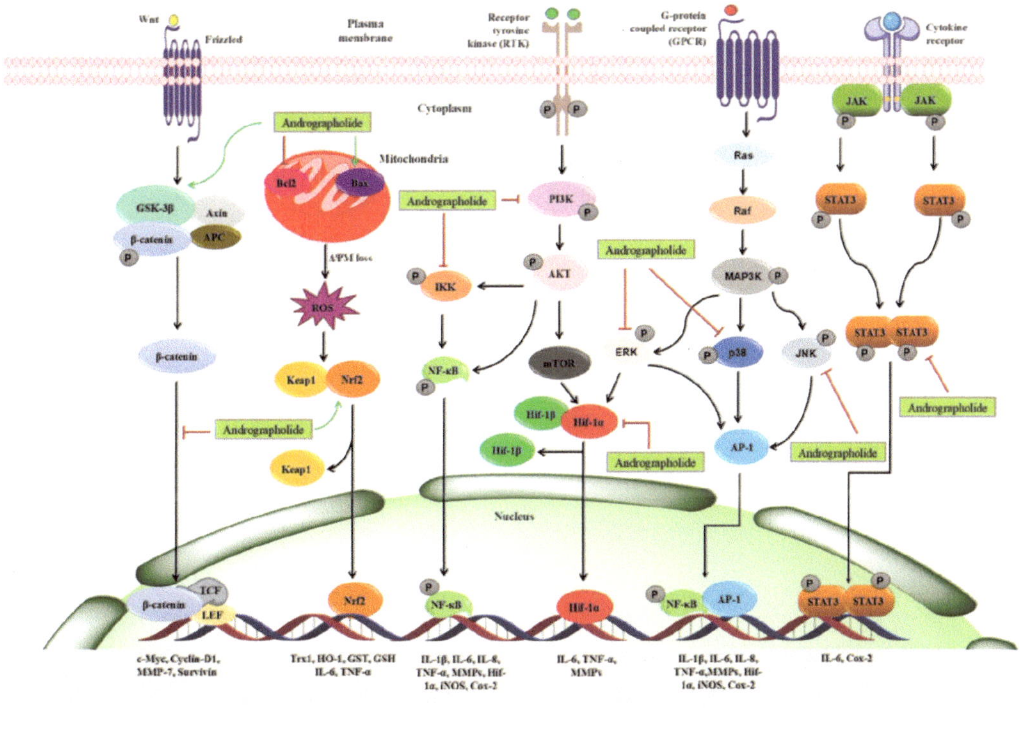

FIGURE 4.2 The signaling networks modulated by andrographolide. Andrographolide exerts anti-inflammatory effects by inhibiting the NF-κB and STAT-3 pathways, preventing the phosphorylation and degradation of IκBα, thereby reducing the inflammatory cytokines (Cox-2, IL-6, IL-1β, and TNF-α). It enhances the antioxidant defense system by activating Nrf2. Nrf2, in turn, shuttles the nucleus and induces antioxidative enzymes like HO-1, NQO1, and GST, thus mitigating oxidative stress. Andrographolide also interferes with the Wnt/β-catenin signaling pathway by ablating β-catenin translocation, resulting in diminished cell proliferation. Andrographolide augments apoptosis favoring protein-Bax and diminishes anti-apoptotic Bcl-2, thereby activating caspases. Additionally, it inhibits the PI3K/Akt/mTOR pathway, leading to reduced cell proliferation.

AP-1, Activator protein-1; APC, Adenomatous polyposis coli; Cox-2, Cyclooxygenase-2; GSH, Glutathione synthetase; GSK-3β, Glycogen synthase kinase-3β; GST, Glutathione S-transferase; Hif-1α, Hypoxia-inducible factor-1 α; HO-1, Heme oxygenase-1; IL, Interleukins; iNOS, Inducible Nitric Oxide Synthase; MMP, Matrix metalloproteinase; Nrf2, Nuclear factor erythroid 2-related factor 2; TNF-α, Tumor Necrosis Factor-α; Trx1, Thioredoxin

4.2.1 Anti-inflammatory Signaling

Andrographolide exerts potent anti-inflammatory effects primarily through the abrogation of the NF-κB signaling axis (Burgos et al. 2020). NF-κB is a pivotal transcription factor modulating inflammatory cytokines, chemokines, and adhesion molecules. Andrographolide stops IκBα from being phosphorylated and degraded and restricts NF-κB from entering the nucleus and starting gene transcription. As a consequence, there is a reduction in the production of inflammation-favoring molecules (TNF-α, IL-6, and IL-1β). Additionally, andrographolide abrogates the NF-κB mediated induction of cyclooxygenase-2 (Cox-2) to circumvent the synthesis of prostaglandins that serve as long-term mediators of inflammation (Li et al. 2022). The inhibition of NF-κB by andrographolide has also been linked to its ability to modulate upstream signaling molecules like inhibitor of kappa B kinase (IKK) and mitogen-activated protein kinases (MAPKs) (Li et al. 2017). The broad-spectrum inhibition reduces the overall inflammatory response and provides a multi-targeted approach to manage neuro-inflammation.

4.2.2 Antioxidant Pathways

Andrographolide enhances the antioxidant defense system by upregulating the expression of Nrf2, a critical modulator of cellular redox homeostasis (Wang et al. 2021). Nrf2 activation leads to the transcription of various antioxidant response elements (AREs) that encode for detoxifying enzymes, examples being heme oxygenase-1 (HO-1), NAD(P)H quinone dehydrogenase 1 (NQO1), and glutathione S-transferases. As a result, antioxidant defense mechanisms are bolstered to counteract oxidative stress, a common underpinning in many chronic diseases. The activation of Nrf2 by andrographolide also involves the disruption of the Keap1/Nrf2 physical interaction (Gong et al. 2022). The disruption allows Nrf2 to translocate to the nucleus and initiate the transcription of antioxidant genes, thus enhancing the cellular defense against oxidative damage.

4.2.3 Apoptotic and Cell Survival Pathways

Andrographolide has demonstrated the ability to induce apoptosis and inhibit cell proliferation. It modulates key components of the apoptotic machinery, including the intrinsic (mitochondrial) and extrinsic (death receptor) pathways. Andrographolide shifts the impetus toward apoptosis by elevating Bax and curtailing Bcl2 levels (Yang et al. 2010). Furthermore, it activates terminal caspases, which are essential for the execution of apoptosis. Andrographolide also interacts with the PI3K/Akt/mTOR pathway, a critical axis that influences cellular growth and survival (Kumar et al. 2015). By inhibiting Akt phosphorylation, andrographolide suppresses mTOR-mediated survival signaling, leading to reduced protein synthesis and cell proliferation. Additionally, andrographolide has been shown to induce cell cycle arrest at the G_2/M phase, thereby preventing the cells from proliferating (Li et al. 2007). The halt in the cell cycle is facilitated by the downregulation of cyclin-dependent kinases and cyclins that are essential for cell cycle progression.

4.2.4 Immune-Modulation

The immunomodulatory effects of andrographolide are evident in its ability to regulate both innate and adaptive immune responses (Cai et al. 2022). It enhances the activity of macrophages, natural killer cells, and cytotoxic T lymphocytes, to mount an effective immune response against pathogens and tumor cells. Additionally, andrographolide modulates the production of cytokines and chemokines, balancing the immune response and preventing excessive inflammation. Andrographolide has been reported to control how T cells and dendritic cells, among other immune cells, differentiate and perform. It can enhance the production of Th1 cytokines (e.g., IFN-γ) while suppressing Th2 cytokines (e.g., IL-4), thereby promoting a more balanced immune response (Burgos et al. 2005).

The maintenance of balance is beneficial in conditions where the immune system is either overactive, as in autoimmune diseases, or underactive, as in chronic infections.

In the succeeding sections of the book chapter, we have delineated the extensive role of andrographolide in mitigating the deleterious effects of various neurological diseases and have highlighted the underlying signaling mechanisms of action as well.

4.3 ANDROGRAPHOLIDE AS NEUROPROTECTIVE AGENT AGAINST NEUROLOGICAL DISORDERS

The multifaceted mechanisms of andrographolide present a holistic approach to addressing the complex pathophysiology underlying debilitating neurological conditions. This section aims to elucidate the current understanding of andrographolide's neuroprotective mechanisms and evaluate its prospective applications in the treatment of specific neurological disorders. Table 4.2 summarizes the therapeutic potential of andrographolide against neurological conditions.

4.3.1 Andrographolide Ameliorates Neuropathic Pain

Neuropathic pain is a prevalent chronic pain disorder that adversely affects quality of life, typically resulting from a lesion or disease of the somatosensory nerve system. Trigeminal neuralgia, excruciating polyneuropathy, postherpetic neuralgia, and central poststroke pain are a few examples (Finnerup et al. 2020). Neuropathic pain manifests as a burning, stabbing, or shooting pain and can be accompanied by abnormal sensations such as tingling or numbness. However, neuropathic pain is not directly linked to tissue damage but rather to nerve injury or dysfunction (Prescott and Ratté 2017). The pathophysiology of neuropathic pain involves several mechanisms, including central and peripheral sensitization. Peripheral sensitization occurs when nerve injury leads to increased

TABLE 4.2
The Effects of Andrographolide on Various Neurological Conditions

Neurological Disease	Mechanism of Action	Key Findings	References
Neuropathic pain	Inhibition of NF-κB pathway, reduction of inflammatory cytokines	Reduced pain behavior	Wang et al. (2018), Yi et al. (2018)
Parkinson's disease	Enhanced activity of PI3K/Akt axis, abrogation of caspase-3 cleavage, modulation of mitochondrial function, reduction of oxidative damage	Enhanced cell survival, reduced dopaminergic neuron loss, and improved motor function	Zhang et al. (2014), Geng et al. (2019), Ahmed et al. (2021)
Alzheimer's disease	Inhibition of Aβ-induced oxidative stress, enhancement of autophagy (via Beclin-1, LC3), and inhibition of amyloid precursor protein (APP) processing	Increased cell viability and decreased Aβ aggregation and production	Geng et al. (2018), Abedi et al. (2021), Patel et al. (2021)
Depression	Modulation of serotonergic and dopaminergic systems and reduction of corticosterone levels	Improved behavioral outcomes and decreased symptoms of depression	Thakur et al. (2014), Geng et al. (2019), Zhang et al. (2019)
Stroke	Inhibition of oxidative stress and inflammation	Improved neurological outcomes and decreased brain damage	Chan et al. (2010), Chern et al. (2011), Yang et al. (2017)
Schizophrenia	Modulation of glutamatergic signaling and reduction of oxidative stress	Improved behavioral outcomes	Wang et al. (2021)
Seizures	Modulation of GABAergic signaling and reduction of neuronal hyperexcitability	Decreased seizure frequency	Baru Venkata et al. (2023)

sensitivity of pain receptors and nerves. Central sensitization involves the amplification of pain signals particularly in the spinal cord and brain (Finnerup et al. 2015). These changes result in heightened pain responses and the persistence of pain despite the absence of ongoing tissue damage. Neuropathic pain can result from diabetes (diabetic neuropathy), postherpetic neuralgia (following shingles), trigeminal neuralgia, and peripheral nerve injuries. Conditions like multiple sclerosis and spinal cord injury can also lead to neuropathic pain due to damage to the central nervous system (Murphy et al. 2017).

Andrographolide has been studied for its role in alleviating neuropathic pain through various biological pathways. Andrographolide reduces neuropathic pain primarily through its anti-inflammatory effects and suppresses the release of cytokines like IL-1β, TNF-α, and IL-6 in neuropathic pain models (Cai et al. 2022). Oxidative stress is also known to contribute to neuropathic pain by damaging nerve tissues and promoting inflammation. Andrographolide possesses antioxidant properties that mitigate oxidative stress. Andrographolide, particularly, enhances the activity of antioxidative enzymes and reduces the levels of ROS in the spinal cord of neuropathic pain models. This reduction in oxidative stress helps to counter pain. Furthermore, andrographolide affects central pain processing by modulating pain signaling pathways. Andrographolide inhibits the NF-κB cascade and reduces the levels of pain-associated proteins, COX-2, and iNOS in the spinal cord (Li et al. 2022). As a result, there is decreased pain perception and improved pain management. Additionally, andrographolide promotes nerve regeneration and reduces neuronal damage, which can be beneficial in neuropathic pain (Wang et al. 2018). Andrographolide has been shown to interact with pain receptors and modulate pain signaling and affects the expression of transient receptor potential vanilloid 1 (TRPV1) channels, which are involved in pain sensation (Smith et al. 2006). Therefore, andrographolide alleviates neuropathic pain through multiple pathways, including the inhibition of inflammatory and oxidative stress pathways, modulation of pain signaling, neuroprotection, and interaction with pain receptors. These mechanisms contribute to its efficacy in managing neuropathic pain (Burgos et al. 2020).

4.3.2 Role of Andrographolide in PD

PD is a neurological illness that results from the degeneration of nerve cells in the substantia nigra, the area of the brain responsible for controlling movement. These nerve cells stop functioning or degenerate, losing their capacity to make the crucial neurotransmitter dopamine (DA) (Bloem et al. 2021). According to studies, people who have lost 80% or more of their substantia nigra's DA-producing cells experience the onset of PD symptoms (Sigcha et al. 2023). PD is more common in the elderly because of the progressive loss of dopaminergic neurons in the substantia nigra. Presently, a more significant pathophysiology is thought to be a rise in neuroinflammatory reactions in the brain, resulting in the production of several messengers of inflammation by overactive microglia, ultimately causing damage to dopaminergic neurons in the midbrain (Qiao et al. 2018).

According to a study by Wang et al., pre- and post-treatment with andrographolide eliminated the decrease in DA uptake induced by lipopolysaccharides (LPSs). Furthermore, andrographolide diminished LPS-mediated induction of inducible NO synthase and Cox-2 in BV-2 microglial cells in a dose-dependent fashion. As a consequence, by preventing stimulation of microglia, andrographolide inhibited inflammation-mediated dopaminergic neurodegeneration in mesencephalic neuroglia cells, therefore, signifying andrographolide's potential against PD (Wang et al. 2004). In a different report, Geng et al. elucidated that in 1-methyl-4-phenyl-1,2,3,6-tetrahydropyridine (MPTP)-prompted PD models, andrographolide slowed dopaminergic neuronal loss, ablated oxidative stress, and conserved mitochondrial morphology (Geng et al. 2019). It also decreased mitochondrial malfunctions and suppressed GTPase activity, resulting in decreased cell death. Of note, andrographolide reduces the activation of the NLRP3 inflammasome in microglia, thereby hampering the loss of dopaminergic neurons (Ahmed et al. 2021). Additionally, synthetic analogs of andrographolide can restore tyrosine hydroxylase-positive neurons in MPTP-induced murine models of PD. As a result, there is an increase in the striatal levels of DA and 3,4-dihydroxyphenylacetic acid to ameliorate MPTP-associated symptoms of PD. Andrographolide also prevents 1-methyl-4-phenylpyridinium-induced neurotoxicity in primary cerebellar granule neurons. Moreover, it raises the expression of superoxide dismutase (SOD) to prevent nerve damage and drastically lower the expression of nitric oxide in the substantia nigra of MPTP mice (Zhang et al. 2014).

4.3.3 Role of Andrographolide in AD

AD is defined by the gradual degeneration and death of brain cells, leading to cognitive impairment and functional decline. AD affects brain regions responsible for language, memory, and thinking processes. The hallmark symptoms include memory loss, confusion, difficulty with language, and problems with spatial awareness. As the disease progresses, individuals may experience changes in personality, mood swings, and difficulty with daily activities (Knopman et al. 2021). The pathophysiology of AD involves several key pathological features. Abnormal clumps of beta-amyloid (Aβ) peptides accumulate between neurons, disrupting cell function (Murphy and LeVine III 2010). Neurofibrillary tangles composed of hyperphosphorylated tau protein disrupt the microtubule function and contribute to neuronal death (Iqbal et al. 1998). Additionally, chronic inflammation in the brain, mediated by glial cells, contributes to neuronal damage and disease progression (Kempuraj et al. 2016). Microglial activation and neuroinflammation play definitive roles particularly in the onset of AD (Serrano-Pozo et al. 2011).

Andrographolide impacts multiple facets of AD, including Aβ plaques, tau phosphorylation, neuroinflammation, and oxidative stress. Andrographolide has been reported to influence Aβ metabolism by mitigating the formation and accumulation of Aβ plaques. Of note, andrographolide enhances Aβ clearance, potentially slowing AD progression (Gu et al. 2018). Furthermore, andrographolide has been found to modulate tau pathology by reducing tau phosphorylation and inhibiting neurofibrillary tangle formation in AD animal models. As a result, tau-related neuronal damage and cognitive deficits are ablated (Serrano et al. 2014). Pathological alterations in AβPPswe/PS-1 transgenic mice demonstrated how andrographolide decreased Aβ plaques in the cortical and hippocampal regions of juvenile animals. In the same model, andrographolide decreased phosphorylation of tau by Aβ oligomers, thereby preserving and restoring synaptic plasticity and cognition (Serrano et al. 2014). Apart from the above-mentioned mechanisms, andrographolide can decelerate AD progression by modulating key signaling networks implicated in AD such as the PI3K/Akt/mTOR, NF-κB, and MAPK pathways (Godoy et al. 2014). Andrographolide can also slow the progression of AD by inhibiting GSK-3β activity and modulating the downstream Wnt/β-catenin pathway linked to AD dysfunction (Tapia-Rojas et al. 2015). Importantly, neuroinflammation was significantly lower in the andrographolide cohort of streptozotocin-induced AD rats, suggesting that inhibiting neuroinflammation is a therapeutically useful avenue for neurodegeneration (Patel et al. 2021). In mice injected with HT22 hippocampal cells, andrographolide treatment activated the Nrf2/Keapl-mediated HO-1 signaling pathway to alleviate oxidative stress. In microglial cell line BV-2, andrographolide treatment resulted in the inhibition of $A\beta_{42}$-overexpressing microglia, resulting in the deactivation of NF-κB and its downstream mediators such as IL-6, IL-1β, PGE2, NO, Cox-2, and inducible nitric oxide (iNOS) (Seo et al. 2017). As a consequence, andrographolide significantly protected neural cells from microglia-mediated Aβ toxicity. Of note, andrographolide-mediated inhibition of NF-κB nuclear translocation hampers IκB phosphorylation and attenuates Aβ-induced JNK-MAPK hyperactivation (Yang et al. 2017).

4.3.4 Role of Andrographolide in Depression

Depression is a mood disorder that affects how individuals think, feel, and handle daily activities. It is characterized by a combination of symptoms that persist for at least 2 weeks and interfere with daily functioning (Beck and Alford 2009). Depression is associated with alterations in brain function and structure and is among the most prevalent neuropsychiatric conditions worldwide (Lépine and Briley 2011). Key neurobiological factors include dysregulation of neurotransmitter systems (e.g., serotonin, norepinephrine, DA) and changes in the areas of the brain involved in mood regulation (e.g., prefrontal cortex, amygdala) (Lang and Borgwardt 2013).

Inflammation in the brain is frequently linked to depression (Hasler 2010). Patients with depression have higher numbers of inflammatory cells, and this increase in inflammatory cells can make depression worse. In animal models, andrographolide counters inflammation through the reduction of cytokines implicated in inflammation as well as the amelioration of depressive symptoms. Its antidepressant effects primarily stem from its ability to reduce neuroinflammation (Zhang et al. 2019). Depression is linked to neurotransmitter abnormalities, specifically those involving serotonin, norepinephrine, and DA. It has been demonstrated that andrographolide affects neurotransmitter levels, which may help reduce feelings of depression. Andrographolide functions similarly to conventional antidepressants by raising serotonin levels in the brain. By readjusting neurotransmitter systems, depression symptoms can be lessened and mood can be elevated (Zhang et al. 2019). Administering andrographolide in chronic unpredictable mild stress (CUMS) animal model ablates depressive behavior and reduces the expression of proinflammatory mediators and cytokines in the prefrontal cortex (Thakur et al. 2014, Zhang et al. 2019). Additionally, in the prefrontal cortex of CUMS rats, andrographolide induced autophagic responses by upregulating Beclin1 expression and abrogating phosphorylated mTORC1 (Geng et al. 2019). Beclin1 promotes autophagy in its early phases, while p-mTOR phosphorylates ATG13 and ULK1/2 to prevent autophagy (Kim et al. 2011). Andrographolide-mediated activation of the autophagy program resulted in minimal symptoms of depression (Geng et al. 2019). In a different study, andrographolide exhibited anti-depressive properties by actuating the brain-derived neurotrophic factor axis in the hippocampus (Zhang et al. 2019). In yet another study, andrographolide reduced the GSK-3β activity and modulated β-catenin accumulation to inhibit long-term depression (Tapia-Rojas et al. 2015).

4.3.5 Role of Andrographolide in Brain Injury/Stroke

Stroke is a disorder that develops when there is a disruption or reduction in blood flow to a portion of the brain, depriving the brain tissue of the necessary nutrients and oxygen it requires. As a consequence, brain cells die within minutes, resulting in various degrees of neurological impairment depending on the location and extent of the brain damage (Iadecola and Anrather 2011). Ischemic stroke occurs when a blood clot obstructs an artery supplying blood to the brain. It accounts for about 87% of all strokes (Feske 2021). Ischemic strokes are often caused by thrombotic or embolic events. Thrombotic strokes are due to blood clots forming in a brain artery, while embolic strokes are caused by clots or debris originating from other parts of the body, such as the heart, traveling to the brain (De Meyer et al. 2016).

Stroke is invariably accompanied by inflammatory conjunctions. Andrographolide reduces microglial activation and neuroinflammation in intracerebral hemorrhage (ICH) mouse models, leading to significantly lower levels of TNF-α and IL-6. It also alleviates brain injury caused by ICH by curtailing both the nuclear export of NF-κB and the assemblage of NLRP3 inflammasomes (Li et al. 2018). In a rodent model of weightlessness-inflicted injury, andrographolide additionally hinders neuroinflammation and stimulation of microglia by blocking the MAPK signaling network. Andrographolide heals brain damage post craniocerebral trauma, improves neurobehavioral functioning, and decreases brain edema in the brain tissue (Tao et al. 2018). A study by Chan et al. indicates that andrographolide significantly decreased microglia cells in rodent models of permanent middle cerebral artery occlusion (pMCAO). There was a significant reduction in the infarction volume, resulting in the remedy of neurological impairments. Conversely, andrographolide demonstrated a neuroprotective effect in the rat model produced by MCAO by lowering inflammation. Neuroinflammatory indicators including NF-κB, TNF-α, and IL-1β were eliminated by andrographolide (Chan et al. 2010). In another study, andrographolide helped ameliorate neuronal cell loss, degeneration, and neurobehavioral impairment in rats with ICH-induced secondary brain injury (Li et al. 2018). Experimental evidence revealed that andrographolide inhibits the activation of microglia in ICH rats by reverse phosphorylating IκBα and p65. Additionally, andrographolide decreases the interaction between NLRP3 inflammasome components in the ICH brain, which prevented the cleavage of caspase-1 (Wang et al. 2004). In an investigation employing a rodent model of middle cerebral artery obstruction, andrographolide boosted HO-1 activity by lowering oxidative stress in the brain. Additionally, the heightened HO-1 activity can be attributed to the modulation of the MAPK-Nrf2-HO-1 axis by andrographolide (Yen et al. 2016). As a result, andrographolide reduces the size of the infarct by up to 50%; inhibits the levels of inflammatory cytokines TNF-α, PGE2, and IL-1β in areas of the ischemic brain; and reduces neurological impairment by inhibiting NF-κB and activity of the microglial cells (Chan et al. 2010). In ischemic animals, andrographolide lessens the production of NADPH oxidase 2 (NOX2) and iNOS and inhibits the activity of NF-κB and HIF-1α to improve neurological impairment (Chern et al. 2011). Moreover, Yang et al. reported that andrographolide treatment can effectively reduce cerebral infarction to improve neurological function possibly through the activation of the Nrf2/ARE signaling pathway and the suppression of the NF-κB signaling pathway (Yang et al. 2019).

4.3.6 Role of Andrographolide in Schizophrenia

One of the most common psychological disorders in the world is schizophrenia, and new research indicates that oxidative stress and inflammation may be important factors in the etiology of the disorder. It is a significant mental illness that alters a person's thoughts, feelings, and behavior (McCutcheon et al. 2020).

A study using a mouse model of schizophrenia produced by phencyclidine was done to investigate whether andrographolide could ameliorate symptoms of schizophrenia-like behaviors by reducing inflammation and oxidative stress (Wang et al. 2021). Andrographolide abrogated the phosphorylation of p65, IκBα, p38, and ERK1/2 in the prefrontal cortex when PCP was present. This, in turn, decreased the levels of inflammatory cytokines in the prefrontal cortex; examples include IL-1, TNF, IL-6, COX-2, and iNOS (Wang et al. 2021). The levels of malondialdehyde and GSH were markedly reduced by andrographolide, while the activities of SOD and catalase were increased. Furthermore, andrographolide boosted the expression of HO-1, NQO-1, and Nrf2. It also facilitated Nrf2's nuclear translocation by obstructing the interaction between Nrf2 and Keapl. Moreover, the application of Nrf2-specific inhibitor ML385 hampered andrographolide's anti-schizophrenia-like behaviors and antioxidant effects. Therefore, these findings indicated that andrographolide, via upregulating the Nrf2 pathway, mitigated oxidative stress and PCP-induced schizophrenia-like behaviors.

4.3.7 Role of Andrographolide in Seizures

Seizures are episodes of abnormal electrical activity in the brain that disrupt normal brain function. The disruption can lead to a range of symptoms depending on the part of the brain affected and the type of seizure (Fisher et al. 2017). Seizures occur due to an imbalance between excitatory and inhibitory neurotransmission, leading to hyperexcitability of neurons because of genetic factors, brain injuries, infections, or metabolic disturbances (Lasoń et al. 2013). Conditions such as cortical malformations, brain tumors, and scar tissue from injuries or surgeries can create a focus for abnormal electrical activity, increasing the risk of seizures.

Neuroprotection involves safeguarding neurons from damage caused by excessive excitatory neurotransmission and neurotoxicity, which are common in seizures. Andrographolide provides neuroprotection in rodent models of epilepsy. The compound reduces neuronal cell death and protects against seizure-induced brain damage (Baru Venkata et al. 2023). Seizures often induce an inflammatory response in the brain, contributing to neuronal damage and the perpetuation of seizures. In animal models, andrographolide curtails inflammation-related cytokines. This decrease in inflammation has a critical role in reducing seizure activity and averting more brain damage (Low et al. 2024). Imbalances in neurotransmitter systems, particularly glutamate and gamma-aminobutyric acid (GABA), are central to seizure pathophysiology. Andrographolide influences neurotransmitter levels and increases GABAergic activity, while reducing excessive glutamatergic activity to stabilize neuronal excitability and prevent seizures (Baru Venkata et al. 2023). Seizures have the potential to weaken the blood–brain barrier (BBB), which can exacerbate inflammation and damage to neurons. Andrographolide supports the maintenance of the BBB in epilepsy models. Andrographolide lessens the entry of inflammatory cells and chemicals into the brain, which can cause or exacerbate seizures, by shielding the BBB (Owen et al. 2023). In conclusion, andrographolide impacts seizures through multiple pathways, including neuroprotection, anti-inflammatory effects, modulation of neurotransmitter systems, antioxidant properties, and enhancement of BBB integrity (Huang et al. 2022).

REFERENCES

Abedi, Z., Basri, H., Hassan, Z., Mat, L. N. I., Khaza'ai, H. and Mohamad, N. A. (2021). "A review of the neuroprotective effects of andrographolide in Alzheimer's disease." *Advances in Traditional Medicine* **21**: pp. 253–266.

Ahmed, S., Kwatra, M., Ranjan Panda, S., Murty, U. S. N. and Naidu, V. G. M. (2021). "Andrographolide suppresses NLRP3 inflammasome activation in microglia through induction of parkin-mediated mitophagy in in-vitro and in-vivo models of Parkinson disease." *Brain, Behavior, and Immunity* **91**: pp. 142–158.

Akbar, S. (2011). "*Andrographis paniculata*: a review of pharmacological activities and clinical effects." *Alternative Medicine Review* **16**(1): pp. 66–77.

Banerjee, M., Parai, D., Chattopadhyay, S. and Mukherjee, S. K. (2017). "Andrographolide: antibacterial activity against common bacteria of human health concern and possible mechanism of action." *Folia Microbiologica* **62**: pp. 237–244.

Baru Venkata, R., Prasanth, D. S. N. B. K., Pasala, P. K., Panda, S. P., Tatipamula, V. B., Mulukuri, S., Kota, R. K., Rudrapal, M., Khan, J., Aldosari, S., Alshehri, B., Banawas, S., Challa, M. C. and Kammili, J. K. (2023). "Utilizing *Andrographis paniculata* leaves and roots by effective usage of the bioactive andrographolide and its nanodelivery: investigation of antikindling and antioxidant activities through in silico and in vivo studies." *Frontiers in Nutrition* **10**: p. 1185236.

Bayazid, A. B., Hwang, U. K., Jang, Y. A., Jeong, Y. H., Jo, Y. C. and Lim, B. O. (2024). "*Andrographis paniculata* leaves extract alleviates UVB-induced HaCaT cells through suppressing mitogen-activated protein kinases activation." *Natural Product Communications* **19**(4): p. 1934578X241238137.

Beck, A. T. and B. A. Alford (2009). Depression: Causes and Treatment. University of Pennsylvania Press: Philadelphia, PA.

Bharati, B. D., Sharma, P. K., Kumar, N., Dudhe, R., and Bansal, V. (2011). "Pharmacological activity of Andrographis paniculata: a brief review." *Pharmacologyonline* **2**(1): p. 10.

Bloem, B. R., Okun, M. S. and Klein, C. (2021). "Parkinson's disease." *The Lancet* **397**(10291): pp. 2284–2303.

Brahmachari, G. (2016). "Andrographolide: a molecule of antidiabetic promise." In Brahmachari, G. (ed.), *Discovery and Development of Antidiabetic Agents from Natural Products*, Elsevier: Amsterdam, Netherlands (pp. 1–27).

Burgos, R. A., Seguel, K., Perez, M., Meneses, A., Ortega, M., Guarda, M. I., Loaiza, A. and Hancke, J. L. (2005). "Andrographolide inhibits IFN-γ and IL-2 cytokine production and protects against cell apoptosis." Planta Medica **71**(05): pp. 429–434.

Burgos, R. A., Alarcón, P., Quiroga, J., Manosalva, C. and Hancke, J. (2020). "Andrographolide, an anti-inflammatory multitarget drug: all roads lead to cellular metabolism." *Molecules* **26**(1): p. 5.

Cai, Q., Zhang, W., Sun, Y., Xu, L., Wang, M., Wang, X., Wang, S. and Ni, Z. (2022). "Study on the mechanism of andrographolide activation." *Frontiers in Neuroscience* **16**: p. 977376.

Chan, S. J., Wong, W. S. F., Wong, P. T. H. and Bian, J. S. (2010). "Neuroprotective effects of andrographolide in a rat model of permanent cerebral ischaemia." *British Journal of Pharmacology* **161**(3): pp. 668–679.

Chao, W.-W. and B.-F. Lin (2010). "Isolation and identification of bioactive compounds in Andrographis paniculata (Chuanxinlian)." *Chinese Medicine* **5**: pp. 1–15.

Chern, C.-M., Liou, K.-T., Wang, Y.-H., Liao, J.-F., Yen, J.-C. and Shen, Y.-C. (2011). "Andrographolide inhibits PI3K/AKT-dependent NOX2 and iNOS expression protecting mice against hypoxia/ischemia-induced oxidative brain injury." *Planta Medica* **77**(15): pp. 1669–1679.

Dai, Y., Chen, S. R., Chai, L., Zhao, J., Wang, Y. and Wang, Y. (2019). "Overview of pharmacological activities of Andrographis paniculata and its major compound andrographolide." *Critical Reviews in Food Science and Nutrition* **59(sup1)**: pp. S17–S29.

De Meyer, S. F., Denorme, F., Langhauser, F., Geuss, E., Fluri, F. and Kleinschnitz, C. (2016). "Thromboinflammation in stroke brain damage." *Stroke* **47**(4): pp. 1165–1172.

Feske, S. K. (2021). "Ischemic stroke." *The American Journal of Medicine* **134**(12): pp. 1457–1464.

Finnerup, N. B., Attal, N., Haroutounian, S., McNicol, E., Baron, R., Dworkin, R. H., Gilron, I., Haanpää, M., Hansson, P., Jensen, T. S. and Kamerman, P. R. (2015). "Pharmacotherapy for neuropathic pain in adults: a systematic review and meta-analysis." *The Lancet Neurology* **14**(2): pp. 162–173.

Finnerup, N. B., Kuner, R. and Jensen, T. S. (2020). "Neuropathic pain: from mechanisms to treatment." *Physiological Reviews* **101**(1):259–301.

Fisher, R. S., Cross, J. H., French, J. A., Higurashi, N., Hirsch, E., Jansen, F. E., Lagae, L., Moshé, S. L., Peltola, J., Roulet Perez, E., Scheffer, I. E. and Zuberi, S. M. (2017). "Operational classification of seizure types by the International League Against Epilepsy: position paper of the ILAE commission for classification and terminology." *Epilepsia* **58**(4): pp. 522–530.

Geng, J., Liu, W., Xiong, Y., Ding, H., Jiang, C., Yang, X., Li, X., Elgehama, A., Sun, Y., Xu, Q., Guo, W. and Gao, J. (2018). "Andrographolide sulfonate improves Alzheimer-associated phenotypes and mitochondrial dysfunction in APP/PS1 transgenic mice." *Biomedicine & Pharmacotherapy* **97**: pp. 1032–1039.

Geng, J., Liu, W., Gao, J., Jiang, C., Fan, T., Sun, Y., Qin, Z. H., Xu, Q., Guo, W. and Gao, J. (2019). "Andrographolide alleviates Parkinsonism in MPTP-PD mice via targeting mitochondrial fission mediated by dynamin-related protein 1." *British Journal of Pharmacology* **176**(23): pp. 4574–4591.

Geng, J., Liu, J., Yuan, X., Liu, W. and Guo, W. (2019). "Andrographolide triggers autophagy-mediated inflammation inhibition and attenuates chronic unpredictable mild stress (CUMS)-induced depressive-like behavior in mice." *Toxicology and Applied Pharmacology* **379**: p. 114688.

Godoy, J. A., Rios, J. A., Zolezzi, J. M., Braidy, N. and Inestrosa, N. C. (2014). "Signaling pathway cross talk in Alzheimer's disease." *Cell Communication and Signaling* **12**: pp. 1–12.

Gong, P., Zhang, W., Zou, C., Han, S., Tian, Q., Wang, J., He, P., Guo, Y. and Li, M. (2022). "Andrographolide attenuates blood–brain barrier disruption, neuronal apoptosis, and oxidative stress through activation of Nrf2/HO-1 signaling pathway in subarachnoid hemorrhage." *Neurotoxicity Research* **40**(2): pp. 508–519.

Gu, L., Yu, Q., Li, Q., Zhang, L., Lu, H. and Zhang, X. (2018). "Andrographolide protects PC12 cells against β-amyloid-induced autophagy-associated cell death through activation of the Nrf2-mediated p62 signaling pathway." *International Journal of Molecular Sciences* **19**(9): p. 2844.

Gupta, S., Mishra, K. P. and Ganju, L. (2017). "Broad-spectrum antiviral properties of andrographolide." *Archives of Virology* **162**(3): pp. 611–623.

Gupta, S., Mishra, K., Dash, P., Parida, M., Ganju, L. and Singh, S. (2018). "Andrographolide inhibits chikungunya virus infection by up-regulating host innate immune pathways." *Asian Pacific Journal of Tropical Medicine* **11**(3): pp. 214–221.

Hasler, G. (2010). "Pathophysiology of depression: do we have any solid evidence of interest to clinicians?" *World Psychiatry* **9**(3): p. 155.

Hossain, M. S., Urbi, Z., Sule, A. and Hafizur Rahman, K. M. (2014). "Andrographis paniculata (Burm. f.) Wall. ex Nees: a review of ethnobotany, phytochemistry, and pharmacology." *The Scientific World Journal* **2014**(1): p. 274905.

Hossain, R., Quispe, C., Herrera-Bravo, J., Beltrán, J. F., Islam, M. T., Shaheen, S., Cruz-Martins, N., Martorell, M., Kumar, M., Sharifi-Rad, J., Ozdemir, F. A., Setzer, W. N., Alshehri, M. M., Calina, D. and Cho, W. C. (2022). "Neurobiological promises of the bitter diterpene lactone andrographolide." *Oxidative Medicine and Cellular Longevity* **2022**(1): p. 3079577.

Hossain, S., Urbi, Z., Karuniawati, H., Mohiuddin, R. B., Moh Qrimida, A., Allzrag, A. M. M., Ming, L. C., Pagano, E. and Capasso, R. (2021). "Andrographis paniculata (burm. F.) wall. Ex nees: an updated review of phytochemistry, antimicrobial pharmacology, and clinical safety and efficacy." *Life* **11**(4): p. 348.

Huang, Y., Xu, H., Wang, P., Gu, R., Li, X., Xu, Y., Wang, J., Qiao, S., Shi, D., Gao, Z. and Li, J. (2022). "Identification of Guaifenesin–Andrographolide as a novel combinatorial drug therapy for epilepsy using network virtual screening and experimental validation." *ACS Chemical Neuroscience* **13**(7): pp. 978–986.

Iadecola, C. and J. Anrather (2011). "Stroke research at a crossroad: asking the brain for directions." *Nature Neuroscience* **14**(11): pp. 1363–1368.

Iqbal, K., del C. Alonso, A., Gong, C. X., Khatoon, S., Pei, J. J., Wang, J. Z. and Grundke-Iqbal, I. (1998). Mechanisms of Neurofibrillary Degeneration and the Formation of Neurofibrillary Tangles. Springer: Vienna.

Jain, P., Satija, J. and Sudandiradoss, C. (2023). "Discovery of andrographolide hit analog as a potent cyclooxygenase-2 inhibitor through consensus MD-simulation, electrostatic potential energy simulation and ligand efficiency metrics." *Scientific Reports* **13**(1): p. 8147.

Kempuraj, D., Thangavel, R., Natteru, P. A., Selvakumar, G. P., Saeed, D., Zahoor, H., Zaheer, S., Iyer, S. S. and Zaheer, A. (2016). "Neuroinflammation induces neurodegeneration." *Journal of Neurology, Neurosurgery and Spine* **1**(1), p. 1003.

Kim, J., Kundu, M., Viollet, B. and Guan, K. L. (2011). "AMPK and mTOR regulate autophagy through direct phosphorylation of Ulk1." *Nature Cell Biology* **13**(2): pp. 132–141.

Knopman, D. S., Amieva, H., Petersen, R. C., Chételat, G., Holtzman, D. M., Hyman, B. T., Nixon, R. A. and Jones, D. T. (2021). "Alzheimer disease." *Nature Reviews Disease Primers* **7**(1): p. 33.

Kumar, D., Das, B., Sen, R., Kundu, P., Manna, A., Sarkar, A., Chowdhury, C., Chatterjee, M. and Das, P. (2015). "Andrographolide analogue induces apoptosis and autophagy mediated cell death in U937 cells by inhibition of PI3K/Akt/mTOR pathway." *Plos One* **10**(10): p. e0139657.

Kumar, G., Singh, D., Tali, J. A., Dheer, D. and Shankar, R. (2020). "Andrographolide: chemical modification and its effect on biological activities." *Bioorganic Chemistry* **95**: p. 103511.

Lang, U. E. and S. Borgwardt (2013). "Molecular mechanisms of depression: perspectives on new treatment strategies." *Cellular Physiology and Biochemistry* **31**(6): pp. 761–777.

Lasoń, W., Chlebicka, M. and Rejdak, K. (2013). "Research advances in basic mechanisms of seizures and antiepileptic drug action." *Pharmacological Reports* **65**(4): pp. 787–801.

Lépine, J.-P. and M. Briley (2011). "The increasing burden of depression." *Neuropsychiatric Disease and Treatment* **7**(**sup1**): pp. 3–7.

Li, J., Cheung, H. Y., Zhang, Z., Chan, G. K. L. and Fong, W. F. (2007). "Andrographolide induces cell cycle arrest at G2/M phase and cell death in HepG2 cells via alteration of reactive oxygen species." *European Journal of Pharmacology* **568**(**1–3**): pp. 31–44.

Li, X., Wang, T., Zhang, D., Li, H., Shen, H., Ding, X. and Chen, G. (2018). "Andrographolide ameliorates intracerebral hemorrhage induced secondary brain injury by inhibiting neuroinflammation induction." *Neuropharmacology* **141**: pp. 305–315.

Li, X., Yuan, W., Wu, J., Zhen, J., Sun, Q. and Yu, M. (2022). "Andrographolide, a natural anti-inflammatory agent: an update." *Frontiers in Pharmacology* **13**: p. 920435.

Li, Y., He, S., Tang, J., Ding, N., Chu, X., Cheng, L., Ding, X., Liang, T., Feng, S., Rahman, S. U., Wang, X. and Wu, J. (2017). "Andrographolide inhibits inflammatory cytokines secretion in LPS-stimulated RAW264. 7 cells through suppression of NF-κB/MAPK signaling pathway." *Evidence-Based Complementary and Alternative Medicine* **2017**(1): p. 8248142.

Liu, G.-Z. and H.-W. Xu (2008). "Development in synthesis of andrographolide derivatives." *Chinese Journal of Organic Chemistry* **28**(2): p. 201.

Low, M., Suresh, H., Zhou, X., Bhuyan, D. J., Alsherbiny, M. A., Khoo, C., Münch, G. and Li, C. G. (2024). "The wide spectrum anti-inflammatory activity of andrographolide in comparison to NSAIDs: a promising therapeutic compound against the cytokine storm." *Plos One* **19**(7): p. e0299965.

Lu, J., Ma, Y., Wu, J., Huang, H., Wang, X., Chen, Z., Chen, J., He, H. and Huang, C. (2019). "A review for the neuroprotective effects of andrographolide in the central nervous system." *Biomedicine & Pharmacotherapy* **117**: p. 109078.

Maiti, K., Mukherjee, K., Murugan, V., Saha, B. P. and Mukherjee, P. K. (2010). "Enhancing bioavailability and hepatoprotective activity of andrographolide from Andrographis paniculata, a well-known medicinal food, through its herbosome." *Journal of the Science of Food and Agriculture* **90**(1): pp. 43–51.

McCutcheon, R. A., Reis Marques, T. and Howes, O. D. (2020). "Schizophrenia—an overview." *JAMA Psychiatry* **77**(2): pp. 201–210.

Meng, W., Xiaoliang, R., Xiumei, G., Vincieri, F. F. and Bilia, A. R. (2009). "Stability of active ingredients of traditional Chinese medicine (TCM)." *Natural Product Communications* **4**(12): doi:10.1177/1934578X0900401229.

Murphy, K. L., Bethea, J. R. and Fischer, R. (2017). "Neuropathic pain in multiple sclerosis–current therapeutic intervention and future treatment perspectives." In: Zagon IS, McLaughlin PJ, editors. *Multiple Sclerosis: Perspectives in Treatment and Pathogenesis [Internet]*. Brisbane (AU): Codon Publications.

Murphy, M. P. and H. LeVine III (2010). "Alzheimer's disease and the amyloid-β peptide." *Journal of Alzheimer's Disease* **19**(1): pp. 311–323.

Mussard, E., Cesaro, A., Lespessailles, E., Legrain, B., Berteina-Raboin, S. and Toumi, H. (2019). "Andrographolide, a natural antioxidant: an update." *Antioxidants* **8**(12): p. 571.

Naomi, R., Bahari, H., Ong, Z. Y., Keong, Y. Y., Embong, H., Rajandram, R., Teoh, S. H., Othman, F., Hasham, R., Yin, K. B., Kaniappan, P., Yazid, M. D. and Zakaria, Z. A. (2022). "Mechanisms of natural extracts of andrographis paniculata that target lipid-dependent cancer pathways: a view from the signaling pathway." *International Journal of Molecular Sciences* **23**(11): p. 5972.

Okhuarobo, A., Ehizogie Falodun, J., Erharuyi, O., Imieje, V., Falodun, A. and Langer, P. (2014). "Harnessing the medicinal properties of Andrographis paniculata for diseases and beyond: a review of its phytochemistry and pharmacology." *Asian Pacific Journal of Tropical Disease* **4**(3): pp. 213–222.

Owen, A. E., Louis, H., Ejiofor, E. U., Emori, W., Gber, T. E., Benjamin, I., Cheng, C. R., Orosun, M. M., Ling, L. and Adeyinka, A. S. (2023). "Natural andrographolide isolated from andrographis paniculata as potent epileptic agent: spectroscopy, molecular structure, and molecular docking investigation." *Chemistry Africa* **6**(5): pp. 2445–2461.

Patel, R., Kaur, K. and Singh, S. (2021). "Protective effect of andrographolide against STZ induced Alzheimer's disease in experimental rats: possible neuromodulation and Aβ (1–42) analysis." *Inflammopharmacology* **29**(4): pp. 1157–1168.

Peng, Y., Wang, Y., Tang, N., Sun, D., Lan, Y., Yu, Z., Zhao, X., Feng, L., Zhang, B., Jin, L., Yu, F., Ma, X. and Lv, C. (2018). "Andrographolide inhibits breast cancer through suppressing COX-2 expression and angiogenesis via inactivation of p300 signaling and VEGF pathway." *Journal of Experimental & Clinical Cancer Research* **37**: pp. 1–14.

Prescott, S. A. and S. Ratté (2017). Somatosensation and pain. In P. Michael Conn, editor. *Conn's Translational Neuroscience*. Elsevier: Cambridge, MA, USA (pp. 517–539).

Qiao, C., Zhang, Q., Jiang, Q., Zhang, T., Chen, M., Fan, Y., Ding, J., Lu, M. and Hu, G. (2018). "Inhibition of the hepatic Nlrp3 protects dopaminergic neurons via attenuating systemic inflammation in a MPTP/p mouse model of Parkinson's disease." *Journal of Neuroinflammation* **15**: pp. 1–11.

Rajagopal, S., Kumar, R. A., Deevi, D. S., Satyanarayana, C. and Rajagopalan, R. (2003). "Andrographolide, a potential cancer therapeutic agent isolated from Andrographis paniculata." *Journal of Experimental Therapeutics and Oncology* **3**(3): pp. 147–158.

Seo, J. Y., Pyo, E., An, J. P., Kim, J., Sung, S. H. and Oh, W. K. (2017). "Andrographolide activates Keap1/Nrf2/ARE/HO-1 pathway in HT22 cells and suppresses microglial activation by Aβ42 through Nrf2-related inflammatory response." *Mediators of Inflammation*. 2017;2017:5906189. doi: 10.1155/2017/5906189

Serrano, F. G., Tapia-Rojas, C., Carvajal, F. J., Hancke, J., Cerpa, W. and Inestrosa, N. C. (2014). "Andrographolide reduces cognitive impairment in young and mature AβPPswe/PS-1 mice." *Molecular Neurodegeneration* **9**: pp. 1–18.

Serrano-Pozo, A., Frosch, M. P., Masliah, E. and Hyman, B. T. (2011). "Neuropathological alterations in Alzheimer disease." *Cold Spring Harbor Perspectives in Medicine* **1**(1): p. a006189.

Sigcha, L., Borzì, L., Amato, F., Rechichi, I., Ramos-Romero, C., Cárdenas, A., Gascó, L. and Olmo, G. (2023). "Deep learning and wearable sensors for the diagnosis and monitoring of Parkinson's disease: a systematic review." *Expert Systems with Applications* **229**: p. 120541.

Smith, P. L., Maloney, K. N., Pothen, R. G., Clardy, J. and Clapham, D. E. (2006). "Bisandrographolide from Andrographis paniculata activates TRPV4 channels." *Journal of Biological Chemistry* **281**(40): pp. 29897–29904.

Sule, A., Ahmed, Q. U., Latip, J., Samah, O. A., Omar, M. N., Umar, A. and Dogarai, B. B. S. (2012). "Antifungal activity of Andrographis paniculata extracts and active principles against skin pathogenic fungal strains in vitro." *Pharmaceutical Biology* **50**(7): pp. 850–856.

Tao, L., Zhang, L., Gao, R., Jiang, F., Cao, J. and Liu, H. (2018). "Andrographolide alleviates acute brain injury in a rat model of traumatic brain injury: possible involvement of inflammatory signaling." *Frontiers in Neuroscience* **12**: p. 657.

Tapia-Rojas, C., Schüller, A., Lindsay, C. B., Ureta, R. C., Mejías-Reyes, C., Hancke, J., Melo, F. and Inestrosa, N. C. (2015). "Andrographolide activates the canonical Wnt signalling pathway by a mechanism that implicates the non-ATP competitive inhibition of GSK-3β: autoregulation of GSK-3β in vivo." *Biochemical Journal* **466**(2): pp. 415–430.

Thakur, A. K., Soni, U. K., Rai, G., Chatterjee, S. S. and Kumar, V. (2014). "Protective effects of Andrographis paniculata extract and pure andrographolide against chronic stress-triggered pathologies in rats." *Cellular and Molecular Neurobiology* **34**: pp. 1111–1121.

Wang, H.-C., Tsay, H.-S., Shih, H.-N., Chen, Y.-A., Chang, K.-M., Agrawal, D. C., Huang, S., Lin, Y.-L. and Lee, M.-J. (2018). "Andrographolide relieved pathological pain generated by spared nerve injury model in mice." *Pharmaceutical Biology* **56**(1): pp. 124–131.

Wang, T., Liu, B., Zhang, W., Wilson, B. and Hong, J. S. (2004). "Andrographolide reduces inflammation-mediated dopaminergic neurodegeneration in mesencephalic neuron-glia cultures by inhibiting microglial activation." *Journal of Pharmacology and Experimental Therapeutics* **308**(3): pp. 975–983.

Wang, W., Wang, J., Dong, S. F., Liu, C. H., Italiani, P., Sun, S. H., Xu, J., Boraschi, D., Ma, S. P. and Qu, D. (2010). "Immunomodulatory activity of andrographolide on macrophage activation and specific antibody response." *Acta Pharmacologica Sinica* **31**(2): pp. 191–201.

Wang, X., Liu, J., Dai, Z. and Sui, Y. (2021). "Andrographolide improves PCP-induced schizophrenia-like behaviors through blocking interaction between NRF2 and KEAP1." *Journal of Pharmacological Sciences* **147**(1): pp. 9–17.

Wong, S. K., Chin, K. Y. and Ima-Nirwana, S. (2021). "A review on the molecular basis underlying the protective effects of Andrographis paniculata and andrographolide against myocardial injury." *Drug Design, Development and Therapy* **15**, pp. 4615–4632.

Wu, B., Chen, X., Zhou, Y., Hu, P., Wu, D., Zheng, G. and Cai, Y. (2018). "Andrographolide inhibits proliferation and induces apoptosis of nasopharyngeal carcinoma cell line C666-1 through LKB1-AMPK-dependent signaling pathways." *Die Pharmazie-An International Journal of Pharmaceutical Sciences* **73**(10): pp. 594–597.

Yang, C.-H., Yen, T.-L., Hsu, C.-Y., Thomas, P.-A., Sheu, J.-R. and Jayakumar, T. (2017a). "Multi-targeting andrographolide, a novel NF-κB inhibitor, as a potential therapeutic agent for stroke." *International Journal of Molecular Sciences* **18**(8): p. 1638.

Yang, M.-Y., Yu, Q.-L., Huang, Y.-S. and Yang, G. (2019). "Neuroprotective effects of andrographolide derivative CX-10 in transient focal ischemia in rat: involvement of Nrf2/AE and TLR/NF-κB signaling." *Pharmacological Research* **144**: pp. 227–234.

Yang, R., Liu, S., Zhou, J., Bu, S. and Zhang, J. (2017b). "Andrographolide attenuates microglia-mediated Aβ neurotoxicity partially through inhibiting NF-κB and JNK MAPK signaling pathway." *Immunopharmacology and Immunotoxicology* **39**(5): pp. 276–284.

Yang, S., Evens, A. M., Prachand, S., Singh, A. T. K., Bhalla, S., David, K. and Gordon, L. I. (2010). "Mitochondrial-mediated apoptosis in lymphoma cells by the diterpenoid lactone andrographolide, the active component of Andrographis paniculata." *Clinical Cancer Research* **16**(19): pp. 4755–4768.

Yen, T.-L., Chen, R.-J., Jayakumar, T., Lu, W.-J., Hsieh, C.-Y., Hsu, M.-J., Yang, C.-H., Chang, C.-C., Lin, Y.-K., Lin, K.-H. and Sheu, J.-R. (2016). "Andrographolide stimulates p38 mitogen-activated protein kinase–nuclear factor erythroid-2-related factor 2–heme oxygenase 1 signaling in primary cerebral endothelial cells for definite protection against ischemic stroke in rats." *Translational Research* **170**: pp. 57–72.

Yi, Z., Ouyang, S., Zhou, C., Xie, L., Fang, Z., Yuan, H., Yang, J., Zou, L., Jia, T., Zhao, S., Li, L., Shi, L., Gao, Y., Li, G., Liu, S., Xu, H., Xu, C., Zhang, C. and Liang, S. (2018). "Andrographolide inhibits mechanical and thermal hyperalgesia in a rat model of HIV-induced neuropathic pain." *Frontiers in Pharmacology* **9**: pp. 593.

Yu, B., Zhang, Z., Liu, W., Yang, T. and Wang, P. (1992). "Study on stability in vitro of andrographolide." *Chinese Traditional Patent Medicine.*

Zhang, J.-J., Gao, T.-T., Wang, Y., Wang, J.-L., Guan, W., Wang, Y.-J., Wang, C.-N., Liu, J.-F. and Jiang, B. (2019). "Andrographolide exerts significant antidepressant-like effects involving the hippocampal BDNF system in mice." *International Journal of Neuropsychopharmacology* **22**(9): pp. 585–600.

Zhang, Z., Lai, D., Wang, L., Yu, P., Zhu, L., Guo, B., Xu, L., Zhou, L., Sun, Y., Lee, S. M. Y. and Wang, Y., (2014). "Neuroprotective effects of the andrographolide analogue AL-1 in the MPP+/MPTP-induced Parkinson's disease model in vitro and in mice." *Pharmacology Biochemistry and Behavior* **122**: pp. 191–202.

5 Natural Products as Neuroprotective Agents

Dipak Kumar and Abhishek Kumar Mishra

ABBREVIATIONS

AD	Alzheimer's Disease
ALS	Amyotrophic Lateral Sclerosis
APP	Amyloid Precursor Protein
Aβ	Amyloid Beta
BDNF	Brain-Derived Neuroptrophic Factor
CBD	Cannabidiol
CNS	Central Nervous System
COX	Cyclooxyegenase
DHA	Docosahexaenoic Acid
EPA	Eicosapentaenoic Acid
HTT	Huntngtin
MPTP	1-Methyl-4-Phenyl-1,2,3,6-tetrapyrodoxine
NAC	N-acetylcystine.
NGF	Nerve Growth Factor
PD	Parkinson's Disease
PSD95	Postsynaptic Density Protein 95
PUFA	Polyunsaturated Fatty Acid
SIRT1	Sirtuin1
6OHDA	6-hydroxydopamine
SNAP23	Synaptosome-Associated Protein 23
SOD	Superoxide Dismutase
TH	Tyrosine Hydrolase

5.1 INTRODUCTION

Degenerative brain disorders, like Alzheimer's disease (AD), Parkinson's disease (PD), and amyotrophic lateral sclerosis (ALS), pose significant challenges to global health, affecting millions of individuals worldwide. These disorders, illustrated by progressive neuronal deterioration and loss of function, result in severe cognitive and motor dysfunction, leading to increased morbidity and mortality. Current pharmacological treatments often focus on alleviating symptoms rather than addressing the underlying neurodegenerative processes, which can limit their long-term effectiveness and result in adverse side effects. As a result, researchers are increasingly investigating alternative strategies to combat these devastating conditions. Among these strategies, natural products' usage as neuroprotective agents has obtained considerable attention. Natural products, encompassing a diverse range of bioactive substances derived from various plant parts, fungi, and other organisms, have been a cornerstone of traditional medicine across various cultures. The rich biodiversity of the natural world offers a vast reservoir of potential therapeutic agents, many of which have evolved to protect organisms from environmental stressors, such as oxidative stress, inflammation,

DOI: 10.1201/9781003520610-5

and neurotoxicity. These compounds often exhibit multifaceted pharmacological properties that enable them to modulate complex biological pathways critical in neuronal survival and function.

Recent advancements in scientific techniques, including high-throughput screening and molecular biology, have facilitated the identification and characterization of numerous natural compounds with neuroprotective properties. This diverse group of compounds includes alkaloids, flavonoids, terpenoids, carotenoids, alkanes, fatty acids, vitamins, valproate, saponins, and monoterpenes, all of which are sourced from several parts of plants, including fruits, seeds, vegetables, leaves, roots, rhizomes, and stems. Research is increasingly focusing on these bioactive compounds due to their potential for cognitive enhancement and their ability to help prevent the pathological progression of numerous diseases, including cardiovascular disease, cancer, diabetes, and neurological disorders (Asgharian et al., 2022). Phytochemicals exhibit diverse biological roles and employ multiple mechanisms to promote cellular health.

These mechanisms include scavenging free radicals, modulating cell signaling pathways, and influencing enzyme activity (Küpeli Akkol et al., 2021; Mohd Sairazi & Sirajudeen, 2020). One of the most compelling aspects of phytochemicals is their potential to target multiple pathways involved in neurodegeneration. For example, curcumin, a polyphenol compound derived from *Curcuma longa*, exhibits potent antioxidant, anti-inflammatory, and amyloid plaque-clearing potentials, making it a preferred candidate for AD treatment. Also, resveratrol, found in red wine, has been linked to improved cognitive behavior and neuroprotection through its effects on sirtuins and brain-derived neurotrophic factor (BDNF) regulation. Moreover, the holistic nature of natural products aligns well with contemporary understandings of the disease, which emphasize the importance of a multi-targeted approach. This is mainly relevant in the context of neurodegenerative diseases, where multiple interrelated biological pathways contribute to disease progression. By harnessing the complexity and diversity of natural products, researchers aim to develop more effective and safer therapeutic strategies that not only slow the progression of these diseases but also enhance the quality of life for affected human beings. This chapter provides a comprehensive overview of natural neuroprotective agents, their sources, cellular mechanism of action, neuroprotective/neurogenesis roles in neurological disorders, and mental health benefits.

5.2 POLYPHENOLS

Polyphenols are structurally diverse plant-based organic compounds with a complex chemical structure. Most polyphenols contain repeating phenolic moieties connected by esters or more stable C–C bonds. These phytochemicals are found in berries, cocoa, coffee, tea, spices, grapes, and olives, exclusively in their fruit skins (Arias-Sánchez et al., 2023). Polyphenols are known for their anti-inflammatory and anti-oxidative properties, and they are natural scavengers of free radicals (Uddin et al., 2020). A study revealed that polyphenols protect brain tissues by reducing oxidative stress, following the alternation of the protein kinase B (AKT)/cAMP response element binding protein (CREB)/BDNF signaling pathway and increasing nuclear translocation of Nrf2 (Uddin et al., 2020). A clinical study showed that polyphenol supplementation promotes neurogenesis by increasing pro-cognitive neurotrophic factors within the brain (Ammar et al., 2020). It also fosters synaptic plasticity and survivability, enhances vascularization, and regulates neurotransmitter systems by interacting with several brain proteins (Rebas et al., 2020; Vauzour, 2017).

5.2.1 Resveratrol

Peanuts, grapes, blueberries, raspberries, and mulberries contain resveratrol. However, a higher concentration of resveratrol is in the peel and seeds of grapes grown in cooler climates. It is also significantly present in red wine, often a suggestive dietary source for this phytochemical. Resveratrol is a prominent free radical scavenger and a potent natural anti-inflammatory polyphenol (Andrade et al., 2018; Repossi et al., 2020). In neuronal cells, it inhibits the activity of cyclooxygenase

(COX) enzymes and modulates the NF-κß pathway to reduce neuroinflammation (Xu et al., 2018). Resveratrol improves mitochondrial health and function, thereby neuronal longevity, by interacting and activating sirtuin 1 (SIRT1) (Ungvari et al., 2011; Wang et al., 2021). Its dietary administration in mice induces mitochondrial biogenesis in the brain (Robb et al., 2008; Ungvari et al., 2011). An animal study on resveratrol demonstrated that the compound strengthens synaptic plasticity and maintains BDNF levels (Wang et al., 2021). Research showed that resveratrol reduces amyloid-β (Aβ) plaques in AD models and α-synuclein levels in PD (Karuppagounder et al., 2009; Wang et al., 2021). The multifaceted approach of resveratrol to inhibit neurodegeneration also makes it a promising bioactive compound for future research to cure neurodegenerative diseases.

5.2.2 Curcumin

It is a major curcuminoid of *C. longa* (turmeric plant). It is derived from the plant's rhizomes and is commonly used as a spice and medicinal herb. Curcumin actively crosses the blood–brain barrier and shows antioxidant competency by neutralizing free radicals and amplifying the activity of endogenous antioxidant enzymes in neurons (Askarizadeh et al., 2020). It blocks pro-inflammatory cytokines and enzymes like COX-2 and lipoxygenase (Del Prado-Audelo et al., 2019). In AD models, curcumin disrupts preformed Aβ plaques and resists further aggregation and formation in AD-related neurons (Mishra & Palanivelu, 2008). Curcumin supplementation improves cognitive health and function in both animal models and human studies (Kuszewski et al., 2018; Tsai et al., 2021). It promotes neurogenesis, specifically in the hippocampus, and enhances mood, memory, and learning (Mishra & Palanivelu, 2008). The wide range of action mechanisms makes curcumin a potent neuroprotective agent for neurodegenerative research.

5.2.3 Epigallocatechin Gallate

Epigallocatechin gallate (EGCG) is a primary catechin in tea and is predominantly found in *Camellia sinensis* (green tea). It is also found in berries, kiwis, cherries, apples, pistachios, etc. EGCG decreases pro-inflammatory cytokine levels and inhibits NF-κß signaling pathway to reduce neuroinflammation (Payne et al., 2023). It chelates transitional metals to inhibit oxidative stress and ferroptosis (Mandel et al., 2006). EGCG has also been investigated to safeguard neurons against intrinsic programmed cell death through CREB/BDNF/TrkB, PI3K/AKT/mTOR, and PI3K/AKT/eNOS signaling (Abdelmeguid et al., 2022). EGCG-induced modulation of these pathways stimulates neuronal growth and synaptogenesis within the central nervous system (CNS) (Ding et al., 2017; Esvald et al., 2020). Recent research showed that EGCG modulates dopamine production and suppresses α-synuclein aggregation (Wang et al., 2022). It protects cholinergic neurons from deterioration as EGCG inhibits tau aggregation by regulating tau phosphorylation and oligomerization (Soto-Mercado et al., 2021; Valverde-Salazar et al., 2023). Moreover, it strengthens synaptic plasticity and potentially improves cognition and memory impairment (Ahn et al., 2022).

5.2.4 Pterostilbene

It is a plant-based stilbenoid compound. Stilbenoid is a family of naturally occurring phenolic phytochemicals having a C6–C2–C6 skeleton and two aromatic rings separated by an ethylene linker. Plants often produce stilbenoid compounds, including pterostilbene, in response to fungal infection (Li et al., 2014). Several plants, including blueberries, grapes, peanuts, almonds, and some widely used Ayurvedic medicinal plants, such as *Pterocarpus marsupium* and *Pterocarpus santalinm*, are rich sources of pterostilbene. In cellular and animal models, pterostilbene showed neuroprotection against cerebral ischemia by attenuating astrocytic inflammation via suppressing COX2 protein and NF-κß signaling pathway; it also inhibits neuronal oxidative stress via SIRT1 (Koh et al., 2023; Yan et al., 2021; Zhu et al., 2022). Research showing the role of pterostilbene on an Aβ-induced AD model indicates that pterostilbene upregulates SIRT1, nuclear factor erythroid 2-related factor 2 (Nrf2), and superoxide dismutase (SOD) levels and inhibits mitochondria-dependent programmed cell death to protect the primary cortical neurons against insults and restores memory-learning declines in animals (Koh et al., 2023; Xu et al., 2020). Fan et al. (2024) demonstrated that pterostilbene exerts neuroprotective effects in 1-methyl-4-phenyl-1,2,3,6-tetrahydropyridine (MPTP)-induced PD animal models by inhibiting neuroinflammation and redox imbalance in related dopamine-releasing neurons (Fan et al., 2024). Moreover, reports validate its involvement in enhancing cognition, synaptic wiring, and neuronal longevity (Fan et al., 2024; Li et al., 2018; Xu et al., 2020).

5.2.5 Anthocyanins

Anthocyanins restrain Aβ formation, attenuate tau phosphorylation, and preserve cholinergic neurons from apoptotic loss (Ali et al., 2017; Noi et al., 2021). Anthocyanin supplementation regulates p-PI3K/Akt/GSK3β and Nrf2/HO-1 pathways to halt programmed neuronal degeneration by inhibiting neurons' apoptotic and other neurodegenerative biomarker (Ali et al., 2017). It also improves cognitive impairments and synaptic structure and function by elevating the expressions of several pre- and postsynaptic proteins, including synaptophysin, postsynaptic density protein 95 (PSD95), and synaptosome-associated protein 23 (SNAP23) (Baek et al., 2023). These polyphenols are predominantly found in fruits and flowers and possess antioxidant, anti-inflammatory, and anti-apoptotic benefits (Ali et al., 2017; Baek et al., 2023).

5.3 FLAVONOIDS

They are a class of phenolic bioactive agents with a 15-carbon skeleton (C6–C3–C6 chemical structure) consisting of two aromatic rings and a heterocyclic ring. Tea, citrus fruits, berries, apples, and legumes are some primary dietary sources of plant flavonoids. Importantly, flavonoids in animals and humans come from plants as they do not synthesize the phytochemical. Like polyphenols, flavonoids are exceptional antioxidants, free radical scavengers, metal chelators, and anti-inflammatory phytochemicals. Reports showed that flavonoid treatment decreases Aβ formation and tau pathology in AD models and prevents α-synuclein accumulation in models of PD (de Andrade Teles et al., 2018; Hole & Williams, 2021). Most flavonoids interact with neurotransmitter systems in the brain and optimize their synthesis, release, and uptake, including acetylcholine and dopamine.

5.3.1 Quercetin

Quercetin is reported to inhibit many enzymes linked to neurodegeneration. It inhibits tyrosine protein kinase, phospholipase, phosphodiesterase, mitochondrial ATPase, and protein kinase C (Singh et al., 2003; Wang et al., 2011). It is naturally present in onions, apples, and broccoli. A study showed that it reduces oxidative stress, iron overload, and inflammation to protect neurons from apoptotic degeneration (Meng-Zhen et al., 2022). The research identified that pretreatment of quercetin improves cognitive decline in AD models by reducing Aβ levels and redox damage through Nrf2/HO-1 pathways (Meng-Zhen et al., 2022). In transgenic AD model models, pretreatment of quercetin protects related neurons from degeneration by correcting mitochondrial dysfunction and activating AMPK (Wang et al., 2014). Pretreatment of quercetin in 6-hydroxydopamine (6-OHDA)-exposed PD model mice demonstrated that it ameliorates mitochondrial dysfunction and endogenous antioxidant enzymatic function and increases the expression of crucial proteins, including CREB and BDNF in dopaminergic nerve cells (Chen et al., 2009; Haleagrahara et al., 2013).

5.3.2 Silymarin

Silymarin is present in *Silybum marianum* and is extracted from the dried fruits and seeds of the plant. It has multifaceted neuroprotective benefits, including anti-aging, antioxidant, anti-inflammatory, anti-apoptotic, and pro-autophagy functions (Lim et al., 2020; Tripathi et al., 2019). This natural flavonoid obstructs neurodegeneration in AD and PD models by modulating pro-inflammatory cytokines and correcting aberrant autophagy (Song et al., 2016). Also, silymarin supplements enhance synaptic wiring and alleviates cognitive decline, and memory deficits (Liu et al., 2020).

5.3.3 Baicalein

The flavonoid is isolated from the roots of *Scutellaria baicalensis*. This phytochemical showed dopaminergic neuroprotection against several toxicant-induced neurotoxicity, including 6-OHDA, MPTP, and methamphetamine (AlSaad et al., 2020; Jiang et al., 2010; Zhu et al., 2004). Studies showed that baicalein prevents dopaminergic loss by reducing α-synuclein aggregation and disaggregating mature α-synuclein fibrils. Moreover, baicalein restores mitochondrial function, reduces microglia and astrocyte activity, and increases neurotransmitter levels (Jiang et al., 2010).

5.3.4 Nobiletin

It is a polymethoxylated flavone present in the outer layer of numerous citrus fruits. Research showed that nobiletin protects dopaminergic neurons by inhibiting neuroinflammation and regulating apoptotic signaling (Cui et al., 2010). In the MPTP-exposed PD model, nobiletin protects neurons via increasing tyrosine hydroxylase (TH) activity through the Ca++/calmodulin pathway; it ameliorates motor impairment and memory-learning deficits (Jeong et al., 2015). In the AD model, pretreatment of nobiletin restores acetylcholinesterase-positive neuron density by ameliorating free radical generation (Lee et al., 2019; Onozuka et al., 2008). It also decreases Aβ plaques levels and hyperphosphorylation of tau.

5.3.5 Naringenin

The anti-inflammatory and anti-oxidative benefits of naringenin increase Nrf2 protein expression and trigger the antioxidant response system to protect degenerating nigrostriatal neurons in the 6-OHDA-exposed PD model (Wang et al., 2019). Moreover, naringenin pretreatment restores dopamine in the midbrain. Pretreatment of naringenin upregulates AMPK-mediated autophagy to digest aberrant proteins (Ahsan et al., 2020). It also ameliorates Aβ plaques-induced cognitive imbalance by mitigating lipid peroxidation and apoptosis (Ghofrani et al., 2015). The antioxidant and anti-inflammatory activities of naringenin to shield acetylcholinergic neurons modulate several proteins in neurons, including Nrf2, NF-κB, COX2, and TNFα (Khajevand-Khazaei et al., 2018). Pretreatment of naringenin protects against intracerebroventricular-streptozotocin-induced redox imbalance, neuronal injury, and cognitive impairment (Khan et al., 2012).

5.3.6 Naringin

Naringin is another flavonoid with antioxidant, anti-inflammatory, anti-apoptotic, and free radical scavenging benefits and is common in citrus fruits (Zhao & Liu, 2021). It penetrates the blood–brain barrier and potentially rescues neurons from Aβ-induced neurotoxicity by restoring mitochondrial membrane potential and resisting lipid peroxidation, apoptosis, and ROS production in cholinergic neurons (Maratha & Mahadevan, 2012). Moreover, the anti-inflammatory activity of naringin increases BDNF levels and protects the neurons against 1-methyl-4-phenylpyridinium (MPP)+-induced neurodegeneration in cellular models of PD (Leem et al., 2014).

5.3.7 Hesperidin

Hesperidin enhances memory by inducing synapse formation between hippocampal and cortex neurons. The antioxidant nature of hesperidin significantly decreases oxidative stress in 6-OHDA exposed neurons by obstructing the activity of glutathione peroxidase and catalase upon co-treatment (Antunes et al., 2014). Hesperidin is obtained from unripe citrus fruits. In the transgenic animal model of AD, hesperidin restores cognitive and non-cognitive ability and social interaction by mitigating Aβ aggregation, β-amyloid precursor protein expression, oxidative stress, and mitochondrial dysfunction (Li et al., 2015; Wang et al., 2014).

5.3.8 Kaempferol

Kaempferol is a flavonoid obtained from the rhizome of *Kaempferol galanga*. Research showed that kaempferol possesses neuroprotective potency as it reduces behavioral impairment and restores striatal dopamine levels in the PD model (Kouhestani et al., 2018). It is also reported to induce autophagy and preserve striatal glutamatergic response in PD-related neurons (Filomeni et al., 2012).

5.3.9 Fisetin

It is abundant in strawberries but is also present in apples. This phytochemical protects neurons against oxidative stress and apoptosis. Fisetin protects dopamine-containing neurons against MPTP/MPP+-induced neuronal loss and alleviates α-synuclein deposition and inflammation in deteriorating neurons (Patel et al., 2012).

5.4 ALKALOIDS

These are nitrogen-containing natural products, typically with at least one uncharged nitrogen atom and heterocyclic ring in their structure. These bioactive compounds are widely found across kingdoms: Monera, Fungi, Plantae, and Animalia. However, only the plant-based alkaloids play a medicinal role in humans. In plants, alkaloids play different roles: nitrogen reservoirs, waste products, protective agents against predators, and growth stimulators/inhibitors (Heinrich et al., 2021). Several alkaloids directly act on the CNS, while others modulate neurotransmitter systems. Some alkaloids have potential therapeutic effects against neurodegenerative disorders and chronic inflammatory conditions.

5.4.1 Caffeine

This natural product is a phytochemical of the methylxanthine alkaloid family. It is exclusively found in several plants' leaves, seeds, and fruits, such as tea, cocoa, coffee, and kola. Caffeine is a popular psychostimulant with antioxidant, anti-inflammatory, anti-aging, and anti-apoptosis properties (Gonçalves et al., 2020). Chronic caffeine intake inhibits lipid peroxidation and ROS generation, ameliorates oxidative stress, improves mitochondrial health, and attenuates neuroinflammation. Research showed that chronic caffeine treatment reduces Aβ formation, improves memory, learning, and recognition, and protects hippocampal neurons from degeneration in AD animal models (Cao et al., 2009). In animals, it also restores levels of BDNF and its receptor (TrkB), which combinedly regulate growth, survival, and neuron differentiation (Lao-Peregrín et al., 2017). Furthermore, increased consumption of caffeine preserves dopamine levels in the human striatum. Caffeine co-treatment inhibits microglial proliferation in the MPTP-induced PD model and blocks nigral neurodegeneration (Bagga et al., 2016). Caffeine consumption as a supplement also reduces the risk of ALS and improves functional outcomes of Huntington's disease (HD) (Huin et al., 2024; Ruggiero et al., 2022).

5.4.2 Huperzine A

Huperzine A is a novel alkaloid in *Huperzia serrata*, the Chinese club moss plant. It is a potent modulator of acetylcholinesterase and is therefore, often consumed to enhance memory-learning function in AD or other related forms of dementia (Mak et al., 2021). Preclinical studies showed that huperzine A potentially modifies Aβ peptide processing, reduces redox imbalance, and regulates the formation and secretion of nerve growth factor (NGF) and NGF signaling (Wang et al., 2006). Researchers investigated that it inhibits GSK3α/β activity and enhances β-catenin levels in animals, suggesting that the neuroprotective role of huperzine A in AD brain includes targeting the Wnt/β-catenin signaling pathways (Friedli & Inestrosa, 2021; Wang et al., 2011). The phytochemical huperzine A also interacts with several synaptic proteins, including synaptotagmin and neuroligins (Friedli & Inestrosa, 2021).

5.4.3 Berberine

Berberine is potentially beneficial for several health issues, for example, diabetes, high cholesterol, high blood pressure, and depression. The genus *Berberis*, comprising ~500 species, is a natural source of berberine; some major berberine-containing plants are goldenseal, goldthread, *Phellodendron*, and European barberry. It belongs to benzylisoquinoline alkaloid. Berberine shows neuroprotective effects against several CNS disorders, like AD, PD, HD, cerebral ischemia, anxiety, and depression. In mice models of AD, berberine alleviates cognitive deficits by reducing tau hyperphosphorylation and promoting autophagy clearance of redundant proteins (Wu et al., 2021). In addition, it inhibits NF-κB activity, GSK3β activity, and lipid peroxidation in the hippocampus (Wu et al., 2021). In 6-OHDA-exposed PD models, berberine increases striatal dopamine, norepinephrine, 3,4-dihydroxyphenylacetic acid, and homovanillic acid levels and rescues TH-positive neurons from degeneration. Pretreatment of berberine upregulates HO-1 expression to suppress oxidative stress and induces PI3K/Akt and p38 activation to promote Nrf2 expression and neuroprotection (Bae et al., 2013). Berberine promotes the degradation of misfolded and aggregated huntingtin (HTT) proteins in HTT-induced mice models of HD (Jiang et al., 2015).

5.5 TERPENOIDS

Chemically, terpenoids naturally occur in unsaturated biomolecules composed of linked isoprene units. Plants, algae, marine phytoplankton, and a few fungi species are sources of terpenoids. Plant-based terpenoids constitute a diversified category of naturally occurring organic bioactive agents that stem from terpenes. They are recognized in various plants for their extensive biological functions, notably neuroprotection. Significant instances involve ginkgolides and bilobalide sourced from *Ginkgo biloba*, demonstrating antioxidant and anti-inflammatory features, thereby enhancing cognitive abilities, and shielding neurons from oxidative stress. The EGb761, a comprehensively studied standard extract of *G. biloba* leaves with ginkgolides and bilobalide as key constituents, has demonstrated neuroprotective properties against cerebral injury (Feng et al., 2019). Another well-known terpenoid, cannabidiol, originating from *Cannabis sativa*, helps regulate neurotransmitter systems, reduce neuroinflammation, and protect against oxidative damage (De Gregorio et al., 2019; Pretzsch et al., 2019). Terpenoids such as these fulfill vital roles in upholding brain well-being, fostering synaptic adaptability, and bolstering neurogenesis in neurodegenerative disorders (Yoo & Park, 2012). Their ability to eliminate free radicals, stabilize cellular membranes, and modulate signaling pathways positions them as auspicious contenders for preventing and managing neurodegenerative ailments. Ongoing investigations persist in delving into their therapeutic capacities to leverage their advantages in augmenting cognitive function and overall neurological welfare.

5.6 SAPONINS

Saponins represent a distinct category of naturally occurring glycosides in various plants, renowned for their distinct foaming characteristics and wide-ranging biological functions. These substances have earned considerable recognition for their potential neuroprotective attributes.

Predominantly, saponins are in plant sources such as soybeans, quinoa, ginseng, and licorice. The neuroprotective attributes of saponins are linked to multiple mechanisms. Initially, they demonstrate robust antioxidant features by eliminating free radicals and diminishing oxidative stress, a pivotal element in neurodegenerative ailments (Sun et al., 2014). By safeguarding neurons from oxidative harm, saponins aid in upholding cellular structure and operation. Moreover, saponins possess anti-inflammatory traits. They impede increased pro-inflammatory cytokines and modulate inflammatory pathways, thereby reducing neuroinflammation (Sun et al., 2014). This proves critical in ailments like AD wherein persistent inflammation significantly influences disease advancement. Furthermore, saponins stimulate neurogenesis and synaptic adaptability (Zhang et al., 2023). They boost the development and distinction of neural stem cells, contributing to the mending and regeneration of impaired neurons. Additionally, saponins from *Panax ginseng* regulate neurotransmitter systems, enhancing cognitive capacities like memory and learning (Jin et al., 2019). Research has indicated that saponins derived from *P. ginseng*, particularly ginsenosides, deliver notable neuroprotective effects. These compounds elevate cognitive performance, shield against neuronal harm, and enhance overall brain well-being (Zheng et al., 2018). As investigations progress, saponins persist, displaying potential as therapeutic agents for preventing and managing neurodegenerative conditions.

5.7 PHENOLIC ACIDS

They are a class of natural products distinguished by a phenol ring containing at least one carboxylic acid group (–COOH). They are diversely distributed in plants and contribute to their taste and color. Berries, apples, carrots, spinach, grains, seeds, and coffee are rich sources of phenolic acids. Chemically, hydroxybenzoic acids and hydroxycinnamic acids are two categories of phenolic acids, having benzoic acid and cinnamic acid structures, respectively. These bioactive chemicals exhibit antioxidant, anti-inflammatory, and neuroprotective properties.

5.7.1 Gallic Acid

It is a phenolic acid of the hydroxybenzoic acid category. Gallic acid is a common tea antioxidant that reduces triglyceride levels and improves cholesterol. Apart from antioxidants, phytomedicine possesses many other biotic and therapeutic activities, including anti-inflammatory, anti-apoptotic, and interfering with cell signaling pathways. Gallic acid acetylates NF-κB to stabilize and decrease the production of pro-inflammatory cytokines and protects neurons against Aβ-induced neurotoxicity (Kim et al., 2011). The systemic administration of gallic acid in rats attenuates LPS-induced neuroinflammation and decreases neuro-apoptosis biomarkers in nigral neurons (Shabani et al., 2020). Moreover, research showed that it alters serotonin and dopamine levels (Ojo et al., 2023).

5.7.2 Ferulic Acid

It is a phenolic acid of the hydroxycinnamic acid category, abundant in commelinid plants, including rice, wheat, and oats. It is also present in beans, coffee seeds, nuts, and peanuts. Like many other phenolic acids, ferulic acids possess antioxidant, anti-inflammatory, and anti-apoptotic properties. Ferulic acid supplementation has been found to enhance survival factors like BDNF levels in the brain (Singh et al., 2022).

5.7.3 Rosameric Acid

Rosameric acid is traditionally used in culinary herbs for its phytomedicinal properties. It possesses multimodal actions on numerous signaling pathways to ameliorate neurodegenerative and CNS pathologies (Nabavi et al., 2015; Ravaria et al., 2023). Besides its antioxidant and anti-inflammatory roles, it modulates synaptic signaling to elevate the levels of synaptic proteins and ameliorate neurotransmission (Ravaria et al., 2023).

5.7.4 Salvianolic Acid

Salvianolic acid protects brain tissues and improves neurological deficits by attenuating cerebral edema and downregulating inflammatory mediators levels: TLR4, p-p38 MAPK, p-JNK, IL-1β, and NF-κB (Zheng et al., 2023). Salvianolic acid is reported to ameliorate pathologies of PD and AD (Liu et al., 2020; Tian et al., 2008).

5.8 CAROTENOIDS

Carotenoids are a family of natural pigments in plants, algae, and photosynthetic bacteria. They are the determinants of organisms' vivid colors. These pigments are crucial in maintaining plant health since they contribute to photosynthesis and protect against photooxidative damage. Carrots, sweet potatoes, spinach, kale, tomatoes, mangoes, and papayas are significant sources of carotenoids in our diet. Chemically, carotenoids are tetraterpenoids that consist of 40 carbon atoms organized in a conjugated double-bond structure. This arrangement determines distinct color and antioxidant effects. They are categorized into two primary classes: carotenes and xanthophylls. Carotenes, such as beta-carotene and lycopene, are hydrocarbons consisting only of carbon and hydrogen atoms, but xanthophylls, like lutein, zeaxanthin, and astaxanthin, contain oxygen atoms. These phytochemicals offer neuroprotective advantages because of their antioxidant and anti-inflammatory characteristics (Manochkumar et al., 2021). They function as antioxidants by counteracting free radicals, diminishing oxidative stress, and safeguarding neurons from harm. It is imperative to prevent and alleviate the advancement of degenerative conditions in CNS, such as AD and PD. Carotenoids possess anti-inflammatory characteristics that regulate major inflammatory pathways, i.e., PI3K/Akt, mTOR, and MAPK pathways, and reduce neuroinflammation, a crucial element in many disorders (Manochkumar et al., 2021).

Particular carotenoids, such as lutein and zeaxanthin, have notable advantages for brain function. In animal models of vision-threatening diseases, lutein reduces oxidative stress and restores rhodopsin, synaptophysin, and BDNF levels to preserve visual functions (Iyer et al., 2024; Ozawa et al., 2012). Dietary zeaxanthin supplementation delays Parkinsonian signs in older adults (Agarwal et al., 2022). These substances are highly concentrated in the retina and brain, specifically to enhance visual and cognitive capabilities. They prevent age-dependent macular degeneration and improve cognitive performance, particularly in older people (Agarwal et al., 2022; Ozawa et al., 2012).

5.9 FATTY ACIDS

They are essential in protecting the brain, owing to their structure and functions. Omega-3 and omega-6 polyunsaturated fatty acids hold substantial importance. Omega-3 fatty acids, specifically eicosapentaenoic acid (EPA) and docosahexaenoic acid (DHA), are essential for maintaining the structure of neuronal membranes, promoting the ability of synapses to change and adapt, and supporting the generation of new neurons. Both EPA and DHA decrease IL-1β, IL-6, and IFN-α and promote hippocampal neurogenesis (Borsini et al., 2021; Michael-Titus & Priestley, 2014). DHA, specifically, is found in high levels in the brain and retina, where it promotes cognitive function and visual acuity (Michael-Titus & Priestley, 2014). Omega-3 fatty acids are obtained from numerous sources, including fatty fish like salmon, mackerel, sardines, flaxseeds, chia seeds, walnuts, and algae. Omega-6 fatty acids, such as linoleic acid and arachidonic acid, are necessary for maintaining brain function. These fatty acids are found in vegetable oils, nuts, and seeds. It is essential to maintain a proper balance between the ingesting of omega-3 and omega-6, as an excessive amount of omega-6 can stimulate inflammation.

5.10 VITAMINS

Vitamins are vital organic substances that are crucial in maintaining proper physiological activities and promoting general health. They are necessary in small quantities and must be acquired from diet since the human body is incapable of either synthesizing or manufacturing them in inadequate amounts. Vitamins are vital for many essential biochemical roles, such as metabolism, immunological function, and cellular maintenance.

5.10.1 Vitamin D

Vitamin D is vital for brain metabolism because of its neuroprotective, anti-inflammatory, and antioxidant qualities. It promotes neuronal development and differentiation by increasing the expression of neurotrophic factors: NGF, neurotrophin3, and glial cell line-derived neurotrophic factor, critical for neuron survival and function (Latimer et al., 2014; Neveu et al., 1994). Vitamin D also affects calcium homeostasis, vital for synaptic development and neurotransmission (Latimer et al., 2014; Ye et al., 2023). It also affects the immune system, lowering neuroinflammation, which significantly contributes to illnesses such as AD and PD (Ye et al., 2023). Its sources include sunlight, fatty fish, fortified dairy products, and supplements.

5.11 AMINO ACIDS AND PEPTIDES

Amino acids are chemical molecules that function as the fundamental constituents of proteins, comprising an amino group, a carboxyl group, and a distinctive side chain. Peptides are concise sequences of amino acids connected by peptide bonds, serving as fundamental frameworks for proteins and fulfilling crucial tasks in biological processes. Amino acids and peptides exhibit neuroprotective characteristics, playing key roles in maintaining brain health and safeguarding against neurodegeneration.

5.11.1 *N*-acetylcysteine

N-acetylcysteine (NAC) is a precursor to glutathione, one of the body's most critical antioxidants. NAC safeguards neurons from injury and mitigates oxidative stress by replenishing glutathione levels. It also possesses anti-inflammatory properties, which modulate pathways that reduce neuroinflammation, a critical cause of the degeneration of neurons in PD and AD (Faghfouri et al., 2020; Fan et al., 2020). Additionally, it has been demonstrated that NAC contributes to the brain's overall health by supporting mitochondrial function and enhancing synaptic plasticity (Faghfouri et al., 2020).

5.11.2 L-Theanine

It is an amino acid primarily found in green tea that can promote relaxation and reduce stress without producing drowsiness. After crossing the blood–brain barrier, L-theanine modulates neurotransmitter levels, emphasizing increasing GABA, serotonin, and dopamine, essential for cognitive function and mood regulation (Nathan et al., 2006). L-Theanine also demonstrates neuroprotective effects by decreasing inflammation and redox imbalance. It regulates oxidative damage-related enzymes: SOD-1, GST, and NOX-4. L-Theanine also reduces TNF-α, IL-1β, and IL-6 expression levels, reducing neuroinflammation and protecting neurons from degeneration (Sumathi et al., 2016). Research indicates that L-theanine can improve learning, memory, and attention by promoting alpha brain wave activity associated with a relaxed yet vigilant mental state (Baba et al., 2021).

5.12 OTHER NATURAL NEUROPROTECTIVE AGENTS

5.12.1 *Bacopa monnieri* (Brahmi)

Brahmi is an Ayurvedic herb that has long been recognized for its ability to safeguard and support the health of the nervous system. It is widely recognized for improving cognitive function, memory, and learning. Brahmi demonstrates antioxidant properties, which safeguard neurons against oxidative stress and harm. Additionally, it possesses anti-inflammatory characteristics, which diminish neuroinflammation, a crucial element in degenerative disorders, including AD (Dubey & Chinnathambi, 2019). In addition, Brahmi stimulates the development of dendrites and nascent neurons and the ability of synapses to change, which helps develop and restore nerve cells and improves the transmission of signals between them (Sarda et al., 2022). The neuroprotective benefits of this substance are ascribed to its active components, such as bacosides, which enhance the functioning of neurotransmitters and safeguard against neuronal degeneration. Brahmi is a powerful natural solution for preserving brain function and averting cognitive deterioration (Aguiar & Borowski, 2013).

5.12.2 *Withania somnifera* (Ashwagandha)

Ashwagandha is a renowned adaptogenic plant highly acknowledged for its neuroprotective qualities. The compound has numerous advantages for brain health, including its antioxidant, anti-inflammatory, and anti-stress properties. Ashwagandha aids in counteracting oxidative stress by neutralizing free radicals, safeguarding neurons from harm, and promoting overall brain well-being. It decreases neuroinflammation, which is involved in many neurological illnesses like AD and PD, by regulating inflammatory pathways and levels of cytokines (Mikulska et al., 2023; Zahiruddin et al., 2020). Ashwagandha reduces cortisol levels and improves coping with stress, leading to beneficial roles in brain function and mental well-being (Lopresti et al., 2019). This herb stimulates the development of dendrites and synaptic connections, which aids in restoring and rejuvenating neurons and improves cognitive abilities such as memory and learning. Ashwagandha modulates the equilibrium of crucial neuro-messengers, such as serotonin and dopamine, enhancing mood and mental performance (Singh et al., 2011). Ashwagandha's neuroprotective properties make it a helpful natural treatment for preserving brain well-being, improving cognitive function, and reducing the likelihood of neurodegenerative diseases.

5.12.3 *Hericium erinaceus* (Lion's Mane Mushroom)

Hericium erinaceus, frequently referred to as Lion's Mane mushroom, is a medicinal fungus well-known for its ability to protect the nervous system. Lion's Mane is known for its neuroprotective effects, which can be linked to its distinct components, including hericenones and erinacines. These chemicals provide various advantages for brain health. *H. erinaceus* extracts activate JNK pathways and promote the synthesis of NGF, a vital protein for neurons' development, upkeep, and viability (Hwang et al., 2020). It stimulates the development of new neurons and aids in the restoration of impaired neurons, hence improving cognitive abilities and decreasing the likelihood of neurodegenerative disorders. Mushrooms provide potent antioxidants that effectively counteract free radicals, diminishing oxidative stress and safeguarding neurons against harm (Hwang et al., 2020). It possesses anti-inflammatory characteristics that diminish neuroinflammation, a factor linked with certain neurodegenerative disorders like AD and PD (Brandalise et al., 2023). Research has demonstrated that *H. erinaceus* can enhance memory, concentration, and other cognitive functions, especially in aged individuals. It improves the ability of synapses to change and supports the effective operation of interconnected groups of neurons. Mushrooms can potentially inhibit the development of beta-amyloid plaques linked to AD, safeguarding against cognitive deterioration

(Mori et al., 2011). Hence, *H. erinaceus* has notable neuroprotective advantages by stimulating neurogenesis, diminishing oxidative stress and inflammation, improving cognitive function, and protecting against neurodegenerative disorders.

5.12.4 *Centella asiatica*

Centella asiatica, also called Gotu Kola, is a herb widely recognized for its neuroprotective characteristics, especially in the context of traditional Ayurvedic and Chinese medicine. The neuroprotective properties of *C. asiatica* are mainly due to its bioactive constituents, including asiaticoside, madecassoside, and asiatic acid. *C. asiatica* demonstrates potent antioxidant activity, which aids in neutralizing free radicals and mitigating oxidative stress. Preventing oxidative damage is essential for safeguarding neurons and preserving brain health. The plant possesses notable anti-inflammatory qualities that aid in reducing neuroinflammation, a crucial element in neurodegenerative conditions such as AD and PD (Wong et al., 2021).

C. asiatica aids in preserving neurons against inflammatory damage by altering inflammatory pathways and cytokine levels (Fujimori et al., 2022; Wong et al., 2021). *C. asiatica* stimulates new neurons' growth and improves synaptic plasticity. It enhances the generation of BDNF, a protein that promotes the development and specialization of new neurons and synapses, hence enhancing cognitive abilities such as memory and learning (Kim et al., 2015). Studies have demonstrated that the plant *C. asiatica* can enhance cognition, memory, attention, and concentration. It is advantageous for persons with cognitive impairments and those aiming to enhance overall brain function (Kim et al., 2015). *C. asiatica* has demonstrated promise in safeguarding against beta-amyloid toxicity, a condition linked to AD (Fujimori et al., 2022). It aids in the reduction of plaque development and neuronal damage. *C. asiatica* provides substantial neuroprotective advantages because of its antioxidant, anti-inflammatory, and neurogenic qualities. This makes it a viable natural solution for preserving brain health and preventing cognitive decline.

5.13 CONCLUSION

In conclusion, the findings of natural products as neuroprotective agents represent a promising confluence of ancient wisdom and contemporary science. The synergy between traditional knowledge and modern research methods holds the potential to foster innovative strategies for preserving and advancing neurological health. As research progresses, it continues to reveal the extensive promise of these compounds in addressing one of the most urgent health challenges of our era. The ongoing integration of natural product knowledge into clinical practice may lead to groundbreaking therapies for neurodegeneration, ultimately improving patient outcomes and deepening our understanding of neuronal well-being. This chapter has examined the specific mechanisms of action, key natural compounds, and their emerging roles in neuroprotection, laying a foundation for future advancements in this critical field. The potential of natural products as neuroprotective agents offers an exciting prospect for developing novel treatments for neurodegenerative diseases, potentially leading to more effective and safer therapeutic strategies. Phytochemicals, primarily derived from the Plantae kingdom, showcase a vast array of origins, biosynthetic pathways, chemical properties, mechanisms of action, and biological functions (Table 5.1). These bioactive compounds are categorized in various categories on the basis of their distinct effects on the CNS.

Flavonoids in citrus fruits, apples, cocoa, berries, and teas are secondary metabolites with a C15 benzene ring structure originating from the phenylpropanoid pathway. They exhibit antioxidant and anti-inflammatory properties that support synaptogenesis, neurogenesis, and synaptic plasticity. Additionally, flavonoids enhance cerebrovascular function and modulate neurotransmitter systems, including norepinephrine, dopamine, and serotonin, thereby improving cognitive function and neuroprotection. Phenolic acids in berries, nuts, and whole grains possess potent anti-inflammatory effects that reduce neuronal oxidative stress and inflammation. For instance, ferulic acid preserves

TABLE 5.1
Table Showing Sources, Chemical Families, and Role in Neurodegenerative Disorder of Natural Products Having Neuroprotective/Neurogenesis Properties

Sr. No.	Natural Products	Source	Chemical Family	Neuroprotective/ Neurogenesis Properties	Role in Neurodegenerative Disorders
1	Resveratrol	Peanuts, grapes, blueberries, raspberries, and mulberries	Polyphenol	Antioxidant, free radicals scavenger, anti-inflammatory	AD and PD
2	Curcumin	*Curcuma longa*	Polyphenol	Antioxidant, free radicals scavenger, anti-inflammatory, neuro-rejuvenator	AD
3	Epigallocatechin gallate (EGCG)	*Camellia sinensis*	Polyphenol	Antioxidant, free radicals scavenger, anti-inflammatory, anti-apoptotic, synaptogenesis	AD and PD
4	Pterostilbene	Blueberries, peanuts, almonds, and grapes	Polyphenol	Antioxidant, free radicals scavenger, anti-inflammatory, anti-apoptotic	Cerebral ischemia, AD, and PD
5	Anthocyanin	Mostly in fruits and flowers	Polyphenol	Antioxidant, anti-inflammatory, anti-apoptotic	AD
6	Quercetin	Onions, apples, and broccoli	Flavonoid	Antioxidant, anti-inflammatory, anti-apoptotic	HD, AD, and PD
7	Silymarin	*Silybum marianum*	Flavonoid	Antioxidant, anti-inflammatory, anti-apoptotic, anti-aging	AD and PD
8	Baicalein	*Scutellaria baicalensis*	Flavonoid	Antioxidant, anti-inflammatory	PD
9	Nobiletin	Citrus fruits	Flavonoid	Antioxidant, anti-inflammatory, anti-apoptotic	AD and PD
10	Naringenin	Citrus fruits	Flavonoid	Antioxidant, anti-inflammatory	AD and PD
11	Naringin	Citrus fruits	Flavonoid	Antioxidant, anti-inflammatory	AD and PD
12	Hesperidin	Unripe citrus fruits	Flavonoid	Antioxidant	AD and PD
13	Kaempferol	*Kaempferol galanga*	Flavonoid	Antioxidant	PD and sporadic dementia
14	Fistein	Strawberries	Flavonoid	Antioxidant, anti-apoptotic	PD
15	Caffeine	Tea, cocoa, coffee, and kola	Alkaloid	Antioxidant, anti-inflammatory, anti-apoptotic, anti-aging, synaptic plasticity	AD, PD, ALS, and HD
16	Huperzine A	*Huperzia serrata*	Alkaloid	Antioxidant, anti-inflammatory	AD and cerebral ishemia

(*Continued*)

TABLE 5.1 (*Continued*)
Table Showing Sources, Chemical Families, and Role in Neurodegenerative Disorder of Natural Products Having Neuroprotective/Neurogenesis Properties

Sr. No.	Natural Products	Source	Chemical Family	Neuroprotective/ Neurogenesis Properties	Role in Neurodegenerative Disorders
17	Berberine	Genus: Berberis	Alkaloid	Antioxidant, anti-inflammatory	AD, PD, HD, cerebral ischemia, and anxiety
18	Ginkgolides	*Ginkgo biloba*	Terpenoid	Antioxidant, anti-inflammatory	Cerebral ischemia
19	Bilobalides	*Ginkgo biloba*	Terpenoid	Antioxidant, anti-inflammatory	Cerebral ischemia
20	Cannabidiol	*Cannabis sativa*	Terpenoid	Antioxidant, anti-inflammatory	Autism spectrum disorder
21	Ginsenosides	*Panax ginseng*	Saponin	Antioxidant, anti-inflammatory	AD
22	Rosmarinic acid	Lamiaceae herbs	Polyphenolic acid	Antioxidant, anti-inflammatory	AD
23	Salvianolic acid	*Salvia miltiorrhiza*	Polyphenolic acid	Antioxidant, anti-inflammatory	AD and PD
24	Lycopene	Tomatoes, watermelon, etc.	Carotenoid	Antioxidant	PD
25	Lutein	Carrots, tomatoes, spinach, mangoes, etc.	Carotenoid	Antioxidant, anti-inflammatory	–
26	Zeaxanthin	Carrots, tomatoes, spinach, mangoes, etc.	Carotenoid	Antioxidant, anti-inflammatory	–
27	EPA and DHA	Salmon, mackerel, walnuts, chia seeds, and algae	Omega-3 Fatty Acids	Antioxidant, anti-inflammatory	–
28	L-theanine	Green tea	Amino acid derivative	Antioxidant, anti-inflammatory	–
29	N-acetylcysteine (NAC)	Legumes, eggs, and poultry	Amino acid derivative	Antioxidant, anti-inflammatory	–
30	Vitamin D	Milk, egg yolk, oily fish, and red meat	Vitamin	Antioxidant, anti-inflammatory	Neurodevelopmental disorders, AD, and PD

neuronal integrity and cognitive function, while gallic acid enhances cognition and dendritic spine density in the hippocampus. Rosmarinic acid, in particular, restores dopamine levels in AD models by inhibiting monoamine oxidase. Carotenoids, like lutein and zeaxanthin, which impart color to fruits and vegetables, accumulate in the CNS and retina. They protect neuronal membranes from oxidative damage and inflammation, enhancing visual episodic memory and cognitive function, though their precise mechanisms remain under investigation. Polyphenols in cocoa, grapes, and olive oil act as free radical scavengers, modulating neurotrophic factor circulation and enhancing synaptic plasticity and vascularization. They interact with brain proteins and neurotransmitter systems, including acetylcholine, GABA, and glutamate, thereby promoting cognitive health.

Terpenoids, accountable for the aromas and flavors of spices and herbs such as sage, peppermint, ginkgo biloba, and ginseng, show promise in improving cognitive function, alertness, and psychological well-being. They are currently being explored for their potential in treating AD. Nitrogen-containing compounds, like alkaloids including caffeine, influence neurotransmitter systems, enhancing cognitive function through modulation of synaptic activity and receptor functions. Glycosides in various plants offer neuroprotective benefits by modulating neuroinflammation and oxidative stress, thereby supporting cognitive health in neurodegenerative conditions. These phytochemicals interact with neurotransmitter systems, including cholinergic, dopaminergic, serotonergic, and GABAergic pathways, providing potential therapeutic strategies for cognitive enhancement and neuroprotection.

REFERENCES

Abdelmeguid, N. E., Hammad, T. M., Abdel-Moneim, A. M., & Salam, S. A. (2022). Effect of Epigallocatechin-3-gallate on stress-induced depression in a mouse model: role of interleukin-1β and brain-derived neurotrophic factor. *Neurochemical Research*, *47*(11), pp. 3464–3475. https://doi.org/10.1007/s11064-022-03707-9

Agarwal, P., Wang, Y., Buchman, A. S., Holland, T. M., Bennett, D. A., & Morris, M. C. (2022). Dietary antioxidants associated with slower progression of parkinsonian signs in older adults. *Nutritional Neuroscience*, *25*(3), pp. 550–557. https://doi.org/10.1080/1028415X.2020.1769411

Aguiar, S., & Borowski, T. (2013). Neuropharmacological review of the nootropic herb Bacopa monnieri. *Rejuvenation Research*, *16*(4), pp. 313–326. https://doi.org/10.1089/rej.2013.1431

Ahn, J.-W., Kim, S., Ko, S., Kim, Y.-H., Jeong, J.-H., & Chung, S. (2022). Modified (-)-gallocatechin gallate-enriched green tea extract rescues age-related cognitive deficits by restoring hippocampal synaptic plasticity. *Biochemistry and Biophysics Reports*, *29*, p. 101201. https://doi.org/10.1016/j.bbrep.2022.101201

Ahsan, A. U., Sharma, V. L., Wani, A., & Chopra, M. (2020). Naringenin upregulates AMPK-mediated autophagy to rescue neuronal cells from β-Amyloid (1–42) evoked neurotoxicity. *Molecular Neurobiology*, *57*(8), pp. 3589–3602. https://doi.org/10.1007/s12035-020-01969-4

Ali, T., Kim, T., Rehman, S. U., Khan, M. S., Amin, F. U., Khan, M., Ikram, M., & Kim, M. O. (2017). Natural Dietary supplementation of Anthocyanins via PI3K/Akt/Nrf2/HO-1 pathways mitigate oxidative stress, neurodegeneration, and memory impairment in a mouse model of Alzheimer's disease. *Molecular Neurobiology*, *55*(7), pp. 6076–6093. https://doi.org/10.1007/s12035-017-0798-6

AlSaad, A. M., Mohany, M., Almalki, M. S., Almutham, I., Alahmari, A. A., AlSulaiman, M., & Al-Rejaie, S. S. (2020). Baicalein neutralizes hypercholesterolemia-induced aggravation of oxidative injury in rats. *International Journal of Medical Sciences*, *17*(9), pp. 1156–1166. https://doi.org/10.7150/ijms.46108

Ammar, A., Trabelsi, K., Boukhris, O., Bouaziz, B., Müller, P., M Glenn, J., Bott, N. T., Müller, N., Chtourou, H., Driss, T., & Hökelmann, A. (2020). Effects of polyphenol-rich interventions on cognition and brain health in healthy young and middle-aged adults: systematic review and meta-analysis. *Journal of Clinical Medicine*, *9*(5), p. 1598. https://doi.org/10.3390/jcm9051598

Andrade, S., Ramalho, M. J., Pereira, M. do C., & Loureiro, J. A. (2018). Resveratrol brain delivery for neurological disorders prevention and treatment. *Frontiers in Pharmacology*, *9*, p. 1261. https://doi.org/10.3389/fphar.2018.01261

Antunes, M. S., Goes, A. T. R., Boeira, S. P., Prigol, M., & Jesse, C. R. (2014). Protective effect of hesperidin in a model of Parkinson's disease induced by 6-hydroxydopamine in aged mice. *Nutrition*, *30*(11–12), pp. 1415–1422. https://doi.org/10.1016/j.nut.2014.03.024

Arias-Sánchez, R. A., Torner, L., & Fenton Navarro, B. (2023). Polyphenols and neurodegenerative diseases: potential effects and mechanisms of neuroprotection. *Molecules (Basel, Switzerland)*, *28*(14), p. 5415. https://doi.org/10.3390/molecules28145415

Asgharian, P., Quispe, C., Herrera-Bravo, J., Sabernavaei, M., Hosseini, K., Forouhandeh, H., Ebrahimi, T., Sharafi-Badr, P., Tarhriz, V., Soofiyani, S. R., Helon, P., Rajkovic, J., Durna Daştan, S., Docea, A. O., Sharifi-Rad, J., Calina, D., Koch, W., & Cho, W. C. (2022). Pharmacological effects and therapeutic potential of natural compounds in neuropsychiatric disorders: an update. *Frontiers in Pharmacology*, *13*, p. 926607. https://doi.org/10.3389/fphar.2022.926607

Askarizadeh, A., Barreto, G. E., Henney, N. C., Majeed, M., & Sahebkar, A. (2020). Neuroprotection by curcumin: a review on brain delivery strategies. *International Journal of Pharmaceutics*, *585*, p. 119476. https://doi.org/10.1016/j.ijpharm.2020.119476

Baba, Y., Inagaki, S., Nakagawa, S., Kaneko, T., Kobayashi, M., & Takihara, T. (2021). Effects of l-theanine on cognitive function in middle-aged and older subjects: a randomized placebo-controlled study. *Journal of Medicinal Food*, *24*(4), pp. 333–341. https://doi.org/10.1089/jmf.2020.4803

Bae, J., Lee, D., Kim, Y. K., Gil, M., Lee, J.-Y., & Lee, K. J. (2013). Berberine protects 6-hydroxydopamine-induced human dopaminergic neuronal cell death through the induction of heme oxygenase-1. *Molecules and Cells*, *35*(2), pp. 151–157. https://doi.org/10.1007/s10059-013-2298-5

Baek, H., Sanjay, Park, M., & Lee, H.-J. (2023). Cyanidin-3-O-glucoside protects the brain and improves cognitive function in APPswe/PS1ΔE9 transgenic mice model. *Journal of Neuroinflammation*, *20*(1), p. 268. https://doi.org/10.1186/s12974-023-02950-3

Bagga, P., Chugani, A. N., & Patel, A. B. (2016). Neuroprotective effects of caffeine in MPTP model of Parkinson's disease: a 13C NMR study. *Neurochemistry International*, *92*, pp. 25–34. https://doi.org/10.1016/j.neuint.2015.11.006

Borsini, A., Nicolaou, A., Camacho-Muñoz, D., Kendall, A. C., Di Benedetto, M. G., Giacobbe, J., Su, K.-P., & Pariante, C. M. (2021). Omega-3 polyunsaturated fatty acids protect against inflammation through production of LOX and CYP450 lipid mediators: relevance for major depression and for human hippocampal neurogenesis. *Molecular Psychiatry*, *26*(11), pp. 6773–6788. https://doi.org/10.1038/s41380-021-01160-8

Brandalise, F., Roda, E., Ratto, D., Goppa, L., Gargano, M. L., Cirlincione, F., Priori, E. C., Venuti, M. T., Pastorelli, E., Savino, E., & Rossi, P. (2023). Hericium erinaceus in neurodegenerative diseases: from bench to bedside and beyond, how far from the shoreline? *Journal of Fungi (Basel, Switzerland)*, *9*(5), p. 551. https://doi.org/10.3390/jof9050551

Cao, C., Cirrito, J. R., Lin, X., Wang, L., Verges, D. K., Dickson, A., Mamcarz, M., Zhang, C., Mori, T., Arendash, G. W., Holtzman, D. M., & Potter, H. (2009). Caffeine suppresses amyloid-beta levels in plasma and brain of Alzheimer's disease transgenic mice. *Journal of Alzheimer's Disease: JAD*, *17*(3), pp. 681–697. https://doi.org/10.3233/JAD-2009-1071

Chen, P.-Y., Vittal, R., Nien, P.-C., & Ho, K.-C. (2009). Enhancing dopamine detection using a glassy carbon electrode modified with MWCNTs, quercetin, and Nafion®. *Biosensors and Bioelectronics*, *24*(12), pp. 3504–3509. https://doi.org/10.1016/j.bios.2009.05.003

Cui, Y., Wu, J., Jung, S.-C., Park, D.-B., Maeng, Y.-H., Hong, J. Y., Kim, S.-J., Lee, S.-R., Kim, S.-J., Kim, S. J., & Eun, S.-Y. (2010). Anti-neuroinflammatory activity of nobiletin on suppression of microglial activation. *Biological and Pharmaceutical Bulletin*, *33*(11), pp. 1814–1821. https://doi.org/10.1248/bpb.33.1814

de Andrade Teles, R. B., Diniz, T. C., Costa Pinto, T. C., de Oliveira Júnior, R. G., Gama E Silva, M., de Lavor, É. M., Fernandes, A. W. C., de Oliveira, A. P., de Almeida Ribeiro, F. P. R., da Silva, A. A. M., Cavalcante, T. C. F., Quintans Júnior, L. J., & da Silva Almeida, J. R. G. (2018). Flavonoids as therapeutic agents in Alzheimer's and Parkinson's diseases: a systematic review of preclinical evidences. *Oxidative Medicine and Cellular Longevity*, *2018*, p. 7043213. https://doi.org/10.1155/2018/7043213

De Gregorio, D., McLaughlin, R. J., Posa, L., Ochoa-Sanchez, R., Enns, J., Lopez-Canul, M., Aboud, M., Maione, S., Comai, S., & Gobbi, G. (2019). Cannabidiol modulates serotonergic transmission and reverses both allodynia and anxiety-like behavior in a model of neuropathic pain. *Pain*, *160*(1), pp. 136–150. https://doi.org/10.1097/j.pain.0000000000001386

Del Prado-Audelo, M. L., Caballero-Florán, I. H., Meza-Toledo, J. A., Mendoza-Muñoz, N., González-Torres, M., Florán, B., Cortés, H., & Leyva-Gómez, G. (2019). Formulations of curcumin nanoparticles for brain diseases. *Biomolecules*, *9*(2), p. 56. https://doi.org/10.3390/biom9020056

Ding, M., Ma, H., Man, Y., & LV, H. (2017). Protective effects of a green tea polyphenol, epigallocatechin-3-gallate, against sevoflurane-induced neuronal apoptosis involve regulation of CREB/BDNF/TrkB and PI3K/Akt/mTOR signalling pathways in neonatal mice. *Canadian Journal of Physiology and Pharmacology*, *95*(12), pp. 1396–1405. https://doi.org/10.1139/cjpp-2016-0333

Dubey, T., & Chinnathambi, S. (2019). Brahmi (Bacopa monnieri): an ayurvedic herb against the Alzheimer's disease. *Archives of Biochemistry and Biophysics*, *676*, p. 108153. https://doi.org/10.1016/j.abb.2019.108153

Esvald, E.-E., Tuvikene, J., Sirp, A., Patil, S., Bramham, C. R., & Timmusk, T. (2020). CREB family transcription factors are major mediators of BDNF transcriptional autoregulation in cortical neurons. *The Journal of Neuroscience: The Official Journal of the Society for Neuroscience*, *40*(7), pp. 1405–1426. https://doi.org/10.1523/JNEUROSCI.0367-19.2019

Faghfouri, A. H., Zarezadeh, M., Tavakoli-Rouzbehani, O. M., Radkhah, N., Faghfuri, E., Kord-Varkaneh, H., Tan, S. C., & Ostadrahimi, A. (2020). The effects of N-acetylcysteine on inflammatory and oxidative stress biomarkers: a systematic review and meta-analysis of controlled clinical trials. *European Journal of Pharmacology*, *884*, p. 173368. https://doi.org/10.1016/j.ejphar.2020.173368

Fan, C., Long, Y., Wang, L., Liu, X., Liu, Z., Lan, T., Li, Y., & Yu, S. Y. (2020). N-Acetylcysteine rescues hippocampal oxidative stress-induced neuronal injury via suppression of p38/JNK signaling in depressed rats. *Frontiers in Cellular Neuroscience*, *14*, p. 554613. https://doi.org/10.3389/fncel.2020.554613

Fan, Y., He, X., Chen, M., Guo, S., & Dong, Z. (2024). Pterostilbene alleviates MPTP-induced neurotoxicity by targeting neuroinflammation and oxidative stress. *Biochemical and Biophysical Research Communications*, *729*, p. 150358. https://doi.org/10.1016/j.bbrc.2024.150358

Feng, Z., Sun, Q., Chen, W., Bai, Y., Hu, D., & Xie, X. (2019). The neuroprotective mechanisms of ginkgolides and bilobalide in cerebral ischemic injury: a literature review. *Molecular Medicine (Cambridge, Mass.)*, *25*(1), p. 57. https://doi.org/10.1186/s10020-019-0125-y

Filomeni, G., Graziani, I., De Zio, D., Dini, L., Centonze, D., Rotilio, G., & Ciriolo, M. R. (2012). Neuroprotection of kaempferol by autophagy in models of rotenone-mediated acute toxicity: possible implications for Parkinson's disease. *Neurobiology of Aging*, *33*(4), pp. 767–785. https://doi.org/10.1016/j.neurobiolaging.2010.05.021

Friedli, M. J., & Inestrosa, N. C. (2021). Huperzine A and its neuroprotective molecular signaling in Alzheimer's disease. *Molecules (Basel, Switzerland)*, *26*(21), p. 6531. https://doi.org/10.3390/molecules26216531

Fujimori, H., Ohba, T., Mikami, M., Nakamura, S., Ito, K., Kojima, H., Takahashi, T., Iddamalgoda, A., Shimazawa, M., & Hara, H. (2022). The protective effect of Centella asiatica and its constituent, araliadiol on neuronal cell damage and cognitive impairment. *Journal of Pharmacological Sciences*, *148*(1), pp. 162–171. https://doi.org/10.1016/j.jphs.2021.11.001

Ghofrani, S., Joghataei, M.-T., Mohseni, S., Baluchnejadmojarad, T., Bagheri, M., Khamse, S., & Roghani, M. (2015). Naringenin improves learning and memory in an Alzheimer's disease rat model: insights into the underlying mechanisms. *European Journal of Pharmacology*, *764*, pp. 195–201. https://doi.org/10.1016/j.ejphar.2015.07.001

Gonçalves, D. F., Tassi, C. C., Amaral, G. P., Stefanello, S. T., Dalla Corte, C. L., Soares, F. A., Posser, T., Franco, J. L., & Carvalho, N. R. (2020). Effects of caffeine on brain antioxidant status and mitochondrial respiration in acetaminophen-intoxicated mice. *Toxicology Research*, *9*(5), pp. 726–734. https://doi.org/10.1093/toxres/tfaa075

Haleagrahara, N., Siew, C. J., & Ponnusamy, K. (2013). Effect of quercetin and desferrioxamine on 6-hydroxydopamine (6-OHDA) induced neurotoxicity in striatum of rats. *The Journal of Toxicological Sciences*, *38*(1), pp. 25–33. https://doi.org/10.2131/jts.38.25

Heinrich, M., Mah, J., & Amirkia, V. (2021). Alkaloids used as medicines: structural phytochemistry meets biodiversity-an update and forward look. *Molecules (Basel, Switzerland)*, *26*(7), p. 1836. https://doi.org/10.3390/molecules26071836

Hole, K. L., & Williams, R. J. (2021). Flavonoids as an intervention for Alzheimer's disease: progress and hurdles towards defining a mechanism of action. *Brain Plasticity (Amsterdam, Netherlands)*, *6*(2), 167–192. https://doi.org/10.3233/BPL-200098

Huin, V., Blum, D., Delforge, V., Cailliau, E., Djeziri, S., Dujardin, K., Genet, A., Viard, R., Attarian, S., Bruneteau, G., Cassereau, J., Genestet, S., Kaminsky, A.-L., Soriani, M.-H., Lefilliatre, M., Couratier, P., Pittion-Vouyovitch, S., Esselin, F., De La Cruz, E., … Rolland, A.-S. (2024). Caffeine consumption outcomes on amyotrophic lateral sclerosis disease progression and cognition. *Neurobiology of Disease*, *199*, 106603. https://doi.org/10.1016/j.nbd.2024.106603

Hwang, J.-H., Chen, C.-C., Lee, L.-Y., Chiang, H.-T., Wang, M.-F., & Chan, Y.-C. (2020). Hericium erinaceus enhances neurotrophic factors and prevents cochlear cell apoptosis in senescence accelerated mice. *Journal of Functional Foods*, *66*, 103832. https://doi.org/10.1016/j.jff.2020.103832

Iyer, S., Bhat, I., & Bangera Sheshappa, M. (2024). Lutein and the underlying neuroprotective promise against neurodegenerative diseases. *Molecular Nutrition & Food Research*, *68*(13). https://doi.org/10.1002/mnfr.202300409

Jeong, K. H., Jeon, M.-T., Kim, H. D., Jung, U. J., Jang, M. C., Chu, J. W., Yang, S. J., Choi, I. Y., Choi, M.-S., & Kim, S. R. (2015). Nobiletin protects dopaminergic neurons in the 1-methyl-4-phenylpyridinium-treated rat model of Parkinson's disease. *Journal of Medicinal Food*, *18*(4), 409–414. https://doi.org/10.1089/jmf.2014.3241

Jiang, M., Porat-Shliom, Y., Pei, Z., Cheng, Y., Xiang, L., Sommers, K., Li, Q., Gillardon, F., Hengerer, B., Berlinicke, C., Smith, W. W., Zack, D. J., Poirier, M. A., Ross, C. A., & Duan, W. (2010). Baicalein reduces E46K alpha-synuclein aggregation in vitro and protects cells against E46K alpha-synuclein toxicity in cell models of familiar Parkinsonism. *Journal of Neurochemistry*, *114*(2), 419–429. https://doi.org/10.1111/j.1471-4159.2010.06752.x

Jiang, W., Wei, W., Gaertig, M. A., Li, S., & Li, X.-J. (2015). Therapeutic effect of berberine on Huntington's disease transgenic mouse model. *PloS One*, *10*(7), e0134142–e0134142. https://doi.org/10.1371/journal.pone.0134142

Jin, Y., Cui, R., Zhao, L., Fan, J., & Li, B. (2019). Mechanisms of Panax ginseng action as an antidepressant. *Cell Proliferation*, *52*(6), e12696. https://doi.org/10.1111/cpr.12696

Karuppagounder, S. S., Pinto, J. T., Xu, H., Chen, H.-L., Beal, M. F., & Gibson, G. E. (2009). Dietary supplementation with resveratrol reduces plaque pathology in a transgenic model of Alzheimer's disease. *Neurochemistry International*, *54*(2), 111–118. https://doi.org/10.1016/j.neuint.2008.10.008

Khajevand-Khazaei, M.-R., Ziaee, P., Motevalizadeh, S.-A., Rohani, M., Afshin-Majd, S., Baluchnejadmojarad, T., & Roghani, M. (2018). Naringenin ameliorates learning and memory impairment following systemic lipopolysaccharide challenge in the rat. *European Journal of Pharmacology*, *826*, 114–122. https://doi.org/10.1016/j.ejphar.2018.03.001

Khan, M. B., Khan, M. M., Khan, A., Ahmed, M. E., Ishrat, T., Tabassum, R., Vaibhav, K., Ahmad, A., & Islam, F. (2012). Naringenin ameliorates Alzheimer's disease (AD)-type neurodegeneration with cognitive impairment (AD-TNDCI) caused by the intracerebroventricular-streptozotocin in rat model. *Neurochemistry International*, *61*(7), 1081–1093. https://doi.org/10.1016/j.neuint.2012.07.025

Kim, H., Hong, J. T., & Park, M. H. (2015). Centella asiatica enhances neurogenesis and protects neuronal cells against H_2O_2-induced oxidative injury. *Journal of Biomedical Research*, *16*(3), 121–128. https://doi.org/10.12729/jbr.2015.16.3.121

Kim, M., Seong, A., Yoo, J., Jin, C., Lee, Y., Kim, Y. J., Lee, J., Jun, W. J., & Yoon, H. (2011). Gallic acid, a histone acetyltransferase inhibitor, suppresses β-amyloid neurotoxicity by inhibiting microglial-mediated neuroinflammation. *Molecular Nutrition & Food Research*, *55*(12), 1798–1808. https://doi.org/10.1002/mnfr.201100262

Koh, Y., Lin, S., Hsu, K., Nagabhushanam, K., Ho, C., & Pan, M. (2023). Pterostilbene enhances thermogenesis and mitochondrial biogenesis by activating the SIRT1/PGC-1α/SIRT3 pathway to prevent western diet-induced obesity. *Molecular Nutrition & Food Research*, *67*(18). https://doi.org/10.1002/mnfr.202300370

Kouhestani, S., Jafari, A., & Babaei, P. (2018). Kaempferol attenuates cognitive deficit via regulating oxidative stress and neuroinflammation in an ovariectomized rat model of sporadic dementia. *Neural Regeneration Research*, *13*(10), 1827–1832. https://doi.org/10.4103/1673-5374.238714

Küpeli Akkol, E., Tatlı Çankaya, I., Şeker Karatoprak, G., Carpar, E., Sobarzo-Sánchez, E., & Capasso, R. (2021). Natural compounds as medical strategies in the prevention and treatment of psychiatric disorders seen in neurological diseases. *Frontiers in Pharmacology*, *12*, 669638. https://doi.org/10.3389/fphar.2021.669638

Kuszewski, J. C., Wong, R. H. X., & Howe, P. R. C. (2018). Can curcumin counteract cognitive decline? clinical trial evidence and rationale for combining ω-3 fatty acids with curcumin. *Advances in Nutrition (Bethesda, Md.)*, *9*(2), 105–113. https://doi.org/10.1093/advances/nmx013

Lao-Peregrín, C., Ballesteros, J. J., Fernández, M., Zamora-Moratalla, A., Saavedra, A., Gómez Lázaro, M., Pérez-Navarro, E., Burks, D., & Martín, E. D. (2017). Caffeine-mediated BDNF release regulates long-term synaptic plasticity through activation of IRS2 signaling. *Addiction Biology*, *22*(6), 1706–1718. https://doi.org/10.1111/adb.12433

Latimer, C. S., Brewer, L. D., Searcy, J. L., Chen, K.-C., Popović, J., Kraner, S. D., Thibault, O., Blalock, E. M., Landfield, P. W., & Porter, N. M. (2014). Vitamin D prevents cognitive decline and enhances hippocampal synaptic function in aging rats. *Proceedings of the National Academy of Sciences of the United States of America*, *111*(41), E4359–E4366. https://doi.org/10.1073/pnas.1404477111

Lee, H., Lee, S., Lee, D., Choi, B., Le, B., & Yang, S. (2019). Ameliorating effect of Citrus aurantium extracts and nobiletin on β-amyloid (1-42)-induced memory impairment in mice. *Molecular Medicine Reports*. https://doi.org/10.3892/mmr.2019.10582

Leem, E., Nam, J. H., Jeon, M.-T., Shin, W.-H., Won, S.-Y., Park, S.-J., Choi, M.-S., Jin, B. K., Jung, U. J., & Kim, S. R. (2014). Naringin protects the nigrostriatal dopaminergic projection through induction of GDNF in a neurotoxin model of Parkinson's disease. *The Journal of Nutritional Biochemistry*, *25*(7), 801–806. https://doi.org/10.1016/j.jnutbio.2014.03.006

Li, C., Zug, C., Qu, H., Schluesener, H., & Zhang, Z. (2015). Hesperidin ameliorates behavioral impairments and neuropathology of transgenic APP/PS1 mice. *Behavioural Brain Research*, *281*, 32–42. https://doi.org/10.1016/j.bbr.2014.12.012

Li, D.-D., Zhao, L.-X., Mylonakis, E., Hu, G.-H., Zou, Y., Huang, T.-K., Yan, L., Wang, Y., & Jiang, Y.-Y. (2014). In vitro and in vivo activities of pterostilbene against Candida albicans biofilms. *Antimicrobial Agents and Chemotherapy*, *58*(4), 2344–2355. https://doi.org/10.1128/AAC.01583-13

Li, Q., Chen, L., Liu, X., Li, X., Cao, Y., Bai, Y., & Qi, F. (2018). Pterostilbene inhibits amyloid-β-induced neuroinflammation in a microglia cell line by inactivating the NLRP3/caspase-1 inflammasome pathway. *Journal of Cellular Biochemistry*, *119*(8), 7053–7062. https://doi.org/10.1002/jcb.27023

Lim, J.-O., Shin, N.-R., Seo, Y.-S., Nam, H.-H., Ko, J.-W., Jung, T.-Y., Lee, S.-J., Kim, H.-J., Cho, Y.-K., Kim, J.-C., Lee, I.-C., Kim, J.-S., & Shin, I.-S. (2020). Silibinin attenuates silica dioxide nanoparticles-induced inflammation by suppressing TXNIP/MAPKs/AP-1 signaling. *Cells*, *9*(3), 678. https://doi.org/10.3390/cells9030678

Liu, J., Wang, Y., Guo, J., Sun, J., & Sun, Q. (2020). Salvianolic acid B improves cognitive impairment by inhibiting neuroinflammation and decreasing Aβ level in porphyromonas gingivalis-infected mice. *Aging*, *12*(11), 10117–10128. https://doi.org/10.18632/aging.103306

Liu, P., Cui, L., Liu, B., Liu, W., Hayashi, T., Mizuno, K., Hattori, S., Ushiki-Kaku, Y., Onodera, S., & Ikejima, T. (2020). Silibinin ameliorates STZ-induced impairment of memory and learning by up- regulating insulin signaling pathway and attenuating apoptosis. *Physiology & Behavior*, *213*, 112689. https://doi.org/10.1016/j.physbeh.2019.112689

Lopresti, A. L., Smith, S. J., Malvi, H., & Kodgule, R. (2019). An investigation into the stress-relieving and pharmacological actions of an ashwagandha (Withania somnifera) extract: a randomized, double-blind, placebo-controlled study. *Medicine*, *98*(37), p. e17186. https://doi.org/10.1097/MD.0000000000017186

Mak, S., Li, W., Fu, H., Luo, J., Cui, W., Hu, S., Pang, Y., Carlier, P. R., Tsim, K. W., Pi, R., & Han, Y. (2021). Promising tacrine/huperzine A-based dimeric acetylcholinesterase inhibitors for neurodegenerative disorders: from relieving symptoms to modifying diseases through multitarget. *Journal of Neurochemistry*, *158*(6), pp. 1381–1393. https://doi.org/10.1111/jnc.15379

Mandel, S., Weinreb, O., Reznichenko, L., Kalfon, L., & Amit, T. (2006). Green tea catechins as brain-permeable, non toxic iron chelators to "iron out iron" from the brain. *Journal of Neural Transmission*, Supplementa, 249–257. https://doi.org/10.1007/978-3-211-33328-0_26

Manochkumar, J., Doss, C. G. P., El-Seedi, H. R., Efferth, T., & Ramamoorthy, S. (2021). The neuroprotective potential of carotenoids in vitro and in vivo. *Phytomedicine*, *91*, p. 153676. https://doi.org/10.1016/j.phymed.2021.153676

Maratha, S. R., & Mahadevan, N. (2012). Memory enhancing activity of naringin in unstressed and stressed mice: possible cholinergic and nitriergic modulation. *Neurochemical Research*, *37*(10), pp. 2206–2212. https://doi.org/10.1007/s11064-012-0844-8

Meng-Zhen, S., Ju, L., Lan-Chun, Z., Cai-Feng, D., Shu-da, Y., Hao-Fei, Y., & Wei-Yan, H. (2022). Potential therapeutic use of plant flavonoids in AD and PD. *Heliyon*, *8*(11), p. e11440. https://doi.org/10.1016/j.heliyon.2022.e11440

Michael-Titus, A. T., & Priestley, J. V. (2014). Omega-3 fatty acids and traumatic neurological injury: from neuroprotection to neuroplasticity? *Trends in Neurosciences*, *37*(1), pp. 30–38. https://doi.org/10.1016/j.tins.2013.10.005

Mikulska, P., Malinowska, M., Ignacyk, M., Szustowski, P., Nowak, J., Pesta, K., Szeląg, M., Szklanny, D., Judasz, E., Kaczmarek, G., Ejiohuo, O. P., Paczkowska-Walendowska, M., Gościniak, A., & Cielecka-Piontek, J. (2023). Ashwagandha (Withania somnifera)-current research on the health-promoting activities: a narrative review. *Pharmaceutics*, *15*(4), p. 1057. https://doi.org/10.3390/pharmaceutics15041057

Mishra, S., & Palanivelu, K. (2008). The effect of curcumin (turmeric) on Alzheimer's disease: an overview. *Annals of Indian Academy of Neurology*, *11*(1), pp. 13–19. https://doi.org/10.4103/0972-2327.40220

Mohd Sairazi, N. S., & Sirajudeen, K. N. S. (2020). Natural products and their bioactive compounds: neuroprotective potentials against neurodegenerative diseases. *Evidence-Based Complementary and Alternative Medicine: ECAM*, *2020*, p. 6565396. https://doi.org/10.1155/2020/6565396

Mori, K., Obara, Y., Moriya, T., Inatomi, S., & Nakahata, N. (2011). Effects of Hericium erinaceus on amyloid β(25–35) peptide-induced learning and memory deficits in mice. *Biomedical Research*, *32*(1), pp. 67–72. https://doi.org/10.2220/biomedres.32.67

Nabavi, S., Tenore, G., Daglia, M., Tundis, R., Loizzo, M., & Nabavi, S. (2015). The cellular protective effects of rosmarinic acid: from bench to bedside. *Current Neurovascular Research*, *12*(1), pp. 98–105. https://doi.org/10.2174/1567202612666150109113638

Nathan, P. J., Lu, K., Gray, M., & Oliver, C. (2006). The neuropharmacology of L-Theanine(*N*-Ethyl-L-Glutamine). *Journal of Herbal Pharmacotherapy*, *6*(2), pp. 21–30. https://doi.org/10.1080/j157v06n02_02

Neveu, I., Naveilhan, P., Jehan, F., Baudet, C., Wion, D., De Luca, H. F., & Brachet, P. (1994). 1,25–Dihydroxyvitamin D3 regulates the synthesis of nerve growth factor in primary cultures of glial cells. *Molecular Brain Research*, *24*(1–4), pp. 70–76. https://doi.org/10.1016/0169-328x(94)90119-8

Noi, K., Ikenaka, K., Mochizuki, H., Goto, Y., & Ogi, H. (2021). Disaggregation behavior of amyloid β Fibrils by anthocyanins studied by total-internal-reflection-fluorescence microscopy coupled with a wireless quartz-crystal microbalance biosensor. *Analytical Chemistry*, *93*(32), pp. 11176–11183. https://doi.org/10.1021/acs.analchem.1c01720

Ojo, O. A., Rotimi, D. E., Ojo, A. B., Ogunlakin, A. D., & Ajiboye, B. O. (2023). Gallic acid abates cadmium chloride toxicity via alteration of neurotransmitters and modulation of inflammatory markers in Wistar rats. *Scientific Reports*, *13*(1), p. 1577. https://doi.org/10.1038/s41598-023-28893-6

Onozuka, H., Nakajima, A., Matsuzaki, K., Shin, R.-W., Ogino, K., Saigusa, D., Tetsu, N., Yokosuka, A., Sashida, Y., Mimaki, Y., Yamakuni, T., & Ohizumi, Y. (2008). Nobiletin, a citrus flavonoid, improves memory impairment and Aβ pathology in a transgenic mouse model of Alzheimer's disease. *Journal of Pharmacology and Experimental Therapeutics*, *326*(3), pp. 739–744. https://doi.org/10.1124/jpet.108.140293

Ozawa, Y., Sasaki, M., Takahashi, N., Kamoshita, M., Miyake, S., & Tsubota, K. (2012). Neuroprotective effects of lutein in the retina. *Current Pharmaceutical Design*, *18*(1), pp. 51–56. https://doi.org/10.2174/138161212798919101

Patel, M. Y., Panchal, H. V, Ghribi, O., & Benzeroual, K. E. (2012). The neuroprotective effect of Fisetin in the MPTP model of Parkinson's disease. *Journal of Parkinson's Disease*, *2*(4), pp. 287–302. https://doi.org/10.3233/jpd-012110

Payne, A., Taka, E., Adinew, G. M., & Soliman, K. F. A. (2023). Molecular mechanisms of the anti-inflammatory effects of Epigallocatechin 3-Gallate (EGCG) in LPS-activated BV-2 microglia cells. *Brain Sciences*, *13*(4), p. 632. https://doi.org/10.3390/brainsci13040632

Pretzsch, C. M., Freyberg, J., Voinescu, B., Lythgoe, D., Horder, J., Mendez, M. A., Wichers, R., Ajram, L., Ivin, G., Heasman, M., Edden, R. A. E., Williams, S., Murphy, D. G. M., Daly, E., & McAlonan, G. M. (2019). Effects of cannabidiol on brain excitation and inhibition systems; a randomised placebo-controlled single dose trial during magnetic resonance spectroscopy in adults with and without autism spectrum disorder. *Neuropsychopharmacology: Official Publication of the American College of Neuropsychopharmacology*, *44*(8), pp. 1398–1405. https://doi.org/10.1038/s41386-019-0333-8

Ravaria, P., Saxena, P., Laksmi BS, S., Ranjan, V., Abidi, S. W. F., Saha, P., Ramamoorthy, S., Ahmad, F., & Rana, S. S. (2023). Molecular mechanisms of neuroprotective offerings by rosmarinic acid against neurodegenerative and other CNS pathologies. *Phytotherapy Research*, *37*(5), pp. 2119–2143. https://doi.org/10.1002/ptr.7825

Rebas, E., Rzajew, J., Radzik, T., & Zylinska, L. (2020). Neuroprotective polyphenols: a modulatory action on neurotransmitter pathways. *Current Neuropharmacology*, *18*(5), pp. 431–445. https://doi.org/10.2174/1570159X18666200106155127

Repossi, G., Das, U. N., & Eynard, A. R. (2020). Molecular basis of the beneficial actions of resveratrol. *Archives of Medical Research*, *51*(2), pp. 105–114. https://doi.org/10.1016/j.arcmed.2020.01.010

Robb, E. L., Winkelmolen, L., Visanji, N., Brotchie, J., & Stuart, J. A. (2008). Dietary resveratrol administration increases MnSOD expression and activity in mouse brain. *Biochemical and Biophysical Research Communications*, *372*(1), pp. 254–259. https://doi.org/10.1016/j.bbrc.2008.05.028

Ruggiero, M., Calvello, R., Porro, C., Messina, G., Cianciulli, A., & Panaro, M. A. (2022). Neurodegenerative diseases: can caffeine be a powerful ally to weaken neuroinflammation? *International Journal of Molecular Sciences*, *23*(21), p. 12958. https://doi.org/10.3390/ijms232112958

Sarda, R. K., Sinha, R., Sherpa, M., & Gupta, A. (2022). Crude extracts of Bacopa Monnieri induces dendrite formation in rodent neural stem cell cultures-a possible use in neuronal injury. *Journal of Neurosciences in Rural Practice*, *13*(2), pp. 254–260. https://doi.org/10.1055/s-0042-1743215

Shabani, S., Rabiei, Z., & Amini-Khoei, H. (2020). Exploring the multifaceted neuroprotective actions of gallic acid: a review. *International Journal of Food Properties*, *23*(1), pp. 736–752. https://doi.org/10.1080/10942912.2020.1753769

Singh, A., Naidu, P. S., & Kulkarni, S. K. (2003). Quercetin potentiates <i>L</i>-Dopa reversal of drug-induced catalepsy in rats: possible COMT/MAO inhibition. *Pharmacology*, *68*(2), pp. 81–88. https://doi.org/10.1159/000069533

Singh, N., Bhalla, M., de Jager, P., & Gilca, M. (2011). An overview on ashwagandha: a Rasayana (rejuvenator) of Ayurveda. *African Journal of Traditional, Complementary, and Alternative Medicines: AJTCAM*, *8*(5 Suppl), pp. 208–213. https://doi.org/10.4314/ajtcam.v8i5S.9

Singh, S., Arthur, R., Upadhayay, S., & Kumar, P. (2022). Ferulic acid ameliorates neurodegeneration via the Nrf2/ARE signalling pathway: a review. *Pharmacological Research - Modern Chinese Medicine*, *5*, p. 100190. https://doi.org/10.1016/j.prmcm.2022.100190

Song, X., Zhou, B., Cui, L., Lei, D., Zhang, P., Yao, G., Xia, M., Hayashi, T., Hattori, S., Ushiki-Kaku, Y., Tashiro, S., Onodera, S., & Ikejima, T. (2016). Silibinin ameliorates Aβ25–35-induced memory deficits in rats by modulating autophagy and attenuating neuroinflammation as well as oxidative stress. *Neurochemical Research*, *42*(4), pp. 1073–1083. https://doi.org/10.1007/s11064-016-2141-4

Soto-Mercado, V., Mendivil-Perez, M., Velez-Pardo, C., & Jimenez-Del-Rio, M. (2021). (-)-Epigallocatechin-3-Gallate diminishes intra-and extracellular amyloid-induced cytotoxic effects on cholinergic-like neurons from familial Alzheimer's disease PSEN1 E280A. *Biomolecules*, *11*(12), p. 1845. https://doi.org/10.3390/biom11121845

Sumathi, T., Asha, D., Nagarajan, G., Sreenivas, A., & Nivedha, R. (2016). l-Theanine alleviates the neuropathological changes induced by PCB (Aroclor 1254) via inhibiting upregulation of inflammatory cytokines and oxidative stress in rat brain. *Environmental Toxicology and Pharmacology*, *42*, pp. 99–117. https://doi.org/10.1016/j.etap.2016.01.008

Sun, A., Xu, X., Lin, J., Cui, X., & Xu, R. (2014). Neuroprotection by Saponins. *Phytotherapy Research*, *29*(2), pp. 187–200. https://doi.org/10.1002/ptr.5246

Tian, L.-L., Wang, X.-J., Sun, Y.-N., Li, C.-R., Xing, Y.-L., Zhao, H.-B., Duan, M., Zhou, Z., & Wang, S.-Q. (2008). Salvianolic acid B, an antioxidant from Salvia miltiorrhiza, prevents 6-hydroxydopamine induced apoptosis in SH-SY5Y cells. *The International Journal of Biochemistry & Cell Biology*, *40*(3), pp. 409–422. https://doi.org/10.1016/j.biocel.2007.08.005

Tripathi, M. K., Rasheed, M. S. U., Mishra, A. K., Patel, D. K., & Singh, M. P. (2019). Silymarin protects against impaired autophagy associated with 1-Methyl-4-phenyl-1,2,3,6-tetrahydropyridine-induced Parkinsonism. *Journal of Molecular Neuroscience*, *70*(2), pp. 276–283. https://doi.org/10.1007/s12031-019-01431-8

Tsai, I.-C., Hsu, C.-W., Chang, C.-H., Tseng, P.-T., & Chang, K.-V. (2021). The effect of curcumin differs on individual cognitive domains across different patient populations: a systematic review and meta-analysis. *Pharmaceuticals (Basel, Switzerland)*, *14*(12), p. 1235. https://doi.org/10.3390/ph14121235

Uddin, M. S., Al Mamun, A., Kabir, M. T., Ahmad, J., Jeandet, P., Sarwar, M. S., Ashraf, G. M., & Aleya, L. (2020). Neuroprotective role of polyphenols against oxidative stress-mediated neurodegeneration. *European Journal of Pharmacology*, *886*, p. 173412. https://doi.org/10.1016/j.ejphar.2020.173412

Ungvari, Z., Sonntag, W. E., de Cabo, R., Baur, J. A., & Csiszar, A. (2011). Mitochondrial protection by resveratrol. *Exercise and Sport Sciences Reviews*, *39*(3), p. 128–132. https://doi.org/10.1097/JES.0b013e3182141f80

Valverde-Salazar, V., Ruiz-Gabarre, D., & García-Escudero, V. (2023). Alzheimer's disease and green tea: epigallocatechin-3-gallate as a modulator of inflammation and oxidative stress. *Antioxidants (Basel, Switzerland)*, *12*(7), p. 1460. https://doi.org/10.3390/antiox12071460

Vauzour, D. (2017). Polyphenols and brain health. *OCL*, *24*(2), p. A202. https://doi.org/10.1051/ocl/2017008

Wang, C.-Y., Zheng, W., Wang, T., Xie, J.-W., Wang, S.-L., Zhao, B.-L., Teng, W.-P., & Wang, Z.-Y. (2011). Huperzine A activates Wnt/β-catenin signaling and enhances the nonamyloidogenic pathway in an Alzheimer transgenic mouse model. *Neuropsychopharmacology: Official Publication of the American College of Neuropsychopharmacology*, *36*(5), pp. 1073–1089. https://doi.org/10.1038/npp.2010.245

Wang, D.-M., Li, S.-Q., Wu, W.-L., Zhu, X.-Y., Wang, Y., & Yuan, H.-Y. (2014). Effects of long-term treatment with quercetin on cognition and mitochondrial function in a mouse model of Alzheimer's disease. *Neurochemical Research*, *39*(8), pp. 1533–1543. https://doi.org/10.1007/s11064-014-1343-x

Wang, D., Liu, L., Zhu, X., Wu, W., & Wang, Y. (2014). Hesperidin alleviates cognitive impairment, mitochondrial dysfunction and oxidative stress in a mouse model of Alzheimer's disease. *Cellular and Molecular Neurobiology*, *34*(8), pp. 1209–1221. https://doi.org/10.1007/s10571-014-0098-x

Wang, G.-Q., Zhang, B., He, X.-M., Li, D.-D., Shi, J.-S., & Zhang, F. (2019). Naringenin targets on astroglial Nrf2 to support dopaminergic neurons. *Pharmacological Research*, *139*, pp. 452–459. https://doi.org/10.1016/j.phrs.2018.11.043

Wang, R. E., Hunt, C. R., Chen, J., & Taylor, J.-S. (2011). Biotinylated quercetin as an intrinsic photoaffinity proteomics probe for the identification of quercetin target proteins. *Bioorganic & Medicinal Chemistry*, *19*(16), pp. 4710–4720. https://doi.org/10.1016/j.bmc.2011.07.005

Wang, R., Wu, Z., Liu, M., Wu, Y., Li, Q., Ba, Y., Zhang, H., Cheng, X., Zhou, G., & Huang, H. (2021). Resveratrol reverses hippocampal synaptic markers injury and SIRT1 inhibition against developmental Pb exposure. *Brain Research*, 1767, p. 147567. https://doi.org/10.1016/j.brainres.2021.147567

Wang, Y., Wu, S., Li, Q., Lang, W., Li, W., Jiang, X., Wan, Z., Chen, J., & Wang, H. (2022). Epigallocatechin-3-gallate: a phytochemical as a promising drug candidate for the treatment of Parkinson's disease. *Frontiers in Pharmacology*, *13*, p. 977521. https://doi.org/10.3389/fphar.2022.977521

Wang, Z., Tang, L., Yan, H., Wang, Y., & Tang, X. (2006). Effects of huperzine A on memory deficits and neurotrophic factors production after transient cerebral ischemia and reperfusion in mice. *Pharmacology Biochemistry and Behavior*, *83*(4), pp. 603–611. https://doi.org/10.1016/j.pbb.2006.03.027

Wong, J. H., Barron, A. M., & Abdullah, J. M. (2021). Mitoprotective effects of Centella asiatica (L.) Urb.: anti-inflammatory and neuroprotective opportunities in neurodegenerative disease. *Frontiers in Pharmacology*, *12*, p. 687935. https://doi.org/10.3389/fphar.2021.687935

Wu, Y., Chen, Q., Wen, B., Wu, N., He, B., & Chen, J. (2021). Berberine reduces Aβ(42) deposition and tau hyperphosphorylation via ameliorating endoplasmic reticulum stress. *Frontiers in Pharmacology*, *12*, p. 640758. https://doi.org/10.3389/fphar.2021.640758

Xu, J., Liu, J., Li, Q., Mi, Y., Zhou, D., Meng, Q., Chen, G., Li, N., & Hou, Y. (2020). Pterostilbene alleviates $A\beta_{1-42}$-induced cognitive dysfunction via inhibition of oxidative stress by activating Nrf2 signaling pathway. *Molecular Nutrition & Food Research*, *65*(2). https://doi.org/10.1002/mnfr.202000711

Xu, L., Botchway, B. O. A., Zhang, S., Zhou, J., & Liu, X. (2018). Inhibition of NF-κB signaling pathway by resveratrol improves spinal cord injury. *Frontiers in Neuroscience*, *12*, p. 690. https://doi.org/10.3389/fnins.2018.00690

Yan, W., Ren, D., Feng, X., Huang, J., Wang, D., Li, T., & Zhang, D. (2021). Neuroprotective and anti-inflammatory effect of pterostilbene against cerebral ischemia/reperfusion injury via suppression of COX-2. *Frontiers in Pharmacology*, *12*, p. 770329. https://doi.org/10.3389/fphar.2021.770329

Ye, X., Zhou, Q., Ren, P., Xiang, W., & Xiao, L. (2023). The synaptic and circuit functions of Vitamin D in neurodevelopment disorders. *Neuropsychiatric Disease and Treatment*, *19*, pp. 1515–1530. https://doi.org/10.2147/NDT.S407731

Yoo, K.-Y., & Park, S.-Y. (2012). Terpenoids as potential anti-Alzheimer's disease therapeutics. *Molecules (Basel, Switzerland)*, *17*(3), pp. 3524–3538. https://doi.org/10.3390/molecules17033524

Zahiruddin, S., Basist, P., Parveen, A., Parveen, R., Khan, W., Gaurav, & Ahmad, S. (2020). Ashwagandha in brain disorders: a review of recent developments. *Journal of Ethnopharmacology*, *257*, p. 112876. https://doi.org/10.1016/j.jep.2020.112876

Zhang, J., Liu, Q., Su, D., Li, L., Xiao, C., He, H., You, Z., & Zhou, T. (2023). Akebia saponin D acts via the PPAR-gamma pathway to reprogramme a pro-neurogenic microglia that can restore hippocampal neurogenesis in mice exposed to chronic mild stress. *CNS Neuroscience & Therapeutics*, *29*(9), pp. 2555–2571. https://doi.org/10.1111/cns.14196

Zhao, Y., & Liu, S. (2021). Bioactivity of naringin and related mechanisms. *Die Pharmazie*, *76*(8), pp. 359–363. https://doi.org/10.1691/ph.2021.1504

Zheng, M., Xin, Y., Li, Y., Xu, F., Xi, X., Guo, H., Cui, X., Cao, H., Zhang, X., & Han, C. (2018). Ginsenosides: a potential neuroprotective agent. *BioMed Research International*, *2018*, p. 8174345. https://doi.org/10.1155/2018/8174345

Zheng, X.-F., Zhang, X.-J., Dong, L.-P., Zhao, J.-R., Zhang, C., & Chen, R. (2023). Neuroprotective mechanism of salvianolic acid B against cerebral ischemia-reperfusion injury in mice through downregulation of TLR4, p-p38MAPK, p-JNK, NF-κB, and IL-1β. *Immunity, Inflammation and Disease*, *11*(10), p. e1030. https://doi.org/10.1002/iid3.1030

Zhu, L., Lu, F., Zhang, X., Liu, S., & Mu, P. (2022). SIRT1 is involved in the neuroprotection of pterostilbene against amyloid β 25–35-induced cognitive deficits in mice. *Frontiers in Pharmacology*, *13*, p. 877098. https://doi.org/10.3389/fphar.2022.877098

Zhu, M., Rajamani, S., Kaylor, J., Han, S., Zhou, F., & Fink, A. L. (2004). The flavonoid baicalein inhibits fibrillation of α-Synuclein and disaggregates existing fibrils. *Journal of Biological Chemistry*, *279*(26), pp. 26846–26857. https://doi.org/10.1074/jbc.m403129200

6 Medicinal Chemistry and the Design of Small Molecules

Faheem Rasool and Nawab John Dar

ABBREVIATIONS

FDA	Food and Drug Administration
SAR	Structure-Activity Relationship
PK	Pharmacokinetics
PD	Pharmacodynamics
HTS	High-Throughput Screening
ADME	Absorption, Distribution, Metabolism, and Excretion
HMG-CoA	3-hydroxy-3-methyl glutaryl coenzyme A
GABA	Gamma-AminoButyric Acid
COPD	Chronic Obstructive Pulmonary Disease
NSAID	Non-Steroidal Anti-Inflammatory Drug
bRo5	Beyond Rule of 5
DNA	Deoxyribonucleic Acid
RNA	Ribonucleic Acid
ACE	Angiotensin-Converting Enzyme
EU-OPENSCREEN	European Open Screening Network

6.1 INTRODUCTION

Medicinal chemistry is a multidisciplinary field that combines principles from chemistry, biology, and pharmacology to design, discover, and develop new drugs and improve existing ones to better treat diseases. The primary goal is to design and optimize compounds that are effective, safe, and suitable for therapeutic use. Medicinal chemistry emerged in a familiar form toward the end of the 19th century as a discipline exploring the relationship between chemical structure and observed biological activity via chemical modification and structural mimicry of nature's materials. Its roots have been said to be in the fertile mix of ancient folk medicine and early awareness of the properties of natural products (Erhardt, 2002). "Medicinal chemistry is a scientific discipline deep rooted in organic chemistry concerning the discovery, development, identification, and interpretation of the mode of action of bioactive compounds at the molecular and cellular level" (Brenk and Rauh, 2012). Some core aspects of medicinal chemistry are drug design, chemical synthesis, structure–activity relationship (SAR), pharmacokinetics (PK), pharmacodynamics (PD), toxicology, computational chemistry, and drug development and testing (Imming, 2015).

Drug discovery, the main focus of medicinal chemistry, is the long and backbreaking process that can finally bring molecules from the laboratories to the market. Although the number of newly approved drugs has shown a substantial increase in recent years (Mullard, 2019), in general only 1 molecule out of 5,000 hit candidates can reach the market (Cong et al., 2012). The process of discovering, testing, and gaining approval for a new drug has changed a lot during the last century. From the isolation of active ingredients from natural products, drug discovery has evolved into a multidisciplinary and complex process that brings together the efforts of chemists, biologists, and pharmacologists. Many different approaches nowadays can be applied in drug discovery.

DOI: 10.1201/9781003520610-6

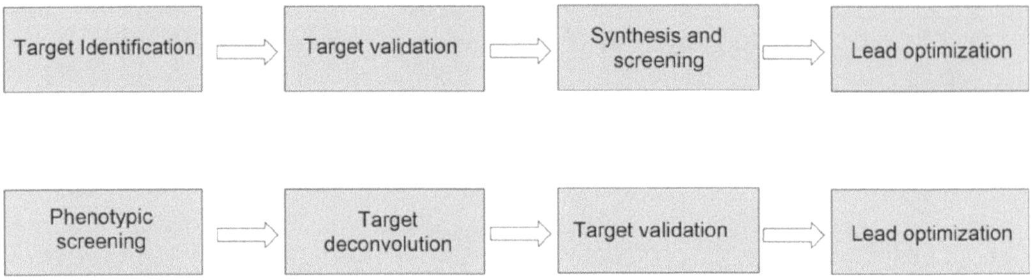

FIGURE 6.1 Conventional target-based and chemical genetics drug discovery.

On the one hand, the rational design of ligands remains the "gold standard" in medicinal chemistry, especially when the biological target is well defined and structurally known (Figure 6.1a) (Lenci and Trabocchi, 2020). On the other hand, there are fields like cancer and neurodegenerative disorders, where the biological target or the mode of binding is not well known (The State of the Art of Chemical Biology, 2009; Walters and Namchuk, 2003) or difficult to study in traditional drug discovery programs (Walters and Namchuk, 2003). In such cases, the discovery of small molecule new chemical entities enables the application of large small molecule libraries in high-throughput screening (HTS), phenotypic assays, and chemical genetics studies (Figure 6.1b) (Cong et al., 2012; Schreiber, 2003; Spandl et al., 2008; Stockwell, 2000; Walsh and Chang, 2006). The relevance of this approach is also highlighted by the emergence of international screening initiatives, such as EU-OPENSCREEN (http://www.eu-openscreen.eu/) or the European Lead Factory (https://www.europeanleadfactory.eu) (Karawajczyk et al., 2015).

In both approaches, synthetic organic chemistry plays a key role in generating high-quality small molecule new libraries. In fact, despite the vast success of biological drugs (monoclonal antibodies or recombinant proteins), the favorable PK properties of small molecule chemical entities allowed them to remain the gold standard for the development of new medications, especially in the case of enzyme inhibitors. Indeed small molecule drugs have been the mainstay of the pharmaceutical industry for nearly a century, contributing more than 90% of approved FDA drugs (Figure 6.2) (Mullard, 2019).

Thus, to address this demand, very powerful synthetic methods are necessary for the generation of large small molecule libraries. Several efforts have been devoted to improving the quality and quantity of small molecules representing a library. In particular, during the last decades, organic chemists have taken advantage of HTS methods, such as solid-phase techniques (Bunin and Ellman, 1992; DeWitt et al., 1993; Merrifield, 1963) and combinatorial chemistry (Balkenhohl et al., 1996; Borrell, 2000). Unfortunately, despite the success achieved, these chemistry approaches have not fulfilled the desired expectations as the automation of discovery processes has proven to be inefficient (O' Connor et al., 2011). Thus, new frontiers in the synthesis of small molecule libraries are being explored, with the aim of improving the quality of the small molecules representing a library, where the synthetic efforts are not guided by a specific core structure but rather by concepts like molecular diversity (i.e., diversity-oriented synthesis [DOS]) and bioactivity or biosynthetic pathway (i.e., biology-oriented synthesis [BIOS]). This chapter focuses on the main principles and techniques of medicinal chemistry that are crucial in designing effective small molecule drugs. This includes SARs, optimization of PK, and the use of computational methods in drug discovery.

6.2 SMALL MOLECULES IN MEDICINAL CHEMISTRY

Small molecules are organic compounds with low molecular weights usually below 1000 Da that can be used as medications to treat various diseases (Arkin and Wells, 2004). They rapidly diffuse across cell membranes and reach the intracellular sites of action (Arkin and Wells, 2004). Unlike their biological counterparts, which are large and complex molecules like proteins or antibodies, small molecules are typically organic compounds with well-defined structures. They are usually

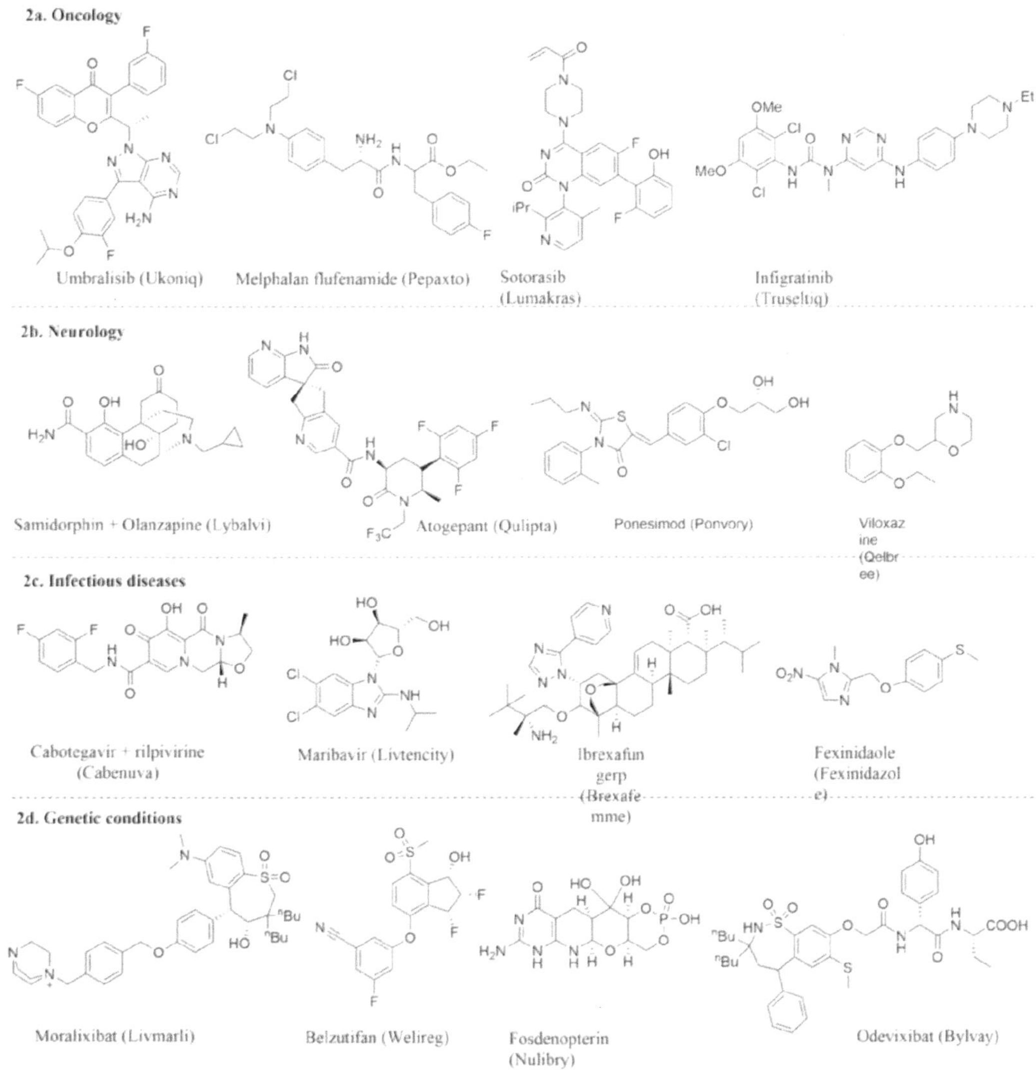

FIGURE 6.2 FDA-approved novel small molecule drugs.

administered orally in the form of pills or tablets, although other routes like intravenous injection can also be used. They can also be designed to engage biological targets by various modes of action, and their distribution can further be tailored, for example, to allow for systemic exposure with or without brain penetration. Small molecule drugs can be developed from leads derived from rational drug design or isolated from natural resources.

Owing to the advances of biotechnology and analytical techniques, biologic drugs have been increasing over the last decade. Despite the fact they have some advantages over small molecules, such as their high specificity and biocompatibility, they often suffer from poor absorption, distribution, metabolism, and excretion (ADME) properties, and the oral delivery route remains practically unattainable, as most of them are still delivered using subcutaneous injections. Also, they are much more expensive compared to low-molecular-weight drugs, and their structural characterization and quality control are more challenging (Oo and Kalbag, 2016).

Small molecules dominate the market, as more than 95% of most prescribed drugs are small molecules (Fuentes et al., 2018). Table 6.1 shows some of the small molecule drugs that have been most prescribed in recent times, highlighting the evolution of their structural complexity and their

TABLE 6.1
Most Prescribed Small Molecule Drugs with Their Biological Activity

Name of the Drug with Structure	Biological Activity
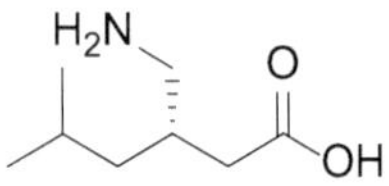 **Pregabalin**	Inhibitor of neuronal activity, used in the treatment of epilepsy and generalized anxiety disorder
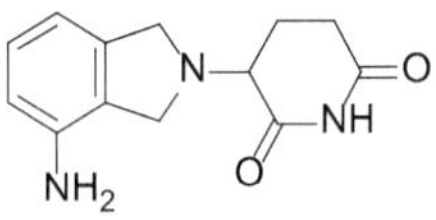 **Lenalidomide**	Chemotherapeuticagent, used in the treatment of multiple myeloma and other cancers
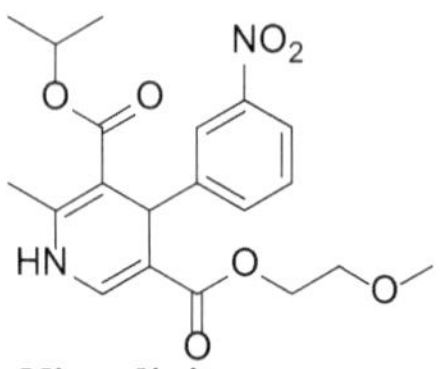 **Nimodipine**	Calcium channel blocker, used in the treatment of high blood pressure
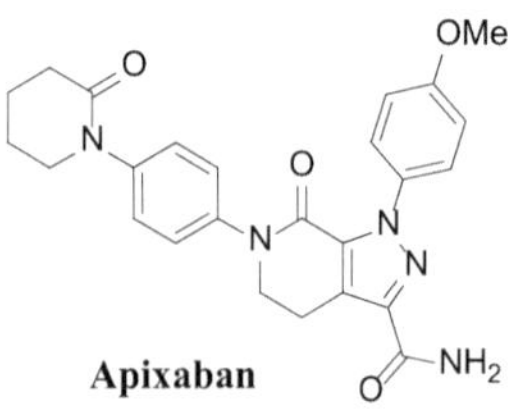 **Apixaban**	Anticoagulant, used in the treatment and for the prevention of strokes and blood clots
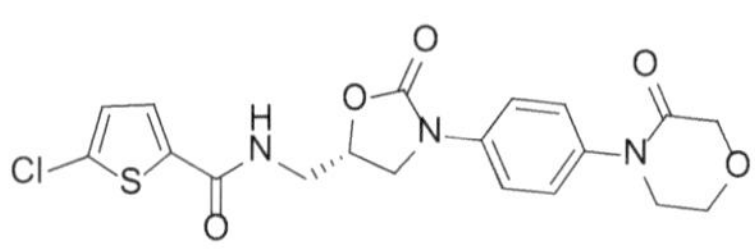 **Rivaroxaban**	Anticoagulant, used in the treatment of blood clots
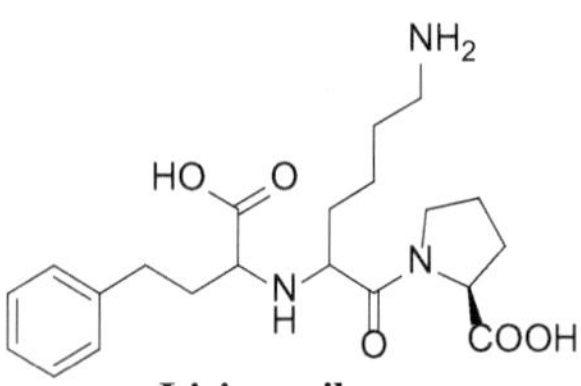 **Lisinopril**	Angiotensin-converting enzyme inhibitor, used in the treatment of hypertension and symptomatic congestive heart failure
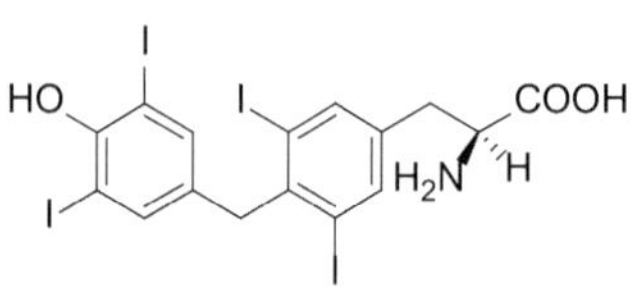 **Levothyroxine**	Natural thyroxine analog, used in the treatment of hypothyroidism

(Continued)

TABLE 6.1 (*Continued*)
Most Prescribed Small Molecule Drugs with Their Biological Activity

Name of the Drug with Structure	Biological Activity
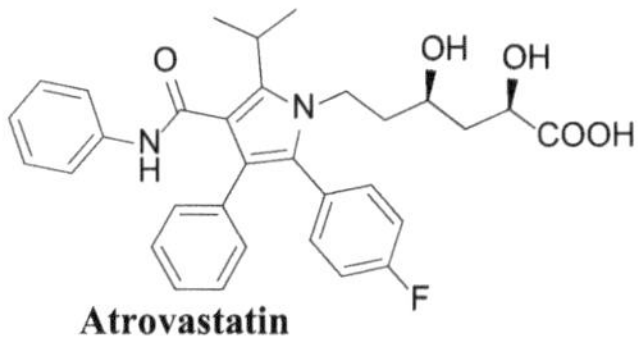 **Atrovastatin**	3-Hydroxy-3-methyl glutaryl coenzyme A (HMG-CoA) reductase inhibitor, used in the treatment of hyperlipidemic diseases
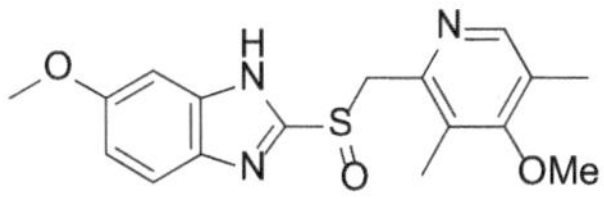 **Omperazole**	Proton-pump (potassium-transporting ATPase alpha chain 1) inhibitor, used in the treatment of gastric acid-related disorders
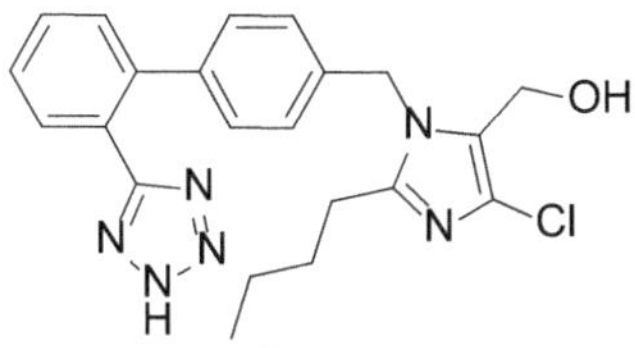 **Losartan**	Angiotensin-receptor blocker, used in the treatment of hypertension
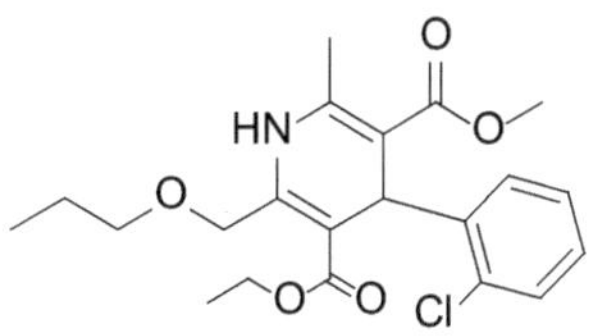 **Amlodipine**	Calcium channel blocker, used in the treatment of high blood pressure and angina
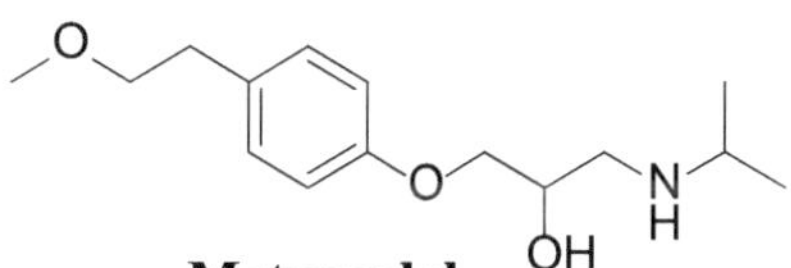 **Metoprolol**	Beta-1 blocker, used in the treatment of angina, heart failure, and hypertension
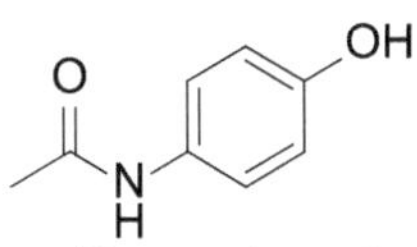 **Paracetamol**	Antipyretic and analgesic
 **Gabapentin**	Increases the synaptic concentration of gamma-aminobutyric acid, an anticonvulsant, used in the treatment of epilepsy
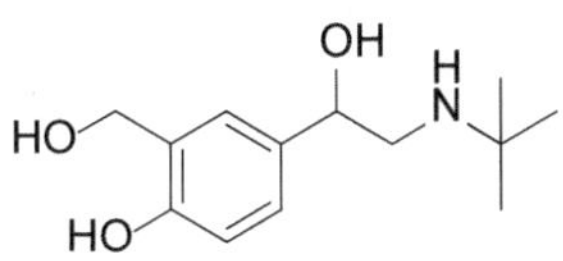 **Salbutamol**	Beta-2 adrenergic receptor agonist, bronchodilator, used in the treatment of asthma and chronic obstructive pulmonary disease (COPD)

(*Continued*)

TABLE 6.1 (*Continued*)
Most Prescribed Small Molecule Drugs with Their Biological Activity

Name of the Drug with Structure	Biological Activity
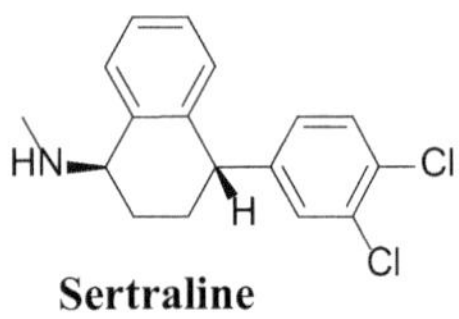 **Sertraline**	Serotonin-reuptake inhibitor, an antidepressant
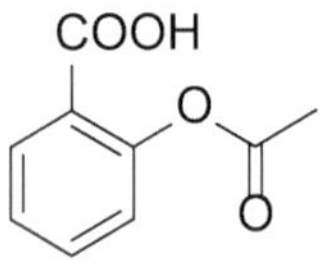 **Aspirin**	Non-steroidal anti-inflammatory drug (NSAID)
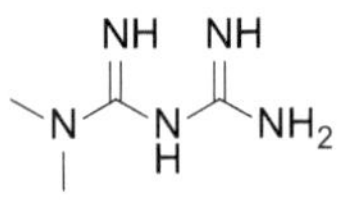 **Metformin**	Antihyperglycemic agent, used in the treatment of type II diabetes
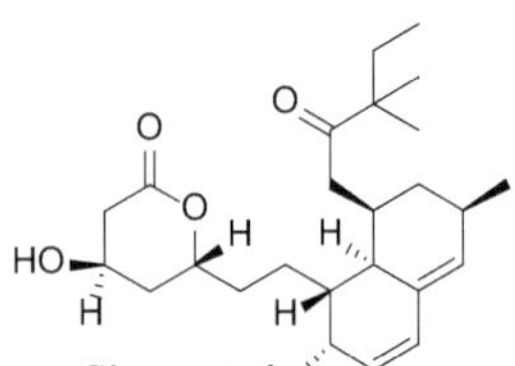 **Simvastatin**	3-Hydroxy-3-methyl glutaryl coenzyme A (HMG-CoA) reductase inhibitor, used in the treatment of hypercholesterolemia
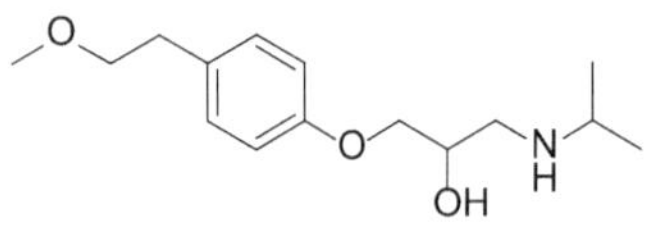 **Metoprolol**	Beta-1 blocker, used in the treatment of angina, heart failure, and hypertension
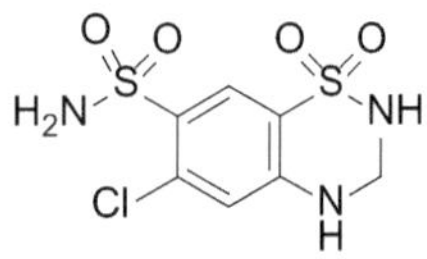 **Hydrochlorothiazide**	Diuretic, used in the treatment of edema, hypertension, and hypoparathyroidism

mechanisms of action. Small molecule drugs usually follow the Lipinski rule-of-five criteria or their variants, which allows for transcellular transport through intestinal epithelial cells. Usually they have no more than five hydrogen bond donors and no more than 10 hydrogen bond acceptors, an octanol–water partition coefficient (log *P*) below 5, and a molecular mass less than 500 Da. Such molecular weight cutoff is not a strict condition (Zhang and Wilkinson, 2007) for oral bioavailability, as some antibiotics with a molecular weight of up to 900 Da have been found to show digestive absorption ability. Also, in recent years, there has been an average increase in the molecular weight and the number of nitrogen atoms of new chemical entities approved for marketing, so scientists are claiming that conducting drug discovery campaigns in the "beyond rule of 5" chemical space can increase the chance of finding new medications (DeGoey et al., 2018). All these properties allow small molecules to have better drug-like properties (solubility, permeability, metabolic stability, transportation, PD, and PK properties) and more importantly higher stability (which reflects in easier and longer conservation), cheaper preparation, and oral availability, although some of them are only absorbed if given as prodrugs (DeGoey et al., 2018).

Small active chemical entities mimic, enhance, or diminish the behavior of natural substances or products within the body. They have relatively simple structures, customizable to meet specific therapeutic goals (Southey and Brunavs, 2023). They are generally stable and rarely need specialized storage conditions. Their behavior in the body, or in vivo, is usually predictable, leading to straightforward, often oral, dosing protocols that patients find easy to manage. They can treat a wide variety of diseases because they can move through the body easily, transferring from the gut via the bloodstream to the site of action and permeating through cell membranes to reach intracellular targets (Southey and Brunavs, 2023). They can be administered as pills, inhalers, suppositories, or injectables, making them very flexible. The chemical structure of small molecules can be designed to interact selectively with specific biological targets. By altering the atomic composition of small molecules, their overall properties can be fine-tuned to a particular purpose, eliciting only the desired response. The flexibility afforded by being able to explore all "chemical space" in this way offers small molecule approaches a marked advantage over other modalities. Once patent life expires, non-branded generic forms of the medicine will increase availability to patients (Southey and Brunavs, 2023).

Small molecules are compounds that alter the activity or the function of a biological target, by interacting with a biological macromolecule, such as DNA, RNA, and proteins (Schreiber, 2005), often in a selective and dose-dependent manner, showing a beneficial effect against a disease or a detrimental one (such as in the case of teratogens and carcinogens) (Schreiber, 2005).

Small molecule drugs can be isolated from naturally occurring resources or can be developed from leads derived from rational drug design (Chen et al., 2018; Ibarrola-Villava et al., 2018; Ma et al., 2021; Raju et al., 2018). In fact, they work rapidly, reversibly, and in tunable conditions depending on the concentration, in contrast with genetic approaches, so they are better probes to analyze complex biological networks.

6.3 HOW DO SMALL MOLECULES INTERACT WITH BIOLOGICAL TARGETS

Over the past few decades, numerous small molecules have been identified that modify the activity of a wide range of proteins, and there is a growing demand for high-quality chemical probes with clearly defined structure, potency, selectivity, mechanism of action, and availability. Small molecules interact with biological targets through various mechanisms, and their effects depend on their specific structure and the nature of the target (Li and Kang, 2020). Here are the key ways small molecules interact with biological targets:

i. **Binding to receptors**: Small molecules can bind to specific receptors on the surface of cells or within cells, usually in one of two ways: agonists which activate the receptor, mimicking the natural signaling molecule, or antagonists which block the receptor, inhibiting the binding of the natural signaling molecule and reducing activation (Southey and Brunavs, 2023). This binding can trigger a cascade of intracellular events. For example, neurotransmitters like dopamine bind to receptors in the brain to transmit signals. Albuterol is a receptor agonist prescribed for the treatment of asthma, which activates the receptor responsible for opening the airways in the lung (Ahrens and Smith, 1984).
ii. **Enzyme inhibition or activation**: Small molecules can act as inhibitors or activators of enzymes. An inhibitor binds to the enzyme and prevents it from catalyzing its usual reaction. For instance, many drugs are enzyme inhibitors, such as statins, which inhibit HMG-CoA reductase in cholesterol synthesis. Conversely, some small molecules can activate enzymes, enhancing their activity (Jiang et al., 2018).

iii. **Modulation of protein–protein interactions (PPIs)**: Small molecules can influence how proteins interact with each other. This can alter cellular processes and signaling pathways. For example, small molecules can disrupt the interaction between a protein and its partner, potentially affecting downstream signaling pathways (Davis and Sali, 2010).
iv. **Binding to DNA/RNA**: Small molecules can interact directly with nucleic acids, influencing gene expression or stabilizing or destabilizing specific structures. For example, some chemotherapeutic agents intercalate into DNA and interfere with replication (Rescifina et al., 2014).
v. **Alteration of cellular membrane properties**: Small molecules can affect the fluidity and permeability of cellular membranes, which can impact the transport of other molecules or ions across the membrane.
vi. **Allosteric modulation**: Small molecules can bind to sites other than the active site of an enzyme or receptor (allosteric sites). This binding changes the shape of the protein, which can enhance or inhibit its activity (Shen, 2010).
vii. **Competitive binding**: Small molecules can compete with endogenous substrates or ligands for binding to a biological target. For example, a drug might compete with a natural hormone to bind to its receptor (Payandeh and Volgraf, 2021).
viii. **Structural changes**: Small molecules can induce conformational changes in proteins, altering their function. This is common in drugs that target protein kinases or other regulatory proteins (Zacharioudakis and Gavathiotis, 2022).
ix. **Ion channel modulators**: Ion channels are proteins embedded in cell membranes that are responsible for regulating the flow of ions into and out of those cells. They play a key role in a wide variety of physiological processes including the regulation of heartbeat and neurotransmission. Small molecule drugs can modulate the opening and closing of these channels to treat diseases such as epilepsy (Dai, 2023).

These interactions are foundational in drug development and understanding cellular mechanisms, as small molecules are often used to probe biological systems or to treat diseases by specifically targeting these interactions. Many of the mechanisms of action described above involve a well-defined region on the protein into which a small molecule can fit and bind. These regions are known as active sites, and the geometric arrangement of their amino acids is such that they only have affinity for a few naturally occurring molecules, or substrates, within the body. This mechanism is often referred to as the "lock and key" theory (Southey and Brunavs, 2023). By understanding the requirements of the lock, researchers can create the best small molecule "key" to fit it and thereby generate the desired response. The stronger and the more specific the compound's interaction with the amino acids of the targeted active site is, the less likely that compound will be to bind to different proteins. In turn, the fewer side effects it will have. Thus, the design of small molecule drugs is highly specialized.

As a result of better disease understanding and the development of innovative technologies, more diverse approaches for disease modulation by small molecules have evolved that exploit different mechanisms of action. Examples include the modulation of PPIs (Wells and McClendon, 2007), bifunctional protein degraders (Sun et al., 2019), and stabilizers (Dong et al., 2021; Henning et al., 2022; Mullard, 2023). Some of these other modalities are explored later in the chapter.

6.4 DRUG DISCOVERY PROCESS OF SMALL MOLECULE

Medicinal chemistry involves the design and synthesis of novel biologically active compounds that have the potential to become drugs. It typically involves cycles of iterative optimization of compounds to enhance their efficacy, safety, and PK characteristics, leading to the selection of a candidate drug that ultimately will need to be manufactured on a large scale if it progresses toward regulatory approval. The process of inventing a small molecule drug and ensuring that it performs precisely as it should, minimizing unwanted side effects, involves meticulous design and synthetic mastery from researchers, often over several years reproducibly, an advantage for researchers seeking a return on their investment (Stocks, 2013). On average, it takes 10–15 years to develop a new medicine from initial discovery through to regulatory approval, and the total cost is often in the billions (Berdigaliyev and Aljofan, 2020). For every drug that makes it to the market, there are many more that do not, and it is the outlay associated with abortive efforts that account for most of this expense (Berdigaliyev and Aljofan, 2020). The discovery of new drugs remains a significant challenge, involving teams of researchers from chemistry, biology, drug development, computer science, and informatics. Drug discovery encompasses the discovery, development, identification, and interpretation of the mode of action of bioactive compounds at the molecular and cellular levels (Brenk and Rauh, 2012). Starting from the 1980s, the attention began to shift away from trial-and-error searches to a more rational and predictive computer-aided drug design, thanks to the growth of computer processing power and the advances in structural biology (Munos, 2009; Ruiz-Garcia et al., 2008).

To synthesize only molecules with a higher chance of possessing biological potency, good PK behavior, and low prediction of cytotoxic effects, medicinal chemists began to follow a multidisciplinary approach. However, this rational approach showed to be not as productive as expected (Hann and Keserü, 2012; Paul et al., 2010), and, therefore, the research community returned to empirical methods, with the so-called reverse pharmacology (Swinney and Anthony, 2011) or phenotypic drug discovery, which are the most frequently used approaches still today. The resurgence of phenotypic drug discovery was driven by the advent of robotic and miniaturized techniques for the HTS of many compounds and also by the development of new synthetic approaches for the generation of large compound collections, such as combinatorial chemistry (Balkenhohl et al., 1996; Plunkett and Ellman, 1995) and solid-phase synthesis (Bunin and Ellman, 1992; DeWitt et al., 1993; Merrifield, 1963). Moreover, strategies like the split-and-pool techniques made it possible to generate up to millions of compounds in a short period of time. However, the lack of real chemical diversity in the small molecule libraries that have been synthesized and screened has led to many failures in drug discovery (Krauss, 2014). Combinatorial libraries often showed limited diversity, as most of them only vary for substituents around a common scaffold, which was also "flat" and structurally simple, whereas it has been demonstrated how a higher scaffold complexity and a more three-dimensional character are generally associated with a more successful outcome in drug discovery and development (Flagstad et al., 2018; Young and Ge, 2004).

New strategies and new ideas were necessary to develop better high-quality new chemical libraries, and in this view, DOS, BIOS, and activity-directed synthesis are providing new tools to explore underpinned and underrepresented areas of the chemical space (Wetzel et al., 2011). Also, developing the therapeutic agents of the future may require keeping up to date with the landscape and the techniques involved in the chemistry aspects, also including entirely new approaches, by introducing new disciplines and new technologies in the process. Today, an important part of modern medicinal chemistry is represented by computer-aided drug design, where the structure of the potential drug candidate is predicted de novo by a virtual screening of potential chemical substructures interacting with a target protein (Rester, 2008; Rollinger et al., 2008). Also, computational methods, such as molecular modeling and dynamics simulations (Barcellos et al., 2008; Durrant and McCammon, 2011), are used to improve the potency and the drug-likeness, in replacement of experimental SAR studies (Borhani and Shaw, 2012; Ciemny et al., 2018). One remarkable example of the evolution of drug discovery is presented by the history of the development of angiotensin-converting enzyme inhibitors (Smith and Vane, 2003).

The drug discovery process is a complex, multistep journey that involves identifying and developing new therapeutic compounds. Paths toward the identification of a small molecule drug candidate that successfully reaches the market are complicated and depend on a variety of factors including the complexity of the disease and the rigorous validation and testing required to meet regulatory approval requirements. The general process is outlined in Figure 6.2. In the following sections we will discuss each stage of the process in more detail.

6.4.1 Target Identification and Validation

First, a biological target (often a protein) involved in the disease must be identified and validated. This target validation is a time-consuming process that involves demonstrating the relevance of the target in a particular biological pathway, process, or disease of interest (Schenone et al., 2013). Upon validation, the assumption is made that the modulation of that target will bring out the desired therapeutic effect. The most represented class of therapeutic targets are proteins (Santos et al., 2017), but other biological molecules can also be targeted by drugs, such as nucleic acids (DNA and RNA) (Kulkarni et al., 2021). Therapeutic targets are chosen based on the disease being treated and the potential to interfere with the mechanisms of disease progression. For example, many anti-cancer drugs target proteins responsible for abnormal cell growth and division while for Alzheimer's disease, a common target is amyloid beta, a protein that forms plaques (Ramanan and Day, 2023). Drugs can be developed to prevent plaque formation or degrade those that have already formed.

6.4.2 Lead Discovery

Once the physiological role played by the target is established, it can be predicted how its modulation will affect the disease and begin their search for a chemical entity to achieve the desired outcome. Lead can be developed by following approaches such as HTS (Carnero, 2006), fragment identification (Erlanson et al., 2016; Singh et al., 2018), structure-based drug design (Anderson, 2003; Moitessier et al., 2016; Ou-Yang et al., 2012), artificial intelligence (AI)-based drug design (Schneider et al., 2020), known compounds from the published literature (Mullard, 2018), the repurposing of approved drugs (Corsello et al., 2017), and DNA-encoded library screening (Chan et al., 2015; Gerry et al., 2019; Goodnow et al., 2017).

In HTS a very large library of structurally diverse compounds will be tested against the target, in search of those that elicit the desired biological response. The method has the advantage that no prior knowledge of the target is required. These compounds can be natural products or synthetic. Although traditional HTS often results in hit compounds, some of which are capable of being modified into a lead, hit rates are often extremely very low, which has limited the usage of HTS in research programs.

Medicinal chemists try very hard to ensure that the compound collections they use for screening are fit for purpose. It is estimated that the total number of compounds that make up oral drug-like chemical space is extremely small (Reymond et al., 2012). Even a screening set containing a million compounds—which is rare—barely scratches the surface of available drug-like space. Rational and predictive drug design methods done by virtual screening are a computational strategy that has played a crucial role in the development of medicinally important small bioactive molecules (Carpenter and Huang, 2018; Lavecchia and Giovanni, 2013; Shoichet, 2004; Walters et al., 1998). These methods are broadly classified as either structure-based or ligand-based methods.

A typical virtual screening of a small molecule library is a multistep process (Figure 6.3). It starts with small molecules for screening with a careful preparation of the compounds by removing the unwanted molecules. Depending on the prior knowledge of the target structure and active molecules, ligand-based virtual screening and structure-based virtual screening (SBVS) can be employed independently or together. Due to the imperfections of the current virtual screening methods, a post-screening process is usually conducted on carefully selecting the hits for further experimental testing.

FIGURE 6.3 Overview of the drug discovery process.

6.4.2.1 Structure-Based Method

In cases where more is known about a target (perhaps the binding site of the agents has already been ascertained), the screening set can be more focused. In this approach both target and ligand structure information is imperative and includes ligand docking, pharmacophore, and ligand design methods. To create that focus, virtual screening can be undertaken. Docking experiments, in which compounds are fitted into the known binding site using a software package, can be carried out ahead of the physical testing (Ali et al., 2023). In this way, researchers can eliminate compounds that have no hope of binding and promote those that have a better chance. This process is known as SBVS because it is enabled by the knowledge of the protein structure. This type of screening is designed to evaluate specific compound-binding hypotheses, unlike the random HTS experiments described earlier.

6.4.2.2 Ligand-Based Method

Where the binding site is not known, it may still be possible to create a focused screening set if agents active against the target have been identified in previous efforts. Ligand-based methods use only ligand information for predicting activity depending on its similarity/dissimilarity to previously known active ligands. In ligand-based virtual screen, compounds with similar structures to the known actives are selected, in the hope that these will provide a best-in-class solution. These might contain motifs structurally similar to those in the comparator actives or different while still bearing similar protein interaction properties. These scaffold-hopping searches can be conducted in two- or three dimensions, and success often opens new intellectual property space. Further a synthetic effort will start from a validated active discovered during a previous effort, to address residual target selectivity, drug stability, or toxicological concerns as rapidly as possible. Often, this can result in the discovery of one or more new chemical series.

6.4.3 Lead Optimization

Having identified a suitable set of compounds for the test, whether that be a random HTS with many compounds or a smaller focused set informed by prior knowledge, the search is then performed. After the structure of a lead compound is determined, it is also confirmed that initial hits are genuinely interacting with the target and not just artifacts of the assay. Still, it is often found to have poor PK properties and/or substantial side effects (Ylä-Herttuala, 2017). These adverse reactions often make lead compounds clinically unusable. Therefore, it is often necessary to optimize the structure of the lead compound to improve biological activity, reduce toxicity, increase specificity, and/or improve PK behavior, to improve the drug-forming properties of the lead compound and apply them in the clinical setting (Chang and Ding, 2018; Wang et al., 2017; Wu et al., 2018; Zheng et al., 2019). Chemists modify hit compounds to improve their efficacy, selectivity, and safety. This involves iterative testing and refinement to enhance the compound's properties.

The primary strategies for lead compound optimization are as follows: (1) changing the metabolic pathway to improve metabolic stability, (2) structural optimization to reduce toxicity risks in drug design, (3) structural modifications to improve water solubility, (4) promoting the passage of compounds through the blood–brain barrier, (5) reducing cardiac toxicity, and (6) improving plasma stability (Igase et al., 2019; Ma et al., 2018; Nejadnik et al., 2020; Ratnayake et al., 2018). The engagement of a compound with the target of interest is investigated using structural biology techniques, through the computational creation and exploration of 3D models of the target. These structural models are central to the further design and optimization of compounds during the lead optimization phase, as researchers endeavor to improve the "fit" of the compound to the active site.

6.4.4 Preclinical Development

Preclinical trials are an essential phase in the development of new drugs. They occur before clinical trials involving human subjects. The main goal of preclinical trials is to gather data on the safety and efficacy of a new optimized lead. This includes understanding how the intervention works, its potential side effects, and how it is metabolized in the body and involves the testing phase where in vitro and in vivo studies are done. In vitro studies are conducted in a controlled environment outside a living organism, such as in cell cultures to understand the drug's mechanism and its effects on cells. They help in understanding the biological activity of the drug or treatment. In vivo studies are conducted in living organisms, often in animals like rodents, to assess the effects of the intervention in a more complex biological system. They help determine the PK, i.e., how the drug is absorbed, distributed, metabolized, and excreted (ADME), and PD (the drug's effects on the body) of the new drug. PK helps in optimizing drug therapy by informing dosing regimens and understanding potential interactions and side effects. It is a critical aspect of drug development and clinical practice. PK influences drug levels and kinetics and hence the pharmacological effect of a compound in vivo (Pellegatti, 2012). As a consequence, a significantly better correlation of in vitro–in vivo parameters and an improved extrapolation from nonhuman to human PK parameters can be achieved (Jones and Rowland-Yeo, 2013; Smith, 2013). The consequent improvement of ADME assays and their application in drug discovery has led to a significant reduction of failure in clinical trials due to poor ADME properties. These assessments drive decisions during the discovery process, combining lab-based experimentation with the use of computational modeling and machine learning methods to make early predictions of various DMPK outcomes (Obrezanova, 2023). They help to determine the optimal dosage, side effect profile, and toxicity risks. Researchers will make modifications to the structures of their leading compounds to arrive at a better balance of DMPK properties. They will seek to bring about the desired clinical outcome from the smallest possible dose—another good strategy for the elimination of side effects. At the same time, they will seek to ensure that any structural changes they make will neither diminish potency against their target nor reduce selectivity. Drug discovery researchers often find themselves optimizing multiple, often competing parameters to find the "sweet spot" that will deliver the optimal properties for a given indication.

Researchers then collect the data on various parameters including efficacy, dosage, side effects, toxicity, and potential interactions with other drugs or substances. Preclinical trials must adhere to strict ethical guidelines and regulations to ensure the humane treatment of animals and the accuracy of the results. Successful preclinical trials provide the necessary evidence to move on to human clinical trials. If significant safety or efficacy issues are discovered, the intervention may need to be modified or abandoned. Preclinical trials are crucial for minimizing risks before new treatments are tested in humans, ultimately aiming to ensure that new therapies are both effective and safe.

6.4.5 Clinical Development

Clinical development is the last phase in drug development where a new drug or treatment is tested in humans to evaluate its safety, efficacy, and overall effectiveness. This phase follows successful preclinical trials and involves several stages, each with specific goals and regulatory requirements. Clinical development is a rigorous process designed to ensure that new drugs or treatments are both safe and effective before they become available to the public. It involves a collaborative effort between researchers, regulatory agencies, and various stakeholders to bring new medical innovations to market. Clinical development involves three phases: phase I, phase II, and phase III, each with different objectives. In phase I, safety, dosage, and PK of the drug are assessed in healthy volunteers or sometimes in patients with the target disease to determine the drug's metabolism and excretion, identify the maximum tolerated dose, and observe for any adverse effects. This phase lasts for a few months and involves 20–100 healthy volunteers or patients. In phase II, the drug is evaluated for its efficacy, optimal dosing, and further assess safety in patients with a larger group of patients. Further drug's effectiveness for a specific indication is determined by refining the dosing regimens. The duration of this phase involves several months to 2 years and involves 100–300 patients. Phase III clinical trials involve large-scale trials, which are conducted to confirm efficacy, monitor side effects, and compare the new drug to standard treatments or a placebo in a larger patient population. In this phase large-scale studies are conducted to gather comprehensive data on efficacy and safety. This important phase provides the evidence needed for regulatory approval. Time duration for this phase usually takes 1–4 years and involves 1,000–3,000 or more patients. This phase is crucial for regulatory approval.

Before starting clinical trials, sponsors must submit an investigational new drug application to regulatory agencies (e.g., the FDA in the United States) to gain approval to test the drug in humans. Clinical trials are carefully designed to answer specific research questions and minimize biases. Common designs include randomized controlled trials, double-blind studies, and placebo-controlled studies. Clinical trials define primary and secondary endpoints to measure the drug's effects. Primary endpoints are the main outcomes that the study is designed to assess, while secondary endpoints are additional effects or benefits. Collecting and analyzing data accurately is critical. Data management systems are used to ensure data integrity and compliance with regulatory standards. Ensuring the ethical treatment of participants is paramount in the clinical development of the drug candidate. Informed consent must be obtained from all participants, ensuring they understand the risks, benefits, and nature of the trial. Continuous monitoring for adverse events and side effects is crucial. Data safety monitoring boards may be established to review data and ensure participant safety. During clinical trials, medicinal chemists face various challenges like finding and keeping participants, particularly for rare diseases or specific populations. Further clinical development is often expensive and time-consuming, with trials taking several years to complete. In addition, navigating regulatory requirements and obtaining approval can be complex and vary by region.

6.5 EMERGING TECHNOLOGIES IN SMALL MOLECULE DRUG DISCOVERY

Despite the large amount of money invested in drug discovery, it suffers from a high failure rate (https://ncats.nih.gov/director/january-2021) (accessed October 30, 2023). High rates of late-stage attrition due to the lack of efficacy or compound-related safety issues make drug discovery expensive and time-consuming. Hence, exploring new technologies is crucial if the industry has to improve its success rate. The use of innovative modern methodologies such as AI and the development of new chemical modalities to target "undruggable" targets are two ways the industry is evolving to improve the chances of producing new therapies and better patient outcomes.

6.5.1 Artificial Intelligence

AI techniques improve the speed and efficiency and therefore reduce the cost of the drug discovery process. Computers are more efficient at analyzing and processing large amounts of data to help researchers make less biased, more informed decisions (Brown et al., 2020). An important example of an AI application is compound screening (Sadybekov and Katritch, 2023). AI can be used to screen millions of compounds virtually, solving the chemical space conundrum in a much more cost-effective way to identify potential novel drug candidates, unlike traditional THS. The ability of AI to explore broader chemical space improves the chances of success in finding potential leads that can give rise to clinically approved drug candidates that reside outside traditional "oral drug-like chemical space" (Doak et al., 2014). Advanced machine learning models are trained to utilize neural networks, imitating the biological network of the human brain, and used to predict the efficacy of compounds and safety endpoints (Cavasotto and Scardino, 2022). Thus, better-informed decisions are made at an earlier stage in the process, mitigating the risk of more costly late-stage failure (Vamathevan et al., 2019). AI tools can design molecules from first principles (Vanhaelen et al., 2020), repurpose (Prasad and Kumar, 2021; Roessler et al., 2021) known drugs for the treatment of other diseases, and even help plan synthetic routes (Thakkar et al., 2021) based on prior knowledge. To be well informed enough to make accurate decisions, AI tools require large volumes of data from which they are trained. In the absence of sufficient quality and volume of data, these tools can be unreliable and inaccurate. DeepMind's AlphaFold, an AI tool which uses deep learning technology to predict three-dimensional protein structures from amino acid sequences (Jumper et al., 2021), has transformed the field of life science. The neural network system used was trained on a vast dataset of protein structures and sequences, learning the intricate relationship between amino acid sequences and their preferred spatial arrangements with high accuracy. While this has revolutionized the field of protein structure prediction, its full impact on the field of drug discovery is only just starting to be realized (Arnold, 2023). AI holds much promise and has achieved some notable successes so far but human researchers are clearly far from being replaced; their experience is a key to the validation of AI tools and the interpretation of their predictions (Jiménez-Luna et al., 2020). The synergistic integration of AI technologies, traditional screening approaches, and human wisdom remains essential for modern drug discovery programs to successfully deliver new treatments (Griffen et al., 2020).

6.5.2 New Chemical Modalities

Novel chemical modalities often work through unique mechanisms of action. Undruggable targets non-enzymatic and/or intrinsically disordered proteins do not possess well-defined active sites to which small molecules can bind. However, innovative strategies to chemically modulate these proteins are emerging and have attracted huge interest in drug discovery, offering the potential to treat many more diseases and address at least some of the gaps in unmet patient needs (Blanco and Gardinier, 2020). For example, covalent drugs are capable of bonding chemically to certain amino acids present either on the surface of a target protein or in featureless active sites. Thus, such compounds directly attach to target proteins, rather than interact transiently, thereby increasing efficacy and specificity as well as extended duration of action. An example of a marketed covalent drug is ibrutinib (Figure 6.4), approved by the FDA in 2013 for the treatment of various blood cancers, which works by blocking the activity of a protein involved in the growth and survival of cancer cells (https://www.forbes.com/sites/davidshaywitz/2013/04/05/the-wild-story-behind-a-promising-experimental-cancerdrug/).

Another new modality for small molecules is their use in the disruption/modulation of the interactions between two or more target proteins. Such PPIs are crucial for many biological processes, so directly targeting them can bring about therapeutic benefits. Venetoclax (Figure 6.4) is an approved PPI drug prescribed for the treatment of acute myeloid leukemia and chronic lymphocytic leukemia. Its mode of action involves disrupting the interaction of two proteins that would otherwise work together to promote cancer cell survival (Roberts and Huang, 2017).

Protein degrader therapies take advantage of the body's natural processes to schedule disease-causing proteins for degradation. The degrader recruits undesirable proteins to the cellular machinery responsible for protein breakdown, leading to their elimination. Small molecules can induce targeted degradation of specific proteins through cellular machinery like the ubiquitin-proteasome

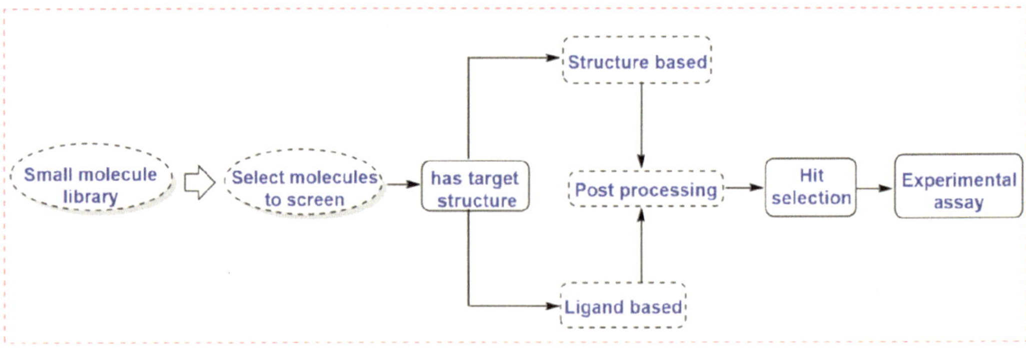

FIGURE 6.4 Workflow of virtual screening of a small molecule library.

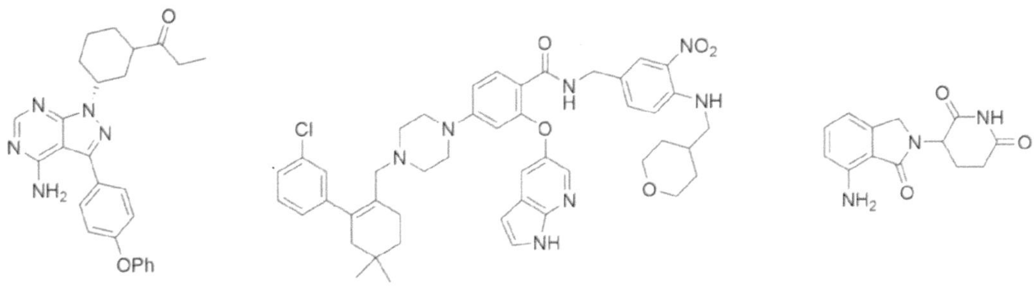

Ibrutinib (Antineoplastiv agent) — Venetoclax (BCL-2 inhibitor) — Lenalidomide (Angiogenesis inhibitor)

FIGURE 6.5 Examples of new chemical modality drugs.

system. Lenalidomide (Figure 6.4) (Armoiry et al., 2008) is a degrader that promotes the removal of a protein promoting cancer cell growth, FDA-approved for the treatment of multiple myeloma in 2005. This approach has been developed further in recent years, giving rise to a new class of drugs called proteolysis-targeting chimeras.

6.6 CONCLUSION

Small molecules are central to medicinal chemistry due to their ability to be precisely designed and optimized, their diverse mechanisms of action, their favorable pharmacological properties, and their cost-effective synthesis and development. They play a crucial role in developing new and effective therapeutic agents for a wide range of diseases. However, due to the low success rate and difficulties in finding new lead compounds for undruggable targets, the whole drug discovery process has changed a lot during the last decades. Through careful design and rigorous testing in both preclinical and clinical settings, medicinal chemists are advancing our understanding of disease biology. Rational design of ligands in combination with computer-aided methods is a powerful approach, especially when the biological target is well defined and structurally known. In addition, new synthetic methods are able to generate high-quality chemical libraries and are being developed to meet the need to improve the quality and quantity of small molecules for biological screenings. Also, biology-oriented synthesis has been conceived with the purpose of taking inspiration from nature to select promising molecular scaffolds being related to natural products in terms of biological output.

Compounds entering the clinic in recent years have successfully challenged previous dogma regarding the molecular properties required for an oral drug and the methodology by which disease progression can be arrested. Ongoing investment and advances in biology, computational technologies, and innovative synthetic chemistry are providing researchers with increasingly efficient and precise tools for small molecule discovery and design. The synergistic partnership between scientific expertise and technological progress still holds great promise for the discovery and development of small molecule medications and treatments in the future.

REFERENCES

Ahrens, R.C., Smith, G.D., 1984. Albuterol: an adrenergic agent for use in the treatment of asthma pharmacology, pharmacokinetics and clinical use. *Pharmacother. J. Hum. Pharmacol. Drug Ther.* 4, pp. 105–120. https://doi.org/10.1002/j.1875-9114.1984.tb03330.x

Ali, A., Wani, A.B., Malla, B.A., Poyya, J., Dar, N.J., Ali, F., Ahmad, S.B., Rehman, M.U., Nadeem, A., 2023. Network pharmacology integrated molecular docking and dynamics to elucidate saffron compounds targeting human COX-2 protein. *Medicina (B. Aires).* 59, p. 2058. https://doi.org/10.3390/medicina59122058

Anderson, A.C., 2003. The process of structure-based drug design. *Chem. Biol.* 10, pp. 787–797. https://doi.org/10.1016/j.chembiol.2003.09.002

Arkin, M.R., Wells, J.A., 2004. Small-molecule inhibitors of protein–protein interactions: progressing towards the dream. *Nat. Rev. Drug Discov.* 3, pp. 301–317. https://doi.org/10.1038/nrd1343

Armoiry, X., Aulagner, G., Facon, T., 2008. Lenalidomide in the treatment of multiple myeloma: a review. *J. Clin. Pharm. Ther.* 33, pp. 219–226. https://doi.org/10.1111/j.1365-2710.2008.00920.x

Arnold, C., 2023. AlphaFold touted as next big thing for drug discovery — but is it? *Nature* 622, pp. 15–17. https://doi.org/10.1038/d41586-023-02984-w

Balkenhohl, F., von dem Bussche-Hünnefeld, C., Lansky, A., Zechel, C., 1996. Combinatorial synthesis of small organic molecules. *Angew. Chemie Int. Ed. English* 35, pp. 2288–2337. https://doi.org/10.1002/anie.199622881

Barcellos, G., Pauli, I., Caceres, R., Macedo Timmers, L.F., Dias, R., de Azevedo Jr, W., 2008. Molecular modeling as a tool for drug discovery. *Curr. Drug Targets* 9, pp. 1084–1091. https://doi.org/10.2174/138945008786949388

Berdigaliyev, N., Aljofan, M., 2020. An overview of drug discovery and development. *Future Med. Chem.* 12, pp. 939–947. https://doi.org/10.4155/fmc-2019-0307

Blanco, M.-J., Gardinier, K.M., 2020. New chemical modalities and strategic thinking in early drug discovery. *ACS Med. Chem. Lett.* 11, pp. 228–231. https://doi.org/10.1021/acsmedchemlett.9b00582

Borhani, D.W., Shaw, D.E., 2012. The future of molecular dynamics simulations in drug discovery. *J. Comput. Aided. Mol. Des.* 26, pp. 15–26. https://doi.org/10.1007/s10822-011-9517-y

Borrell, J.I., 2000. Editorial: past, present and future of the combinatorial chemistry in Spain. *Mol. Divers.* 6, pp. 73–74. https://doi.org/10.1023/B:MODI.0000006988.67158.6d

Brenk, R., Rauh, D., 2012. Change or be changed. *Bioorg. Med. Chem.* 20, pp. 3695–3697. https://doi.org/10.1016/j.bmc.2012.04.038

Brown, N., Ertl, P., Lewis, R., Luksch, T., Reker, D., Schneider, N., 2020. Artificial intelligence in chemistry and drug design. *J. Comput. Aided. Mol. Des.* 34, pp. 709–715. https://doi.org/10.1007/s10822-020-00317-x

Bunin, B.A., Ellman, J.A., 1992. A general and expedient method for the solid-phase synthesis of 1,4-benzodiazepine derivatives. *J. Am. Chem. Soc.* 114, pp. 10997–10998. https://doi.org/10.1021/ja00053a067

Carnero, A., 2006. High throughput screening in drug discovery. *Clin. Transl. Oncol.* 8, pp. 482–490. https://doi.org/10.1007/s12094-006-0048-2

Carpenter, K.A., Huang, X., 2018. Machine learning-based virtual screening and its applications to Alzheimer's drug discovery: a review. *Curr. Pharm. Des.* 24, pp. 3347–3358. https://doi.org/10.2174/1381612824666180607124038

Cavasotto, C.N., Scardino, V., 2022. Machine learning toxicity prediction: latest advances by toxicity end point. *ACS Omega* 7, pp. 47536–47546. https://doi.org/10.1021/acsomega.2c05693

Chan, A.I., McGregor, L.M., Liu, D.R., 2015. Novel selection methods for DNA-encoded chemical libraries. *Curr. Opin. Chem. Biol.* 26, pp. 55–61. https://doi.org/10.1016/j.cbpa.2015.02.010

Chang, S.C., Ding, J.L., 2018. Ubiquitination and SUMOylation in the chronic inflammatory tumor microenvironment. *Biochim. Biophys. Acta - Rev. Cancer* 1870, pp. 165–175. https://doi.org/10.1016/j.bbcan.2018.08.002

Chen, F., Long, Q., Fu, D., Zhu, D., Ji, Y., Han, L., Zhang, B., Xu, Q., Liu, B., Li, Y., Wu, S., Yang, C., Qian, M., Xu, J., Liu, S., Cao, L., Chin, Y.E., Lam, E.W.-F., Coppé, J.-P., Sun, Y., 2018. Targeting SPINK1 in the damaged tumour microenvironment alleviates therapeutic resistance. *Nat. Commun.* 9, p. 4315. https://doi.org/10.1038/s41467-018-06860-4

Ciemny, M., Kurcinski, M., Kamel, K., Kolinski, A., Alam, N., Schueler-Furman, O., Kmiecik, S., 2018. Protein–peptide docking: opportunities and challenges. *Drug Discov. Today* 23, pp. 1530–1537. https://doi.org/10.1016/j.drudis.2018.05.006

Cong, F., Cheung, A.K., Huang, S.-M.A., 2012. Chemical genetics–based target identification in drug discovery. *Annu. Rev. Pharmacol. Toxicol.* 52, pp. 57–78. https://doi.org/10.1146/annurev-pharmtox-010611-134639

Corsello, S.M., Bittker, J.A., Liu, Z., Gould, J., McCarren, P., Hirschman, J.E., Johnston, S.E., Vrcic, A., Wong, B., Khan, M., Asiedu, J., Narayan, R., Mader, C.C., Subramanian, A., Golub, T.R., 2017. The drug repurposing hub: a next-generation drug library and information resource. *Nat. Med.* 23, pp. 405–408. https://doi.org/10.1038/nm.4306

Dai, G., 2023. Signaling by ion channels: pathways, dynamics and channelopathies. *Mo. Med.* 120, pp. 367–373.

Davis, F.P., Sali, A., 2010. The overlap of small molecule and protein binding sites within families of protein structures. *PLoS Comput. Biol.* 6, p. e1000668. https://doi.org/10.1371/journal.pcbi.1000668

DeGoey, D.A., Chen, H.-J., Cox, P.B., Wendt, M.D., 2018. Beyond the rule of 5: lessons learned from AbbVie's drugs and compound collection. *J. Med. Chem.* 61, pp. 2636–2651. https://doi.org/10.1021/acs.jmedchem.7b00717

DeWitt, S.H., Kiely, J.S., Stankovic, C.J., Schroeder, M.C., Cody, D.M., Pavia, M.R., 1993. "Diversomers": an approach to nonpeptide, nonoligomeric chemical diversity. *Proc. Natl. Acad. Sci.* 90, pp. 6909–6913. https://doi.org/10.1073/pnas.90.15.6909

Doak, B.C., Over, B., Giordanetto, F., Kihlberg, J., 2014. Oral druggable space beyond the rule of 5: insights from drugs and clinical candidates. *Chem. Biol.* 21, pp. 1115–1142. https://doi.org/10.1016/j.chembiol.2014.08.013

Dong, G., Ding, Y., He, S., Sheng, C., 2021. Molecular glues for targeted protein degradation: from serendipity to rational discovery. *J. Med. Chem.* 64, pp. 10606–10620. https://doi.org/10.1021/acs.jmedchem.1c00895

Durrant, J.D., McCammon, J.A., 2011. Molecular dynamics simulations and drug discovery. *BMC Biol.* 9, p. 71. https://doi.org/10.1186/1741-7007-9-71

Erhardt, P.W., 2002. Medicinal chemistry in the new millennium. A glance into the future. *Pure Appl. Chem.* 74, pp. 703–785. https://doi.org/10.1351/pac200274050703

Erlanson, D.A., Fesik, S.W., Hubbard, R.E., Jahnke, W., Jhoti, H., 2016. Twenty years on: the impact of fragments on drug discovery. *Nat. Rev. Drug Discov.* 15, pp. 605–619. https://doi.org/10.1038/nrd.2016.109

Flagstad, T., Azevedo, C.M.G., Min, G., Willaume, A., Morgentin, R., Nielsen, T.E., Clausen, M.H., 2018. Petasis/Diels–Alder/Cyclization cascade reactions for the generation of scaffolds with multiple stereogenic centers and orthogonal handles for library production. *European J. Org. Chem.* 2018, p. 6596. https://doi.org/10.1002/ejoc.201801730

Fuentes, A., Pineda, M., Venkata, K., 2018. Comprehension of top 200 prescribed drugs in the US as a resource for pharmacy teaching, training and practice. *Pharmacy* 6, p. 43. https://doi.org/10.3390/pharmacy6020043

Gerry, C.J., Wawer, M.J., Clemons, P.A., Schreiber, S.L., 2019. DNA barcoding a complete matrix of stereoisomeric small molecules. *J. Am. Chem. Soc.* 141, pp. 10225–10235. https://doi.org/10.1021/jacs.9b01203

Goodnow, R.A., Dumelin, C.E., Keefe, A.D., 2017. DNA-encoded chemistry: enabling the deeper sampling of chemical space. *Nat. Rev. Drug Discov.* 16, pp. 131–147. https://doi.org/10.1038/nrd.2016.213

Griffen, E.J., Dossetter, A.G., Leach, A.G., 2020. Chemists: AI is here; unite to get the benefits. *J. Med. Chem.* 63, pp. 8695–8704. https://doi.org/10.1021/acs.jmedchem.0c00163

Hann, M.M., Keserü, G.M., 2012. Finding the sweet spot: the role of nature and nurture in medicinal chemistry. *Nat. Rev. Drug Discov.* 11, pp. 355–365. https://doi.org/10.1038/nrd3701

Henning, N.J., Boike, L., Spradlin, J.N., Ward, C.C., Liu, G., Zhang, E., Belcher, B.P., Brittain, S.M., Hesse, M.J., Dovala, D., McGregor, L.M., Valdez Misiolek, R., Plasschaert, L.W., Rowlands, D.J., Wang, F., Frank, A.O., Fuller, D., Estes, A.R., Randal, K.L., Panidapu, A., McKenna, J.M., Tallarico, J.A., Schirle, M., Nomura, D.K., 2022. Deubiquitinase-targeting chimeras for targeted protein stabilization. *Nat. Chem. Biol.* 18, pp. 412–421. https://doi.org/10.1038/s41589-022-00971-2

Ibarrola-Villava, M., Cervantes, A., Bardelli, A., 2018. Preclinical models for precision oncology. *Biochim. Biophys. Acta - Rev. Cancer* 1870, pp. 239–246. https://doi.org/10.1016/j.bbcan.2018.06.004

Igase, M., Shibutani, S., Kurogouchi, Y., Fujiki, N., Hwang, C.C., Coffey, M., Noguchi, S., Nemoto, Y., Mizuno, T., 2019. Combination therapy with Reovirus and ATM inhibitor enhances cell death and virus replication in Canine Melanoma. *Mol. Ther. - Oncolytics* 15, pp. 49–59. https://doi.org/10.1016/j.omto.2019.08.003

Imming, P., 2015. Medicinal chemistry. In: *The Practice of Medicinal Chemistry*. Elsevier, pp. 3–13. https://doi.org/10.1016/B978-0-12-417205-0.00001-8

Jiang, S.-Y., Li, H., Tang, J.-J., Wang, J., Luo, J., Liu, B., Wang, J.-K., Shi, X.-J., Cui, H.-W., Tang, J., Yang, F., Qi, W., Qiu, W.-W., Song, B.-L., 2018. Discovery of a potent HMG-CoA reductase degrader that eliminates statin-induced reductase accumulation and lowers cholesterol. *Nat. Commun.* 9, p. 5138. https://doi.org/10.1038/s41467-018-07590-3

Jiménez-Luna, J., Grisoni, F., Schneider, G., 2020. Drug discovery with explainable artificial intelligence. *Nat. Mach. Intell.* 2, pp. 573–584. https://doi.org/10.1038/s42256-020-00236-4

Jones, H., Rowland-Yeo, K., 2013. Basic concepts in physiologically based pharmacokinetic modeling in drug discovery and development. *CPT Pharmacometrics Syst. Pharmacol.* 2, pp. 1–12. https://doi.org/10.1038/psp.2013.41

Jumper, J., Evans, R., Pritzel, A., Green, T., Figurnov, M., Ronneberger, O., Tunyasuvunakool, K., Bates, R., Žídek, A., Potapenko, A., Bridgland, A., Meyer, C., Kohl, S.A.A., Ballard, A.J., Cowie, A., Romera-Paredes, B., Nikolov, S., Jain, R., Adler, J., Back, T., Petersen, S., Reiman, D., Clancy, E.,

Zielinski, M., Steinegger, M., Pacholska, M., Berghammer, T., Bodenstein, S., Silver, D., Vinyals, O., Senior, A.W., Kavukcuoglu, K., Kohli, P., Hassabis, D., 2021. Highly accurate protein structure prediction with AlphaFold. *Nature* 596, pp. 583–589. https://doi.org/10.1038/s41586-021-03819-2

Karawajczyk, A., Giordanetto, F., Benningshof, J., Hamza, D., Kalliokoski, T., Pouwer, K., Morgentin, R., Nelson, A., Müller, G., Piechot, A., Tzalis, D., 2015. Expansion of chemical space for collaborative lead generation and drug discovery: the European lead factory perspective. *Drug Discov. Today* 20, pp. 1310–1316. https://doi.org/10.1016/j.drudis.2015.09.009

Krauss, G. (Ed.), 2014. *Biochemistry of Signal Transduction and Regulation*. Wiley. https://doi.org/10.1002/9783527667475

Kulkarni, J.A., Witzigmann, D., Thomson, S.B., Chen, S., Leavitt, B.R., Cullis, P.R., van der Meel, R., 2021. The current landscape of nucleic acid therapeutics. *Nat. Nanotechnol.* 16, pp. 630–643. https://doi.org/10.1038/s41565-021-00898-0

Lavecchia, A., Giovanni, C., 2013. Virtual screening strategies in drug discovery: a critical review. *Curr. Med. Chem.* 20, pp. 2839–2860. https://doi.org/10.2174/09298673113209990001

Lenci, E., Trabocchi, A., 2020. Synthetic approaches toward small molecule libraries, In: *Small Molecule Drug Discovery*. Elsevier, pp. 1–34. https://doi.org/10.1016/B978-0-12-818349-6.00001-7

Li, Q., Kang, C., 2020. Mechanisms of action for small molecules revealed by structural biology in drug discovery. *Int. J. Mol. Sci.* 21, p. 5262. https://doi.org/10.3390/ijms21155262

Ma, Y.-S., Xin, R., Yang, X.-L., Shi, Y., Zhang, D.-D., Wang, H.-M., Wang, P.-Y., Liu, J.-B., Chu, K.-J., Fu, D., 2021. Paving the way for small-molecule drug discovery. *Am. J. Transl. Res.* 13, pp. 853–870. https://doi.org/PMC8014367

Ma, Y.-S., Yu, F., Zhong, X.-M., Lu, G.-X., Cong, X.-L., Xue, S.-B., Xie, W.-T., Hou, L.-K., Pang, L.-J., Wu, W., Zhang, W., Cong, L.-L., Liu, T., Long, H.-D., Sun, R., Sun, H.-Y., Lv, Z.-W., Wu, C.-Y., Fu, D., 2018. miR-30 Family reduction maintains self-renewal and promotes tumorigenesis in NSCLC-initiating cells by targeting oncogene TM4SF1. *Mol. Ther.* 26, pp. 2751–2765. https://doi.org/10.1016/j.ymthe.2018.09.006

Merrifield, R.B., 1963. Solid phase peptide synthesis. I. The synthesis of a tetrapeptide. *J. Am. Chem. Soc.* 85, pp. 2149–2154. https://doi.org/10.1021/ja00897a025

Moitessier, N., Pottel, J., Therrien, E., Englebienne, P., Liu, Z., Tomberg, A., Corbeil, C.R., 2016. Medicinal chemistry projects requiring imaginative structure-based drug design methods. *Acc. Chem. Res.* 49, pp. 1646–1657. https://doi.org/10.1021/acs.accounts.6b00185

Mullard, A., 2023. Glue-based KRAS inhibitors make their debut cancer trial mark. *Nat. Rev. Drug Discov.* 22, p. 942. https://doi.org/10.1038/d41573-023-00169-8

Mullard, A., 2019. 2018 FDA drug approvals. *Nat. Rev. Drug Discov.* 18, pp. 85–89. https://doi.org/10.1038/d41573-019-00014-x

Mullard, A., 2018. A snapshot of lead-generation strategies. *Nat. Rev. Drug Discov.* 17, p. 534. https://doi.org/10.1038/nrd.2018.129

Munos, B., 2009. Lessons from 60 years of pharmaceutical innovation. *Nat. Rev. Drug Discov.* 8, pp. 959–968. https://doi.org/10.1038/nrd2961

Nejadnik, H., Jung, K.O., Theruvath, A.J., Kiru, L., Liu, A., Wu, W., Sulchek, T., Pratx, G., Daldrup-Link, H.E., 2020. Instant labeling of therapeutic cells for multimodality imaging. *Theranostics* 10, pp. 6024–6034. https://doi.org/10.7150/thno.39554

O' Connor, C.J., Laraia, L., Spring, D.R., 2011. Chemical genetics. *Chem. Soc. Rev.* 40, p. 4332. https://doi.org/10.1039/c1cs15053g

Obrezanova, O., 2023. Artificial intelligence for compound pharmacokinetics prediction. *Curr. Opin. Struct. Biol.* 79, p. 102546. https://doi.org/10.1016/j.sbi.2023.102546

Oo, C., Kalbag, S.S., 2016. Leveraging the attributes of biologics and small molecules, and releasing the bottlenecks: a new wave of revolution in drug development. *Expert Rev. Clin. Pharmacol.* 9, pp. 747–749. https://doi.org/10.1586/17512433.2016.1160778

Ou-Yang, S., Lu, J., Kong, X., Liang, Z., Luo, C., Jiang, H., 2012. Computational drug discovery. *Acta Pharmacol. Sin.* 33, pp. 1131–1140. https://doi.org/10.1038/aps.2012.109

Paul, S.M., Mytelka, D.S., Dunwiddie, C.T., Persinger, C.C., Munos, B.H., Lindborg, S.R., Schacht, A.L., 2010. How to improve R&D productivity: the pharmaceutical industry's grand challenge. *Nat. Rev. Drug Discov.* 9, pp. 203–214. https://doi.org/10.1038/nrd3078

Payandeh, J., Volgraf, M., 2021. Ligand binding at the protein–lipid interface: strategic considerations for drug design. *Nat. Rev. Drug Discov.* 20, pp. 710–722. https://doi.org/10.1038/s41573-021-00240-2

Pellegatti, M., 2012. Preclinical in vivo ADME studies in drug development: a critical review. *Expert Opin. Drug Metab. Toxicol.* 8, pp. 161–172. https://doi.org/10.1517/17425255.2012.652084

Plunkett, M.J., Ellman, J.A., 1995. Solid-phase synthesis of structurally diverse 1,4-benzodiazepine derivatives using the stille coupling reaction. *J. Am. Chem. Soc.* 117, pp. 3306–3307. https://doi.org/10.1021/ja00116a049

Prasad, K., Kumar, V., 2021. Artificial intelligence-driven drug repurposing and structural biology for SARS-CoV-2. *Curr. Res. Pharmacol. Drug Discov.* 2, p. 100042. https://doi.org/10.1016/j.crphar.2021.100042

Raju, G.S.R., Pavitra, E., Merchant, N., Lee, H., Prasad, G.L.V., Nagaraju, G.P., Huh, Y.S., Han, Y.-K., 2018. Targeting autophagy in gastrointestinal malignancy by using nanomaterials as drug delivery systems. *Cancer Lett.* 419, pp. 222–232. https://doi.org/10.1016/j.canlet.2018.01.044

Ramanan, V.K., Day, G.S., 2023. Anti-amyloid therapies for Alzheimer disease: finally, good news for patients. *Mol. Neurodegener.* 18, p. 42. https://doi.org/10.1186/s13024-023-00637-0

Ratnayake, G., Bain, A.L., Fletcher, N., Howard, C.B., Khanna, K.K., Thurecht, K.J., 2018. RNA interference to enhance radiation therapy: targeting the DNA damage response. *Cancer Lett.* 439, pp. 14–23. https://doi.org/10.1016/j.canlet.2018.09.011

Rescifina, A., Zagni, C., Varrica, M.G., Pistarà, V., Corsaro, A., 2014. Recent advances in small organic molecules as DNA intercalating agents: synthesis, activity, and modeling. *Eur. J. Med. Chem.* 74, pp. 95–115. https://doi.org/10.1016/j.ejmech.2013.11.029

Rester, U., 2008. From virtuality to reality - Virtual screening in lead discovery and lead optimization: a medicinal chemistry perspective. *Curr. Opin. Drug Discov. Devel.* 11, pp. 559–568.

Reymond, J., Ruddigkeit, L., Blum, L., van Deursen, R., 2012. The enumeration of chemical space. *WIREs Comput. Mol. Sci.* 2, pp. 717–733. https://doi.org/10.1002/wcms.1104

Roberts, A., Huang, D., 2017. Targeting BCL2 with BH3 mimetics: basic science and clinical application of venetoclax in chronic lymphocytic leukemia and related B cell malignancies. *Clin. Pharmacol. Ther.* 101, pp. 89–98. https://doi.org/10.1002/cpt.553

Roessler, H.I., Knoers, N.V.A.M., van Haelst, M.M., van Haaften, G., 2021. Drug repurposing for rare diseases. *Trends Pharmacol. Sci.* 42, pp. 255–267. https://doi.org/10.1016/j.tips.2021.01.003

Rollinger, J.M., Stuppner, H., Langer, T., 2008. Virtual screening for the discovery of bioactive natural products, In: *Natural Compounds as Drugs Volume I.* Birkhäuser Basel, Basel, pp. 211–249. https://doi.org/10.1007/978-3-7643-8117-2_6

Ruiz-Garcia, A., Bermejo, M., Moss, A., Casabo, V.G., 2008. Pharmacokinetics in drug discovery. *J. Pharm. Sci.* 97, pp. 654–690. https://doi.org/10.1002/jps.21009

Sadybekov, A. V., Katritch, V., 2023. Computational approaches streamlining drug discovery. *Nature* 616, pp. 673–685. https://doi.org/10.1038/s41586-023-05905-z

Santos, R., Ursu, O., Gaulton, A., Bento, A.P., Donadi, R.S., Bologa, C.G., Karlsson, A., Al-Lazikani, B., Hersey, A., Oprea, T.I., Overington, J.P., 2017. A comprehensive map of molecular drug targets. *Nat. Rev. Drug Discov.* 16, pp. 19–34. https://doi.org/10.1038/nrd.2016.230

Schenone, M., Dančík, V., Wagner, B.K., Clemons, P.A., 2013. Target identification and mechanism of action in chemical biology and drug discovery. *Nat. Chem. Biol.* 9, pp. 232–240. https://doi.org/10.1038/nchembio.1199

Schneider, P., Walters, W.P., Plowright, A.T., Sieroka, N., Listgarten, J., Goodnow, R.A., Fisher, J., Jansen, J.M., Duca, J.S., Rush, T.S., Zentgraf, M., Hill, J.E., Krutoholow, E., Kohler, M., Blaney, J., Funatsu, K., Luebkemann, C., Schneider, G., 2020. Rethinking drug design in the artificial intelligence era. *Nat. Rev. Drug Discov.* 19, pp. 353–364. https://doi.org/10.1038/s41573-019-0050-3

Schreiber, S.L., 2005. Small molecules: the missing link in the central dogma. *Nat. Chem. Biol.* 1, pp. 64–66. https://doi.org/10.1038/nchembio0705-64

Schreiber, S.L., 2003. The small-molecule approach to biology. *Chem. Eng. News* 81, pp. 51–61.

Shen, A., 2010. Allosteric regulation of protease activity by small molecules. *Mol. Biosyst.* 6, p. 1431. https://doi.org/10.1039/c003913f

Shoichet, B.K., 2004. Virtual screening of chemical libraries. *Nature* 432, pp. 862–865. https://doi.org/10.1038/nature03197

Singh, M., Tam, B., Akabayov, B., 2018. NMR-fragment based virtual screening: a brief overview. *Molecules* 23, p. 233. https://doi.org/10.3390/molecules23020233

Smith, C.G., Vane, J.R., 2003. The discovery of Captopril. *FASEB J.* 17, pp. 788–789. https://doi.org/10.1096/fj.03-0093life

Smith, D.A., 2013. Evolution of ADME science: where else can modeling and simulation contribute? *Mol. Pharm.* 10, pp. 1162–1170. https://doi.org/10.1021/mp3005319

Southey, M.W.Y., Brunavs, M., 2023. Introduction to small molecule drug discovery and preclinical development. *Front. Drug Discov.* 3. https://doi.org/10.3389/fddsv.2023.1314077

Spandl, R.J., Bender, A., Spring, D.R., 2008. Diversity-oriented synthesis; a spectrum of approaches and results. *Org. Biomol. Chem.* 6, p. 1149. https://doi.org/10.1039/b719372f

Stocks, M., 2013. The small molecule drug discovery process – from target selection to candidate selection, In: *Introduction to Biological and Small Molecule Drug Research and Development*. Elsevier, pp. 81–126. https://doi.org/10.1016/B978-0-12-397176-0.00003-0

Stockwell, B.R., 2000. Chemical genetics: ligand-based discovery of gene function. *Nat. Rev. Genet.* 1, pp. 116–125. https://doi.org/10.1038/35038557

Sun, X., Gao, H., Yang, Y., He, M., Wu, Y., Song, Y., Tong, Y., Rao, Y., 2019. PROTACs: great opportunities for academia and industry. *Signal Transduct. Target. Ther.* 4, p. 64. https://doi.org/10.1038/s41392-019-0101-6

Swinney, D.C., Anthony, J., 2011. How were new medicines discovered? *Nat. Rev. Drug Discov.* 10, pp. 507–519. https://doi.org/10.1038/nrd3480

Thakkar, A., Chadimová, V., Bjerrum, E.J., Engkvist, O., Reymond, J.-L., 2021. Retrosynthetic accessibility score (RAscore) – rapid machine learned synthesizability classification from AI driven retrosynthetic planning. *Chem. Sci.* 12, pp. 3339–3349. https://doi.org/10.1039/D0SC05401A

The State of the Art of Chemical Biology, 2009. *ChemBioChem* 10, pp. 16–29. https://doi.org/10.1002/cbic.200800758

Vamathevan, J., Clark, D., Czodrowski, P., Dunham, I., Ferran, E., Lee, G., Li, B., Madabhushi, A., Shah, P., Spitzer, M., Zhao, S., 2019. Applications of machine learning in drug discovery and development. *Nat. Rev. Drug Discov.* 18, pp. 463–477. https://doi.org/10.1038/s41573-019-0024-5

Vanhaelen, Q., Lin, Y.-C., Zhavoronkov, A., 2020. The advent of generative chemistry. *ACS Med. Chem. Lett.* 11, pp. 1496–1505. https://doi.org/10.1021/acsmedchemlett.0c00088

Walsh, D.P., Chang, Y.-T., 2006. Chemical genetics. *Chem. Rev.* 106, pp. 2476–2530. https://doi.org/10.1021/cr0404141

Walters, W.P., Namchuk, M., 2003. Designing screens: how to make your hits a hit. *Nat. Rev. Drug Discov.* 2, pp. 259–266. https://doi.org/10.1038/nrd1063

Walters, W.P., Stahl, M.T., Murcko, M.A., 1998. Virtual screening—an overview. *Drug Discov. Today* 3, pp. 160–178. https://doi.org/10.1016/S1359-6446(97)01163-X

Wang, D., Hu, G., Du, Y., Zhang, C., Lu, Q., Lv, N., Luo, S., 2017. Aberrant activation of hedgehog signaling promotes cell proliferation via the transcriptional activation of forkhead Box M1 in colorectal cancer cells. *J. Exp. Clin. Cancer Res.* 36, p. 23. https://doi.org/10.1186/s13046-017-0491-7

Wells, J.A., McClendon, C.L., 2007. Reaching for high-hanging fruit in drug discovery at protein–protein interfaces. *Nature* 450, pp. 1001–1009. https://doi.org/10.1038/nature06526

Wetzel, S., Bon, R.S., Kumar, K., Waldmann, H., 2011. Biology-oriented synthesis. *Angew. Chemie Int. Ed.* 50, pp. 10800–10826. https://doi.org/10.1002/anie.201007004

Wu, G., Ma, Z., Cheng, Y., Hu, W., Deng, C., Jiang, S., Li, T., Chen, F., Yang, Y., 2018. Targeting Gas6/TAM in cancer cells and tumor microenvironment. *Mol. Cancer* 17, p. 20. https://doi.org/10.1186/s12943-018-0769-1

Ylä-Herttuala, S., 2017. The pharmacology of gene therapy. *Mol. Ther.* 25, pp. 1731–1732. https://doi.org/10.1016/j.ymthe.2017.07.007

Young, S.S., Ge, N., 2004. Design of diversity and focused combinatorial libraries in drug discovery. *Curr. Opin. Drug Discov. Devel.* 7, pp. 318–324.

Zacharioudakis, E., Gavathiotis, E., 2022. Targeting protein conformations with small molecules to control protein complexes. *Trends Biochem. Sci.* 47, pp. 1023–1037. https://doi.org/10.1016/j.tibs.2022.07.002

Zhang, M.-Q., Wilkinson, B., 2007. Drug discovery beyond the 'rule-of-five.' *Curr. Opin. Biotechnol.* 18, pp. 478–488. https://doi.org/10.1016/j.copbio.2007.10.005

Zheng, M., Huang, J., Tong, A., Yang, H., 2019. Oncolytic viruses for cancer therapy: barriers and recent advances. *Mol. Ther. - Oncolytics* 15, pp. 234–247. https://doi.org/10.1016/j.omto.2019.10.007

7 Neurotrophic Factors and Small Molecule Mimetics

Ishtiaq Jeelani, Aamir Sharif, Urmilla John, Alisha Sheikh, Saeed Ahmed, and Allah Nawaz

ABBREVIATIONS

AD Alzheimer's Disease
ALS Amyotrophic Lateral Sclerosis
BBB Blood–Brain Barrier
BDNF Brain-Derived Neurotrophic Factor
CNTF Ciliary Neurotrophic Factor
CREB cAMP Response Element-Binding Protein
CSF Cerebrospinal Fluid
GDNF Glial Cell Line-Derived Neurotrophic Factor
HD Huntington's Disease
ICV Intracerebroventricular
IGFs Insulin-like Growth Factors
LTD Long-Term Depression
LTP Long-Term Potentiation
NGF Nerve Growth Factor
NT-3 Neurotrophin-3
PD Parkinson's Disease

7.1 INTRODUCTION

Neurodegenerative diseases, characterized by the progressive deterioration of neuron structure and function, ultimately lead to neuronal death. Prominent among these conditions are Alzheimer's disease (AD), Parkinson's disease (PD), Huntington's disease (HD), and Amyotrophic lateral sclerosis (ALS), each presenting significant challenges to modern medicine and society. The global burden of neurodegenerative diseases is substantial and escalating. Recent estimates suggest that over 50 million people worldwide are currently living with dementia, with AD accounting for 60–70% of these cases. This number is projected to nearly triple by 2050, underscoring the urgent need for effective therapeutic interventions (Nichols and Vos, 2021).

At the molecular level, these neurodegenerative disorders share several pathological hallmarks. Wilson et al. (2023) proposed a comprehensive framework identifying key mechanisms underlying neurodegeneration (Wilson et al., 2023). A central mechanism is protein misfolding and aggregation, where abnormal protein aggregates accumulate within the brain. In AD, this is evidenced by beta-amyloid plaques and tau tangles, while, in PD, alpha-synuclein aggregates form Lewy bodies (Sweeney et al., 2017). Recent research has further elucidated the role of these aggregates. Studies have demonstrated that beta-amyloid oligomers are not only toxic to neurons but also contribute to the spread of tau pathology, suggesting a synergistic relationship between these two proteins in AD progression (Henstridge et al., 2019). In PD, novel findings revealed that alpha-synuclein aggregates might induce mitochondrial DNA damage, leading to cellular energy deficits and exacerbating neurodegeneration (Wong and Krainc, 2017).

DOI: 10.1201/9781003520610-7

Mitochondrial dysfunction remains a critical factor, with impaired energy metabolism and increased oxidative stress, resulting from mitochondrial abnormalities and contributing significantly to neuronal death (Golpich et al., 2017). Recent studies have highlighted the potential of targeting mitochondrial biogenesis as a therapeutic strategy in PD, showing that enhancing mitochondrial function could mitigate neurodegeneration (Bose and Beal, 2016). Neuroinflammation, characterized by the chronic activation of the brain's immune cells (microglia and astrocytes), also exacerbates neuronal damage and accelerates disease progression (Heneka et al., 2015). New insights have shown that microglial activation is not merely a bystander effect but plays an active role in synaptic pruning and neuronal death in AD (Hansen et al., 2018). The recognition of these shared pathological mechanisms, combined with insights from recent research, has led to a more holistic approach to understanding these disorders and developing targeted therapies.

7.1.1 Importance of Neurotrophic Factors in Neuroprotection

Neurotrophic factors play a crucial role in neuroprotection, offering significant potential for treating neurodegenerative diseases. These endogenous proteins promote neuronal survival, growth, and plasticity, making them attractive candidates for therapeutic interventions. Recent studies have highlighted several key aspects of neurotrophic factors in neuroprotection:

i. **Neuroprotective mechanisms**: Neurotrophic factors, particularly brain-derived neurotrophic factor (BDNF), glial cell line-derived neurotrophic factor (GDNF), and nerve growth factor (NGF), activate pro-survival signaling pathways in neurons. For instance, Bothwell (2019) demonstrated that BDNF binding to its the high-affinity tropomyosin-related kinase receptor (TrkB) initiates signaling cascades that promote neuronal survival and synaptic plasticity (Bothwell, 2016).

ii. **Mitochondrial function**: A study by Hong et al. (2008) revealed that GDNF enhances mitochondrial function in dopaminergic neurons, potentially offering protection against oxidative stress in PD models (Hong et al., 2008).

iii. **Neuroinflammation modulation**: Neurotrophic factors have been shown to modulate neuroinflammatory responses. Fang et al. (2019) found that BDNF suppresses microglial activation and reduces pro-inflammatory cytokine production in a mouse model of AD (Fang et al., 2019).

iv. **Synaptic plasticity:** BDNF, in particular, plays a crucial role in synaptic plasticity. Leal et al. (2023) demonstrated that BDNF-TrkB signaling enhances long-term potentiation (LTP) in the hippocampus, which is essential for learning and memory processes (Leal et al., 2017).

v. **Neurogenesis promotion**: Riberioand Xapelli (2021) showed that neurotrophic factors stimulate adult neurogenesis in the hippocampus and subventricular zone, potentially offering a regenerative approach to neurodegenerative diseases (Ribeiro and Xapelli, 2021).

vi. **Axonal regeneration**: In spinal cord injury models, Keefe et al. (2017) found that a combination of BDNF and NT-3 promotes axonal regeneration and functional recovery (Keefe et al., 2017).

vii. **Protection against excitotoxicity**: Neurotrophic factors have been shown to protect neurons against glutamate-induced excitotoxicity. A study by Nicole et al. (2001) demonstrated that GDNF reduces calcium influx and subsequent neuronal death in an in vitro model of stroke (Nicole et al., 2001; Rehman et al., 2023).

viii. **Epigenetic regulation**: Emerging evidence suggests that neurotrophic factors influence epigenetic mechanisms. Karpova (2014) showed that BDNF regulates histone acetylation in the prefrontal cortex, affecting gene expression related to neuroplasticity (Karpova, 2014).

Despite these promising findings, translating the neuroprotective effects of neurotrophic factors into clinical treatments remains challenging. Issues such as blood–brain barrier (BBB) penetration and potential side effects have hindered progress. However, novel delivery methods are being explored. For instance, Bondarenko and Saarma, (2021) developed a nanoparticle-based delivery system for BDNF that showed improved brain penetration and neuroprotective effects in a PD model (Bondarenko and Saarma, 2021). In conclusion, the importance of neurotrophic factors in neuroprotection is well established and continues to be a focus of intense research. Their multifaceted effects on neuronal survival, plasticity, and regeneration make them promising candidates for treating various neurodegenerative diseases. Ongoing research is focused on overcoming delivery challenges and optimizing their therapeutic potential.

7.1.2 Challenges in Current Neurodegenerative Therapies

Current therapies for neurodegenerative diseases face several significant challenges, which underscore the need for novel approaches such as small molecules and neurotrophic factors:

i. **Limited disease-modifying effects**: Most existing treatments for neurodegenerative diseases focus on symptom management rather than addressing the underlying pathology. For instance, in AD, cholinesterase inhibitors like donepezil improve cognitive symptoms but do not halt disease progression (Cummings et al., 2021). This limitation highlights the need for therapies that can modify disease course, a potential strength of neurotrophic factors.

ii. **BBB penetration**: Many potential therapeutic agents, including large protein molecules, struggle to cross the BBB effectively. This challenge is particularly relevant for neurotrophic factors. However, small molecule mimetics of neurotrophic factors offer a promising solution. For example, Jiang et al. (2013) developed small molecule TrkB agonists that showed improved BBB penetration and neuroprotective effects in animal models of HD (Jiang et al., 2013).

iii. **Targeting multiple pathological mechanisms:** Neurodegenerative diseases often involve multiple pathological processes. Current single-target approaches have shown limited success. Neurotrophic factors, with their pleiotropic effects on neuronal survival, synaptic plasticity, and neuroinflammation, offer a multifaceted approach to neuroprotection (Bothwell, 2016).

iv. **Long-term safety and efficacy:** The chronic nature of neurodegenerative diseases necessitates long-term treatment. However, many current therapies have side effects that limit their long-term use. Small molecules designed to mimic neurotrophic factors could potentially offer improved safety profiles for long-term administration (Longo and Massa, 2013).

v. **Delivery and distribution:** Ensuring adequate distribution of therapeutic agents throughout affected brain regions remains a significant challenge. Small molecules and engineered delivery systems for neurotrophic factors are being explored to overcome this hurdle. For instance, Géral et al. (2013) developed a nanoparticle-based delivery system for BDNF that showed improved brain penetration and distribution in a PD model (Géral et al., 2013).

vi. **Timing of intervention:** Many neurodegenerative diseases are diagnosed after significant neuronal loss has occurred, limiting the effectiveness of current therapies. Neurotrophic factors and their small molecule mimetics offer the potential for both neuroprotection and neuroregeneration, potentially expanding the therapeutic window (Sampaio et al., 2017).

vii. **Personalized medicine approaches:** Current treatments often follow a one-size-fits-all approach, overlooking individual variations in disease pathology. Small molecules and neurotrophic factor-based therapies could potentially be tailored to specific genetic or molecular profiles of patients, aligning with the growing trend toward precision medicine in neurodegenerative disease treatment (Hampel et al., 2017).

These challenges collectively highlight the limitations of current therapeutic approaches and justify the exploration of small molecules and neurotrophic factors as promising avenues for neurodegenerative disease treatment. Their potential to address multiple pathological mechanisms, overcome BBB penetration issues, and offer both neuroprotective and neurodegenerative effects positions them as valuable candidates for next-generation therapies.

7.2 NEUROTROPHIC FACTORS: MECHANISMS AND FUNCTIONS

Neurotrophic factors are a family of endogenous proteins that play crucial roles in the development, maintenance, and survival of neurons in both the central nervous system (CNS) and the peripheral nervous system. These molecular guardians of neuronal health have captivated researchers for decades, offering tantalizing possibilities for treating neurodegenerative diseases and promoting neuronal repair. The story of neurotrophic factors began with the groundbreaking discovery of NGF by Rita Levi-Montalcini and Stanley Cohen in the 1950s, a feat that earned them the Nobel Prize in Physiology or Medicine in 1986. Since then, our understanding of these remarkable proteins has expanded dramatically, revealing a complex network of molecular interactions that regulate neuronal life and death. At their core, neurotrophic factors function by binding to specific receptor tyrosine kinases on the surface of neurons. This binding initiates a cascade of intracellular signaling events that ultimately influence gene expression, protein synthesis, and cellular metabolism. The result is a powerful neuroprotective effect that can promote neuronal survival, stimulate neurite outgrowth, enhance synaptic plasticity, and even induce the birth of new neurons through neurogenesis. Recent research has shed light on the intricate mechanisms by which neurotrophic factors exert their effects. For instance, Bothwell (2016) elucidated the complex interplay between neurotrophic factors and their receptors, revealing how different receptor conformations can lead to diverse cellular responses (Bothwell, 2016). This work has highlighted the potential for developing targeted therapies that can activate specific signaling pathways while minimizing unwanted effects. One of the most fascinating aspects of neurotrophic factor function is their role in synaptic plasticity. Park and Poo (2012) demonstrated how BDNF, one of the most well-studied neurotrophic factors, modulates synaptic transmission and promotes the formation of new synapses (Park and Poo, 2012). This process is crucial for learning and memory, and disruptions in BDNF signaling have been implicated in various cognitive disorders. Beyond their role in normal neuronal function, neurotrophic factors have shown remarkable potential in neuroprotection and neuroregeneration. In a groundbreaking study, Keefe et al. (2017) used a combination of BDNF and NT-3 to promote axonal regeneration after spinal cord injury in animal models, demonstrating functional recovery that was previously thought impossible (Keefe et al., 2017). The neuroprotective effects of neurotrophic factors extend to various neurodegenerative diseases. For example, Palasz et al. (2020) reviewed the potential of GDNF in PD, highlighting its ability to protect dopaminergic neurons from degeneration and even restore function in damaged neurons (Palasz et al., 2020).

7.2.1 Types of Neurotrophic Factors

Neurotrophic factors can be broadly classified into several families, each with unique characteristics and functions:

i. **Neurotrophins:** This family includes some of the most well-known neurotrophic factors:
   a. **NGF:** The first discovered neurotrophic factor, NGF is crucial for the survival and maintenance of sympathetic and sensory neurons. A study by Minnone et al. (2017) has revealed its unexpected role in immune system regulation, opening new avenues for treating inflammatory disorders (Minnone et al., 2017).
   b. **BDNF:** BDNF is widely expressed in the brain and plays a vital role in synaptic plasticity, learning, and memory. Notaras and van den Buuse (2020) provided a

comprehensive review of BDNF's role in psychiatric and neurodegenerative disorders, highlighting its potential as a biomarker and therapeutic target (Notaras and van den Buuse, 2020).

   c. **Neurotrophin-3 (NT-3):** NT-3 is crucial for the survival and differentiation of new neurons during neurogenesis. A study by Jiang et al. (2011) demonstrated NT-3's potential in promoting functional recovery after stroke, showcasing its regenerative capabilities (Jiang et al., 2011).
   d. **Neurotrophin-4/5 (NT-4/5):** While less studied than other neurotrophins, NT-4/5 has shown promise in protecting retinal ganglion cells. Guo et al. (2021) revealed its potential in treating glaucoma, a leading cause of irreversible blindness (Guo et al., 2021).

ii. **GDNF Family Ligands (GFLs):** This family includes:
   a. **GDNF:** GDNF is particularly important for the survival of dopaminergic neurons, making it a key focus in PD research. Gash et al. (2020) provided insights into the complex biology of GDNF, including its potential off-target effects that need to be considered in therapeutic development (Gash et al., 2020).
   b. **Neurturin, Artemin, and Persephin**: These GFLs have shown neuroprotective effects in various contexts. For instance, Wong et al. (2015) demonstrated the potential of artemin in treating neuropathic pain, offering hope for new analgesic strategies (Wong et al., 2015).

iii. **Neurokines:** This family includes:
   a. **Ciliary Neurotrophic Factor (CNTF):** CNTF has shown promise in protecting motor neurons. A recent study by Pasquin et al. (2015) revealed its potential in treating ALS, highlighting the complex interplay between CNTF and neuroinflammation (Pasquin et al., 2015).
   b. **Leukemia Inhibitory Factor (LIF):** LIF has demonstrated neuroprotective effects in multiple sclerosis models. Janssens et al. (2015) elucidated its role in promoting oligodendrocyte survival and myelination, opening new therapeutic possibilities for demyelinating diseases (Janssens et al., 2015).

iv. **Other Neurotrophic Factors:**
   a. **Insulin-Like Growth Factors (IGFs):** IGFs play crucial roles in neuronal survival and have shown potential in treating neurodegenerative diseases. Wrigley et al. (2017) provided evidence for IGF-1's neuroprotective effects in AD models, suggesting a potential therapeutic avenue (Wrigley et al., 2017).
   b. **Vascular Endothelial Growth Factor (VEGF):** While primarily known for its role in angiogenesis, VEGF has also demonstrated neurotrophic properties. Mackenzie and Ruhrberg (2012) reviewed VEGF's diverse roles in nervous system development and function, highlighting its potential in treating neurological disorders (Mackenzie and Ruhrberg, 2012).

This diverse array of neurotrophic factors, each with its unique properties and functions, offers a rich landscape for therapeutic exploration in neurodegenerative diseases. As our understanding of these molecular guardians continues to grow, so too does the potential for developing targeted, effective treatments for some of the most challenging neurological disorders of our time.

7.2.2 Mechanisms of Action in Neuronal Survival and Repair

Neurotrophic factors play crucial roles in neuronal survival, growth, and repair through several key mechanisms. These endogenous proteins, which include neurotrophins and other growth factors, act via complex signaling pathways to promote neuronal health and function. One of the primary mechanisms of neurotrophic factor action is through receptor binding and subsequent activation of intracellular signaling cascades. The neurotrophin family, NGF, BDNF,

NT-3, and NT-4, binds to two types of receptors: the high-affinity tropomyosin-related kinase (Trk) receptors and the low-affinity p75 neurotrophin receptor (Skaper, 2012). This binding initiates various signaling pathways that promote neuronal survival and growth. The activation of Trk receptors leads to the stimulation of several critical pathways. The Ras/MAPK pathway promotes neurite outgrowth and differentiation, while the PI3K/Akt pathway supports neuronal survival. Additionally, the PLCγ pathway plays a role in regulating synaptic plasticity (Huang and Reichardt, 2001). These pathways collectively contribute to the neuroprotective and neurotrophic effects of these factors. Retrograde signaling is another crucial mechanism by which neurotrophic factors exert their effects. Typically, these factors are released from target cells or postsynaptic neurons and are taken up by presynaptic axon terminals. They are then retrogradely transported to the neuronal soma, providing information about the quality of the neuron's connection with its target (Castrén, 2014). This retrograde signaling is vital for neuronal survival during development and for maintaining neuronal connections in adulthood. Upon reaching the cell body, neurotrophic factors can influence gene expression, leading to the synthesis of proteins necessary for neuronal survival, growth, and differentiation. For example, BDNF activation of TrkB receptors can lead to the phosphorylation and activation of the cAMP response element-binding protein (CREB), a transcription factor that regulates the expression of genes involved in neuronal plasticity and survival (Park and Poo, 2012). Neurotrophic factors, particularly BDNF, play a crucial role in synaptic plasticity. They can modulate both LTP and long-term depression, processes fundamental to learning and memory. This modulation occurs through the regulation of synaptic strength and number (Lu et al., 2014). In the context of neuroprotection, neurotrophic factors have been shown to protect neurons from various insults. For instance, BDNF has demonstrated protective effects against excitotoxicity, oxidative stress, and apoptosis (Bothwell, 2016). This neuroprotective function is particularly relevant in the context of neurodegenerative diseases and brain injuries. Some neurotrophic factors, like BDNF, can promote neurogenesis in specific brain regions, even in the adult brain. They also play crucial roles in neuronal differentiation during development (Castrén, 2014). This ability to stimulate the growth of new neurons and support their differentiation has significant implications for regenerative therapies in neurological disorders. Neurotrophic factors also stimulate the growth and branching of both axons and dendrites. This function is crucial not only during development but also in neuronal repair processes following injury (Huang and Reichardt, 2001). Lastly, these factors can modulate neurotransmitter release at synapses, thereby influencing synaptic transmission and plasticity (Park and Poo, 2012). This modulation of neurotransmitter dynamics contributes to the overall impact of neurotrophic factors on neuronal function and communication. Understanding these mechanisms has significant implications for developing therapies for neurodegenerative diseases, psychiatric disorders, and brain injuries. By targeting these pathways, researchers aim to enhance neuronal survival, promote brain repair, and improve brain plasticity in various pathological conditions.

7.2.3 Role in Synaptic Plasticity and Cognitive Function

Neurotrophic factors, especially BDNF, play crucial roles in synaptic plasticity and cognitive function. BDNF has emerged as a key mediator of synaptic plasticity, influencing both the structure and the function of synapses in ways that are fundamental to learning and memory processes. At the synaptic level, BDNF acts as a powerful modulator of both short-term and long-term plasticity. In the hippocampus, a brain region critical for memory formation, BDNF has been shown to enhance LTP, a cellular model of learning and memory (Cunha et al., 2010). This enhancement occurs through multiple mechanisms, including the facilitation of synaptic transmission and the modulation of *N*-Methyl-D-Aspartate receptor function (Leal et al., 2017). BDNF's effects on synaptic plasticity are both rapid and long-lasting. Acute application of BDNF can quickly potentiate synaptic transmission, while longer-term exposure leads to more persistent changes in synaptic strength (Panja and Bramham, 2014). These effects are mediated through BDNF's primary receptor, TrkB, which activates multiple intracellular signaling cascades that ultimately influence both pre- and postsynaptic function. On the presynaptic side, BDNF enhances neurotransmitter release by increasing the number of docked vesicles at the active zone and by facilitating vesicle fusion (Jovanovic et al., 2000). Post-synaptically, BDNF promotes the insertion of alpha-amino-3-hydroxy-5-nethyl-4-isoxazolepropionic acid (AMPA) receptors into the synaptic membrane, thereby increasing the sensitivity of the postsynaptic neuron to neurotransmitter release (Caldeira et al., 2007). Beyond its acute effects on synaptic transmission, BDNF also plays a crucial role in structural plasticity. It promotes the growth and branching of dendrites and axons and increases the formation and maturation of dendritic spines, the primary sites of excitatory synapses in the brain (Horch and Katz, 2002). These structural changes provide a physical substrate for long-term information storage in neural circuits. The importance of BDNF in cognitive function is further underscored by studies linking BDNF levels to cognitive performance. In humans, a common polymorphism in the BDNF gene (Val66Met) has been associated with impaired episodic memory and abnormal hippocampal activation patterns (Egan et al., 2003). This polymorphism affects the activity-dependent secretion of BDNF, highlighting the critical nature of BDNF's regulated release in cognitive processes. Animal studies have provided direct evidence for the role of BDNFs in learning and memory. Mice with reduced BDNF expression show impairments in spatial learning and memory (Linnarsson et al., 1997), while increasing BDNF levels can enhance cognitive performance (Bekinschtein et al., 2008). Interestingly, environmental enrichment and physical exercise, both known to improve cognitive function have been shown to increase BDNF levels in the brain (Vaynman et al., 2004). The cognitive effects of BDNF are not limited to the hippocampus. In the prefrontal cortex, BDNF signaling is crucial for working memory and cognitive flexibility (Sakata et al., 2013). In the amygdala, BDNF modulates fear learning and extinction, processes that are relevant to anxiety disorders and post-traumatic stress disorder (Andero and Ressler, 2012). Understanding the role of BDNF in synaptic plasticity and cognitive function has important implications for the treatment of neurological and psychiatric disorders. For example, many antidepressants have been shown to increase BDNF levels, and this increase is thought to be crucial for their therapeutic effects (Castrén, 2014). Similarly, strategies to enhance BDNF signaling are being explored as potential treatments for neurodegenerative disorders such as AD and PD. In conclusion, BDNF stands out as a critical regulator of synaptic plasticity and cognitive function. Its multifaceted effects on synaptic transmission, structural plasticity, and gene expression make it a key player in the brain's ability to adapt and learn from experience. As our understanding of roles of BDNF continues to grow, so too does the potential for developing new therapeutic strategies targeting this crucial neurotrophic factor.

7.2.4 Clinical Implications of Neurotrophic Factor Deficiency

Neurotrophic factor deficiency has been implicated in various neurological and psychiatric disorders. Understanding these deficiencies and their clinical implications is crucial for developing targeted therapies and improving patient outcomes.

i. **Neurodegenerative diseases**: Neurotrophic factor deficiency plays a significant role in several neurodegenerative diseases:
   a. **AD:** Reduced levels of BDNF have been observed in the hippocampus and neocortex of AD patients (Ferrer et al., 1999). This deficiency may contribute to the loss of synaptic plasticity and neuronal death characteristic of AD. Furthermore, the BDNF Val66Met polymorphism has been associated with an increased risk of AD and more rapid cognitive decline (Lim et al., 2013).

   b. **PD:** GDNF deficiency has been implicated in the pathogenesis of PD. Post-mortem studies have shown reduced GDNF levels in the substantia nigra of PD patients (Chauhan et al., 2001). This deficiency may contribute to the progressive loss of dopaminergic neurons observed in PD.
   c. **HD:** Both BDNF and GDNF levels are reduced in the striatum and cortex of HD patients (Zuccato and Cattaneo, 2007). This deficiency may exacerbate the neuronal loss and cognitive decline associated with HD.

ii. **Psychiatric disorders:** Neurotrophic factor deficiency has been linked to various psychiatric conditions:
   a. **Depression:** Reduced serum BDNF levels have been consistently observed in patients with major depressive disorder (MDD) (Sen et al., 2008). This deficiency may contribute to the reduced neuroplasticity and neurogenesis observed in depression.
   b. **Schizophrenia:** Several studies have reported altered BDNF levels in schizophrenia patients, with most showing a decrease in serum BDNF (Green et al., 2010). This deficiency may contribute to the cognitive deficits and altered neuroplasticity observed in schizophrenia.

iii. **Neurodevelopmental disorders:** Neurotrophic factor deficiency has been implicated in various neurodevelopmental disorders:
   a. **Autism spectrum disorders (ASDs):** Altered BDNF levels have been reported in individuals with ASD, with some studies showing increased levels and others showing decreased levels (Saghazadeh and Rezaei, 2017). These alterations may contribute to the abnormal brain development and function observed in ASD.
   b. **Rett syndrome:** BDNF deficiency has been implicated in the pathogenesis of Rett syndrome, a rare genetic disorder affecting brain development in girls (Katz, 2014). BDNF supplementation has shown promise in animal models of Rett syndrome.

iv. **Peripheral neuropathies:** Neurotrophic factor deficiency has been associated with various peripheral neuropathies:
   a. **Diabetic neuropathy:** Reduced levels of NGF and NT-3 have been observed in diabetic patients with peripheral neuropathy (Pittenger and Vinik, 2003). This deficiency may contribute to the axonal degeneration and impaired regeneration observed in diabetic neuropathy.
   b. **Chemotherapy-induced peripheral neuropathy:** Deficiency in BDNF and other neurotrophic factors has been implicated in the development of chemotherapy-induced peripheral neuropathy (Hoke and Ray, 2014).

v. **Implications for therapy:** Understanding neurotrophic factor deficiencies has led to the development of potential therapeutic strategies:
   a. **Exogenous neurotrophic factor administration:** Direct administration of neurotrophic factors or their mimetics has shown promise in preclinical studies for various disorders (Allen et al., 2013).
   b. **Enhancing endogenous neurotrophic factor production:** Strategies to increase endogenous neurotrophic factor production, such as exercise and environmental enrichment, have shown beneficial effects in various neurological and psychiatric disorders (Cotman, 2002).
   c. **Gene therapy:** Viral vector-mediated delivery of neurotrophic factor genes has shown promise in preclinical studies for neurodegenerative diseases (Kordower and Bjorklund, 2013).

In conclusion, neurotrophic factor deficiency has a wide range of clinical implications across various neurological, psychiatric, and neurodevelopmental disorders. Understanding these deficiencies provides valuable insights into disease pathogenesis and opens up new avenues for therapeutic interventions.

7.3 CHALLENGES IN NEUROTROPHIC FACTOR THERAPY

7.3.1 BBB as a Limitation

The BBB represents one of the most formidable obstacles in the field of neurotrophic factor therapy. This highly selective semipermeable border of endothelial cells acts as a gatekeeper, meticulously regulating the passage of substances from the bloodstream into the CNS. While this barrier is crucial for maintaining brain homeostasis and protecting the CNS from potentially harmful substances, it also presents a significant challenge for the delivery of therapeutic agents, including neurotrophic factors. The structure and function of BBB are complex and multifaceted. Tight junctions between endothelial cells form a nearly impermeable barrier to the paracellular diffusion of most molecules. Additionally, the presence of efflux transporters, such as P-glycoprotein, actively pumps many substances out of the brain, further limiting drug penetration (Abbott et al., 2010). This intricate defense system effectively prevents approximately 98% of small molecule drugs and nearly 100% of large molecule therapeutics from reaching the brain parenchyma (Pardridge, 2005). Neurotrophic factors, being large protein molecules typically ranging from 13 to 24 kDa in size, face particular difficulty in crossing the BBB. For instance, BDNF, a 27 kDa protein, has been shown to have minimal BBB penetration when administered systemically. Studies have estimated that less than 0.1% of peripherally administered BDNF reaches the CNS (Poduslo and Curran, 1996). The challenge posed by the BBB has led to numerous attempts to circumvent or overcome this barrier. Strategies have included direct intracerebral delivery, temporary disruption of the BBB, and the development of BBB-penetrating drug delivery systems. However, each of these approaches comes with its own set of limitations and potential risks. Direct intracerebral delivery, while bypassing the BBB, is highly invasive and carries risks associated with neurosurgical procedures. Temporary BBB disruption techniques, such as the use of hyperosmolar solutions or focused ultrasound, can potentially allow increased drug penetration but also risk allowing entry of unwanted substances into the CNS (Burgess and Hynynen, 2013). The development of BBB-penetrating delivery systems has shown promise but faces challenges in terms of efficiency and specificity. Approaches such as the use of cell-penetrating peptides, nanoparticles, and receptor-mediated transcytosis have demonstrated some success in preclinical studies but have yet to be fully translated to clinical applications (Gabathuler, 2010). In conclusion, the BBB remains a significant limitation in neurotrophic factor therapy, necessitating continued research and innovation in drug delivery strategies. Overcoming this barrier is crucial for realizing the full therapeutic potential of neurotrophic factors in treating various neurological disorders.

7.3.2 Side Effects and Complications in Current Therapeutic Approaches

Current therapeutic approaches for delivering neurotrophic factors face several challenges related to side effects and complications. These issues arise from both the nature of the neurotrophic factors themselves and the delivery methods employed.

i. **Systemic administration**: Systemic administration of neurotrophic factors, such as intravenous or subcutaneous injections has been associated with significant side effects:
   a. **Peripheral effects:** Neurotrophic factors can bind to receptors throughout the body, leading to unwanted effects in non-target tissues. For example, systemic administration of NGF has been associated with hyperalgesia and weight loss in clinical trials (Apfel, 2002).
   b. **Immunogenicity:** Repeated administration of protein-based neurotrophic factors can trigger immune responses, leading to the production of neutralizing antibodies. This not only reduces the efficacy of the treatment but can also cause allergic reactions (Pardridge, 2012).

c. **Limited CNS penetration:** Due to the BBB, only a small fraction of systemically administered neurotrophic factors reach the CNS. This necessitates the use of high doses, which exacerbates peripheral side effects (Pardridge, 2012).

ii. **Direct CNS delivery:** Methods that bypass the BBB by delivering neurotrophic factors directly to the CNS also come with their own set of complications:

a. **Surgical risks:** Intracerebroventricular (ICV) infusion and intraparenchymal injection require invasive neurosurgical procedures, which carry risks such as infection, hemorrhage, and neurological deficits (Kordower et al., 1995).

b. **Catheter-related complications**: Chronic infusion systems can lead to catheter obstruction, infection, or misplacement, potentially interrupting treatment or causing local tissue damage (Nutt et al., 2003).

c. **Limited distribution:** Even when delivered directly to the CNS, neurotrophic factors often have limited diffusion within brain tissue. This can result in high concentrations near the delivery site but subtherapeutic levels in more distant regions (Salvatore et al., 2006).

iii. **Gene therapy approaches**: While gene therapy offers the potential for long-term, localized production of neurotrophic factors, it also presents unique challenges:

a. **Vector-related toxicity:** Viral vectors used for gene delivery can trigger immune responses or cause direct cellular toxicity (Marks et al., 2010).

b. **Uncontrolled expression:** Continuous, unregulated expression of neurotrophic factors may lead to adverse effects over time, such as aberrant neuronal sprouting or tumor formation (Kordower et al., 2000).

c. **Irreversibility:** Unlike pharmacological approaches, gene therapy is not easily reversible if adverse effects occur (Marks et al., 2010).

iv. **Off-target effects**: Neurotrophic factors often have pleiotropic effects, which can lead to unintended consequences:

a. **Neuroplasticity-related issues:** While promoting neuronal growth and plasticity is generally beneficial, excessive or uncontrolled neuroplasticity can potentially lead to adverse outcomes such as neuropathic pain or epileptogenesis (Pezet and McMahon, 2006).

b. **Tumor promotion:** Some neurotrophic factors, particularly those involved in cell survival signaling, have been implicated in tumor growth and metastasis when overexpressed (Hellweg et al., 2008).

v. **Dosing and timing challenges**: The efficacy and safety of neurotrophic factor therapies can be highly dependent on dosing and timing:

a. **Narrow therapeutic window:** The dose–response relationship for neurotrophic factors is often non-linear, with both insufficient and excessive levels potentially causing adverse effects (Allen et al., 2013).

b. **Temporal considerations**: The timing of neurotrophic factor administration can significantly impact outcomes, particularly in acute conditions like stroke or traumatic brain injury (Géral et al., 2013).

These challenges highlight the complexity of developing safe and effective neurotrophic factor therapies. Ongoing research is focused on developing novel delivery strategies, such as nanoparticle-based systems and small molecule mimetics, to overcome these limitations and improve the therapeutic potential of neurotrophic factors while minimizing side effects and complications.

7.3.3 Limitations of Traditional Delivery Methods

Traditional delivery methods for neurotrophic factors face several significant limitations that have hindered their clinical application:

i. **BBB penetration:** The BBB represents a major obstacle to the systemic delivery of neurotrophic factors. These proteins are typically large (5–30kDa) and hydrophilic, making it extremely difficult for them to cross the BBB (Pardridge, 2002). Studies have shown that less than 0.1% of peripherally administered neurotrophic factors reach the CNS, necessitating high systemic doses that can lead to unwanted side effects (Poduslo and Curran, 1996).
ii. **Short half-life and poor pharmacokinetics:** Neurotrophic factors generally have short in vivo half-lives and poor pharmacokinetic profiles when administered systemically. They are subjected to rapid enzymatic degradation and clearance in the bloodstream, further limiting their ability to reach therapeutic concentrations in the CNS (Pardridge, 2012).
iii. **Invasiveness of direct CNS delivery:** Methods that bypass the BBB by delivering neurotrophic factors directly to the CNS, such as ICV infusion or intraparenchymal injection, are highly invasive. These procedures carry significant risks including infection, hemorrhage, and neurological deficits (Kordower et al., 2000).
iv. **Cerebrospinal fluid (CSF) clearance:** Delivery into the CSF, while less invasive than direct parenchymal injection, is limited by significant clearance of CSF into the venous and lymphatic circulation. This rapid clearance reduces the effective concentration of neurotrophic factors reaching target tissues (Pardridge, 2012).
v. **Lack of targeted delivery**: Traditional delivery methods often lack specificity in targeting particular brain regions or cell types. This can lead to off-target effects and reduce the overall efficacy of the treatment (Allen et al., 2013).
vi. **Difficulty in maintaining therapeutic levels**: Due to the short half-life and poor distribution of neurotrophic factors, maintaining consistent therapeutic levels over time is challenging with traditional delivery methods. This is particularly problematic for chronic neurodegenerative conditions that require long-term treatment (Jain, 2012).
vii. **Potential for immune responses**: Repeated administration of protein-based neurotrophic factors can trigger immune responses, leading to the production of neutralizing antibodies. This not only reduces the efficacy of the treatment but can also cause allergic reactions (Pardridge, 2012).
viii. **Dosing challenges:** The dose–response relationship for neurotrophic factors is often non-linear, with both insufficient and excessive levels potentially causing adverse effects. Traditional delivery methods make it difficult to achieve and maintain optimal therapeutic concentrations (Allen et al., 2013).

These limitations have led to the exploration of novel delivery strategies, including nanoparticle-based systems, gene therapy approaches, and the development of small molecule mimetics of neurotrophic factors. Overcoming these challenges is crucial for realizing the full therapeutic potential of neurotrophic factors in treating neurodegenerative disorders.

7.4 SMALL MOLECULE MIMETICS OF NEUROTROPHIC FACTORS

7.4.1 Introduction to Small Molecule Mimetics

Small molecule mimetics of neurotrophic factors represent a promising therapeutic strategy to overcome the limitations of using full-length neurotrophic proteins as drugs. Neurotrophic factors like BDNF and NGF play crucial roles in neuronal survival, growth, and plasticity, making them attractive candidates for treating neurodegenerative diseases (Longo and Massa, 2013). However, the clinical application of neurotrophic factors has been hampered by their poor pharmacokinetic properties, including limited BBB penetration, short half-lives, and potential for unwanted systemic effects (Allen et al., 2013). To address these challenges, researchers have focused on developing small molecule compounds that can mimic the beneficial effects of neurotrophic factors while possessing more favorable drug-like properties. These mimetics are designed to modulate neurotrophic factor receptors and downstream signaling pathways in ways that replicate key aspects of neurotrophic factor activity (Longo et al., 2007). Some key advantages of small molecule mimetics include:

i. Improved BBB penetration
ii. Enhanced metabolic stability and longer half-lives
iii. Potential for oral bioavailability
iv. More specific targeting of desired receptor subtypes or signaling pathways
v. Reduced risk of immunogenicity compared to protein therapeutics

Small molecule mimetics can be broadly classified into two main categories:

i. **Peptide mimetics:** Short peptide sequences derived from neurotrophic factor loop regions that interact with receptors
ii. **Non-peptide small molecules:** Synthetic compounds identified through screening or rational drug design approaches

Early efforts to develop mimetics focused on peptide-based compounds corresponding to specific domains of neurotrophic factors like NGF (Longo et al., 1997). More recently, high-throughput screening and computational approaches have enabled the discovery of diverse non-peptide small molecules that can activate neurotrophic signaling (Jang et al., 2010). Some notable examples of small molecule mimetics that have shown promise in preclinical studies include:

i. 7,8-dihydroxyflavone: A BDNF receptor (TrkB) agonist (Jang et al., 2010)
ii. LM22A compounds: Small molecules that activate TrkB signaling (Massa et al., 2010)
iii. P021: A CNTF peptide mimetic (Li et al., 2010)

These compounds have demonstrated neuroprotective and neurorestorative effects in animal models of various neurodegenerative conditions, highlighting the therapeutic potential of the small molecule mimetic approach (Longo and Massa, 2013). As research in this area progresses, small molecule mimetics of neurotrophic factors offer an exciting avenue for developing novel treatments for AD, PD, and other neurological disorders where promoting neuronal health and plasticity could have significant benefits (Allen et al., 2013).

7.4.2 Development and Design of Neurotrophic Mimetics

The development of neurotrophic mimetics represents a significant advancement in the field of neurodegenerative disease research. These small molecules or peptide-like compounds are designed to modulate the action of neurotrophin receptors, mimicking the beneficial effects of neurotrophic factors like NGF and BDNF while overcoming their limitations as therapeutic agents (Longo and Massa, 2013). The rationale behind developing neurotrophic mimetics stems from the poor pharmacokinetic properties of full-length neurotrophic proteins. These proteins have limited BBB penetration, short half-lives, and potential for unwanted systemic effects, which have hindered their clinical use (Allen et al., 2013). Small molecule mimetics aim to overcome these issues while retaining the therapeutic benefits of neurotrophic factors. Several design strategies have been employed in the development of neurotrophic mimetics:

i. **Peptide mimetics:** These are short peptide sequences derived from neurotrophic factor loop regions that interact with receptors. For example, O'Leary and Hughes (2003) designed potent peptide mimetics of BDNF by focusing on the loop 2 region of the protein (O'Leary and Hughes, 2003).

ii. **Non-peptide small molecules**: These are synthetic compounds identified through screening or rational drug design approaches. Jang et al. (2010) discovered 7,8-dihydroxyflavone, a selective TrkB agonist, through this method (Jang et al., 2010).
iii. **Natural product-inspired**: Some natural compounds have been found to have neurotrophic-like activity and have inspired the design of synthetic mimetics (Longo and Massa, 2013). Early approaches focused on synthesizing peptide compounds corresponding to specific domains of neurotrophins like NGF. The first was a cyclized dimeric peptide (P7) that showed neurotrophic effects (Longo et al., 1997). More recent efforts use high-throughput screening and computational approaches to discover diverse non-peptide small molecules that can activate neurotrophin signaling (Jang et al., 2010). Key goals in the design of neurotrophic mimetics include:
    a. Improving BBB penetration
    b. Enhancing metabolic stability and half-life
    c. Allowing for oral bioavailability
    d. Providing more specific targeting of desired receptor subtypes/pathways
    e. Reducing immunogenicity compared to protein therapeutics

Examples of promising mimetics that have emerged from these design efforts include:

a. **7,8-dihydroxyflavone**: A BDNF receptor (TrkB) agonist (Jang et al., 2010)
b. **LM22A compounds**: Small molecules that activate TrkB signaling (Massa et al., 2010)
c. **P021**: A CNTF peptide mimetic (Li et al., 2010)

These mimetics are designed to modulate neurotrophin receptors (e.g., TrkA, TrkB, p75NTR) and downstream signaling pathways in ways that replicate key aspects of neurotrophic factor activity. The goal is to develop compounds that promote neurogenesis, neuroprotection, and synaptic plasticity for treating neurodegenerative and psychiatric disorders (Longo et al., 2007). Recent advancements in the field include the development of conformationally defined and proteolytically stable circular mimetics of BDNF (Fletcher et al., 2008) and the design of small molecules that can selectively activate specific Trk receptor subtypes (Longo et al., 1997). In conclusion, the development and design of neurotrophic mimetics represent a promising approach to harnessing the therapeutic potential of neurotrophic factors while overcoming the limitations that have hindered their clinical application. Ongoing research aims to optimize the design of these compounds for improved efficacy and drug-like properties, potentially leading to novel treatments for neurodegenerative diseases.

7.4.3 Mechanisms of Action: Mimicking Neurotrophic Factor Signaling

Small molecule mimetics of neurotrophic factors have emerged as a promising approach to harness the beneficial effects of neurotrophic signaling in the treatment of neurodegenerative diseases and other neurological disorders. These mimetics are designed to activate the same signaling pathways as endogenous neurotrophic factors, such as BDNF and NGF, but with improved pharmacokinetic properties and reduced side effects.

i. **Activation of Trk Receptors**

The primary mechanism by which neurotrophic factors exert their effects is through the activation of Trk receptors. For instance, BDNF primarily signals through TrkB, while NGF signals through TrkA (Rahman et al., 2023). Small molecule mimetics are designed to selectively bind to these receptors, mimicking the action of their larger counterparts. For example, compounds like 7,8-dihydroxyflavone have been shown to act as TrkB agonists, promoting neuronal survival and differentiation (Jang et al., 2010).

ii. **Modulation of Intracellular Signaling Pathways**

Upon binding to their respective receptors, neurotrophic factors activate several intracellular signaling cascades that lead to various biological responses, including neuronal survival, growth, and synaptic plasticity. These pathways include the Ras/MAPK pathway, the PI3K/Akt pathway, and the PLCγ pathway (Yoshii and Constantine-Paton, 2010). Small molecule mimetics aim to activate these pathways in a manner similar to neurotrophic factors. For instance, studies have shown that small molecule TrkB agonists can activate downstream signaling pathways, leading to neuroprotective effects comparable to those of BDNF (Massa et al., 2010).

iii. **Influence on Gene Expression**

Neurotrophic factors also play a crucial role in regulating gene expression through their signaling pathways. Activation of Trk receptors can lead to the phosphorylation of transcription factors such as CREB, which is essential for the transcription of genes involved in neuronal survival and plasticity (Esvald et al., 2020). Small molecule mimetics that effectively activate Trk receptors can similarly influence gene expression patterns, promoting neuroprotective and neurogenic effects.

iv. **Synaptic Plasticity and Neuronal Health**

Small molecule mimetics have been shown to enhance synaptic plasticity, a critical process for learning and memory. For example, BDNF is known to facilitate LTP in the hippocampus, a cellular mechanism underlying memory formation (Leal et al., 2017). Mimetics that activate TrkB can replicate these effects, enhancing synaptic efficacy and promoting the formation of new synapses (Jang et al., 2010).

v. **Neuroprotection in Neurodegenerative Models**

Several studies have demonstrated the neuroprotective effects of small molecule mimetics in preclinical models of neurodegenerative diseases. For instance, a study by Massa et al. (2010) showed that small molecule BDNF mimetics could prevent neuronal degeneration in rodent models of neurodegenerative conditions, demonstrating efficacy comparable to that of BDNF itself (Massa et al., 2010). These findings suggest that small molecule mimetics can effectively mimic the neuroprotective signaling of endogenous neurotrophic factors.

vi. **Therapeutic Potential and Future Directions**

The development of small molecule mimetics offers a novel therapeutic avenue for treating various neurological disorders characterized by neurotrophic factor deficiencies. Given their improved pharmacokinetic profiles and ability to selectively activate specific signaling pathways, these compounds hold promise for enhancing neuronal survival, promoting neurogenesis, and improving synaptic plasticity in conditions such as AD, PD, and depression (Longo and Massa, 2013). The small molecule mimetics of neurotrophic factors represent a significant advancement in the field of neurotherapeutics. By effectively mimicking the signaling mechanisms of endogenous neurotrophic factors, these compounds have the potential to provide new treatment options for a range of neurodegenerative and psychiatric disorders.

7.4.4 Advantages of Traditional Neurotrophic Factors

The development of small molecule mimetics of neurotrophic factors has gained significant attention due to their potential to address the limitations associated with traditional neurotrophic factors such as NGF and BDNF. These advantages can be categorized into several key areas:

i. **Improved Pharmacokinetic Properties**

One of the most significant advantages of small molecule mimetics is their improved pharmacokinetic profiles compared to full-length neurotrophic proteins. Traditional neurotrophic factors often exhibit poor bioavailability, short half-lives, and limited stability in circulation (Gudasheva et al., 2020). In contrast, small molecule mimetics can be designed to have enhanced metabolic stability, allowing for longer-lasting effects and potentially less frequent dosing (Longo and Massa, 2013).

ii. **Enhanced BBB Penetration**

Traditional neurotrophic factors struggle to cross the BBB effectively due to their large size and hydrophilic nature, which limits their therapeutic application in CNS disorders (Pardridge, 2005). Small molecule mimetics, being significantly smaller and often more lipophilic, can penetrate the BBB more readily, facilitating direct access to the CNS and enhancing their therapeutic potential for neurodegenerative diseases (Kazim and Iqbal, 2016).

iii. **Reduced Risk of Systemic Side Effects**

Full-length neurotrophic factors can bind to multiple receptors, leading to pleiotropic effects that may result in unwanted systemic side effects (Kazim and Iqbal, 2016). Small molecule mimetics can be engineered to selectively target specific receptors or signaling pathways, thereby minimizing off-target effects and enhancing the safety profile of the treatment (Gudasheva et al., 2020). This specificity can help mitigate adverse reactions associated with traditional neurotrophic factor therapies.

iv. **Flexibility in Administration Routes**

Small molecule mimetics offer greater flexibility in terms of administration routes compared to traditional neurotrophic factors, which often require invasive delivery methods (e.g., ICV). Many small molecule mimetics can be administered orally or through other non-invasive routes, improving patient compliance and convenience (Longo and Massa, 2013).

v. **Potential for Oral Bioavailability**

The ability to formulate small molecule mimetics for oral administration is a significant advantage. Traditional neurotrophic factors typically require parenteral administration due to their instability and poor absorption when taken orally (Kazim and Iqbal, 2016). Small molecule mimetics can be designed to withstand gastrointestinal conditions and maintain their bioactivity, making oral delivery a viable option.

vi. **Modulation of Multiple Signaling Pathways**

Small molecule mimetics can be designed to modulate various aspects of neurotrophic signaling pathways, allowing for more nuanced therapeutic effects. This ability to influence multiple pathways can be particularly beneficial in treating complex neurodegenerative disorders where multiple signaling mechanisms are involved (Gudasheva et al., 2020).

vii. **Cost-Effectiveness**

The production of small molecule mimetics is generally more cost-effective compared to the large-scale production of recombinant neurotrophic factors, which require complex biotechnological processes (Simić et al., 2021). This cost-effectiveness can facilitate broader access to effective treatments for patients with neurodegenerative diseases.

7.5 CURRENT RESEARCH AND DEVELOPMENT IN NEUROTROPHIC MIMETICS

Recent advances in the field of neurotrophic mimetics have focused on developing small molecules that can effectively mimic the actions of neurotrophic factors such as BDNF and NGF. These mimetics aim to overcome the limitations associated with traditional neurotrophic factors, including poor pharmacokinetics, limited BBB penetration, and systemic side effects (Gudasheva et al., 2020).

i. **Mechanistic Insights and Target Identification**

Research has increasingly focused on understanding the mechanisms by which neurotrophic mimetics activate their targets. For instance, studies have employed pharmacophore modeling and virtual screening to identify small molecules that selectively activate TrkB, the receptor for BDNF. These approaches have led to the discovery of compounds like LM22A-4, which demonstrate high potency and specificity for TrkB signaling, promoting neuronal survival and differentiation (Longo and Massa, 2013).

ii. **Diverse Classes of Mimetics**

Current research has identified various classes of neurotrophic mimetics, including:

a. **Peptidomimetics:** These are synthetic peptides designed to mimic specific domains of neurotrophic factors. For example, the dimeric peptide GSB-106 has shown neuroprotective effects by specifically activating TrkB pathways (Vakhitova et al., 2021).
b. **Non-peptide small molecules:** Compounds such as 7,8-dihydroxyflavone have been identified as selective TrkB agonists, demonstrating neurotrophic activities similar to BDNF (Jang et al., 2010). These small molecules can penetrate the BBB more effectively than full-length proteins.
c. **Natural product-derived mimetics:** Several natural compounds have been found to possess neurotrophic activity, enhancing BDNF expression and promoting neuronal growth. For instance, flavonoids and alkaloids have shown potential in modulating neurotrophic signaling pathways (Kooshki et al., 2023).

iii. **Preclinical and Clinical Studies**

Preclinical studies have demonstrated the efficacy of neurotrophic mimetics in various models of neurodegenerative diseases. For example, the small molecule BDNF mimetic LM22A-4 has been shown to prevent neuronal degeneration in rodent models and improve cognitive function after traumatic brain injury (Massa et al., 2010). Additionally, the CNTF mimetic P021 has demonstrated neurogenic and neuroprotective effects in models of AD, enhancing neurogenesis and improving memory processes (Kazim and Iqbal, 2016)

iv. **Addressing Pharmacokinetic Challenges**

One of the primary focuses of current research is to enhance the pharmacokinetic properties of neurotrophic mimetics. Strategies being explored include:

a. **Chemical modifications:** Alterations to the chemical structure of small molecules can improve their stability, bioavailability, and ability to cross the BBB (Gudasheva et al., 2020).
b. **Nanoparticle delivery systems:** Encapsulating neurotrophic mimetics in nanoparticles can enhance their delivery to the CNS while providing controlled release and reduced systemic toxicity (Angelova et al., 2013).
c. **Combination therapies:** Investigating the synergistic effects of neurotrophic mimetics with other therapeutic agents may enhance their overall efficacy in treating neurodegenerative conditions (El Ouaamari et al., 2023).

The field of neurotrophic mimetics is rapidly evolving, with significant advancements in the design and development of small molecules that can effectively mimic the actions of neurotrophic factors. Ongoing research is focused on understanding the mechanisms of action, improving pharmacokinetic properties, and evaluating the therapeutic potential of these mimetics in preclinical and clinical settings. As these efforts continue, small molecule neurotrophic mimetics hold great promise for providing innovative treatments for neurodegenerative diseases and enhancing neuronal health.

7.6 CLINICAL TRIALS AND EMERGING THERAPIES

The exploration of neurotrophic factor mimetics has gained momentum in recent years, particularly in the context of neurodegenerative diseases such as AD and stroke. Clinical trials and emerging therapies are focusing on the potential of these small molecules to mimic the effects of traditional neurotrophic factors while overcoming their limitations.

i. **Current Clinical Trials**

Clinical trials involving neurotrophic mimetics are still in the early stages, but several promising studies have been initiated:

a. **Intranasal NGF for ischemic stroke:** One notable trial (NCT03686163) is evaluating the effects of intranasal administration of NGF in patients with acute ischemic stroke. This trial aims to assess the safety and efficacy of NGF in enhancing recovery and neuroprotection post-stroke (Gudasheva et al., 2020). Results from this trial are anticipated to provide insights into the potential of NGF as a therapeutic agent in stroke recovery.
b. **BDNF and memantine combination**: Another ongoing trial (NCT02144584) is investigating the use of memantine, a drug known to enhance BDNF levels, in combination with standard care for stroke recovery. This trial aims to evaluate whether enhancing BDNF signaling can improve functional outcomes in stroke patients (Pichardo-Rojas et al., 2023).
c. **Donepezil and transcranial direct current stimulation**: A study (NCT02535611) is examining the effects of donepezil, an acetylcholinesterase inhibitor, in combination with transcranial direct current stimulation and intensive speech therapy. This trial seeks to determine the impact of combined therapies on cognitive recovery and neurotrophic factor levels in stroke patients (Sims et al., 2022).

ii. **Emerging Therapies**

Emerging therapies are increasingly focusing on the development of small molecule mimetics that can effectively mimic the actions of neurotrophic factors:

a. **Small molecule mimetics for AD:** Research has highlighted the potential of small molecule mimetics to provide neuroprotective effects and promote neurogenesis in AD models. For example, compounds such as 7,8-dihydroxyflavone, a TrkB agonist, have demonstrated neurotrophic activities that could enhance cognitive function and synaptic plasticity (Jang et al., 2010). These compounds are currently undergoing preclinical evaluations and may soon enter clinical trials.
b. **Peptide mimetics:** Advances in the design of peptide mimetics, such as the cyclized dimeric peptide P7, have shown promise in activating neurotrophic signaling pathways. P7 has been demonstrated to exert neurotrophic effects through the p75 neurotrophin receptor, providing a novel approach to neuroprotection (Knowles et al., 2009).
c. **Combination therapies:** There is growing interest in combining neurotrophic mimetics with other therapeutic modalities, such as pharmacological agents and rehabilitation techniques. For instance, combining small molecule mimetics with exercise therapy may enhance endogenous neurotrophic factor levels and promote recovery from neurodegenerative conditions (El Ouaamari et al., 2023).

iii. **Case Studies**

Several case studies have highlighted the potential benefits of neurotrophic factor therapies:

a. **Case studies on NGF:** Historical case studies from the 1980s and 1990s reported the administration of NGF in patients with neurodegenerative conditions, demonstrating improvements in symptoms and functional recover (Rocco et al., 2018). Although these studies were limited in scope, they provided early evidence for the therapeutic potential of neurotrophic factors.
b. **BDNF in depression:** Recent case studies have explored the role of BDNF in mood disorders, with findings suggesting that increased BDNF levels are associated with improved treatment outcomes in patients with MDD (Mosiołek et al., 2021). These observations support the hypothesis that enhancing neurotrophic signaling could be beneficial for mental health.

The landscape of clinical trials and emerging therapies involving neurotrophic mimetics is evolving rapidly. While challenges remain in translating preclinical findings into effective clinical treatments, ongoing research and innovative approaches hold promise for developing new therapies that can harness the neuroprotective and regenerative potential of neurotrophic factors. Continued investigation into small molecule mimetics may lead to significant advancements in the treatment of neurodegenerative diseases and other neurological disorders.

7.6.1 Overcoming Delivery Challenges: Novel Approaches

The therapeutic potential of neurotrophic factors, such as BDNF and NGF, has been well established in preclinical studies. However, translating these benefits into clinical applications has been hindered by significant delivery challenges, including poor bioavailability, limited BBB penetration, and systemic side effects (Gudasheva et al., 2020). To address these challenges, researchers are exploring novel approaches to enhance the delivery of neurotrophic mimetics and improve their therapeutic efficacy.

i. **Innovative Delivery Systems**

   Recent advancements in drug delivery systems have shown promise in overcoming the limitations associated with traditional neurotrophic factor therapies:

   a. **Nanoparticle-based delivery:** Nanoparticles can encapsulate neurotrophic mimetics, facilitating targeted delivery to the CNS while providing controlled release and reducing systemic toxicity. Studies have demonstrated that nanoparticles can enhance the stability and bioavailability of neurotrophic factors, improving their therapeutic potential (Bondarenko and Saarma, 2021; Faustino et al., 2017).
   b. **Hydrogel systems:** Biodegradable hydrogels have been developed for sustained local delivery of neurotrophic factors. These systems can provide a controlled release of mimetics over time, maintaining therapeutic levels in the target area and minimizing the need for frequent administration (Liu et al., 2013).
   c. **Viral vector delivery:** Advances in viral vector technology enable the targeted delivery of neurotrophic factors directly to the CNS. This approach allows for long-lasting expression of therapeutic proteins and can overcome some of the challenges associated with systemic delivery (Gudasheva et al., 2020).

ii. **Non-Invasive Delivery Routes**

   Emerging non-invasive delivery routes are being explored to enhance the accessibility of neurotrophic mimetics:

   a. **Intranasal delivery:** The intranasal route has gained attention as a promising method for delivering neurotrophic factors directly to the CNS. This approach bypasses the BBB and allows for rapid transport of therapeutic agents into the CSF (Fan et al., 2018). Studies have shown that intranasal administration of neurotrophic factors can lead to significant neuroprotective effects in animal models of neurodegeneration (Manni et al., 2021).
   b. **Transdermal delivery:** Transdermal patches and microneedle systems are being investigated for their potential to deliver neurotrophic mimetics through the skin, providing a non-invasive alternative to traditional injection methods (Govender et al., 2016).

iii. **Combination Therapies**

   Combining neurotrophic mimetics with other therapeutic modalities may enhance their efficacy and address delivery challenges:

   a. **Synergistic effects:** Research is exploring the potential of combining small molecule mimetics with existing pharmacological agents, such as acetylcholinesterase inhibitors or anti-inflammatory drugs, to enhance neuroprotective effects and improve patient outcomes (Padmakumar et al., 2020).
   b. **Rehabilitation strategies:** Integrating neurotrophic mimetics with rehabilitation therapies, such as physical or cognitive therapy, may promote neuroplasticity and improve functional recovery in patients with neurodegenerative diseases (Gubert and Hannan, 2021).
   c. **Personalized Medicine Approaches**

The identification of biomarkers that can predict patient response to neurotrophic therapies is crucial for optimizing treatment strategies. Personalized medicine approaches may enable clinicians to tailor neurotrophic mimetic therapies based on individual patient profiles, improving efficacy and minimizing adverse effects (Jain, 2020).

Overcoming the delivery challenges associated with neurotrophic mimetics is critical for translating their therapeutic potential into clinical practice. Innovative delivery systems, non-invasive routes, combination therapies, and personalized medicine approaches represent promising strategies to enhance the efficacy of neurotrophic mimetics. Continued research in these areas is essential for developing effective treatments for neurodegenerative diseases and improving patient outcomes.

7.6.2 Safety and Efficacy of Long-Term Treatment

The exploration of neurotrophic factor small molecule mimetics has garnered significant interest due to their potential to treat neurodegenerative diseases, particularly AD. As these compounds progress through clinical trials, understanding their safety and efficacy in long-term treatment regimens is crucial.

i. **Safety Profile of Neurotrophic Mimetics**

   Initial studies have indicated that neurotrophic mimetics, such as 7,8-dihydroxyflavone (7,8-DHF), exhibit a favorable safety profile in preclinical models. For instance, a 3-week treatment of 7,8-DHF at a dosage of 5 mg/kg did not induce pathological changes in murine organs such as the kidney and liver, suggesting that this compound is well-tolerated (Cerquone Perpetuini et al., 2019). Furthermore, the small molecule mimetic P021 has demonstrated stability and a plasma half-life of over 3 hours, with no adverse effects reported after 1 year of administration in 3xTg-AD mice, a model for AD (Kazim and Iqbal, 2016). These findings are encouraging, highlighting the potential for long-term use without significant toxicity.

ii. **Efficacy in Neurodegenerative Models**

   Neurotrophic mimetics have shown promising efficacy in promoting neurogenesis and synaptic repair in various preclinical models. For example, studies have demonstrated that 7,8-DHF can enhance cognitive function and stimulate synaptogenesis in animal models of AD (Zhang et al., 2014). Similarly, the mimetic LM22A-4 has been shown to activate TrkB signaling, leading to neuroprotective effects that may counteract cognitive decline associated with neurodegeneration (Massa et al., 2010). Clinical trials are currently underway to assess the therapeutic efficacy of these compounds in human populations. The rationale for using neurotrophic mimetics is strong, given their ability to replicate the beneficial effects of full-length neurotrophic factors while avoiding the pharmacokinetic limitations and systemic side effects associated with them (Kazim and Iqbal, 2016).

iii. **Long-Term Treatment Considerations**

   While the early data on safety and efficacy are promising, several considerations remain for long-term treatment with neurotrophic mimetics:

   a. **Cumulative effects:** The long-term effects of sustained neurotrophic signaling need to be thoroughly investigated. Continuous activation of neurotrophic pathways could potentially lead to adverse outcomes, such as excessive neuronal growth or maladaptive plasticity (Cerquone Perpetuini et al., 2019).
   b. **Patient variability:** Individual responses to neurotrophic mimetics may vary based on genetic, environmental, and lifestyle factors. Identifying biomarkers that predict response to treatment could enhance personalized medicine approaches (Gotovac et al., 2013).
   c. **Monitoring for adverse effects:** Rigorous monitoring during clinical trials is essential to identify any delayed or cumulative adverse effects that may arise from long-term use of these compounds.

7.6.3 Summary and Future of Neurodegeneration Therapy: Integrating Neurotrophic Factors and Small Molecules

The increasing prevalence of neurodegenerative diseases, such as AD and PD, has prompted extensive research into novel therapeutic strategies. Traditional treatments have proven inadequate, highlighting the need for innovative approaches that can effectively target the underlying mechanisms of these diseases. Neurotrophic factors and their small molecule mimetics represent a promising avenue for developing new therapies aimed at promoting neuronal survival, enhancing neuroplasticity, and mitigating neurodegeneration.

i. **Current Status of Neurotrophic Factors in Therapy**

   Neurotrophic factors, such as BDNF and NGF, play critical roles in neuronal health, survival, and plasticity. They have been shown to promote neurogenesis and protect against neurodegeneration in various preclinical models (El Ouaamari et al., 2023). However, the clinical application of these factors has been limited by challenges such as poor pharmacokinetics, limited BBB penetration, and potential systemic side effects (Thorne and Frey, 2001). Recent advancements have focused on developing small molecule mimetics that can effectively activate neurotrophic signaling pathways without the limitations associated with full-length proteins. These mimetics, such as 7,8-dihydroxyflavone, have demonstrated the ability to cross the BBB and exert neuroprotective effects comparable to those of traditional neurotrophic factors (Jang et al., 2010).

ii. **Emerging Therapies and Innovations**

   Innovative therapeutic strategies are emerging that integrate neurotrophic factors and small molecule mimetics:

   a. **Nanotechnology:** Nanoparticles are being engineered to deliver neurotrophic factors and mimetics directly to the CNS, enhancing their bioavailability and therapeutic efficacy (Krsek and Baticic, 2024). These nanoparticles can be designed to cross the BBB and provide targeted delivery, minimizing systemic exposure and side effects.
   b. **Gene therapy:** Advances in gene therapy techniques are allowing for the delivery of genes encoding neurotrophic factors directly to affected neurons. This approach has the potential to provide sustained expression of neurotrophic factors at the site of injury or degeneration (Blesch et al., 2012).
   c. **Extracellular vesicles (EVs):** Research is increasingly focusing on the use of EVs derived from mesenchymal stem cells as a novel therapeutic strategy. These vesicles can carry bioactive molecules, including neurotrophic factors, and have shown promise in promoting neuroprotection and neurogenesis while effectively crossing the BBB (El Ouaamari et al., 2023).

iii. **Future Directions**

   The future of neurodegeneration therapy lies in the integration of neurotrophic factors and small molecule mimetics into comprehensive treatment strategies. Key areas for future research include:

   a. **Personalized medicine:** Identifying biomarkers that predict patient response to neurotrophic therapies can help tailor treatments to individual needs, enhancing efficacy and minimizing adverse effects.
   b. **Combination therapies:** Exploring the synergistic effects of combining neurotrophic mimetics with existing pharmacological agents or rehabilitation strategies may enhance treatment outcomes and promote recovery from neurodegenerative conditions.
   c. **Longitudinal studies:** Conducting long-term studies to assess the safety and efficacy of neurotrophic mimetics and their impact on disease progression will be essential for translating these therapies into clinical practice.

The integration of neurotrophic factors and small molecule mimetics represents a promising frontier in the treatment of neurodegenerative diseases. As research continues to advance, innovative therapeutic strategies are likely to emerge, offering new hope for patients suffering from these debilitating conditions. By addressing the challenges associated with traditional neurotrophic therapies and leveraging novel approaches, the future of neurodegeneration therapy holds the potential for significant improvements in patient outcomes.

REFERENCES

Abbott, N.J., Patabendige, A.A.K., Dolman, D.E.M., Yusof, S.R., Begley, D.J., 2010. Structure and function of the blood–brain barrier. *Neurobiol. Dis.* 37, pp. 13–25. https://doi.org/10.1016/j.nbd.2009.07.030

Allen, S.J., Watson, J.J., Shoemark, D.K., Barua, N.U., Patel, N.K., 2013. GDNF, NGF and BDNF as therapeutic options for neurodegeneration. *Pharmacol. & Ther.* 138, pp. 155–175. https://doi.org/10.1016/j.pharmthera.2013.01.004

Andero, R., Ressler, K.J., 2012. Fear extinction and BDNF: translating animal models of PTSD to the clinic. *Genes. Brain. Behav.* 11, pp. 503–512. https://doi.org/10.1111/j.1601-183X.2012.00801.x

Angelova, A., Angelov, B., Drechsler, M., Lesieur, S., 2013. Neurotrophin delivery using nanotechnology. *Drug Discov. Today* 18, pp. 1263–1271. https://doi.org/10.1016/j.drudis.2013.07.010

Apfel, S.C., 2002. Nerve growth factor for the treatment of diabetic neuropathy: what went wrong, what went right, and what does the future hold? *Int. Rev. Neurobiol.* https://doi.org/10.1016/S0074-7742(02)50083-0

Bekinschtein, P., Cammarota, M., Katche, C., Slipczuk, L., Rossato, J.I., Goldin, A., Izquierdo, I., Medina, J.H., 2008. BDNF is essential to promote persistence of long-term memory storage. *Proc. Natl. Acad. Sci. U. S. A.* 105, pp. 2711–2716. https://doi.org/10.1073/pnas.0711863105

Blesch, A., Fischer, I., Tuszynski, M.H., 2012. Gene therapy, neurotrophic factors and spinal cord regeneration. *Handb. Clin. Neurol.* https://doi.org/10.1016/b978-0-444-52137-8.00035-8

Bondarenko, O., Saarma, M., 2021. Neurotrophic factors in Parkinson's disease: clinical trials, open challenges and nanoparticle-mediated delivery to the brain. *Front. Cell. Neurosci.* 15, p. 682597. https://doi.org/10.3389/fncel.2021.682597

Bose, A., Beal, M.F., 2016. Mitochondrial dysfunction in Parkinson's disease. *J. Neurochem.* 139, pp. 216–231. https://doi.org/10.1111/jnc.13731

Bothwell, M., 2016. Recent advances in understanding neurotrophin signaling. *F1000Research 5, F1000 Faculty Rev-1885.* https://doi.org/10.12688/f1000research.8434.1

Burgess, A., Hynynen, K., 2013. Noninvasive and targeted drug delivery to the brain using focused ultrasound. *ACS Chem. Neurosci.* 4, 519–526. https://doi.org/10.1021/cn300191b

Caldeira, M. V, Melo, C. V, Pereira, D.B., Carvalho, R., Correia, S.S., Backos, D.S., Carvalho, A.L., Esteban, J.A., Duarte, C.B., 2007. Brain-derived neurotrophic factor regulates the expression and synaptic delivery ofα-amino-3-hydroxy-5-methyl-4-isoxazole propionic acid receptor subunits in hippocampal neurons. *J. Biol. Chem.* 282, pp. 12619–12628. https://doi.org/10.1074/jbc.m700607200

Castrén, E., 2014. Neurotrophins and psychiatric disorders. *Handb. Exp. Pharmacol.* https://doi.org/10.1007/978-3-642-45106-5_17

Cerquone Perpetuini, A., Mathoux, J., Kennedy, B.N., 2019. The potential of small molecule brain-derived neurotrophic factor: mimetics to treat inherited retinal degeneration. *Neural Regen. Res.* 14, pp. 85–86. https://doi.org/10.4103/1673-5374.243711

Chauhan, N.B., Siegel, G.J., Lee, J.M., 2001. Depletion of glial cell line-derived neurotrophic factor in substantia nigra neurons of Parkinson's disease brain. *J. Chem. Neuroanat.* 21, pp. 277–288. https://doi.org/10.1016/s0891-0618(01)00115-6

Cotman, C., 2002. Exercise: a behavioral intervention to enhance brain health and plasticity. *Trends Neurosci.* 25, pp. 295–301. https://doi.org/10.1016/s0166-2236(02)02143-4

Cummings, J., Lee, G., Zhong, K., Fonseca, J., Taghva, K., 2021. Alzheimer's disease drug development pipeline: 2021. *Alzheimer's Dement. (New York, N. Y.)* 7, p. e12179. https://doi.org/10.1002/trc2.12179

Cunha, C., Brambilla, R., Thomas, K.L., 2010. A simple role for BDNF in learning and memory? *Front. Mol. Neurosci.* 3, 1. https://doi.org/10.3389/neuro.02.001.2010

Egan, M.F., Kojima, M., Callicott, J.H., Goldberg, T.E., Kolachana, B.S., Bertolino, A., Zaitsev, E., Gold, B., Goldman, D., Dean, M., Lu, B., Weinberger, D.R., 2003. The BDNF val66met polymorphism affects activity-dependent secretion of BDNF and human memory and hippocampal function. *Cell* 112, pp. 257–269. https://doi.org/10.1016/s0092-8674(03)00035-7

El Ouaamari, Y., Van den Bos, J., Willekens, B., Cools, N., Wens, I., 2023. Neurotrophic factors as regenerative therapy for neurodegenerative diseases: current status, challenges and future perspectives. *Int. J. Mol. Sci.* 24, p. 3866. https://doi.org/10.3390/ijms24043866

Esvald, E.-E., Tuvikene, J., Sirp, A., Patil, S., Bramham, C.R., Timmusk, T., 2020. CREB family transcription factors are major mediators of BDNF transcriptional autoregulation in cortical neurons. *J. Neurosci.* 40, pp. 1405–1426. https://doi.org/10.1523/JNEUROSCI.0367-19.2019

Fan, Y., Chen, M., Zhang, J., Maincent, P., Xia, X., Wu, W., 2018. Updated progress of nanocarrier-based intranasal drug delivery systems for treatment of brain diseases. *Crit. Rev. Ther. Drug Carr. Syst.* 35, pp. 433–467. https://doi.org/10.1615/critrevtherdrugcarriersyst.2018024697

Fang, W., Liao, W., Zheng, Y., Huang, X., Weng, X., Fan, S., Chen, X., Zhang, X., Chen, J., Xiao, S., Thea, A., Luan, P., Liu, J., 2019. Neurotropin reduces memory impairment and neuroinflammation via BDNF/NF-κB in a transgenic mouse model of Alzheimer's disease. *Am. J. Transl. Res.* 11, pp. 1541–1554. PMCID: PMC6456545 PMID: 30972181

Faustino, C., Rijo, P., Reis, C.P., 2017. Nanotechnological strategies for nerve growth factor delivery: therapeutic implications in Alzheimer's disease. *Pharmacol. Res.* 120, pp. 68–87. https://doi.org/10.1016/j.phrs.2017.03.020

Ferrer, I., Marín, C., Rey, M.J., Ribalta, T., Goutan, E., Blanco, R., Tolosa, E., Martí, E., 1999. BDNF and full-length and truncated TrkB expression in Alzheimer disease. Implications in therapeutic strategies. *J. Neuropathol. Exp. Neurol.* 58, pp. 729–739. https://doi.org/10.1097/00005072-199907000-00007

Fletcher, J.M., Morton, C.J., Zwar, R.A., Murray, S.S., O'Leary, P.D., Hughes, R.A., 2008. Design of a conformationally defined and proteolytically stable circular mimetic of brain-derived neurotrophic factor. *J. Biol. Chem.* 283, pp. 33375–33383. https://doi.org/10.1074/jbc.M802789200

Gabathuler, R., 2010. Approaches to transport therapeutic drugs across the blood–brain barrier to treat brain diseases. *Neurobiol. Dis.* 37, pp. 48–57. https://doi.org/10.1016/j.nbd.2009.07.028

Gash, D.M., Gerhardt, G.A., Bradley, L.H., Wagner, R., Slevin, J.T., 2020. GDNF clinical trials for Parkinson's disease: a critical human dimension. *Cell Tissue Res.* 382, pp. 65–70. https://doi.org/10.1007/s00441-020-03269-8

Géral, C., Angelova, A., Lesieur, S., 2013. From molecular to nanotechnology strategies for delivery of neurotrophins: emphasis on brain-derived neurotrophic factor (BDNF). *Pharmaceutics* 5, 127–167. https://doi.org/10.3390/pharmaceutics5010127

Golpich, M., Amini, E., Mohamed, Z., Azman Ali, R., Mohamed Ibrahim, N., Ahmadiani, A., 2017. Mitochondrial dysfunction and biogenesis in neurodegenerative diseases: pathogenesis and treatment. *CNS Neurosci. Ther.* 23, pp. 5–22. https://doi.org/10.1111/cns.12655

Gotovac, K., Hajnšek, S., Pašić, M.B., Pivac, N., Borovečki, F., 2013. Personalized medicine in neurodegenerative diseases: how far away? *Mol. Diagnosis & Ther.* 18, pp. 17–24. https://doi.org/10.1007/s40291-013-0058-z

Govender, T., Choonara, Y.E., Kumar, P., Bijukumar, D., du Toit, L.C., Modi, G., Naidoo, D., Pillay, V., 2016. Implantable and transdermal polymeric drug delivery technologies for the treatment of central nervous system disorders. *Pharm. Dev. Technol.* 22, pp. 476–486. https://doi.org/10.1080/10837450.2016.1189937

Green, M.J., Matheson, S.L., Shepherd, A., Weickert, C.S., Carr, V.J., 2010. Brain-derived neurotrophic factor levels in schizophrenia: a systematic review with meta-analysis. *Mol. Psychiatry* 16, pp. 960–972. https://doi.org/10.1038/mp.2010.88

Gubert, C., Hannan, A.J., 2021. Exercise mimetics: harnessing the therapeutic effects of physical activity. *Nat. Rev. Drug Discov.* 20, pp. 862–879. https://doi.org/10.1038/s41573-021-00217-1

Gudasheva, T.A., Povarnina, P.Y., Tarasiuk, A. V, Seredenin, S.B., 2020. Low-molecular mimetics of nerve growth factor and brain-derived neurotrophic factor: design and pharmacological properties. *Med. Res. Rev.* 41, pp. 2746–2774. https://doi.org/10.1002/med.21721

Guo, X., Zhou, J., Starr, C., Mohns, E.J., Li, Y., Chen, E.P., Yoon, Y., Kellner, C.P., Tanaka, K., Wang, H., Liu, W., Pasquale, L.R., Demb, J.B., Crair, M.C., Chen, B., 2021. Preservation of vision after CaMKII-mediated protection of retinal ganglion cells. *Cell* 184, pp. 4299–4314.e12. https://doi.org/10.1016/j.cell.2021.06.031

Hampel, H., O'Bryant, S.E., Durrleman, S., Younesi, E., Rojkova, K., Escott-Price, V., Corvol, J.-C., Broich, K., Dubois, B., Lista, S., 2017. A Precision Medicine Initiative for Alzheimer's disease: the road ahead to biomarker-guided integrative disease modeling. *Climacteric* 20, pp. 107–118. https://doi.org/10.1080/13697137.2017.1287866

Hansen, D. V., Hanson, J.E., Sheng, M., 2018. Microglia in Alzheimer's disease. *J. Cell Biol.* 217, pp. 459–472. https://doi.org/10.1083/jcb.201709069

Hellweg, R., Schulte-Herbruggen, O., Jockers-Scherubl, M., 2008. Neurotrophins - from pathophysiology to treatment in Alzheimers disease. *Curr. Alzheimer Res.* 5, pp. 38–44. https://doi.org/10.2174/156720508783884620

Heneka, M.T., Carson, M.J., Khoury, J. El, Landreth, G.E., Brosseron, F., Feinstein, D.L., Jacobs, A.H., Wyss-Coray, T., Vitorica, J., Ransohoff, R.M., Herrup, K., Frautschy, S.A., Finsen, B., Brown, G.C., Verkhratsky, A., Yamanaka, K., Koistinaho, J., Latz, E., Halle, A., Petzold, G.C., Town, T., Morgan, D., Shinohara, M.L., Perry, V.H., Holmes, C., Bazan, N.G., Brooks, D.J., Hunot, S., Joseph, B., Deigendesch, N., Garaschuk, O., Boddeke, E., Dinarello, C.A., Breitner, J.C., Cole, G.M., Golenbock, D.T., Kummer, M.P., 2015. Neuroinflammation in Alzheimer's disease. *Lancet Neurol.* 14, pp. 388–405. https://doi.org/10.1016/S1474-4422(15)70016-5

Henstridge, C.M., Hyman, B.T., Spires-Jones, T.L., 2019. Beyond the neuron-cellular interactions early in Alzheimer disease pathogenesis. *Nat. Rev. Neurosci.* 20, pp. 94–108. https://doi.org/10.1038/s41583-018-0113-1

Hoke, A., Ray, M., 2014. Rodent models of chemotherapy-induced peripheral neuropathy. *ILAR J.* 54, pp. 273–281. https://doi.org/10.1093/ilar/ilt053

Hong, M., Mukhida, K., Mendez, I., 2008. GDNF therapy for Parkinson's disease. *Expert Rev. Neurother.* 8, pp. 1125–1139. https://doi.org/10.1586/14737175.8.7.1125

Horch, H.W., Katz, L.C., 2002. BDNF release from single cells elicits local dendritic growth in nearby neurons. *Nat. Neurosci.* 5, pp. 1177–1184. https://doi.org/10.1038/nn927

Huang, E.J., Reichardt, L.F., 2001. Neurotrophins: roles in neuronal development and function. *Annu. Rev. Neurosci.* 24, pp. 677–736. https://doi.org/10.1146/annurev.neuro.24.1.677

Jain, K.K., 2020. Personalized therapy of neurological disorders. *Textb. Pers. Med.* https://doi.org/10.1007/978-3-030-62080-6_11

Jain, K.K., 2012. Neurotrophic factors. *Appl. Biotechnol. Neurol.* https://doi.org/10.1007/978-1-62703-272-8_10

Jang, S.-W., Liu, X., Yepes, M., Shepherd, K.R., Miller, G.W., Liu, Y., Wilson, W.D., Xiao, G., Blanchi, B., Sun, Y.E., Ye, K., 2010. A selective TrkB agonist with potent neurotrophic activities by 7,8-dihydroxyflavone. *Proc. Natl. Acad. Sci. U. S. A.* 107, pp. 2687–2692. https://doi.org/10.1073/pnas.0913572107

Janssens, K., Slaets, H., Hellings, N., 2015. Immunomodulatory properties of the IL-6 cytokine family in multiple sclerosis. *Ann. N. Y. Acad. Sci.* 1351, pp. 52–60. https://doi.org/10.1111/nyas.12821

Jiang, M., Peng, Q., Liu, X., Jin, J., Hou, Z., Zhang, J., Mori, S., Ross, C.A., Ye, K., Duan, W., 2013. Small-molecule TrkB receptor agonists improve motor function and extend survival in a mouse model of Huntington's disease. *Hum. Mol. Genet.* 22, pp. 2462–2470. https://doi.org/10.1093/hmg/ddt098

Jiang, Y., Wei, N., Lu, T., Zhu, J., Xu, G., Liu, X., 2011. Intranasal brain-derived neurotrophic factor protects brain from ischemic insult via modulating local inflammation in rats. *Neuroscience* 172, pp. 398–405. https://doi.org/10.1016/j.neuroscience.2010.10.054

Jovanovic, J.N., Czernik, A.J., Fienberg, A.A., Greengard, P., Sihra, T.S., 2000. Synapsins as mediators of BDNF-enhanced neurotransmitter release. *Nat. Neurosci.* 3, pp. 323–329. https://doi.org/10.1038/73888

Karpova, N.N., 2014. Role of BDNF epigenetics in activity-dependent neuronal plasticity. *Neuropharmacology* 76, pp. 709–718. https://doi.org/10.1016/j.neuropharm.2013.04.002

Katz, D.M., 2014. Brain-derived neurotrophic factor and rett syndrome. *Handb. Exp. Pharmacol.* https://doi.org/10.1007/978-3-642-45106-5_18

Kazim, S.F., Iqbal, K., 2016. Neurotrophic factor small-molecule mimetics mediated neuroregeneration and synaptic repair: emerging therapeutic modality for Alzheimer's disease. *Mol. Neurodegener.* 11, p. 50. https://doi.org/10.1186/s13024-016-0119-y

Keefe, K.M., Sheikh, I.S., Smith, G.M., 2017. Targeting neurotrophins to specific populations of neurons: NGF, BDNF, and NT-3 and their relevance for treatment of spinal cord injury. *Int. J. Mol. Sci.* 18, p. 548. https://doi.org/10.3390/ijms18030548

Knowles, J.K., Rajadas, J., Nguyen, T.-V. V., Yang, T., LeMieux, M.C., Vander Griend, L., Ishikawa, C., Massa, S.M., Wyss-Coray, T., Longo, F.M., 2009. The p75 neurotrophin receptor promotes amyloid-β(1–42)-induced neuritic dystrophy in vitro and in vivo. *J. Neurosci.* 29, pp. 10627–10637. https://doi.org/10.1523/JNEUROSCI.0620-09.2009

Kooshki, L., Zarneshan, S.N., Fakhri, S., Moradi, S.Z., Echeverria, J., 2023. The pivotal role of JAK/STAT and IRS/PI3K signaling pathways in neurodegenerative diseases: mechanistic approaches to polyphenols and alkaloids. *Phytomedicine* 112, p. 154686. https://doi.org/10.1016/j.phymed.2023.154686

Kordower, J.H., Bjorklund, A., 2013. Trophic factor gene therapy for Parkinson's disease. *Mov. Disord.* 28, pp. 96–109. https://doi.org/10.1002/mds.25344

Kordower, J.H., Emborg, M.E., Bloch, J., Ma, S.Y., Chu, Y., Leventhal, L., McBride, J., Chen, E.-Y., Palfi, S., Roitberg, B.Z., Brown, W.D., Holden, J.E., Pyzalski, R., Taylor, M.D., Carvey, P., Ling, Z., Trono, D., Hantraye, P., Déglon, N., Aebischer, P., 2000. Neurodegeneration prevented by lentiviral vector delivery of GDNF in primate models of Parkinson's disease. *Science (80-.).* 290, pp. 767–773. https://doi.org/10.1126/science.290.5492.767

Kordower, J.H., Freeman, T.B., Snow, B.J., Vingerhoets, F.J.G., Mufson, E.J., Sanberg, P.R., Hauser, R.A., Smith, D.A., Nauert, G.M., Perl, D.P., Olanow, C.W., 1995. Neuropathological evidence of graft survival and striatal reinnervation after the transplantation of fetal mesencephalic tissue in a patient with Parkinson's disease. *N. Engl. J. Med.* 332, pp. 1118–1124. https://doi.org/10.1056/nejm199504273321702

Krsek, A., Baticic, L., 2024. Nanotechnology-driven therapeutic innovations in neurodegenerative disorders: a focus on Alzheimer's and Parkinson's disease. *Futur. Pharmacol.* 4, pp. 352–379. https://doi.org/10.3390/futurepharmacol4020020

Leal, G., Bramham, C.R., Duarte, C.B., 2017. BDNF and hippocampal synaptic plasticity. *Vitam. Horm.* https://doi.org/10.1016/bs.vh.2016.10.004

Li, B., Wanka, L., Blanchard, J., Liu, F., Chohan, M.O., Iqbal, K., Grundke-Iqbal, I., 2010. Neurotrophic peptides incorporating adamantane improve learning and memory, promote neurogenesis and synaptic plasticity in mice. *FEBS Lett.* 584, pp. 3359–3365. https://doi.org/10.1016/j.febslet.2010.06.025

Lim, Y.Y., Villemagne, V.L., Laws, S.M., Ames, D., Pietrzak, R.H., Ellis, K.A., Harrington, K.D., Bourgeat, P., Salvado, O., Darby, D., Snyder, P.J., Bush, A.I., Martins, R.N., Masters, C.L., Rowe, C.C., Nathan, P.J., Maruff, P., 2013. BDNF Val66Met, Aβ amyloid, and cognitive decline in preclinical Alzheimer's disease. *Neurobiol. Aging* 34, pp. 2457–2464. https://doi.org/10.1016/j.neurobiolaging.2013.05.006

Linnarsson, S., Björklund, A., Ernfors, P., 1997. Learning deficit in BDNF mutant mice. *Eur. J. Neurosci.* 9, pp. 2581–2587. https://doi.org/10.1111/j.1460-9568.1997.tb01687.x

Liu, H., Zhou, Y., Chen, S., Bu, M., Xin, J., Li, S., 2013. Current sustained delivery strategies for the design of local neurotrophic factors in treatment of neurological disorders. *Asian J. Pharm. Sci.* 8, pp. 269–277. https://doi.org/10.1016/j.ajps.2013.10.003

Longo, F., Yang, T., Knowles, J., Xie, Y., Moore, L., Massa, S., 2007. Small molecule neurotrophin receptor ligands: novel strategies for targeting Alzheimers disease mechanisms. *Curr. Alzheimer Res.* 4, pp. 503–506. https://doi.org/10.2174/156720507783018316

Longo, F.M., Manthorpe, M., Xie, Y.M., Varon, S., 1997. Synthetic NGF peptide derivatives prevent neuronal death via a p75 receptor-dependent mechanism. *J. Neurosci. Res.* 48, pp. 1–17. https://doi.org/10.1002/(sici)1097-4547(19970401)48:1<1::aid-jnr1>3.0.co;2–k

Longo, F.M., Massa, S.M., 2013. Small-molecule modulation of neurotrophin receptors: a strategy for the treatment of neurological disease. *Nat. Rev. Drug Discov.* 12, pp. 507–525. https://doi.org/10.1038/nrd4024

Lu, B., Nagappan, G., Lu, Y., 2014. BDNF and synaptic plasticity, cognitive function, and dysfunction. *Handb. Exp. Pharmacol.* https://doi.org/10.1007/978-3-642-45106-5_9

Mackenzie, F., Ruhrberg, C., 2012. Diverse roles for VEGF-A in the nervous system. *Development* 139, pp. 1371–1380. https://doi.org/10.1242/dev.072348

Manni, L., Conti, G., Chiaretti, A., Soligo, M., 2021. Intranasal delivery of nerve growth factor in neurodegenerative diseases and neurotrauma. *Front. Pharmacol.* 12, p. 754502. https://doi.org/10.3389/fphar.2021.754502

Marks, W.J., Bartus, R.T., Siffert, J., Davis, C.S., Lozano, A., Boulis, N., Vitek, J., Stacy, M., Turner, D., Verhagen, L., Bakay, R., Watts, R., Guthrie, B., Jankovic, J., Simpson, R., Tagliati, M., Alterman, R., Stern, M., Baltuch, G., Starr, P.A., Larson, P.S., Ostrem, J.L., Nutt, J., Kieburtz, K., Kordower, J.H., Olanow, C.W., 2010. Gene delivery of AAV2-neurturin for Parkinson's disease: a double-blind, randomised, controlled trial. *Lancet Neurol.* 9, pp. 1164–1172. https://doi.org/10.1016/S1474-4422(10)70254-4

Massa, S.M., Yang, T., Xie, Y., Shi, J., Bilgen, M., Joyce, J.N., Nehama, D., Rajadas, J., Longo, F.M., 2010. Small molecule BDNF mimetics activate TrkB signaling and prevent neuronal degeneration in rodents. *J. Clin. Invest.* 120, pp. 1774–1785. https://doi.org/10.1172/JCI41356

Minnone, G., De Benedetti, F., Bracci-Laudiero, L., 2017. NGF and its receptors in the regulation of inflammatory response. *Int. J. Mol. Sci.* 18, p. 1028. https://doi.org/10.3390/ijms18051028

Mosiołek, A., Mosiołek, J., Jakima, S., Pięta, A., Szulc, A., 2021. Effects of antidepressant treatment on neurotrophic factors (BDNF and IGF-1) in patients with major depressive disorder (MDD). *J. Clin. Med.* 10, 3377. https://doi.org/10.3390/jcm10153377

Nichols, E., Vos, T., 2021. The estimation of the global prevalence of dementia from 1990-2019 and forecasted prevalence through 2050: an analysis for the Global Burden of Disease (GBD) study 2019. *Alzheimer's & Dement.* 17. https://doi.org/10.1002/alz.051496

Nicole, O., Ali, C., Docagne, F., Plawinski, L., MacKenzie, E.T., Vivien, D., Buisson, A., 2001. Neuroprotection mediated by glial cell line-derived neurotrophic factor: involvement of a reduction of NMDA-induced calcium influx by the mitogen-activated protein kinase pathway. *J. Neurosci.* 21, pp. 3024–3033. https://doi.org/10.1523/JNEUROSCI.21-09-03024.2001

Notaras, M., van den Buuse, M., 2020. Neurobiology of BDNF in fear memory, sensitivity to stress, and stress-related disorders. *Mol. Psychiatry* 25, pp. 2251–2274. https://doi.org/10.1038/s41380-019-0639-2

Nutt, J.G., Burchiel, K.J., Comella, C.L., Jankovic, J., Lang, A.E., Laws, E.R., Lozano, A.M., Penn, R.D., Simpson, R.K., Stacy, M., Wooten, G.F., 2003. Randomized, double-blind trial of glial cell line-derived neurotrophic factor (GDNF) in PD. *Neurology* 60, pp. 69–73. https://doi.org/10.1212/wnl.60.1.69

O'Leary, P.D., Hughes, R.A., 2003. Design of potent peptide mimetics of brain-derived neurotrophic factor. *J. Biol. Chem.* 278, pp. 25738–25744. https://doi.org/10.1074/jbc.m303209200

Padmakumar, S., Taha, M.S., Kadakia, E., Bleier, B.S., Amiji, M.M., 2020. Delivery of neurotrophic factors in the treatment of age-related chronic neurodegenerative diseases. *Expert Opin. Drug Deliv.* 17, pp. 323–340. https://doi.org/10.1080/17425247.2020.1727443

Palasz, E., Wysocka, A., Gasiorowska, A., Chalimoniuk, M., Niewiadomski, W., Niewiadomska, G., 2020. BDNF as a promising therapeutic agent in Parkinson's disease. *Int. J. Mol. Sci.* 21, p. 1170. https://doi.org/10.3390/ijms21031170

Panja, D., Bramham, C.R., 2014. BDNF mechanisms in late LTP formation: a synthesis and breakdown. *Neuropharmacology* 76, pp. 664–676. https://doi.org/10.1016/j.neuropharm.2013.06.024

Pardridge, W.M., 2012. Drug transport across the blood–brain barrier. *J. Cereb. Blood Flow Metab.* 32, pp. 1959–1972. https://doi.org/10.1038/jcbfm.2012.126

Pardridge, W.M., 2005. The blood-brain barrier: bottleneck in brain drug development. *NeuroRX* 2, pp. 3–14. https://doi.org/10.1602/neurorx.2.1.3

Pardridge, W.M., 2002. Drug and gene delivery to the brain. *Neuron* 36, pp. 555–558. https://doi.org/10.1016/s0896-6273(02)01054-1

Park, H., Poo, M., 2012. Neurotrophin regulation of neural circuit development and function. *Nat. Rev. Neurosci.* 14, pp. 7–23. https://doi.org/10.1038/nrn3379

Pasquin, S., Sharma, M., Gauchat, J.-F., 2015. Ciliary neurotrophic factor (CNTF): new facets of an old molecule for treating neurodegenerative and metabolic syndrome pathologies. *Cytokine & Growth Factor Rev.* 26, pp. 507–515. https://doi.org/10.1016/j.cytogfr.2015.07.007

Pezet, S., McMahon, S.B., 2006. Neurotrophins: mediators and modulators of pain. *Annu. Rev. Neurosci.* 29, pp. 507–538. https://doi.org/10.1146/annurev.neuro.29.051605.112929

Pichardo-Rojas, D., Pichardo-Rojas, P.S., Cornejo-Bravo, J.M., Serrano-Medina, A., 2023. Memantine as a neuroprotective agent in ischemic stroke: preclinical and clinical analysis. *Front. Neurosci.* 17, p. 1096372. https://doi.org/10.3389/fnins.2023.1096372

Pittenger, G., Vinik, A., 2003. Nerve growth factor and diabetic neuropathy. *Exp. Diabesity Res.* 4, pp. 271–285. https://doi.org/10.1155/EDR.2003.271

Poduslo, J.F., Curran, G.L., 1996. Permeability at the blood-brain and blood-nerve barriers of the neurotrophic factors: NGF, CNTF, NT-3, BDNF. *Mol. Brain Res.* 36, pp. 280–286. https://doi.org/10.1016/0169-328x(95)00250-v

Rahman, M.M., Islam, M.R., Supti, F.A., Dhar, P.S., Shohag, S., Ferdous, J., shuvo, S. khan, Akter, A., Hossain, M.S., Sharma, R., 2023. Exploring the therapeutic effect of neurotrophins and neuropeptides in neurodegenerative diseases: at a glance. *Mol. Neurobiol.* 60, pp. 4206–4231. https://doi.org/10.1007/s12035-023-03328-5

Rehman, M.U., Sehar, N., Dar, N.J., Khan, A., Arafah, A., Rashid, S., Rashid, S.M., Ganaie, M.A., 2023. Mitochondrial dysfunctions, oxidative stress and neuroinflammation as therapeutic targets for neurodegenerative diseases: an update on current advances and impediments. *Neurosci. Biobehav. Rev.* 144, p. 104961. https://doi.org/10.1016/j.neubiorev.2022.104961

Ribeiro, F.F., Xapelli, S., 2021. Intervention of brain-derived neurotrophic factor and other neurotrophins in adult neurogenesis. pp. 95–115. https://doi.org/10.1007/978-3-030-74046-7_8

Rocco, M.L., Soligo, M., Manni, L., Aloe, L., 2018. Nerve growth factor: early studies and recent clinical trials. *Curr. Neuropharmacol.* 16, pp. 1455–1465. https://doi.org/10.2174/1570159X16666180412092859

Saghazadeh, A., Rezaei, N., 2017. Brain-derived neurotrophic factor levels in autism: a systematic review and meta-analysis. *J. Autism Dev. Disord.* 47, pp. 1018–1029. https://doi.org/10.1007/s10803-016-3024-x

Sakata, K., Martinowich, K., Woo, N.H., Schloesser, R.J., Jimenez, D. V, Ji, Y., Shen, L., Lu, B., 2013. Role of activity-dependent BDNF expression in hippocampal-prefrontal cortical regulation of behavioral perseverance. *Proc. Natl. Acad. Sci. U. S. A.* 110, pp. 15103–15108. https://doi.org/10.1073/pnas.1222872110

Salvatore, M., Ai, Y., Fischer, B., Zhang, A., Grondin, R., Zhang, Z., Gerhardt, G., Gash, D., 2006. Point source concentration of GDNF may explain failure of phase II clinical trial. *Exp. Neurol.* 202, pp. 497–505. https://doi.org/10.1016/j.expneurol.2006.07.015

Sampaio, T.B., Savall, A.S., Gutierrez, M.E.Z., Pinton, S., 2017. Neurotrophic factors in Alzheimer's and Parkinson's diseases: implications for pathogenesis and therapy. *Neural Regen. Res.* 12, pp. 549–557. https://doi.org/10.4103/1673-5374.205084

Sen, S., Duman, R., Sanacora, G., 2008. Serum brain-derived neurotrophic factor, depression, and antidepressant medications: meta-analyses and implications. *Biol. Psychiatry* 64, pp. 527–532. https://doi.org/10.1016/j.biopsych.2008.05.005

Simić, S., Zukić, E., Schmermund, L., Faber, K., Winkler, C.K., Kroutil, W., 2021. Shortening synthetic routes to small molecule active pharmaceutical ingredients employing biocatalytic methods. *Chem. Rev.* 122, pp. 1052–1126. https://doi.org/10.1021/acs.chemrev.1c00574

Sims, S.-K., Wilken-Resman, B., Smith, C.J., Mitchell, A., McGonegal, L., Sims-Robinson, C., 2022. Brain-derived neurotrophic factor and nerve growth factor therapeutics for brain injury: the current translational challenges in preclinical and clinical research. *Neural Plast.* 2022, p. 3889300. https://doi.org/10.1155/2022/3889300

Skaper, S.D., 2012. The neurotrophin family of neurotrophic factors: an overview. *Methods Mol. Biol.* https://doi.org/10.1007/978-1-61779-536-7_1

Sweeney, P., Park, H., Baumann, M., Dunlop, J., Frydman, J., Kopito, R., McCampbell, A., Leblanc, G., Venkateswaran, A., Nurmi, A., Hodgson, R., 2017. Protein misfolding in neurodegenerative diseases: implications and strategies. *Transl. Neurodegener.* 6, 6. https://doi.org/10.1186/s40035-017-0077-5

Thorne, R.G., Frey, W.H., 2001. Delivery of neurotrophic factors to the central nervous system. *Clin. Pharmacokinet.* 40, pp. 907–946. https://doi.org/10.2165/00003088-200140120-00003

Vakhitova, Y. V, Kalinina, T.S., Zainullina, L.F., Lusta, A.Y., Volkova, A. V, Kudryashov, N. V, Gudasheva, T.A., Shimshirt, A.A., Kadnikov, I.A., Voronin, M. V, Seredenin, S.B., 2021. Analysis of antidepressant-like effects and action mechanisms of GSB-106, a small molecule, affecting the TrkB signaling. *Int. J. Mol. Sci.* 22, p. 13381. https://doi.org/10.3390/ijms222413381

Vaynman, S., Ying, Z., Gomez-Pinilla, F., 2004. Hippocampal BDNF mediates the efficacy of exercise on synaptic plasticity and cognition. *Eur. J. Neurosci.* 20, pp. 2580–2590. https://doi.org/10.1111/j.1460-9568.2004.03720.x

Wilson, D.M., Cookson, M.R., Van Den Bosch, L., Zetterberg, H., Holtzman, D.M., Dewachter, I., 2023. Hallmarks of neurodegenerative diseases. *Cell* 186, 693–714. https://doi.org/10.1016/j.cell.2022.12.032

Wong, L.E., Gibson, M.E., Arnold, H.M., Pepinsky, B., Frank, E., 2015. Artemin promotes functional long-distance axonal regeneration to the brainstem after dorsal root crush. *Proc. Natl. Acad. Sci. U. S. A.* 112, pp. 6170–6175. https://doi.org/10.1073/pnas.1502057112

Wong, Y.C., Krainc, D., 2017. α-synuclein toxicity in neurodegeneration: mechanism and therapeutic strategies. *Nat. Med.* 23, pp. 1–13. https://doi.org/10.1038/nm.4269

Wrigley, S., Arafa, D., Tropea, D., 2017. Insulin-like growth factor 1: at the crossroads of brain development and aging. *Front. Cell. Neurosci.* 11, p. 14. https://doi.org/10.3389/fncel.2017.00014

Yoshii, A., Constantine-Paton, M., 2010. Postsynaptic BDNF-TrkB signaling in synapse maturation, plasticity, and disease. *Dev. Neurobiol.* 70, pp. 304–322. https://doi.org/10.1002/dneu.20765

Zhang, Z., Liu, X., Schroeder, J.P., Chan, C.-B., Song, M., Yu, S.P., Weinshenker, D., Ye, K., 2014. 7,8–dihydroxyflavone prevents synaptic loss and memory deficits in a mouse model of Alzheimer's disease. *Neuropsychopharmacology* 39, pp. 638–650. https://doi.org/10.1038/npp.2013.243

Zuccato, C., Cattaneo, E., 2007. Role of brain-derived neurotrophic factor in Huntington's disease. *Prog. Neurobiol.* 81, pp. 294–330. https://doi.org/10.1016/j.pneurobio.2007.01.003

8 Repurposed Drugs
Expanding the Therapeutic Resource

Duilio Cervantes and Aida Adlimoghaddam

ABBREVIATIONS

AD	Alzheimer's Disease
ALS	Amyotrophic Lateral Sclerosis
AMP	Adenosine Monophosphate
AMPK	Adenosine Monophosphate Kinase
APP	Amyloid Precursor Protein
Aβ	Amyloid Beta
BBB	Blood–Brain Barrier
CNS	Central Nervous System
HD	Huntington's Disease
IND	Investigational New Drug Application
ODC	Oligodendrocytes
OPC	Oligodendrocyte Precursor Cells
PCOS	Polycystic Ovarian Syndrome
PMS	Progressive Multiple Sclerosis
PTLD	Post-Transplant Lymphoproliferative Disorder
NFT	Neurofibrillary Tangle

8.1 WHY DRUG REPURPOSING IS IMPORTANT

Ever since the medical disaster of thalidomide in 1962, the FDA's approach to drug approval and regulation has evolved to include stricter supervision and processing (Darrow et al., 2020). Recent reports showcase the timeline from submitting an Investigational New Drug Application, a step occurring just prior to phase 1 of clinical trials, to final approval of the pharmaceutical agent averaging about 9.1 years. Additionally, recent data reported that fees collected under the Prescription Drug User Act have increased from 29 million in 1992 to 908 million in 2018 (Darrow et al., 2020). Although the emergence of new technologies such as high-throughput screening and X-ray crystallography has enabled a more profound way to uncover new therapeutics, research and development from pharmaceutical companies have gone down in productivity in the past 30 years (Scannell et al., 2012). One study conducted by Wong et al. concluded that the probability of success for potential drug targets in fields such as oncology has been estimated to be only 3.4% (Wong et al., 2019). New drug discoveries using traditional methods remain a lengthy, expensive, and complicated procedure. This is especially detrimental for those patients who suffer from rare and untreatable diseases that require little to no medical intervention. Patients who endure common illnesses such as Alzheimer's disease (AD) or rare inherited disorders like Rett syndrome require a complex, multifactorial approach to care, which would benefit the most from new drug developments that may provide an increasing quality of life (Roessler et al., 2021). The sense of urgency for new forms of treatment is universally instilled in patients who put up with incurable diseases daily, so finding a way to speed

DOI: 10.1201/9781003520610-8

up drug discovery is crucial. One approach to overcoming these challenges would be to use the strategies behind drug repurposing, also referred to as drug repositioning (Ashburn & Thor, 2004).

Drug repurposing is a drug-discovering methodology that serves as a promising and efficient way to identify new therapeutic benefits with pre-existing pharmaceuticals (Jourdan et al., 2020). Ever since the introduction of the term in the early 2000s, there has been constant modification to the definition. Drug repurposing may include active chemicals that failed to reach certain steps in their clinical trials due to factors such as toxicity or efficacy. On the contrary, chemicals that haven't undergone any safety observations for clinical trials are excluded from what can be "repurposed." Additionally, the concept excludes any structural modification of the drug but rather "repositions" the drug with a different indication on already observed biological targets. One way to illustrate this is by taking a known drug whose secondary target may play a role in a new disease pathway; this would allow it to serve another purpose than it was intended for. A prime example of a recent life event was the COVID-19 pandemic, in which drug companies used a variety of available antiviral drugs, such as remdesivir, to find potential benefits for infected patients (Kato et al., 2022). Repurposing of drugs may occur accidentally, such as in the discovery of sildenafil, or is a stepwise approach (Nosengo, 2016). A couple of avenues drug repurposing could take would be utilizing either a computational or an experimental way of discovering. This novel approach to drug development can utilize technological advances, such as artificial intelligence-based computational methods, to successfully compare an existing compound with well-understood pharmacokinetic interactions to known disease pathways, gene expression profiles, and so forth (Tan et al., 2023). Drug repurposing via in silico approaches has made modern advances such as screening for potential estrogen receptor antagonists from a list of currently known natural products in one study (Pang et al., 2018). Drug repurposing offers a time-saving, cost-effective, and safe approach to drug manufacturing while also utilizing data from traditional drug discovery methods.

8.2 IMPLICATIONS IN AGING AND NEURODEGENERATIVE DISEASE

Ever since the 1960s, life expectancy globally has increased significantly due to interventions such as state-of-the-art medical innovation and public health initiatives (Woolf & Schoomaker, 2019). This has enabled an increasing prevalence of aging and age-related diseases (Adlimoghaddam et al., 2021; Eshraghi et al., 2022), especially in countries with more healthcare accessibility and resources. In this chapter, we will talk about the repurposing of two important drugs, metformin and rapamycin, for treating aging and age-related diseases.

8.2.1 METFORMIN

Metformin remains the most widely used first-line therapy for the treatment of type 2 diabetes (T2D) (Foretz et al., 2014). Its history dates back to the French lilac (*Galega officinalis*), where its extracts were used to treat patients who had sweet urine and polyuria. It is part of the biguanide class of antihyperglycemic agents that improves glucose tolerance by lowering both basal and postprandial plasma glucose, utilizing mechanisms that differ from other oral antidiabetics. Evidence suggests that one of its main underlying mechanisms is through inhibition of complex 1 of the mitochondrial respiratory chain to lower blood glucose (Owen et al., 2000). Moreover, metformin decreases hepatic gluconeogenesis production, decreases intestinal absorption of glucose, and improves insulin sensitivity through various mechanisms. Although the exact mechanism of metformin is not well understood, it has long been suggested that one of the ways metformin works is through inhibition of gluconeogenesis via the activation of the adenosine monophosphate (AMP)-activated protein kinase (AMPK) pathway, which leads to ATP depletion and an increase in AMP levels (Hawley et al., 2010). Additionally, there are studies that also showcase the inhibition of hepatic glucose output that is independent of the AMPK pathway (Madiraju et al., 2014). Metformin has also been shown to help treat polycystic ovary syndrome patients by increasing success rates of ovulation and lowering complications (Notaro & Neto, 2022). Besides its role in patients with hyperglycemia in patients with T2D, metformin has been implicated in aiding in a variety of pathologies such as myocardial infarction, acute kidney injury, and coronary artery disease (Bell et al., 2017; Han et al., 2019; Legtenberg et al., 2002). Metformin has shown promise in intervening in the transition from non-invasive ductal carcinoma in situ to life-threatening breast cancer and also being a possible antineoplastic drug in other types of cancers (Pollak, 2014; Vazquez-Martin et al., 2011; Yang et al., 2015).

The safety profile of metformin is well understood, and one of the most common adverse effects is gastrointestinal discomfort. This may arise through various means of diarrhea, cramps, nausea, vomiting, increased flatulence, etc. (Nabrdalik et al., 2022). One of the more serious side effects is metformin-associated lactic acidosis. However, it occurs very rarely, at a rate of <10 cases per 100,000 patient-years. Due to the risk of lactic acidosis, metformin is contraindicated in patients with renal or hepatic insufficiency, congestive heart failure, metabolic acidosis, and hypersensitivity to the drug (DeFronzo et al., 2016; Misbin et al., 1998; Rocha et al., 2012). There has been great interest in repurposing metformin for the treatment of a variety of neurodegenerative disorders, especially since it may act directly on neurons by crossing the blood–brain barrier (Łabuzek et al., 2010). It is important to mention that diabetes and insulin resistance are associated with the risk of neurological disorders such as AD and Parkinson's disease (PD) (Arvanitakis et al., 2004; Wrighten et al., 2009). Being an antidiabetic drug, the benefits of metformin against neurodegenerative disorders may arise from hindering the damaging effects of insulin resistance in the brain and central nervous system (CNS) in general. One example is its protective effects against endothelial damage, which may translate to a healthier CNS vasculature (Nafisa et al., 2018). There have been multiple studies that showcase metformin being a positive agent against neurodegeneration with increasing health and lifespan properties. For example, metformin was shown to mitigate amnestic mild cognitive impairment in a clinical trial with AD patients (Luchsinger et al., 2016). In a like manner, an 8-week randomized placebo-controlled study, metformin was associated with improved executive functioning in patients suffering from AD-related dementia (Koenig et al., 2017). Multiple studies showcase similar results with metformin, either reducing the likelihood of being diagnosed with a neurodegenerative disease or offering other neuroprotective effects in humans (Samaras et al., 2021; Shi et al., 2019; Zhang et al., 2022).

In the context of aging and neurodegeneration, mitochondrial dysfunction is critically important (Adlimoghaddam et al., 2019; Adlimoghaddam et al., 2022; Eshraghi et al., 2021). Interestingly, metformin has been linked to modulating mitochondrial function through various methods that are apart from its main antidiabetic properties. One example is that it was able to restore mitophagy (refers to the recycling of damaged or dysfunctional mitochondria using lysosomal degradation) levels via PINK1 and PARKIN in T2D patients as well as increase AMPK activation, which are crucial for mitophagy (de Marañón et al., 2022). Furthermore, an in vitro study utilizing healthy human cell cultures indicated that metformin treatment increased the formation of mitophagosomes, upregulated mitophagy markers, and intensified mitophagic flux through the increased expression of LC3-11 and low p62 protein levels (Bhansali et al., 2019). It has been shown that metformin's use in human neural stem cells activates AMPK pathways, reducing amyloid beta (Aβ)-induced mitochondrial deficits and restoring fragmented mitochondria (Chiang et al., 2016). Metformin holds the potential for treating PD by improving mitochondrial function, particularly in mouse neurons, and decreasing phospho-Ser 129 alpha-synuclein levels (Pérez-Revuelta et al., 2014). Second, metformin's PP2A activity results in a reduction in alpha-syn in an MPTP-induced mouse model of PD (Katila et al., 2017). Similarly, metformin was found to preserve tyrosine-hydroxylase-positive dopamine cells and volume after MPTP exposure in mice (Bayliss et al., 2016). Furthermore, *Caenorhabditis elegans* has also been utilized in PD's models, where a study found that metformin significantly decreased neurodegeneration in 6-hydroxydopamine-induced worms, inhibited alpha-syn aggregation, and aided with food-sensing behavior (Saewanee et al., 2021). Further, metformin as an AMPK activator is able to inhibit tau phosphorylation and aggregation through the activation of SIRT1 and downregulation of GSK3-beta (Greco et al., 2009; Greco et al., 2009). Moreover, it was demonstrated that metformin induced PP2A activity and reduced tau phosphorylation at PP2A-dependent epitopes in vitro and in vivo in human tau transgenic mice (Kickstein et al., 2010). Recent studies suggest metformin inhibits tau pathology and decreases amyloid precursor protein (APP) protein expression in AD models (Matthes et al., 2018; Zhao et al., 2023). In *C. elegans* models, metformin reversed Aβ metabolic defects and reduced protein aggregation (Teo et al., 2019). Studies also link metformin to altered gene expressions like SPP1 and APOE in AD-related contexts (Charpignon et al., 2022; Rujescu & Jung, 2020; Sala Frigerio et al., 2019).

Another critical aspect in the context of aging and neurodegeneration is inflammation, which plays a pivotal role in the progression of various neurological disorders (Adlimoghaddam et al., 2022; Adlimoghaddam & Albensi, 2021). Metformin shows the potential to mitigate aging-related inflammation by downregulating NF-kB in senescent cells (Moiseeva et al., 2013). It also enhances autophagy in CD4+ T cells from older individuals, improving mitochondrial function and reducing inflammation (Bharath et al., 2020). These effects could be beneficial in treating age-related dementias and neurodegenerative diseases by addressing inflammation and protein aggregation (Andronie-Cioara et al., 2023). Metformin has further shown promise in reducing Aβ plaque and reversing AD phenotypes in animal models (Xu et al., 2021). Additionally, metformin targets monoacylglycerol lipase (MglI), potentially influencing neurodegenerative processes (Syal et al., 2020). Metformin shows promise in Huntington's disease (HD) by enhancing cortical neuronal activity and reducing aberrant protein load in animal models (Arnoux et al., 2018). It has also been associated with improved motor and neuropsychiatric symptoms, along with reduced mutant huntingtin aggregates in preclinical studies (Sanchis et al., 2019). Observational research suggests that metformin may correlate with better cognitive function in human HD patients (Hervás et al., 2017).

Some further literature suggests that metformin may be repurposed to battle the pathology behind amyotrophic lateral sclerosis (ALS), multiple toxin protein variants implicated in ALS pathology through repeated non-ATG translation of the C9orf72 gene, and this finding suggests potential drug targets for ALS symptoms. Metformin has shown promise by reducing RAN protein levels and improving behavior and neuropathology in C9 BAC transgenic mice (Zu et al., 2020). Currently, metformin is in phase two clinical trials for C9orf72 ALS and frontotemporal dementia, with ongoing results (Hautbergue et al., 2021). Progressive multiple sclerosis lacks effective treatments despite advances in multiple sclerosis (MS) research. Inducing remyelination through oligodendrocyte precursor cells (OPCs) and oligodendrocytes is a potential strategy. Metformin has shown promise by reversing functional decline in aged OPCs and improving remyelination in MS mouse models (Neumann et al., 2019). Studies in various animal models further support metformin's potential in MS, prompting clinical trials to assess its effects on brain remyelination and neurodegeneration in MS (De Keersmaecker et al., 2024; Nath et al., 2009; Paintlia et al., 2013).

8.2.2 Rapamycin

In the remote areas of Easter Island during the early 1970s, a potential anti-fungal agent produced by native bacterial species was found to inhibit the growth of *Candida albicans*, *Microsporum gypseum*, and *Trichophyton granulosum* (Vézina et al., 1975). This compound that was isolated from strains of *Streptomyces hygroscopicus* possessed properties beyond hindering fungi, such as immunosuppression and anti-cancer properties (Houchens et al., 1983; Spilman et al., 2010). Rapamycin, also known as Sirolimus and brand name Rapamune, is a macrolide that is FDA-approved to help against rejection prophylaxis in kidney transplants and the treatment of lymphangioleiomyomatosis. It often acts with other immunosuppressants, such as cyclosporine, for immunosuppression with little to no overlapping toxicity (Tsai et al., 2009). Additionally, it is used to coat coronary stents to prevent intra-stent neointimal proliferation (Serruys et al., 2002). Rapamycin works by inhibiting T-cell activation and proliferation (Russell et al., 1992). Rapamycin binds to a specific immunophilin protein known as FKBP12. The FKBP12–rapamycin complex is directly associated with the mammalian target of rapamycin (mTOR) by binding to its domain, rendering it inactive. Once inserted into the FKBP12-rapamycin-binding domain, the inactivation of mTOR blocks signal transduction pathways in cell cycle progression (Sehgal, 1998). More specifically, the inhibition of mTOR suppresses inflammatory biomarkers such as IL-2, IL-4, IL-15, and IL-17, which induce proliferation, resulting in cell cycle arrest during the G1 phase. Rapamycin not only enhances its immunosuppressive effects but also inhibits the synthesis of proteins crucial for T-cell progression into S-phase through its action on p70sk6 (Chung et al., 1992). Moreover, inactivating mTOR reduces the kinase activity of cdk4/cyclinD and cdk2/cyclinE complexes, and it suppresses the expression of bcl-2 and BAG-1 while leaving certain proto-oncogenes unaffected (Asnaghi et al., 2004). Rapamycin impacts various cell types beyond the immune system, including smooth muscle cells (Marx et al., 1995). Its immunosuppressive effects can lead to side effects like thrombocytopenia, hyperlipidemia, impaired wound healing, nephrotoxicity, and altered insulin sensitivity. Contraindications include hypersensitivity, immunocompromised states, and pregnancy. Deregulation of the mTOR pathway has been implicated in multiple disease pathologies such as cancer, obesity, T2D, and neurodegeneration, which will be discussed later. mTOR belongs to the PI3K-related kinase family and interacts with various proteins to form mTORC1 and mTORC2. mTORC1 is specifically sensitive to rapamycin, and its pathway has been studied more extensively in the literature. Evidence suggests that rapamycin is mostly insensitive to mTORC2 (Jacinto et al., 2004). Nevertheless, the mTOR pathway serves an imperative role in cellular metabolism and has been shown to be influenced by growth factors, oxygen, energy levels, and amino acids (Szwed et al., 2021). Therefore, there has been much research into the pleiotropic abilities of rapamycin and its influence on mTOR with various disease processes and aging (Li et al., 2014; Selvarani et al., 2021). For instance, rapamycin markedly reduced the development and growth of tobacco carcinogen-induced murine lung tumors in A/J mice (Granville et al., 2007). Furthermore, in cancer-prone mice, rapamycin was able to decrease the number and size of malignancies and increase lifespan by over 12% (Anisimov et al., 2010). A specific type of malignancy has increased incidence after solid organ and hematopoietic stem cell transplantation, known as post-transplant lymphoproliferative disorder (PTLD) (Dierickx & Habermann, 2018; Opelz & Döhler, 2004). In heart transplant patients, rapamycin-based immunosuppression reduced the incidence of PTLD (Asleh et al., 2022). In mouse models, rapamycin showed promise in preserving cardiac and skeletal muscle in lamin A/C-deficient mice, suggesting potential benefits for certain muscular dystrophies (Ramos et al., 2012). The repurposing of rapamycin has not only been limited to advances in cancer but also to possible drugs to combat the debilitating progression of neurodegenerative disorders. In post-mortem AD patient brains, senescent cells were found alongside neurons containing neurofibrillary tangle tau pathology (Dehkordi et al., 2021). Removing senescent cells has been shown to delay aging-related disorders in animal models, suggesting a potential treatment approach for tauopathies in neurodegenerative diseases (Baker et al., 2011). Rapamycin, an mTOR inhibitor, is emerging as a promising candidate to combat senescence-induced neurodegeneration. For instance, in vitro studies have demonstrated that rapamycin delays cellular senescence and promotes clearance through autophagy in Hutchingson-Gilford progeria cells (Cao et al., 2011). Additionally, rapamycin has been found to activate the Nrf2 pathway and protect against hydrogen peroxide-induced premature senescence in mouse fibroblasts (Wang et al., 2017).

Inhibiting mTOR is known to trigger autophagy, facilitating the targeted breakdown of cell components like dysfunctional mitochondria (Kim & Guan, 2015; Twig et al., 2008). This property makes rapamycin a leading candidate for inducing mitophagy and promoting overall cellular health. In vitro studies using cybrid cell lines have demonstrated rapamycin's ability to induce mitochondrial autophagy, particularly in cells with mtDNA mutations (Gilkerson et al., 2012). Additionally, it has been shown to reduce mitochondrial mutations in vitro (Dai et al., 2014). Studies in *Drosophila* have similarly indicated that rapamycin can mitigate mitochondrial decline (Kandul et al., 2016). Recent in vivo research using genetically diverse mice has revealed that long-term rapamycin treatment enhances mtDNA quality (Bielas et al., 2018). Moreover, rapamycin shows promise in reducing phenotypic expression in murine models of mitochondrial diseases like Leigh syndrome (Johnson et al., 2015).

There has been a booming interest in anti-aging therapeutics, resulting in an increased regard for rapamycin as a possible key to longevity. It was first studied in invertebrates such as *C. elegans* and *Drosophila* that certain configurations in the mTOR pathway increased lifespan (Kapahi et al., 2004; Vellai et al., 2003). Rapamycin, known for its mTOR inhibiting properties, has been shown to extend lifespan in genetically diverse mice when administered during a stage of life with 90% mortality (Harrison et al., 2009). Additionally, researchers have explored rapamycin's potential benefits in AD, particularly in relation to the impact of the APOE4 allele, a significant risk factor associated with decreased cerebral function, neuronal circuitry malfunction, tauopathy, and memory loss (Liu et al., 2013; Rujescu & Jung, 2020). Nevertheless, it was reported that rapamycin was able to restore cerebrovascular function and vascular density in mouse models of AD (Lin et al., 2017). Previous in vivo studies showed reduced Aβ aggregation as well as improved spatial learning and memory in these mouse models using rapamycin (Lin et al., 2013; Spilman et al., 2010). Furthermore, rapamycin was observed to decrease tauopathy in mouse models that overexpressed human mutant tau genes, such as P301S mutant tau transgenic mice (Ozcelik et al., 2013). Another study was conducted on an early-stage Alzheimer-type tauopathy mouse model where rapamycin was able to alleviate neurodegeneration in the perforant pathway, an important contributor to the progression of cognitive deficits in AD patients (Liu et al., 2012; Siman et al., 2015). Even more, one study reported that rapamycin increased Parkin-mediated mitophagy and fusion of the mitophagosome and lysosome in APP/PS1-AD mouse models, which shows promise in not only AD but PD patients as well (Wang et al., 2021). Similarly, it was found that mitochondrial function and cognition were enhanced in 5XFAD-AD mice through the activation of the PINK1-Parkin pathway using rapamycin (Zheng et al., 2024). Due to the compelling evidence of rapamycin with AD models, there is currently a phase trial using rapamycin in patients with AD (Svensson et al., 2024).

Further, rapamycin inhibits the expression of REDD1 (RTP801), a protein implicated in stress-induced pathways leading to PD progressions, such as oxidative stress, DNA damage, and mitochondrial dysfunction in PD (Ellisen et al., 2002; Malagelada et al., 2006). Rapamycin has been shown to reduce RTP801 expression in cell cultures and MPTP-treated mice (Malagelada et al., 2010). By inhibiting mTOR, rapamycin facilitates Akt phosphorylation, which supports neuronal survival and exhibits neuroprotective effects in PD models (Dudek et al., 1997). Additionally, rapamycin improved motor function in alpha-synuclein transgenic mice, highlighting its potential against synucleinopathies (Bai et al., 2015). Remarkably, mTOR inhibition by rapamycin also mitigated cognitive and behavioral deficits in 6-OHDA mouse models of PD (Masini et al., 2018).

Possible mechanisms by which rapamycin may aid in HD haven't been fully explored. However, one study observed a reduction of polyglutamine expansions in fly and mouse models of HD. They found that rapamycin was also able to improve performance on different behavioral tests and reduce protein aggregate formation in the mice (Ravikumar et al., 2004). Being the third most common neurodegenerative disease, there has been a constant hunt for therapies to attenuate ALS progression. As discussed before, rapamycin can aid in regulatory T-cell (Treg) expansion and autophagy integrity by inhibiting the mTOR pathway. This pathway is associated with ALS as it is often dysfunctional and reduced (Zhao et al., 2013). Also, it has been reported that the percentage of Tregs in the blood is inversely correlated with the progression rate of ALS (Yazdani et al., 2022). Rapamycin has been observed to reduce TDP-43 fragment aggregation through the means of autophagy in the ALS model (Caccamo et al., 2009). Additionally, rapamycin was able to prolong survival in human stem cell-derived neurons and astrocytes with a mutation of TDP-43 (Barmada et al., 2014). Preclinical evidence inspired rapamycin to be included in human-based research. In fact, rapamycin has been involved in phase two clinical trials for ALS patients, which show it to be a well-tolerated drug that may show promise for future studies in other neurodegenerative disorders (Mandrioli et al., 2023).

In summary, rapamycin has been observed to have ground-breaking capabilities in reducing the progression of tumor growth, increasing lifespan, combating neurodegenerative diseases, and much more (Li et al., 2014; Selvarani et al., 2021). Due to its influence on mTOR, rapamycin has been the most proven molecule to directly mitigate the hallmarks of aging and give way to a possible target for age-related diseases. Additionally, metformin has been mentioned in various sources as a possible drug agent that can be used to slow the progression of various aging and age-related diseases such as neurodegenerative diseases and provide a possible therapeutic agent to the millions of patients suffering from these illnesses.

REFERENCES

Adlimoghaddam, A., & Albensi, B. C. (2021). The nuclear factor kappa B (NF-κB) signaling pathway is involved in ammonia-induced mitochondrial dysfunction. *Mitochondrion*, *57*, pp. 63–75. https://doi.org/10.1016/j.mito.2020.12.008

Adlimoghaddam, A., Albensi, B. C., & Eiser, A. R. (2022). Neurocovid, neuroinflammation, and nuclear factor-κB: the role for micronutrients. *Journal of Integrative and Complementary Medicine*, *28*(12), pp. 916–918. https://doi.org/10.1089/jicm.2022.0736

Adlimoghaddam, A., Benson, T., & Albensi, B. C. (2022). Mitochondrial transfusion improves mitochondrial function through up-regulation of mitochondrial complex II protein subunit SDHB in the hippocampus of aged mice. *Molecular Neurobiology*, *59*(10), pp. 6009–6017. https://doi.org/10.1007/s12035-022-02937-w

Adlimoghaddam, A., Odero, G. G., Glazner, G., Turner, R. S., & Albensi, B. C. (2021). Nilotinib improves bioenergetic profiling in brain astroglia in the 3xTg mouse model of Alzheimer's disease. *Aging and Disease*, *12*(2), pp. 441–465. https://doi.org/10.14336/AD.2020.0910

Adlimoghaddam, A., Snow, W. M., Stortz, G., Perez, C., Djordjevic, J., Goertzen, A. L., Ko, J. H., & Albensi, B. C. (2019). Regional hypometabolism in the 3xTg mouse model of Alzheimer's disease. *Neurobiology of Disease*, *127*, pp. 264–277. https://doi.org/10.1016/j.nbd.2019.03.008

Andronie-Cioara, F. L., Ardelean, A. I., Nistor-Cseppento, C. D., Jurcau, A., Jurcau, M. C., Pascalau, N., & Marcu, F. (2023). Molecular mechanisms of neuroinflammation in aging and Alzheimer's disease progression. *International Journal of Molecular Sciences*, *24*(3), p. 1869. https://doi.org/10.3390/ijms24031869

Anisimov, V. N., Zabezhinski, M. A., Popovich, I. G., Piskunova, T. S., Semenchenko, A. V, Tyndyk, M. L., Yurova, M. N., Antoch, M. P., & Blagosklonny, M. V. (2010). Rapamycin extends maximal lifespan in cancer-prone mice. *The American Journal of Pathology*, *176*(5), pp. 2092–2097. https://doi.org/10.2353/ajpath.2010.091050

Arnoux, I., Willam, M., Griesche, N., Krummeich, J., Watari, H., Offermann, N., Weber, S., Narayan Dey, P., Chen, C., Monteiro, O., Buettner, S., Meyer, K., Bano, D., Radyushkin, K., Langston, R., Lambert, J. J., Wanker, E., Methner, A., Krauss, S., … Stroh, A. (2018). Metformin reverses early cortical network dysfunction and behavior changes in Huntington's disease. *ELife*, *7*. https://doi.org/10.7554/eLife.38744

Arvanitakis, Z., Wilson, R. S., Bienias, J. L., Evans, D. A., & Bennett, D. A. (2004). Diabetes mellitus and risk of Alzheimer disease and decline in cognitive function. *Archives of Neurology*, *61*(5), p. 661. https://doi.org/10.1001/archneur.61.5.661

Ashburn, T. T., & Thor, K. B. (2004). Drug repositioning: identifying and developing new uses for existing drugs. *Nature Reviews Drug Discovery*, *3*(8), pp. 673–683. https://doi.org/10.1038/nrd1468

Asleh, R., Vucicevic, D., Petterson, T. M., Kremers, W. K., Pereira, N. L., Daly, R. C., Edwards, B. S., Steidley, D. E., Scott, R. L., & Kushwaha, S. S. (2022). Sirolimus-based immunosuppression is associated with decreased incidence of post-transplant lymphoproliferative disorder after heart transplantation: a double-center study. *Journal of Clinical Medicine*, *11*(2), p. 322. https://doi.org/10.3390/jcm11020322

Asnaghi, L., Bruno, P., Priulla, M., & Nicolin, A. (2004). mTOR: a protein kinase switching between life and death. *Pharmacological Research*, *50*(6), pp. 545–549. https://doi.org/10.1016/j.phrs.2004.03.007

Bai, X., Wey, M. C.-Y., Fernandez, E., Hart, M. J., Gelfond, J., Bokov, A. F., Rani, S., & Strong, R. (2015). Rapamycin improves motor function, reduces 4-hydroxynonenal adducted protein in brain, and attenuates synaptic injury in a mouse model of synucleinopathy. *Pathobiology of Aging & Age Related Diseases*, *5*, p. 28743. https://doi.org/10.3402/pba.v5.28743

Baker, D. J., Wijshake, T., Tchkonia, T., LeBrasseur, N. K., Childs, B. G., van de Sluis, B., Kirkland, J. L., & van Deursen, J. M. (2011). Clearance of p16Ink4a-positive senescent cells delays ageing-associated disorders. *Nature*, *479*(7372), pp. 232–236. https://doi.org/10.1038/nature10600

Barmada, S. J., Serio, A., Arjun, A., Bilican, B., Daub, A., Ando, D. M., Tsvetkov, A., Pleiss, M., Li, X., Peisach, D., Shaw, C., Chandran, S., & Finkbeiner, S. (2014). Autophagy induction enhances TDP43 turnover and survival in neuronal ALS models. *Nature Chemical Biology*, *10*(8), pp. 677–685. https://doi.org/10.1038/nchembio.1563

Bayliss, J. A., Lemus, M. B., Santos, V. V, Deo, M., Davies, J. S., Kemp, B. E., Elsworth, J. D., & Andrews, Z. B. (2016). Metformin prevents nigrostriatal dopamine degeneration independent of AMPK activation in dopamine neurons. *PloS One*, *11*(7), p. e0159381. https://doi.org/10.1371/journal.pone.0159381

Bell, S., Farran, B., McGurnaghan, S., McCrimmon, R. J., Leese, G. P., Petrie, J. R., McKeigue, P., Sattar, N., Wild, S., McKnight, J., Lindsay, R., Colhoun, H. M., & Looker, H. (2017). Risk of acute kidney injury and survival in patients treated with Metformin: an observational cohort study. *BMC Nephrology*, *18*(1), p. 163. https://doi.org/10.1186/s12882-017-0579-5

Bhansali, S., Bhansali, A., & Dhawan, V. (2019). Metformin promotes mitophagy in mononuclear cells: a potential *in vitro* model for unraveling metformin's mechanism of action. *Annals of the New York Academy of Sciences*, *1463*(1), pp. 23–36. https://doi.org/10.1111/nyas.14141

Bharath, L. P., Agrawal, M., McCambridge, G., Nicholas, D. A., Hasturk, H., Liu, J., Jiang, K., Liu, R., Guo, Z., Deeney, J., Apovian, C. M., Snyder-Cappione, J., Hawk, G. S., Fleeman, R. M., Pihl, R. M. F., Thompson, K., Belkina, A. C., Cui, L., Proctor, E. A., … Nikolajczyk, B. S. (2020). Metformin enhances autophagy and normalizes mitochondrial function to alleviate aging-associated inflammation. *Cell Metabolism*, *32*(1), pp. 44–55.e6. https://doi.org/10.1016/j.cmet.2020.04.015

Bielas, J., Herbst, A., Widjaja, K., Hui, J., Aiken, J. M., McKenzie, D., Miller, R. A., Brooks, S. V, & Wanagat, J. (2018). Long term rapamycin treatment improves mitochondrial DNA quality in aging mice. *Experimental Gerontology*, *106*, pp. 125–131. https://doi.org/10.1016/j.exger.2018.02.021

Caccamo, A., Majumder, S., Deng, J. J., Bai, Y., Thornton, F. B., & Oddo, S. (2009). Rapamycin rescues TDP-43 mislocalization and the associated low molecular mass neurofilament instability. *The Journal of Biological Chemistry*, *284*(40), 27416–27424. https://doi.org/10.1074/jbc.M109.031278

Cao, K., Graziotto, J. J., Blair, C. D., Mazzulli, J. R., Erdos, M. R., Krainc, D., & Collins, F. S. (2011). Rapamycin reverses cellular phenotypes and enhances mutant protein clearance in hutchinson-gilford progeria syndrome cells. *Science Translational Medicine*, *3*(89). https://doi.org/10.1126/scitranslmed.3002346

Charpignon, M.-L., Vakulenko-Lagun, B., Zheng, B., Magdamo, C., Su, B., Evans, K., Rodriguez, S., Sokolov, A., Boswell, S., Sheu, Y.-H., Somai, M., Middleton, L., Hyman, B. T., Betensky, R. A., Finkelstein, S. N., Welsch, R. E., Tzoulaki, I., Blacker, D., Das, S., & Albers, M. W. (2022). Causal inference in medical records and complementary systems pharmacology for metformin drug repurposing towards dementia. *Nature Communications*, *13*(1), p. 7652. https://doi.org/10.1038/s41467-022-35157-w

Chiang, M.-C., Cheng, Y.-C., Chen, S.-J., Yen, C.-H., & Huang, R.-N. (2016). Metformin activation of AMPK-dependent pathways is neuroprotective in human neural stem cells against Amyloid-beta-induced mitochondrial dysfunction. *Experimental Cell Research*, *347*(2), pp. 322–331. https://doi.org/10.1016/j.yexcr.2016.08.013

Chung, J., Kuo, C. J., Crabtree, G. R., & Blenis, J. (1992). Rapamycin-FKBP specifically blocks growth-dependent activation of and signaling by the 70 kd S6 protein kinases. *Cell*, *69*(7), pp. 1227–1236. https://doi.org/10.1016/0092-8674(92)90643-q

Dai, Y., Zheng, K., Clark, J., Swerdlow, R. H., Pulst, S. M., Sutton, J. P., Shinobu, L. A., & Simon, D. K. (2014). Rapamycin drives selection against a pathogenic heteroplasmic mitochondrial DNA mutation. *Human Molecular Genetics*, *23*(3), pp. 637–647. https://doi.org/10.1093/hmg/ddt450

Darrow, J. J., Avorn, J., & Kesselheim, A. S. (2020). FDA approval and regulation of pharmaceuticals, 1983–2018. *JAMA*, *323*(2), p. 164. https://doi.org/10.1001/jama.2019.20288

De Keersmaecker, A.-V., Van Doninck, E., Popescu, V., Willem, L., Cambron, M., Laureys, G., D'Haeseleer, M., Bjerke, M., Roelant, E., Lemmerling, M., D'hooghe, M. B., Derdelinckx, J., Reynders, T., & Willekens, B. (2024). A metformin add-on clinical study in multiple sclerosis to evaluate brain remyelination and neurodegeneration (MACSiMiSE-BRAIN): study protocol for a multi-center randomized placebo controlled clinical trial. *Frontiers in Immunology*, *15*, p. 1362629. https://doi.org/10.3389/fimmu.2024.1362629

de Marañón, A. M., Díaz-Pozo, P., Canet, F., Díaz-Morales, N., Abad-Jiménez, Z., López-Domènech, S., Vezza, T., Apostolova, N., Morillas, C., Rocha, M., & Víctor, V. M. (2022). Metformin modulates mitochondrial function and mitophagy in peripheral blood mononuclear cells from type 2 diabetic patients. *Redox Biology*, *53*, p. 102342. https://doi.org/10.1016/j.redox.2022.102342

DeFronzo, R., Fleming, G. A., Chen, K., & Bicsak, T. A. (2016). Metformin-associated lactic acidosis: current perspectives on causes and risk. *Metabolism*, *65*(2), pp. 20–29. https://doi.org/10.1016/j.metabol.2015.10.014

Dehkordi, S. K., Walker, J., Sah, E., Bennett, E., Atrian, F., Frost, B., Woost, B., Bennett, R. E., Orr, T. C., Zhou, Y., Andhey, P. S., Colonna, M., Sudmant, P. H., Xu, P., Wang, M., Zhang, B., Zare, H., & Orr, M. E. (2021). Profiling senescent cells in human brains reveals neurons with CDKN2D/p19 and tau neuropathology. *Nature Aging*, *1*(12), pp. 1107–1116. https://doi.org/10.1038/s43587-021-00142-3

Dierickx, D., & Habermann, T. M. (2018). Post-transplantation lymphoproliferative disorders in adults. *New England Journal of Medicine*, *378*(6), pp. 549–562. https://doi.org/10.1056/nejmra1702693

Dudek, H., Datta, S. R., Franke, T. F., Birnbaum, M. J., Yao, R., Cooper, G. M., Segal, R. A., Kaplan, D. R., & Greenberg, M. E. (1997). Regulation of neuronal survival by the serine-threonine protein kinase Akt. *Science*, *275*(5300), pp. 661–665. https://doi.org/10.1126/science.275.5300.661

Ellisen, L. W., Ramsayer, K. D., Johannessen, C. M., Yang, A., Beppu, H., Minda, K., Oliner, J. D., McKeon, F., & Haber, D. A. (2002). REDD1, a developmentally regulated transcriptional target of p63 and p53, links p63 to regulation of reactive oxygen species. *Molecular Cell*, *10*(5), pp. 995–1005. https://doi.org/10.1016/s1097-2765(02)00706-2

Eshraghi, M., Adlimoghaddam, A., Mahmoodzadeh, A., Sharifzad, F., Yasavoli-Sharahi, H., Lorzadeh, S., Albensi, B. C., & Ghavami, S. (2021). Alzheimer's disease pathogenesis: role of autophagy and mitophagy focusing in microglia. *International Journal of Molecular Sciences*, *22*(7), p. 3330. https://doi.org/10.3390/ijms22073330

Eshraghi, M., Ahmadi, M., Afshar, S., Lorzadeh, S., Adlimoghaddam, A., Rezvani Jalal, N., West, R., Dastghaib, S., Igder, S., Torshizi, S. R. N., Mahmoodzadeh, A., Mokarram, P., Madrakian, T., Albensi, B. C., Łos, M. J., Ghavami, S., & Pecic, S. (2022). Enhancing autophagy in Alzheimer's disease through drug repositioning. *Pharmacology & Therapeutics*, *237*, p. 108171. https://doi.org/10.1016/j.pharmthera.2022.108171

Foretz, M., Guigas, B., Bertrand, L., Pollak, M., & Viollet, B. (2014). Metformin: from mechanisms of action to therapies. *Cell Metabolism*, *20*(6), pp. 953–966. https://doi.org/10.1016/j.cmet.2014.09.018

Gilkerson, R. W., De Vries, R. L. A., Lebot, P., Wikstrom, J. D., Torgyekes, E., Shirihai, O. S., Przedborski, S., & Schon, E. A. (2012). Mitochondrial autophagy in cells with mtDNA mutations results from synergistic loss of transmembrane potential and mTORC1 inhibition. *Human Molecular Genetics*, *21*(5), pp. 978–990. https://doi.org/10.1093/hmg/ddr529

Granville, C. A., Warfel, N., Tsurutani, J., Hollander, M. C., Robertson, M., Fox, S. D., Veenstra, T. D., Issaq, H. J., Linnoila, R. I., & Dennis, P. A. (2007). Identification of a highly effective rapamycin schedule that markedly reduces the size, multiplicity, and phenotypic progression of tobacco carcinogen–induced murine lung tumors. *Clinical Cancer Research*, *13*(7), pp. 2281–2289. https://doi.org/10.1158/1078-0432.ccr-06-2570

Greco, S. J., Sarkar, S., Casadesus, G., Zhu, X., Smith, M. A., Ashford, J. W., Johnston, J. M., & Tezapsidis, N. (2009). Leptin inhibits glycogen synthase kinase-3beta to prevent tau phosphorylation in neuronal cells. *Neuroscience Letters*, *455*(3), pp. 191–194. https://doi.org/10.1016/j.neulet.2009.03.066

Greco, S. J., Sarkar, S., Johnston, J. M., & Tezapsidis, N. (2009). Leptin regulates tau phosphorylation and amyloid through AMPK in neuronal cells. *Biochemical and Biophysical Research Communications*, *380*(1), pp. 98–104. https://doi.org/10.1016/j.bbrc.2009.01.041

Han, Y., Xie, H., Liu, Y., Gao, P., Yang, X., & Shen, Z. (2019). Effect of metformin on all-cause and cardiovascular mortality in patients with coronary artery diseases: a systematic review and an updated meta-analysis. *Cardiovascular Diabetology*, *18*(1), p. 96. https://doi.org/10.1186/s12933-019-0900-7

Harrison, D. E., Strong, R., Sharp, Z. D., Nelson, J. F., Astle, C. M., Flurkey, K., Nadon, N. L., Wilkinson, J. E., Frenkel, K., Carter, C. S., Pahor, M., Javors, M. A., Fernandez, E., & Miller, R. A. (2009). Rapamycin fed late in life extends lifespan in genetically heterogeneous mice. *Nature*, *460*(7253), pp. 392–395. https://doi.org/10.1038/nature08221

Hautbergue, G. M., Cleary, J. D., Guo, S., & Ranum, L. P. W. (2021). Therapeutic strategies for C9orf72 amyotrophic lateral sclerosis and frontotemporal dementia. *Current Opinion in Neurology*, *34*(5), pp. 748–755. https://doi.org/10.1097/WCO.0000000000000984

Hawley, S. A., Ross, F. A., Chevtzoff, C., Green, K. A., Evans, A., Fogarty, S., Towler, M. C., Brown, L. J., Ogunbayo, O. A., Evans, A. M., & Hardie, D. G. (2010). Use of cells expressing gamma subunit variants to identify diverse mechanisms of AMPK activation. *Cell Metabolism*, *11*(6), pp. 554–565. https://doi.org/10.1016/j.cmet.2010.04.001

Hervás, D., Fornés-Ferrer, V., Gómez-Escribano, A. P., Sequedo, M. D., Peiró, C., Millán, J. M., & Vázquez-Manrique, R. P. (2017). Metformin intake associates with better cognitive function in patients with Huntington's disease. *PloS One*, *12*(6), p. e0179283. https://doi.org/10.1371/journal.pone.0179283

Houchens, D. P., Ovejera, A. A., Riblet, S. M., & Slagel, D. E. (1983). Human brain tumor xenografts in nude mice as a chemotherapy model. *European Journal of Cancer and Clinical Oncology*, *19*(6), pp. 799–805. https://doi.org/10.1016/0277-5379(83)90012-3

Jacinto, E., Loewith, R., Schmidt, A., Lin, S., Rüegg, M. A., Hall, A., & Hall, M. N. (2004). Mammalian TOR complex 2 controls the actin cytoskeleton and is rapamycin insensitive. *Nature Cell Biology*, *6*(11), pp. 1122–1128. https://doi.org/10.1038/ncb1183

Johnson, S. C., Yanos, M. E., Bitto, A., Castanza, A., Gagnidze, A., Gonzalez, B., Gupta, K., Hui, J., Jarvie, C., Johnson, B. M., Letexier, N., McCanta, L., Sangesland, M., Tamis, O., Uhde, L., Van Den Ende, A., Rabinovitch, P. S., Suh, Y., & Kaeberlein, M. (2015). Dose-dependent effects of mTOR inhibition on weight and mitochondrial disease in mice. *Frontiers in Genetics*, *6*, p. 247. https://doi.org/10.3389/fgene.2015.00247

Jourdan, J.-P., Bureau, R., Rochais, C., & Dallemagne, P. (2020). Drug repositioning: a brief overview. *The Journal of Pharmacy and Pharmacology*, *72*(9), pp. 1145–1151. https://doi.org/10.1111/jphp.13273

Kandul, N. P., Zhang, T., Hay, B. A., & Guo, M. (2016). Selective removal of deletion-bearing mitochondrial DNA in heteroplasmic Drosophila. *Nature Communications*, *7*, p. 13100. https://doi.org/10.1038/ncomms13100

Kapahi, P., Zid, B. M., Harper, T., Koslover, D., Sapin, V., & Benzer, S. (2004). Regulation of lifespan in Drosophila by modulation of genes in the TOR signaling pathway. *Current Biology: CB*, *14*(10), pp. 885–890. https://doi.org/10.1016/j.cub.2004.03.059

Katila, N., Bhurtel, S., Shadfar, S., Srivastav, S., Neupane, S., Ojha, U., Jeong, G.-S., & Choi, D.-Y. (2017). Metformin lowers α-synuclein phosphorylation and upregulates neurotrophic factor in the MPTP mouse model of Parkinson's disease. *Neuropharmacology*, *125*, pp. 396–407. https://doi.org/10.1016/j.neuropharm.2017.08.015

Kato, Y., Nishiyama, K., Nishimura, A., Noda, T., Okabe, K., Kusakabe, T., Kanda, Y., & Nishida, M. (2022). Drug repurposing for the treatment of COVID-19. *Journal of Pharmacological Sciences*, *149*(3), pp. 108–114. https://doi.org/10.1016/j.jphs.2022.04.007

Kickstein, E., Krauss, S., Thornhill, P., Rutschow, D., Zeller, R., Sharkey, J., Williamson, R., Fuchs, M., Köhler, A., Glossmann, H., Schneider, R., Sutherland, C., & Schweiger, S. (2010). Biguanide metformin acts on tau phosphorylation via mTOR/protein phosphatase 2A (PP2A) signaling. *Proceedings of the National Academy of Sciences of the United States of America*, *107*(50), pp. 21830–21835. https://doi.org/10.1073/pnas.0912793107

Kim, Y. C., & Guan, K.-L. (2015). mTOR: a pharmacologic target for autophagy regulation. *The Journal of Clinical Investigation*, *125*(1), pp. 25–32. https://doi.org/10.1172/JCI73939

Koenig, A. M., Mechanic-Hamilton, D., Xie, S. X., Combs, M. F., Cappola, A. R., Xie, L., Detre, J. A., Wolk, D. A., & Arnold, S. E. (2017). Effects of the insulin sensitizer metformin in Alzheimer disease: pilot data from a randomized placebo-controlled crossover study. *Alzheimer Disease and Associated Disorders*, *31*(2), pp. 107–113. https://doi.org/10.1097/WAD.0000000000000202

Łabuzek, K., Suchy, D., Gabryel, B., Bielecka, A., Liber, S., & Okopień, B. (2010). Quantification of metformin by the HPLC method in brain regions, cerebrospinal fluid and plasma of rats treated with lipopolysaccharide. *Pharmacological Reports*, *62*(5), pp. 956–965. https://doi.org/10.1016/s1734-1140(10)70357-1

Legtenberg, R. J., Houston, R. J. F., Oeseburg, B., & Smits, P. (2002). Metformin improves cardiac functional recovery after Ischemia in rats. *Hormone and Metabolic Research*, *34*(4), pp. 182–185. https://doi.org/10.1055/s-2002-26705

Li, J., Kim, S. G., & Blenis, J. (2014). Rapamycin: one drug, many effects. *Cell Metabolism*, *19*(3), pp. 373–379. https://doi.org/10.1016/j.cmet.2014.01.001

Lin, A.-L., Jahrling, J. B., Zhang, W., DeRosa, N., Bakshi, V., Romero, P., Galvan, V., & Richardson, A. (2017). Rapamycin rescues vascular, metabolic and learning deficits in apolipoprotein E4 transgenic mice with pre-symptomatic Alzheimer's disease. *Journal of Cerebral Blood Flow and Metabolism: Official Journal of the International Society of Cerebral Blood Flow and Metabolism*, *37*(1), pp. 217–226. https://doi.org/10.1177/0271678X15621575

Lin, A.-L., Zheng, W., Halloran, J. J., Burbank, R. R., Hussong, S. A., Hart, M. J., Javors, M., Shih, Y.-Y. I., Muir, E., Solano Fonseca, R., Strong, R., Richardson, A. G., Lechleiter, J. D., Fox, P. T., & Galvan, V. (2013). Chronic rapamycin restores brain vascular integrity and function through NO synthase activation and improves memory in symptomatic mice modeling Alzheimer's disease. *Journal of Cerebral Blood Flow and Metabolism: Official Journal of the International Society of Cerebral Blood Flow and Metabolism*, *33*(9), pp. 1412–1421. https://doi.org/10.1038/jcbfm.2013.82

Liu, C.-C., Kanekiyo, T., Xu, H., & Bu, G. (2013). Apolipoprotein E and Alzheimer disease: risk, mechanisms and therapy. *Nature Reviews Neurology*, *9*(2), pp. 106–118. https://doi.org/10.1038/nrneurol.2012.263

Liu, L., Drouet, V., Wu, J. W., Witter, M. P., Small, S. A., Clelland, C., & Duff, K. (2012). Trans-synaptic spread of tau pathology in vivo. *PloS One*, *7*(2), p. e31302. https://doi.org/10.1371/journal.pone.0031302

Luchsinger, J. A., Perez, T., Chang, H., Mehta, P., Steffener, J., Pradabhan, G., Ichise, M., Manly, J., Devanand, D. P., & Bagiella, E. (2016). Metformin in amnestic mild cognitive impairment: results of a pilot randomized placebo controlled clinical trial. *Journal, of Alzheimer's Disease: JAD*, *51*(2), pp. 501–514. https://doi.org/10.3233/JAD-150493

Madiraju, A. K., Erion, D. M., Rahimi, Y., Zhang, X.-M., Braddock, D. T., Albright, R. A., Prigaro, B. J., Wood, J. L., Bhanot, S., MacDonald, M. J., Jurczak, M. J., Camporez, J.-P., Lee, H.-Y., Cline, G. W., Samuel, V. T., Kibbey, R. G., & Shulman, G. I. (2014). Metformin suppresses gluconeogenesis by inhibiting mitochondrial glycerophosphate dehydrogenase. *Nature*, *510*(7506), pp. 542–546. https://doi.org/10.1038/nature13270

Malagelada, C., Jin, Z. H., Jackson-Lewis, V., Przedborski, S., & Greene, L. A. (2010). Rapamycin protects against neuron death in in vitro and in vivo models of Parkinson's disease. *The Journal of Neuroscience: The Official Journal of the Society for Neuroscience*, *30*(3), pp. 1166–1175. https://doi.org/10.1523/JNEUROSCI.3944-09.2010

Malagelada, C., Ryu, E. J., Biswas, S. C., Jackson-Lewis, V., & Greene, L. A. (2006). RTP801 is elevated in Parkinson brain substantia nigral neurons and mediates death in cellular models of Parkinson's disease by a mechanism involving mammalian target of rapamycin inactivation. *The Journal of Neuroscience: The Official Journal of the Society for Neuroscience*, *26*(39), pp. 9996–10005. https://doi.org/10.1523/JNEUROSCI.3292-06.2006

Mandrioli, J., D'Amico, R., Zucchi, E., De Biasi, S., Banchelli, F., Martinelli, I., Simonini, C., Lo Tartaro, D., Vicini, R., Fini, N., Gianferrari, G., Pinti, M., Lunetta, C., Gerardi, F., Tarlarini, C., Mazzini, L., De Marchi, F., Scognamiglio, A., Sorarù, G., … Cossarizza, A. (2023). Randomized, double-blind, placebo-controlled trial of rapamycin in amyotrophic lateral sclerosis. *Nature Communications*, *14*(1), p. 4970. https://doi.org/10.1038/s41467-023-40734-8

Marx, S. O., Jayaraman, T., Go, L. O., & Marks, A. R. (1995). Rapamycin-FKBP inhibits cell cycle regulators of proliferation in vascular smooth muscle cells. *Circulation Research*, *76*(3), pp. 412–417. https://doi.org/10.1161/01.res.76.3.412

Masini, D., Bonito-Oliva, A., Bertho, M., & Fisone, G. (2018). Inhibition of mtorc1 signaling reverts cognitive and affective deficits in a mouse model of Parkinson's disease. *Frontiers in Neurology*, *9*, p. 208. https://doi.org/10.3389/fneur.2018.00208

Matthes, F., Hettich, M. M., Schilling, J., Flores-Dominguez, D., Blank, N., Wiglenda, T., Buntru, A., Wolf, H., Weber, S., Vorberg, I., Dagane, A., Dittmar, G., Wanker, E., Ehninger, D., & Krauss, S. (2018). Inhibition of the MID1 protein complex: a novel approach targeting APP protein synthesis. *Cell Death Discovery*, *4*, p. 4. https://doi.org/10.1038/s41420-017-0003-8

Misbin, R. I., Green, L., Stadel, B. V, Gueriguian, J. L., Gubbi, A., & Fleming, G. A. (1998). Lactic acidosis in patients with diabetes treated with metformin. *New England Journal of Medicine*, *338*(4), pp. 265–266. https://doi.org/10.1056/nejm199801223380415

Moiseeva, O., Deschênes-Simard, X., St-Germain, E., Igelmann, S., Huot, G., Cadar, A. E., Bourdeau, V., Pollak, M. N., & Ferbeyre, G. (2013). Metformin inhibits the senescence-associated secretory phenotype by interfering with IKK/NF-κB activation. *Aging Cell*, *12*(3), pp. 489–498. https://doi.org/10.1111/acel.12075

Nabrdalik, K., Skonieczna-Żydecka, K., Irlik, K., Hendel, M., Kwiendacz, H., Łoniewski, I., Januszkiewicz, K., Gumprecht, J., & Lip, G. Y. H. (2022). Gastrointestinal adverse events of metformin treatment in patients with type 2 diabetes mellitus: a systematic review, meta-analysis and meta-regression of randomized controlled trials. *Frontiers in Endocrinology*, *13*, p. 975912. https://doi.org/10.3389/fendo.2022.975912

Nafisa, A., Gray, S. G., Cao, Y., Wang, T., Xu, S., Wattoo, F. H., Barras, M., Cohen, N., Kamato, D., & Little, P. J. (2018). Endothelial function and dysfunction: impact of metformin. *Pharmacology & Therapeutics*, *192*, pp. 150–162. https://doi.org/10.1016/j.pharmthera.2018.07.007

Nath, N., Khan, M., Paintlia, M. K., Singh, I., Hoda, M. N., & Giri, S. (2009). Metformin attenuated the autoimmune disease of the central nervous system in animal models of multiple sclerosis. *Journal of Immunology (Baltimore, Md. : 1950)*, *182*(12), pp. 8005–8014. https://doi.org/10.4049/jimmunol.0803563

Neumann, B., Baror, R., Zhao, C., Segel, M., Dietmann, S., Rawji, K. S., Foerster, S., McClain, C. R., Chalut, K., van Wijngaarden, P., & Franklin, R. J. M. (2019). Metformin restores CNS remyelination capacity by rejuvenating aged stem cells. *Cell Stem Cell*, *25*(4), pp. 473–485.e8. https://doi.org/10.1016/j.stem.2019.08.015

Nosengo, N. (2016). Can you teach old drugs new tricks? *Nature*, *534*(7607), pp. 314–316. https://doi.org/10.1038/534314a

Notaro, A. L. G., & Neto, F. T. L. (2022). The use of metformin in women with polycystic ovary syndrome: an updated review. *Journal of Assisted Reproduction and Genetics*, *39*(3), pp. 573–579. https://doi.org/10.1007/s10815-022-02429-9

Opelz, G., & Döhler, B. (2004). Lymphomas after solid organ transplantation: a collaborative transplant study report. *American Journal of Transplantation*, *4*(2), pp. 222–230. https://doi.org/10.1046/j.1600-6143.2003.00325.x

Owen, M. R., Doran, E., & Halestrap, A. P. (2000). Evidence that metformin exerts its anti-diabetic effects through inhibition of complex 1 of the mitochondrial respiratory chain. *Biochemical Journal*, *348*(3), p. 607. https://doi.org/10.1042/0264-6021:3480607

Ozcelik, S., Fraser, G., Castets, P., Schaeffer, V., Skachokova, Z., Breu, K., Clavaguera, F., Sinnreich, M., Kappos, L., Goedert, M., Tolnay, M., & Winkler, D. T. (2013). Rapamycin attenuates the progression of tau pathology in P301S tau transgenic mice. *PloS One*, *8*(5), pp. e62459. https://doi.org/10.1371/journal.pone.0062459

Paintlia, A. S., Paintlia, M. K., Mohan, S., Singh, A. K., & Singh, I. (2013). AMP-activated protein kinase signaling protects oligodendrocytes that restore central nervous system functions in an experimental autoimmune encephalomyelitis model. *The American Journal of Pathology*, *183*(2), pp. 526–541. https://doi.org/10.1016/j.ajpath.2013.04.030

Pang, X., Fu, W., Wang, J., Kang, D., Xu, L., Zhao, Y., Liu, A.-L., & Du, G.-H. (2018). Identification of Estrogen receptor α antagonists from natural products via in vitro and in silico approaches. *Oxidative Medicine and Cellular Longevity*, *2018*, p. 6040149. https://doi.org/10.1155/2018/6040149

Pérez-Revuelta, B. I., Hettich, M. M., Ciociaro, A., Rotermund, C., Kahle, P. J., Krauss, S., & Di Monte, D. A. (2014). Metformin lowers Ser-129 phosphorylated α-synuclein levels via mTOR-dependent protein phosphatase 2A activation. *Cell Death & Disease*, *5*(5), p. e1209. https://doi.org/10.1038/cddis.2014.175

Pollak, M. (2014). Overcoming drug development bottlenecks with repurposing: repurposing biguanides to target energy metabolism for cancer treatment. *Nature Medicine*, *20*(6), pp. 591–593. https://doi.org/10.1038/nm.3596

Ramos, F. J., Chen, S. C., Garelick, M. G., Dai, D.-F., Liao, C.-Y., Schreiber, K. H., MacKay, V. L., An, E. H., Strong, R., Ladiges, W. C., Rabinovitch, P. S., Kaeberlein, M., & Kennedy, B. K. (2012). Rapamycin reverses elevated mTORC1 signaling in lamin A/C-deficient mice, rescues cardiac and skeletal muscle function, and extends survival. *Science Translational Medicine*, *4*(144), p. 144ra103. https://doi.org/10.1126/scitranslmed.3003802

Ravikumar, B., Vacher, C., Berger, Z., Davies, J. E., Luo, S., Oroz, L. G., Scaravilli, F., Easton, D. F., Duden, R., O'Kane, C. J., & Rubinsztein, D. C. (2004). Inhibition of mTOR induces autophagy and reduces toxicity of polyglutamine expansions in fly and mouse models of Huntington disease. *Nature Genetics*, *36*(6), pp. 585–595. https://doi.org/10.1038/ng1362

Rocha, A., Almeida, M., Santos, J., & Carvalho, A. (2012). Metformin in patients with chronic kidney disease: strengths and weaknesses. *Journal of Nephrology*, *26*(1), pp. 55–60. https://doi.org/10.5301/jn.5000166

Roessler, H. I., Knoers, N. V. A. M., van Haelst, M. M., & van Haaften, G. (2021). Drug repurposing for rare diseases. *Trends in Pharmacological Sciences*, *42*(4), pp. 255–267. https://doi.org/10.1016/j.tips.2021.01.003

Rujescu, D., & Jung, M. (2020). Faculty opinions recommendation of the TREM2-APOE pathway drives the transcriptional phenotype of dysfunctional microglia in neurodegenerative diseases. *Faculty Opinions – Post-Publication Peer Review of the Biomedical Literature. H1 Connect.* https://doi.org/10.3410/f.731249475.793578324

Russell, G., Graveley, R., Seid, J., Al-Humidan, A.-K., & Skjodt, H. (1992). Mechanisms of action of cyclosporine and effects on connective tissues. *Seminars in Arthritis and Rheumatism*, *21*(6), pp. 16–22. https://doi.org/10.1016/0049-0172(92)90009-3

Saewanee, N., Praputpittaya, T., Malaiwong, N., Chalorak, P., & Meemon, K. (2021). Neuroprotective effect of metformin on dopaminergic neurodegeneration and α-synuclein aggregation in C. elegans model of Parkinson's disease. *Neuroscience Research*, *162*, pp. 13–21. https://doi.org/10.1016/j.neures.2019.12.017

Sala Frigerio, C., Wolfs, L., Fattorelli, N., Thrupp, N., Voytyuk, I., Schmidt, I., Mancuso, R., Chen, W.-T., Woodbury, M. E., Srivastava, G., Möller, T., Hudry, E., Das, S., Saido, T., Karran, E., Hyman, B., Perry, V. H., Fiers, M., & De Strooper, B. (2019). The major risk factors for Alzheimer's disease: age, sex, and genes modulate the microglia response to Aβ plaques. *Cell Reports*, *27*(4), pp. 1293–1306.e6. https://doi.org/10.1016/j.celrep.2019.03.099

Samaras, K., Crawford, J. D., Draper, B., Trollor, J. N., Brodaty, H., & Sachdev, P. S. (2021). Response to comment on Samara et al. metformin use is associated with slowed cognitive decline and reduced incident dementia in older adults with Type 2 diabetes: the Sydney memory and ageing study. Diabetes Care 2020;43:2691–2701. *Diabetes Care*, *44*(4), p. e74. https://doi.org/10.2337/dci20-0077

Sanchis, A., García-Gimeno, M. A., Cañada-Martínez, A. J., Sequedo, M. D., Millán, J. M., Sanz, P., & Vázquez-Manrique, R. P. (2019). Metformin treatment reduces motor and neuropsychiatric phenotypes in the zQ175 mouse model of Huntington disease. *Experimental & Molecular Medicine*, *51*(6), pp. 1–16. https://doi.org/10.1038/s12276-019-0264-9

Scannell, J. W., Blanckley, A., Boldon, H., & Warrington, B. (2012). Diagnosing the decline in pharmaceutical R&D efficiency. *Nature Reviews Drug Discovery*, *11*(3), pp. 191–200. https://doi.org/10.1038/nrd3681

Sehgal, S. N. (1998). Rapamune® (RAPA, rapamycin, sirolimus): mechanism of action immunosuppressive effect results from blockade of signal transduction and inhibition of cell cycle progression. *Clinical Biochemistry*, *31*(5), pp. 335–340. https://doi.org/10.1016/s0009-9120(98)00045-9

Selvarani, R., Mohammed, S., & Richardson, A. (2021). Effect of rapamycin on aging and age-related diseases-past and future. *GeroScience*, *43*(3), pp. 1135–1158. https://doi.org/10.1007/s11357-020-00274-1

Serruys, P. W., Regar, E., & Carter, A. J. (2002). Rapamycin eluting stent: the onset of a new era in interventional cardiology. *Heart (British Cardiac Society)*, *87*(4), pp. 305–307. https://doi.org/10.1136/heart.87.4.305

Shi, Q., Liu, S., Fonseca, V. A., Thethi, T. K., & Shi, L. (2019). Effect of metformin on neurodegenerative disease among elderly adult US veterans with type 2 diabetes mellitus. *BMJ Open*, *9*(7), p. e024954. https://doi.org/10.1136/bmjopen-2018-024954

Siman, R., Cocca, R., & Dong, Y. (2015). The mTOR inhibitor rapamycin mitigates perforant pathway neurodegeneration and synapse loss in a mouse model of early-stage Alzheimer-type tauopathy. *PloS One*, *10*(11), p. e0142340. https://doi.org/10.1371/journal.pone.0142340

Spilman, P., Podlutskaya, N., Hart, M. J., Debnath, J., Gorostiza, O., Bredesen, D., Richardson, A., Strong, R., & Galvan, V. (2010). Inhibition of mTOR by rapamycin abolishes cognitive deficits and reduces amyloid-beta levels in a mouse model of Alzheimer's disease. *PloS One*, *5*(4), p. e9979. https://doi.org/10.1371/journal.pone.0009979

Svensson, J. E., Bolin, M., Thor, D., Williams, P. A., Brautaset, R., Carlsson, M., Sörensson, P., Marlevi, D., Spin-Neto, R., Probst, M., Hagman, G., Morén, A. F., Kivipelto, M., & Plavén-Sigray, P. (2024). Evaluating the effect of rapamycin treatment in Alzheimer's disease and aging using in vivo imaging: the ERAP phase IIa clinical study protocol. *BMC Neurology*, *24*(1), p. 111. https://doi.org/10.1186/s12883-024-03596-1

Syal, C., Kosaraju, J., Hamilton, L., Aumont, A., Chu, A., Sarma, S. N., Thomas, J., Seegobin, M., Dilworth, F. J., He, L., Wondisford, F. E., Zimmermann, R., Parent, M., Fernandes, K., & Wang, J. (2020). Dysregulated expression of monoacylglycerol lipase is a marker for anti-diabetic drug metformin-targeted therapy to correct impaired neurogenesis and spatial memory in Alzheimer's disease. *Theranostics*, *10*(14), pp. 6337–6360. https://doi.org/10.7150/thno.44962

Szwed, A., Kim, E., & Jacinto, E. (2021). Regulation and metabolic functions of mTORC1 and mTORC2. *Physiological Reviews*, *101*(3), pp. 1371–1426. https://doi.org/10.1152/physrev.00026.2020

Tan, G. S. Q., Sloan, E. K., Lambert, P., Kirkpatrick, C. M. J., & Ilomäki, J. (2023). Drug repurposing using real-world data. *Drug Discovery Today*, *28*(1), p. 103422. https://doi.org/10.1016/j.drudis.2022.103422

Teo, E., Ravi, S., Barardo, D., Kim, H.-S., Fong, S., Cazenave-Gassiot, A., Tan, T. Y., Ching, J., Kovalik, J.-P., Wenk, M. R., Gunawan, R., Moore, P. K., Halliwell, B., Tolwinski, N., & Gruber, J. (2019). Metabolic stress is a primary pathogenic event in transgenic Caenorhabditis elegans expressing pan-neuronal human amyloid beta. *ELife*, *8*. https://doi.org/10.7554/eLife.50069

Tsai, M.-K., Wu, F.-L. L. I. N., Lai, I -R., Lee, C.-Y., Hu, R.-H., & Lee, P.-H. (2009). Decreased acute rejection and improved renal allograft survival using sirolimus and low-dose Calcineurin inhibitors without induction therapy. *The International Journal of Artificial Organs*, *32*(6), pp. 371–380. https://doi.org/10.1177/039139880903200608

Twig, G., Hyde, B., & Shirihai, O. S. (2008). Mitochondrial fusion, fission and autophagy as a quality control axis: the bioenergetic view. *Biochimica et Biophysica Acta*, *1777*(9), pp. 1092–1097. https://doi.org/10.1016/j.bbabio.2008.05.001

Vazquez-Martin, A., López-Bonetc, E., Cufí, S., Oliveras-Ferraros, C., Del Barco, S., Martin-Castillo, B., & Menendez, J. A. (2011). Repositioning chloroquine and metformin to eliminate cancer stem cell traits in pre-malignant lesions. *Drug Resistance Updates: Reviews and Commentaries in Antimicrobial and Anticancer Chemotherapy*, *14*(4–5), pp. 212–223. https://doi.org/10.1016/j.drup.2011.04.003

Vellai, T., Takacs-Vellai, K., Zhang, Y., Kovacs, A. L., Orosz, L., & Müller, F. (2003). Influence of TOR kinase on lifespan in C. elegans. *Nature*, *426*(6967), p. 620. https://doi.org/10.1038/426620a

Vézina, C., Kudelski, A., & Sehgal, S. N. (1975). Rapamycin (AY-22,989), a new antifungal antibiotic. I. Taxonomy of the producing streptomycete and isolation of the active principle. *The Journal of Antibiotics*, *28*(10), pp. 721–726. https://doi.org/10.7164/antibiotics.28.721

Wang, H., Fu, J., Xu, X., Yang, Z., & Zhang, T. (2021). Rapamycin activates mitophagy and alleviates cognitive and synaptic plasticity deficits in a mouse model of Alzheimer's disease. *The Journals of Gerontology: Series A*, *76*(10), pp. 1707–1713. https://doi.org/10.1093/gerona/glab142

Wang, R., Yu, Z., Sunchu, B., Shoaf, J., Dang, I., Zhao, S., Caples, K., Bradley, L., Beaver, L. M., Ho, E., Löhr, C. V., & Perez, V. I. (2017). Rapamycin inhibits the secretory phenotype of senescent cells by a Nrf2-independent mechanism. *Aging Cell*, *16*(3), pp. 564–574. https://doi.org/10.1111/acel.12587

Wong, C. H., Siah, K. W., & Lo, A. W. (2019). Estimation of clinical trial success rates and related parameters. *Biostatistics (Oxford, England)*, *20*(2), pp. 273–286. https://doi.org/10.1093/biostatistics/kxx069

Woolf, S. H., & Schoomaker, H. (2019). Life expectancy and mortality rates in the United States, 1959–2017. *JAMA*, *322*(20), pp. 1996–2016. https://doi.org/10.1001/jama.2019.16932

Wrighten, S. A., Piroli, G. G., Grillo, C. A., & Reagan, L. P. (2009). A look inside the diabetic brain: contributors to diabetes-induced brain aging. *Biochimica et Biophysica Acta*, *1792*(5), pp. 444–453. https://doi.org/10.1016/j.bbadis.2008.10.013

Xu, X., Sun, Y., Cen, X., Shan, B., Zhao, Q., Xie, T., Wang, Z., Hou, T., Xue, Y., Zhang, M., Peng, D., Sun, Q., Yi, C., Najafov, A., & Xia, H. (2021). Metformin activates chaperone-mediated autophagy and improves disease pathologies in an Alzheimer disease mouse model. *Protein & Cell*, *12*(10), pp. 769–787. https://doi.org/10.1007/s13238-021-00858-3

Yang, T., Yang, Y., & Liu, S. (2015). Association between metformin therapy and breast cancer incidence and mortality: evidence from a meta-analysis. *Journal of Breast Cancer*, *18*(3), pp. 264–270. https://doi.org/10.4048/jbc.2015.18.3.264

Yazdani, S., Seitz, C., Cui, C., Lovik, A., Pan, L., Piehl, F., Pawitan, Y., Kläppe, U., Press, R., Samuelsson, K., Yin, L., Vu, T. N., Joly, A.-L., Westerberg, L. S., Evertsson, B., Ingre, C., Andersson, J., & Fang, F. (2022). T cell responses at diagnosis of amyotrophic lateral sclerosis predict disease progression. *Nature Communications*, *13*(1), p. 6733. https://doi.org/10.1038/s41467-022-34526-9

Zhang, Y., Zhang, Y., Shi, X., Han, J., Lin, B., Peng, W., Mei, Z., & Lin, Y. (2022). Metformin and the risk of neurodegenerative diseases in patients with diabetes: a meta-analysis of population-based cohort studies. *Diabetic Medicine*, *39*(6). https://doi.org/10.1111/dme.14821

Zhao, S., Fan, Z., Zhang, X., Li, Z., Shen, T., Li, K., Yan, Y., Yuan, Y., Pu, J., Tian, J., Liu, Z., Chen, Y., & Zhang, B. (2023). Metformin attenuates tau pathology in tau-seeded PS19 mice. *Neurotherapeutics: The Journal of the American Society for Experimental NeuroTherapeutics*, *20*(2), pp. 452–463. https://doi.org/10.1007/s13311-022-01316-6

Zhao, W., Beers, D. R., & Appel, S. H. (2013). Immune-mediated mechanisms in the pathoprogression of amyotrophic lateral sclerosis. *Journal of Neuroimmune Pharmacology : The Official Journal of the Society on NeuroImmune Pharmacology*, *8*(4), pp. 888–899. https://doi.org/10.1007/s11481-013-9489-x

Zheng, W., Li, K., Zhong, M., Wu, K., Zhou, L., Huang, J., Liu, L., & Chen, Z. (2024). Mitophagy activation by rapamycin enhances mitochondrial function and cognition in 5×FAD mice. *Behavioural Brain Research*, *463*, p. 114889. https://doi.org/10.1016/j.bbr.2024.114889

Zu, T., Guo, S., Bardhi, O., Ryskamp, D. A., Li, J., Khoramian Tusi, S., Engelbrecht, A., Klippel, K., Chakrabarty, P., Nguyen, L., Golde, T. E., Sonenberg, N., & Ranum, L. P. W. (2020). Metformin inhibits RAN translation through PKR pathway and mitigates disease in C9orf72 ALS/FTD mice. *Proceedings of the National Academy of Sciences of the United States of America*, *117*(31), pp. 18591–18599. https://doi.org/10.1073/pnas.2005748117

9 Integrating Small Molecules into Multimodal Treatment Strategies

Sakshi Patel, Prem Shankar Mishra, S. K. Abdul Rahaman, Rahul Kumar, Mudassir Jan Makhdoomi, and Abid Ali Sheikh

ABBREVIATIONS

AD	Alzheimer's Disease
ATP13A2	ATPase Cation-Carrying 13 A2
BBB	Blood–Brain Barrier
CSF	Cerebrospinal Fluid
DML	Designed Multiple Ligand
GBA	Glucosylceramidase B Acid
HDAC6	Histone Deacetylase 6
HTS	High-Throughput Screening
KI	Knock In
LRRK2	Leucin-Rich Repeat Kinase 2
MAO-B	Monoamine Oxidase B
NDD	Neurodegenerative Disease
PD	Parkinson's Disease
ROS	Reactive Oxygen Species
SYT11	Synaptotagmin 11

9.1 INTRODUCTION

Alzheimer's disease and Parkinson's disease (AD & PD), the most common Neurodegenerative Diseases (NDDs), are having a growing influence on society as a greater proportion of the overall population ages. One of the biggest therapeutic difficulties facing medicine today is the management of AD & PD due to their complicated pathogenesis (Morphy & Rankovic, 2005). A reduced proportion of cells in distinct neuronal populations as a result of cellular demise is a characteristic shared by PD and AD (Zimmermann et al., 2007). This condition is clinically represented in the emergence of symptoms unique to each illness. From a pharmacological perspective, several therapeutic modalities are being investigated (Zheng et al., 2009). First, the etiopathogenic strategy is thought to be "disease-modifying," with the goal of preventing cell demise and encouraging the healing of damaged cell counts (Youdim & Buccafusco, 2005).

As the illness process alters neurotransmitter levels, the cognitive approach is now the most commonly used for curing NDDs (Van der Schyf et al., 2007). Its goal is to either avoid or mitigate the onset of symptoms (Shastry, 2003). Thus, maintaining normal function in the face of rising disorder insult—increased NDDs in the context of AD & PD—is the lead goal of the pathophysiological approach (Shimazu et al., 2005). A group of medications that belong to the latter category are

DOI: 10.1201/9781003520610-9

presently employed in clinical settings and serve as the foundation for strategies intended to manage the symptoms experienced by individuals with AD & PD (Sayre et al., 2008). Recent research has provided increasing evidence that the range of therapeutic options that can be identified through the identification of novel pharmacological targets is expected to grow significantly in the future. However, there are currently few novel drug candidates on the market that combine etiopathogenic or pathophysiological agents (Sagi et al., 2003). It is believed that no etiopathogenic drug has been created and introduced to the clinic that may reverse the degeneration process or restore damaged neurons may be the strongest argument for the complexity of these disorders (Hindo et al., 2009). At most, the pharmaceutical investigation has led to the commercialization of drugs that can patho-physiologically regulate the symptoms of those with these disorders (Barbiero et al., 2011).

There is a lack of neural-protective medications for AD and PD, in part because most drug development labs in the academics and the pharmaceutical sector have adopted the single aim for single disorder approach (Amenta & Tayebati, 2008). For many years, Big Pharma has been focused on creating extremely selective molecules due to the fact that there is probably a negative correlation between medication aim selectivity and a manageable side-effect profile (Bordet et al., 2006). This paradigm assumes that the possibility of adverse effects coming from interfering at other biological locations in the body will be reduced by designing a highly selective chemical that addresses a sole receptor or enzyme (Chen et al., 2009). In recent times, there has been a growing recognition of a multi-drug therapeutic approach that aims to target complicated illness etiologies using a multitargeted single ligand strategy (Chimenti et al., 2006). Pharmaceutical terms such as "hybrid," "bifunctional," "multipurpose," "multi-mechanistic," "multimodal," "pan-agonist," "double mechanism," and "dual ligand" have gained popularity recently and are used interchangeably (Chen et al., 2007). The acronym system for these medications is still evolving. Since there are so many different names for the same idea, Morphy and Rankovic (2005) proposed unifying the terminology and naming these compounds "Designed Multiple Ligands," which are substances that are primarily and logically made to aim at each of the processes that underlie the etiology of a particular disorder (Collino et al., 2006).

Essentially, there has been a noticeable shift in emphasis from "magic bullets" that are intended to be extremely powerful and selective to a more promiscuous "dirty drug" strategy that promotes single-molecule modulators of several important elements (also known as "dirty targets") in a certain disease pathway (Fernandez et al., 2010). Therefore, ligands that control the variety of pharmacological targets in AD and PD are an emerging drug design approach for therapies for these illnesses (Foley et al., 2008). The examples of new multimodal medications with a range of central nervous system targets that are being developed for the symptomatic and potentially neuroprotective–neurorescue therapy of AD & PD are provided in this chapter (Monte, 2003).

9.2 MULTIMODAL THERAPY

The goal of multimodal treatments is to combine several forms of therapy to maximize the treatment of brain problems (Dall'lgna et al., 2003). Pharmacotherapy (small molecule medicines and biologics), technologies, and behavioral/psychosocial therapies are some of the various techniques that may be employed (Gaeta & Hider, 2005) (Figure 9.1). One of these modalities may be utilized to activate or amplify a neurological circuit for the other modality to have a therapeutic effect, or they may target separate neural systems (e.g., a drug with neurostimulation) (Geldenhuys et al., 2010). In this text, "multimodal therapy" is used for clarification purposes only, unless it is expressly referring to a product that satisfies the FDA's official description of a mixture product or when describing drug–drug combos (Glennon et al., 2010). Importantly, the workshop covered FDA-defined combination products and drug–drug combinations that do not treat distinct elements of a disease, even if they did not match the operational definition of multimodal therapy (Heneka et al., 2005). Since these methods have undergone more development and present comparable difficulties, the development of multimodal treatment may benefit from the lessons learned.

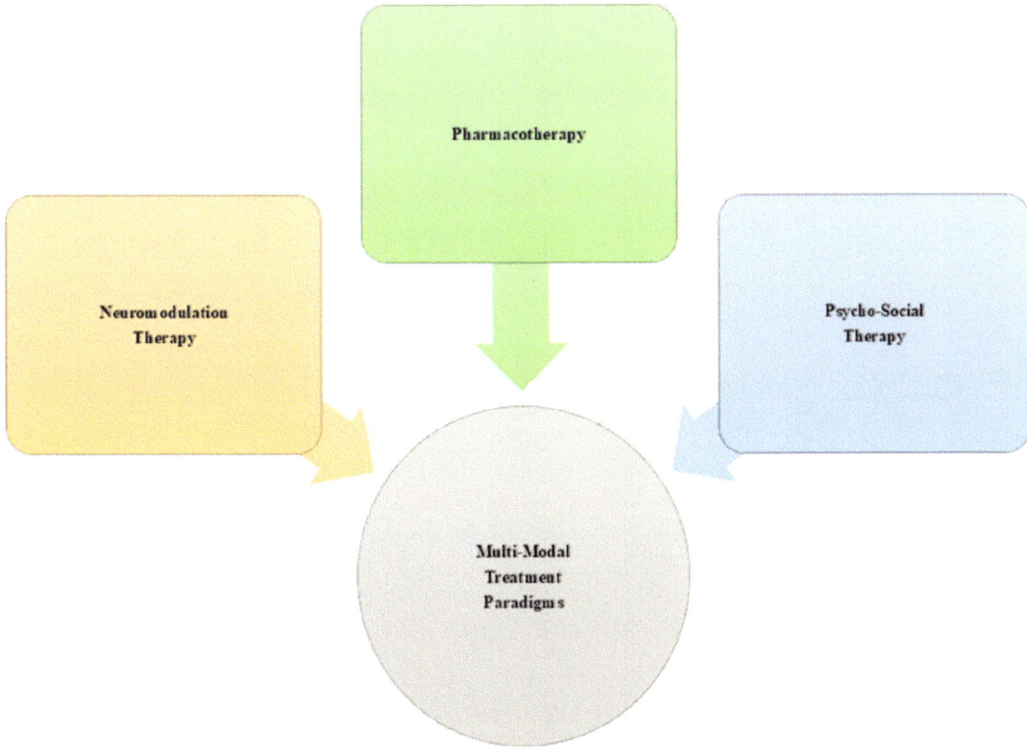

FIGURE 9.1 Multimodal therapy for therapeutic optimization. In treatment paradigms that integrate several therapeutic modalities, such as non-invasive neuromodulation or psychosocial intervention, one treatment modality, like medication, might be joined with another.

9.3 IDEAS FOR CREATING MULTIMODAL DRUG DESIGNS

We've covered a few design techniques for multimodal ligands here. A pharmaceutical chemistry program that focuses on designed multiple ligand (DML) development uses two methods: broad-based screening and knowledge-based drug design. When designing drugs with a knowledge-based approach, biological and pharmacological facts are gathered from previous studies or by examining the literature on medicinal chemistry (Morphy & Rankovic, 2005). Large compound libraries and high-throughput screening (HTS) techniques are typically linked to the use of broad-based screening techniques to identify lead compounds.

Knowing which pharmacophoric properties are necessary for their action at several aims might help guide the creation of a novel medication, as these components can be added to the proposed ligand (Morphy et al., 2004). For instance, a lead optimization program might be designed to maximize the intended ratio of activity at several targets if a lead molecule is currently exhibiting activity at two of them (Maia & De Mendonça, 2002). A few factors that could be used in determining the successful creation of multimodal ligands are depicted in Figure 9.2. Alternatively, multifunctional ligands may also be discovered by an expert medicinal chemist by intuition (Morelli et al., 2009).

9.4 CONVENTIONAL TECHNIQUES A KEY TARGET FOR NDDS: MONOAMINE OXIDASE

The patho-physiologies of both AD & PD are complicated illnesses involving a variety of cascades that are started and sustained by interconnected molecular processes. The monoamine oxidase (MAO) enzyme is one of the significant patho-etiological medication targets in both conditions

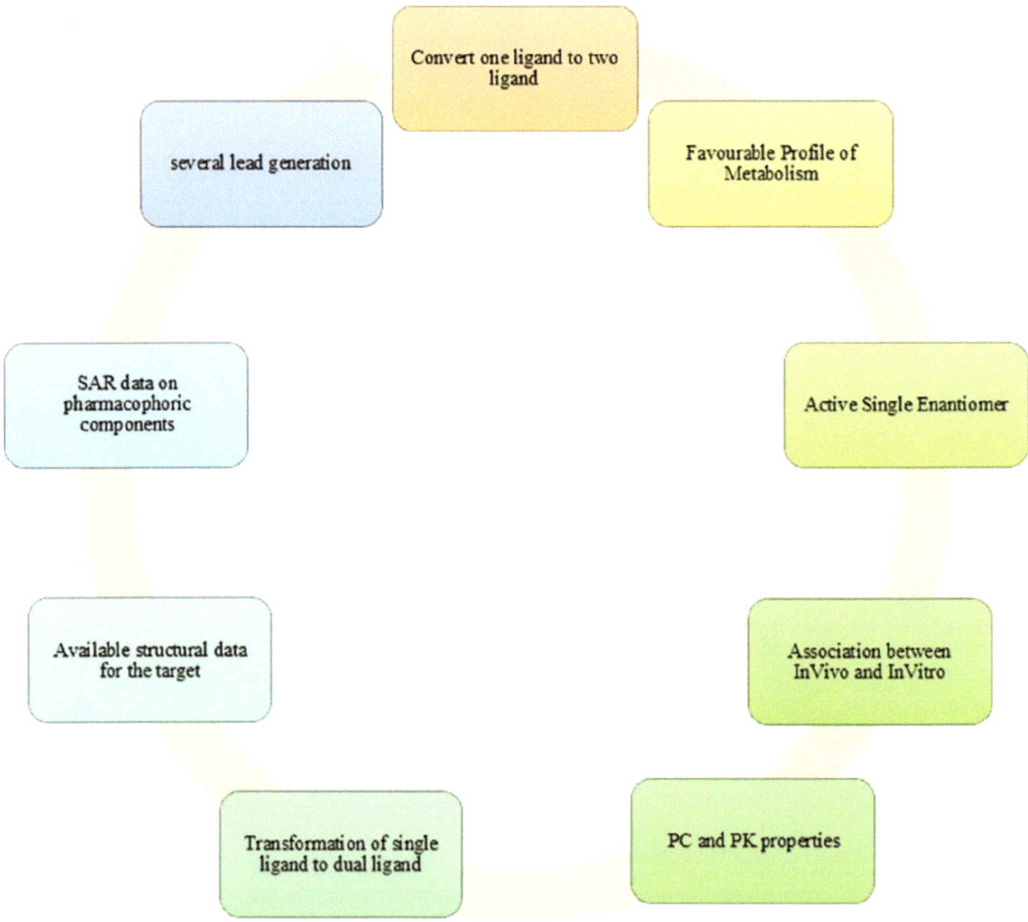

FIGURE 9.2 Important considerations for creating multifunctional drugs are presented. The possibility of discovering a new compound is determined by a number of factors, including the availability of a single potential substance for DML design that exhibits motion at each of the desired therapeutic seeks, the use of HTS techniques for multiple lead generation, the targeting of similar targets, such as receptors in the same superfamily, and the availability of structural data for pharmacological targets. A single agonist can be converted into a dual ligand more easily than many ligands. Lead optimization programs are concerned with pharmacokinetic and physicochemical aspects.

(Weinreb, 2008). It is well known that hydrogen peroxide is produced when monoamines such as brain catecholamines and other monoamines undergo catalytic oxidative deamination by MAO (Weinreb, 2008). Further reactive oxygen species are subsequently produced because of this process (reactive oxygen species, ROS) (Dar et al., 2022; Weinreb et al., 2009).

Research conducted on human brain tissue has demonstrated that MAO-B levels rise with aging and in numerous NDDs (Schapira & Olanow, 2009). The placement of MAO-B among glial cells and the age-associated increases in MAO levels are correlated. Glial cell numbers rise with age in the healthy human brain, as well as in NDDs, which include AD and PD, and following brain trauma (Shadrina et al., 2010). Age-related brain enlargements up to three times the usual amount of MAO-B can be found in the brain, which exacerbates oxidative stress, and raises the risk of age-related NDDs including AD and PD (LeWitt et al., 2008). Therefore, increased MAO movement and catecholamine turnover act as an oxidative stressor, preventing mitochondrial respiration and increasing the likelihood of mitochondrial distortion with aging in the brain and subsequent neurodegeneration in both AD and PD (Li et al., 2008). However, this is undoubtedly only one among numerous related and unrelated contributors (Morelli et al., 2009).

Consequently, monoamine oxidase-B activity rises with age in AD patients and is especially high within the vicinity of aging particles in the brain of AD patients (Joghataie et al., 2004). Oxidative stress subtly raises tau and amyloid-beta (Aβ) hyperphosphorylation and happens well in advance of the formation of neurofibrillary tangles and senile plaques. In addition to decreasing the production of ROS by reducing catalysis, MAO inhibition raises monoamine levels in the CNS (Haycox et al., 2009). When comparing AD individuals to aging- and sex-similarity regulates, the majority of monoamines are reduced (Honda et al., 2008). The finding that selegiline, a specifically targeted monoamine oxidase-B inhibitor, improves memory and cognition deficits in animal models of AD lends credence to the usefulness of MAO suppression in AD (Kaundal et al., 2009).

9.5 SMALL MOLECULES FOR PSYCHIATRY NEUROLOGY DISORDER TREATMENT

Small molecules may recognize evidence of a large variety of biological metabolites and are important participants in various chemical events (Mitchinson & Finkelstein, 2008). Indeed, a number of molecular changes that are normally absent from physiological states can be seen in a wide range of clinical diseases. These primarily relate to elevated free radical production, altered pH, and impaired synthesis of certain particular biomolecules (Shugrue & Miller, 2017). A growing body of research has been published in recent years that focuses on the possible therapeutic application of tiny molecules with potent biological characteristics (Gurevich & Gurevich, 2014).

However, the majority of studies on this topic so far have mostly focused on infectious or proliferative disorders (Kehler & Nielsen, 2011; Law et al., 2012; Leng et al., 2008; Weïwer et al., 2013). Table 9.1 highlights the most important findings on the feasible therapeutic application of tiny molecules in the therapy of NDDs and psychiatry neurology illnesses including anxiety, sadness, schizophrenia, and autism.

9.6 UTILIZING SMALL MOLECULES IN NDD CONDITIONS

9.6.1 Proposed AD Remedies

An essential stage in the pathogenesis of AD is the aggregation of amyloid-β peptide since this starts a cascade of molecular events that conclude with the degradation and killing of nerve cells (Haass & Selkoe, 2007). It is noteworthy that the anomalous metabolism of Aβ can be identified several years

TABLE 9.1
Small Compounds and Psychiatry Neurology Disorders

Psychiatry Neurology Disorders	Small Molecules Used	M.O.A	Effects Induced
Autism	CLK2 inhibitor	It inhibits CLK2.	Decrease in synaptic deficits in neurons from people with signs of autism spectrum disorder
Anxiety	HDAC inhibitor	It inhibits HDAC1 and HDAC 2.	Improvement of the first contextual fear conditioning learning and neuroprotection
Mood disorder	HDAC inhibitors (scriptaid, suberoylanilide hydroxamic acid, TSA, and aliphatic acid derivatives)	It inhibits HDAC1 and HDAC 2.	Producing effects analogous to those of an antidepressant, defense of neurons against the death of neurons brought on by oxidative stress
Schizophrenia	Phosphodiesterase 10A inhibitor	It inhibits phosphodiesterase 10A.	Dopamine D1 receptor signaling is potentiated. Suppression of the signaling from dopamine D2 receptors

prior to the beginning of AD, and this feature serves as a crucial pharmacological target for first treatment measures (Buchhave, 2012). Unfortunately, no medication that primarily targets the Aβ accumulation process has been successfully applied in clinical settings thus far. Animal specimens of the disease were used in the development of these compounds (Karran & Hardy, 2014). To potentially create a new and promising "chapter" in the development of drugs for use in AD, a growing array of research initiatives in this sector have focused on identifying small molecules that act on Aβ buildup and aggregation. A relatively recent paper by Habchi and colleagues represents one of the most comprehensive and uncomplicated works in this sense. The researchers stated on the planned amalgamation of a sequence of small compounds, designated as "set A" (seven molecules exhibiting an effect comparable to or greater than that of bexarotene) and "set B" (five molecules capable of completely inhibiting Aβ42 accumulation), that block particular stages of Aβ42 aggregation (Dar & Glazner, 2020; Habchi et al., 2017). To be clear, however, there have been earlier reports by other researchers that several of the tiny chemicals discovered in this article, such as MM11253 and adapalene, have a significant effect on the onset of the accumulation as well as the propagation of Aβ42 oligomers (Cohen et al., 2013, 2015).

The further data that our research technique uncovered are mostly concerned with the creation of small compounds that target the disruption of metal ion equilibria and their reduced compartmentalization, which is another significant pathogenetic step in the development of AD (Rauk, 2009; Scott & Orvig, 2009). This is particularly true with regard to the reduction and oxidation active Cu (I & II) and Fe (II & III), which are extensively distributed in plaques of senility, within the hippocampal and cortical regions (Barnham & Bush, 2008; Molina-Holgado et al., 2007; Zatta et al., 2009).

Consequently, Aβ species can be attached to these reactive metals. The hydroxyl radical and hydrogen peroxide produced by this mechanism, called the Fenton reaction, are examples of reactive oxygen species that may encourage Aβ aggregation and lead to dementia (Scott & Orvig, 2009). As a result, the metal–Aβ coupling outbreak by metal chelation may offer a useful treatment alternative to reduce neurotoxicity brought on by metals' chemical bonds with Aβ molecules and to restore metal ion balance in the brain (Bush & Tanzi, 2008). Thus far, clioquinol, EDTA, and PBT2, an 8-hydroxyquinoline alternative, have been the most frequently utilized chelators in AD treatment. The latter two substances were also studied in phase 2 clinical examinations to try to convert preclinical results into potential therapeutic applications. The persistent myelo-optic neuropathy that clioquinol causes is one of its major adverse effects, but they have also been shown to markedly enhance cognitive function (Adlard et al., 2008). Persistent attempts have also been directed toward the creation of tiny compounds, such as cyclin, KLVFF peptide, curcumin, IMPY, and p-I-stilbene (Hureau et al., 2010; Wu et al., 2008). These compounds were able to overcome the drawbacks of earlier metal-chelating molecules, especially those pertaining to blood–brain barrier (BBB) permeability, by harmoniously linking both metal ions and Aβ (Hindo et al., 2009).

9.6.2 The AD Amyloid Theory

Currently, the most widely recognized explanation to explain the underlying mechanisms and clinical development of AD is the amyloid cascade hypothesis. According to the theory, the pathology associated with AD is produced by an imbalance in the production and clearance of Aβ, whereas the clinical signs of AD are brought on by a series of anomalies in the breakdown of the amyloid primary protein (Hardy & De Strooper, 2017). Two membrane-bound enzyme complexes, BACE1 and γ-secretase, take part in the double-step protein breakdown procedure of amyloid precursor protein and Aβ synthesis.

Two membrane-bound enzyme complexes, BACE1 and γ-secretase, are engaged in the dual-step proteolysis procedure that produces Aβ and APP. According to the aforementioned theory, peptides Aβ40 and Aβ42 are essential to the pathophysiology of AD (Dorababu, 2019). Five main explanations support the benefits the amyloid present in the pathophysiology of AD.

First, amyloid deposits set the stage for early pathological understanding of AD, and neuropathic plaques are essential elements of the diagnostic standard (Watt et al., 2019). Two, amyloid cause promotes tissue deterioration in peripheral amyloidosis (without any connection to Aβ and AD), presenting the toxicity of CNS amyloid. Three, Aβ fibrils generated from plaques have pro-inflammatory actions and strain neuronal toxicity, whereas Aβ oligomers clarify the features of acute synaptic toxicity.

Four, increased amyloid load is associated with apolipoprotein E4, the most common genetic risk factor for AD. The most astonishing thing is that five alterations that reason to familial early-onset AD also raise Aβ42 generation or its association with the less likely aggregate Aβ40 isoform. The processing of amyloidogenic APP is directly increased by each of these mutations (Dinda et al., 2019).

9.6.2.1 Novel Small Molecule Therapeutics for AD That Are Directed at Multiple Targets

Over the recent few decades-long, a growing number of potential intervention mechanisms have emerged to generate disorder-modifying medicines that will slow or even reverse the advancement of AD. These research endeavors have concentrated on the start and progression of the illness. Preventing the early stages of oligomerization through either direct interference with monomeric Aβ/Aβ interactions or encouraging non-toxic oligomerization appears to be the most promising strategy (Liu et al., 2017). Targeting the disparity in trace element cations like iron, copper, and zinc ions with metal-protein-attenuating compounds or chelators may be the second most viable strategy (Parsons & Rammes, 2017). Lastly, three different sorts of techniques can be used to broadly illustrate a possible therapy notion that involves the manipulation of APP/Aβ pathways: alteration of Aβ amalgamation, curation of Aβ accumulation, and promotion of Aβ approval (Madav et al., 2019).

9.6.2.2 ADs: Emphasizing the Amyloid β Cascading Amyloid Primary Protein

As was previously mentioned, BACE1 and γ-secretase sequentially cleave APP to produce Aβ. There are two ways that APP processing can occur: amyloidogenic and nonamyloidogenic. APP processing has been linked to three secretases: α, β, and γ. The Aβ domain is cleaved in the middle of the α-secretase-based nonamyloidogenic route, yielding soluble amyloid precursor-α and the C-terminal portion C83 (De Strooper & Karran, 2016). γ-secretase then splits the C-terminal portion C83, producing P3 and the amyloid primary protein intrinsic domain. Neurons often eliminate these smaller particles. The Aβ domain's beginning is where the BACE1-based amyloidogenic pathway cleaves to create soluble APP-β and the C-terminal region portion C99.

Following C99 breakage, Aβ and ACID monomers are produced. In AD neurons, the accumulation of Aβ fragments in the subcellular chamber involving mitochondria, lysosomes, and ER-endoplasmic reticulum results in oxidative stress and dysfunction of these compartments (Manzine et al., 2019). A growing corpus of research over the past 20 years has clarified the critical role that Aβ generation and amyloid precursor protein processing play in the development, onset, and risk of AD.

In order to reduce the toxicity of Aβ, research has been done on developing pharmacological molecules that manufacture anti-BACE1 and γ-secretases, reduce Aβ formation, and accelerate the approval of Aβ via Aβ sequesters and α-secretase activators. With ten preliminary phase two or third trials with BACE1 inhibitors at the moment, they have become the majority of the popular class of tiny compounds in the AD pipeline (Watt et al., 2019). However, a variety of serious adverse effects have caused the majority of gamma-secretase blockers to fail in clinical trial investigations.

9.6.3 Proposed PD Remedies

PD is an NDDs emphasized by the lack of dopamine-transmitting neurons in the substantia nigra, a part of the brain. This CNS disorder is explained by a number of pathogenetic ideas. Redox and mitochondrial dysfunctions are two of the most studied topics (Nunnari & Suomalainen, 2012; Rugarli & Langer, 2012). This includes variations in the mitochondria-associated protein Parkin, an E3 ubiquitin ligase, and the mitochondria-specific kinase PTEN-induced kinase 1 (PINK1) (Hertz et al., 2013; Khan et al., 2024). The PINK1/Parkin route has demonstrated therapeutic potential in the therapy of PD, as demonstrated by several researchers (Arena et al., 2013; Kondapalli et al., 2012). However, it has been shown that PINK1 lacks small molecule-binding sites, indicating that other processes control its pharmacological activation. This stands in contrast to the well-established mechanisms of small molecule-mediated kinase stimulation, which are often accomplished by adhering to the allosteric regulatory sites of the kinases (Kondapalli et al., 2012). Conversely, the tiny molecule mildronate, which possesses charged nitrogen as well as oxygen atoms, shields mitochondria. It has been documented to function as a neuro-protective agent in a mouse method of azidothymidine-generated neurotoxicity by inhibiting brain inflammatory response and cell death, additionally by lowering levels of cytochrome oxidase c, caspase, and susceptibility-protein for cell death (Kondapalli et al., 2012). Another study used a rat model method of PD produced via one-sided intrastriatal parenteral preparation of the neurotoxin 6-hydroxydopamine to further confirm the neuroprotection linked to mildronate.

Mildronate prevented the loss of biomarkers called preserve neural and glial integrity, like tyrosine hydroxylase, ubiquitin, and Notch-3, and it also decreased the concentration of certain inflammatory markers, such as inducible nitric oxide synthase in animals given 6-OHDA injections (Klusa et al., 2010). In addition to malfunctioning mitochondria, another significant pathogenetic mechanism associated with the onset of PD is malfunctions of GTPases, specifically those governing the dynamic procedures of cellular mitochondria nuclear fusion and fission, the transportation of organelles along axons, the preservation of axons and neuronal survival, neuroinflammation, and the amplification of oxidative stress (Hong & Sklar, 2014).

Regarding redox imbalance, it has been observed that PD patients' dopaminergic neurons contain micro GTPase Rac1, which is essential for controlling the activity of the free radical generator NOX1 (Choi et al., 2012). Fascinatingly, a collection of 5 million tiny molecules has been examined as an adjuster for several GTPases, including mutant Ras, Cdc42, Rab5, Rab7, wild-type Ras, and Rho, which specifically stimulate GTPase subfamilies (Choi et al., 2012).

A noteworthy addition to our comprehension of the molecule mechanisms behind PD is the function of certain proteins, such as α-synuclein. To quickly and effectively identify possible pharmacological "chaperones," or small compounds that adhere protein molecules and anchor them against deterioration, along with innovative chaperone inhibitors against α-synuclein, the authors of a very interesting paper by Mistook and colleagues worked on creating a library of small molecules (Oh et al., 2014) (Figure 9.3).

9.6.3.1 Mechanisms of the Autolysosome Pathway in the Etiology of PD

The CNS relies on the autophagous mode to keep neurons and glial cells in a state of homeostasis. Sophisticated quality control systems are necessary because highly differentiated neuronal and glial cells are more challenging to fix than other cells and require increased vigor for routine operations as well as autophagic recycling of components (Maria Fimia et al., 2007). As a matter of fact, ALP can

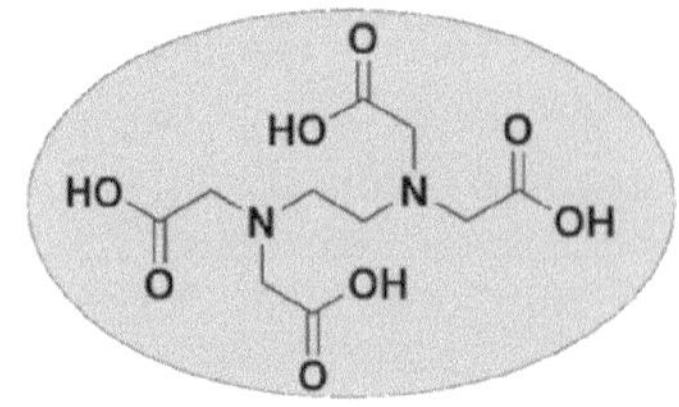

2,2',2'',2'''-(Ethane-1,2-diyldinitrilo)tetraacetic acid

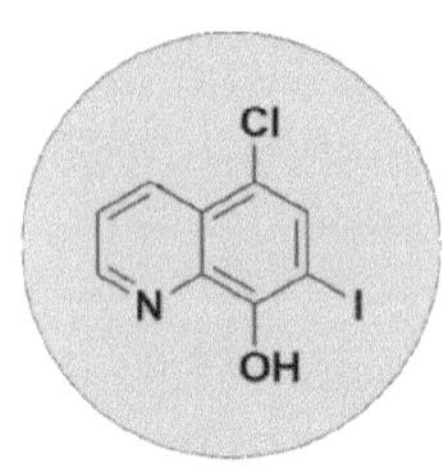

5-chloro-7-iodoquinolin-8-ol

FIGURE 9.3 The chemical structures of tiny compounds with potential therapeutic use in AD and PD.

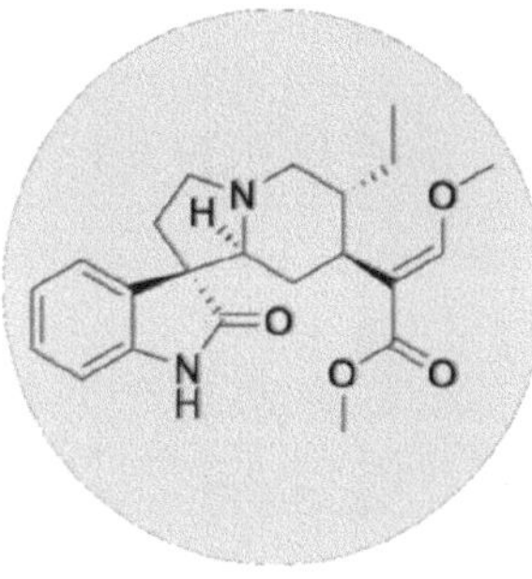

Isorhynchophylline

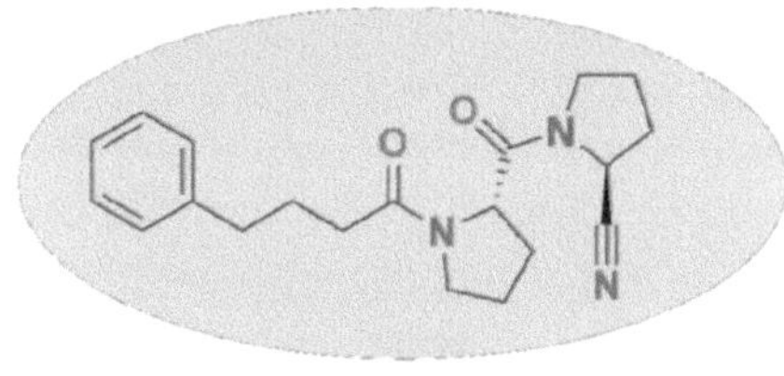

KYP-2047

FIGURE 9.4 Compounds targeting beclin-1 and their structures.

malfunction in the brain throughout extremely dense labor, and some PD risk gene changes, such as those that influence certain autophagous mode networks, such as chaperone-mediated autophagous mode and mitophagy process, can also have a deleterious effect on ALP. Decreased ALP is therefore assumed to be a PD stimulant because it may promote the formation of toxic protein aggregates and cause lysosome failure, delayed approval of defective mitochondria, and an increase in oxidative strain (Figure 9.4) (Levine & Klionsky, 2017). ATG5 was discovered to be lower in a zebra fish PD model treated with MPTP. This finding was linked to inhibited the autophagous mode flux, which resulted in an increase in PD-connected proteins such b-syn, Parkin, and PINK1, as well as the elimination of DA (dopamine) neurons (Son et al., 2012). Recent analysis of genome-wide association study data has identified the 5 PD-related genes: PTEN-induced kinase 1, α -syn, gene leucine-rich continue order kinase 2, Parkin RBR E3 ubiquitin amino acids ligase (PARK2, Parkin), and Parkinson protein 7 (Hu et al., 2017). Notably, ALP deficiencies are one of the pathogenic processes of PD that underpin these genes' mutations, which are thought to be indicative of risk for the illness.

For instance, LRRK2 has drawn a lot of interest in the pathophysiology of PD, and the connection between its genetic variants along with the autophagous mode has been thoroughly studied and extensively updated in recent years (Nuytemans et al., 2010; O'Hara et al., 2020). Though the precise processes are yet unknown, mounting evidence suggests that one of the harmful mechanisms of LRRK2 mutations may be the defective ALP. Generally speaking, dopamine transporter dysfunctions have been found to be triggered by G2019S, the most common mutation in LRRK2. Furthermore, leucyl-tRNA synthetase (LRS) may be phosphorylated and activated by LRRK2 G2019S, which would subsequently activate mTORC1, which would decrease the autophagous mode (Albanese et al., 2019). Furthermore, it has been found that a single deletion of LRRK2 does not increase the amount of α -syn, selective neurodegeneration, or impairment of ALP, indicating the critical function LRRK2 plays in modulating ALP pathways (Cogo et al., 2019).

In addition to the conventional autophagous mode pathways, recent research has shown its function in controlling PD-related selective autophagous mode, including CMA and mitophagy (Wauters et al., 2020). A gathering of CMA-specific lysosome-associated surface proteins 2A, thermal trauma cognates 70kDa protein 8/heat trauma cognate 71kDa amino acids (HSPA8/HSC70), and raised CMA substrate glyceraldehyde 3-phosphate dehydrogenase was observed in the striatum and the cerebral cortex of aged LRRK2 R1441G knock-in (KI) mice. These results imply that LRRK2 alterations would impair CMA's capacity to break down oligomeric a-syn, which would encourage PD (Wauters et al., 2020). It has been observed that upregulation of kinase active LRRK2 inhibits mitophagy, causing mitochondrial malfunction. This dysfunction is exacerbated via the amplification of the kinase excitable LRRK2G2019S variant, which speeds up the progression of PD (Bonello et al., 2019).

Significantly, SNCA mutations lead to aberrant a-syn aggregates (such as oligomer and fibrosis forms), which were found to have a role in the PD mechanism, which is the primary Lewy pathology in Parkinson's disease. Toxic protein aggregates may be exacerbated by these aggregates, which the ALP may be able to break down (Nuytemans et al., 2010). This might also have an adverse effect on autophagic, lysosomal, and mitochondrial functioning. It was shown that the overexpression of α -Syn causes ER stress, lysosomal pH alkalization that decreased LAMP1 amounts, and a disturbance in lysosomal architecture and classification, all of which hinder the autophagous mode (Lamonaca & Volta, 2020). The increase of α -syn inhibits lysosomes' capacity to break down dopaminergic neurons in the middle cerebral cortex dopamine models by eliminating hydrolase trafficking.

Furthermore, ALP impairment brought on by glucosylceramidase B acid (GBA) alleles reduces GCase activity and raises α -syn amounts, which in turn reduce GCase function (Wauters et al., 2020). As part of its pathogenetic involvement in PD, the L444P heterozygous alteration of GBA, a lysosomal protein which convert sugar and a ceramide from glucosylceramide, has been found to cause mitochondrial malfunction by blocking essential processes in mitophagy. More proof has linked ALP malfunction to mutations in additional genes. Mutations in PINK1 and Parkin were thought to impact mitophagy that causes a stalled approval of damaged mitochondria and exacerbates Parkinson's disease (Bonello et al., 2019). Due to the inability of ATPase cation-carrying 13A2 (ATP13A2) to recruit the enzyme histone deacetylase 6 (HDAC6) toward lysosomes, autophagosome–lysosome fusion might be harmed (Neumann et al., 2004). In another study, it was shown that ATP13A2 inhibited the production of TEFB-dependent synaptotagmin 11 (SYT11), which in turn blocked the autophagous mode, and activated mTORC by the stimulation of tuberous sclerosis complication 2 due to its linkage with MYC-connecting protein II.

As previously stated, protein particles and damaged organelles may accumulate, leading to cellular damage, as an outcome of the autophagous mode abnormalities associated with both familial and sporadic Parkinson's disease. It has been demonstrated that genetically or chemically targeting ALP increases neuronal survival, suggesting this as a possible PD therapeutic approach.

9.7 EMPHASIZING THE AUTOPHAGOUS MODE AS A POSSIBLE PD TREATMENT

The identification of toxic aggregates resulting from defective autophagous mode as a pathogenesis of PD in recent years has allowed for the targeting of many autophagous mode regulators, including adenosine monophosphate-activated protein kinase, mammalian target of rapamycin complex 1, Unc-51-like autophagous mode activating kinase 1, IMPase, leucine-rich repeat kinase 2, beclin-1, transcription factor EB, glucosylceramidase, ERRa, and C-Abelson. Numerous tiny molecule inhibitors or activators that have been linked to these potentially druggable domains have been discovered; these compounds may have neuroprotective benefits in various forms of PD models. Nevertheless, a lot of these tiny molecules are solely employed in PD research as tool probes. Developing a comprehensive knowledge of autophagic pathways in the pathophysiology of PD and translating existing tiny molecules ahead of time therapeutic techniques remain a significant challenge.

9.7.1 Beclin-1 as a Possible PD Therapy

Beclin-1, sometimes referred to as ATG6, is a member of the ATG peptide family and is highly conserved throughout eukaryotes. Its overexpression can decrease apoptosis and increase the autophagous mode since it is a crucial molecule for controlling the autophagous mode (Wirawan et al., 2012). Beclin-1 is involved in the induction of the autophagous mode through the nucleation of autophagic vesicles, which are phagocyte production processes. To initiate the nucleation of vesicles, BECN1 connected with its connecting partners, which included Ambra1, Vps15, Atg14L, and PI3KC3, a class III phosphatidylinositol-3 kinase. Phosphatidylinositol 3-phosphate, the distinct result of PI3KC3/Vps34 activation upon complexation with beclin-1, was responsible for triggering the autophagous mode (Savolainen et al., 2014).

According to Spencer's team, beclin-1 activates the autophagous mode, which may lessen the buildup of α -syn in PD. KYP-2047 acts as a serine protease and prolyl oligopeptidase antagonist, which can cause the production of beclin-1 (Figure 9.4). Following this, there was a rise in the amount of the autophagous mode indicators LC3BII, which improved the autophagous mode and the removal of increased M.W. oligomeric a-syn (Savolainen et al., 2015). In comparison to other PREP inhibitors, the 4-phenylbutyryl-L-prolylpyrrolidine skeleton of 28 has been demonstrated to be more effective in reducing the dimerization of a-syn (Song et al., 2014).

Furthermore, by upregulating the concentration of beclin-1 and preserving the functionality of neural cells (Figure 9.4), natural compounds such isorhynchophylline and corynoxine B may additionally reinforce the cytoprotective autophagous mode (Yang et al., 2018). Beclin-1 controls both the autophagous mode and apoptosis. Beclin-1 has significant scientific relevance because of the interplay between the autophagous mode and cell damage, which includes regulating cell living and damage (Palmieri et al., 2011).

9.7.2 GCase as a Possible PD Therapy

In PD models, selectively stimulating ALP through lysosome regulation has emerged as a compelling approach. One of the interesting strategies is to directly aim certain lysosomal enzymes (like glucocerebrosidase or GCase) to promote the lysosomal breakdown of α -syn. Numerous findings point to a reduction in GCase activity and level, as well as an inverse relationship between brain levels of GCase and a-syn in idiopathic PD (Liu & Sabatini, 2020). Thus, increasing GCase activity may likewise be used to treat PD. However, direct enzyme replacement treatment is not successful for PD because GCase cannot cross the BBB (Altarescu et al., 2001). Eventually, the creation of small molecule medication chaperones was suggested as a potential substitute treatment for PD. Certain medications were engineered to attach to altered GCase, rectifying its erratical folding and facilitating its transfer to the lysosomes, therefore enhancing both GCase function and lysosomal function. A tiny molecule chaperone utilized for respiratory disorders was ambroxol (Richter et al., 2014). In PD fibroblast with GBA1 mutations, McNeill et al. discovered that it might lower oxidative stress and increase GCase activity and also increase the amounts of saposin C, LIMP2, and cathepsin D, all of which were essential for GCase function (Sawkar et al., 2006).

This compound has the prospective to not only cross the BBB and dramatically boost GCase activity in lymphocytes but also lower the amount of glycosphingosine in CSF fluid, treat myoclonus, and lessen seizures in patients. The FDA authorized a phase II non-random order, noncontrolled clinical investigation of participants (NCT02941822), which the business PRO.MED.CSA just concluded (Mullin et al., 2020). An oral drug that was detected in the blood and cerebrospinal fluid (CSF) of Parkinson's disease (PD) patients after a prolonged period (186 days), without notable adverse effects. (Blanz & Saftig, 2016). Additionally, a few additional small molecule chaperones showed comparable results (Figure 9.5), such as isofagomine (Jiang et al., 2013).

However, these chaperones have the drawback of blocking GCase action by attaching to the enzyme's catalytic site. It implies, then, that, when employing these compounds, the appropriate equilibrium among chaperone function and inhibitory properties should be carefully evaluated, necessitating the creation of those medicines without affinity to the GCase catalyst site (Weikum et al., 2018). As a result, HTS revealed NCGC607 as a noninhibitory GCase micromolecular chaperone (Figure 9.5). It has been demonstrated that in dopaminergic neurons produced from generated pluripotent stem cells, this chaperone decreases α -syn levels while increasing GCase functionality and protein levels (Giguère, 2008). Tiny molecular chaperone improvers of GCase have been shown in these PD trials to increase lysosomal function and a-syn approval. This suggests that targeting GCase to specifically boost the ALP pathway may be a viable treatment option for PD.

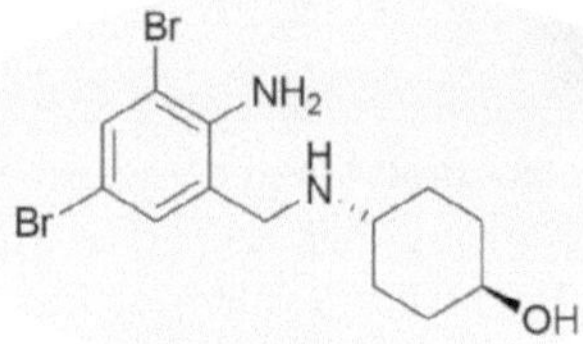

(1r,4r)-4-((2-amino-3,5-dibromobenzyl)amino)cyclohexan
1-ol
(Ambroxol)

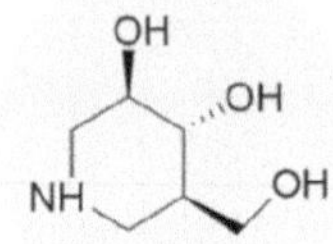

(3R,4R,5R)-5-(hydroxymethyl)piperidine-3,4-diol
(Isofagomine)

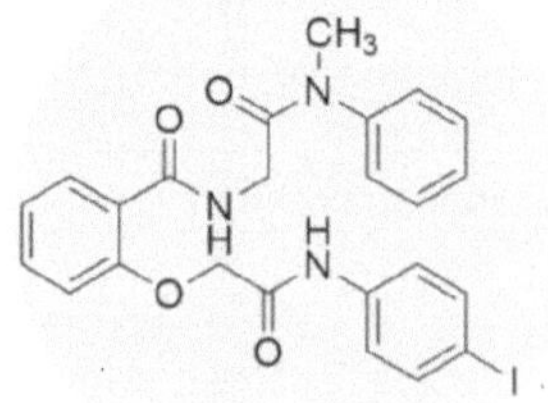

2-(2-((4-iodophenyl)amino)-2-oxoethoxy)-N-(2-(methyl(phenyl)amino)-2-oxoethyl)benzamide
NCGC607

FIGURE 9.5 The configurations of substances that target GCase.

9.8 TECHNIQUE ADVANCEMENTS FOR MULTI-TARGET MEDICATION DESIGN

Conventional approaches to drug discovery depend on creating molecules that are selective for a certain target to alter its function. However, multi-target ligands that might seek a system that engages receptors must be taken into consideration due to various metabolic dysfunctions of CNS illnesses (Gervasio & Spiwok, 2018). The creation of an effective process for determining the structure of multi-potent medications is the primary obstacle in the drug development of multitarget-directed ligands. Consequently, the creation of various computational techniques has made a substantial contribution to the creation of novel medications.

9.8.1 NDDs Syndrome QSAR

Quantitative structure-activity relationship (QSAR) is a method used in chemistry and biology to predict the effects, properties, or biological activity of chemical compounds based on their chemical structure. The relationship between a chemical entry's molecular structure and its accompanying biological activity is used by QSAR (Chen et al., 2020). This method creates a mathematical connection between the molecular descriptors, or structural characteristics, of agonists and biological targets. In the field of AD, several QSAR models were developed and effectively predicted a wide range of biological processes (Srour et al., 2019).

9.8.2 NDD Virtual Screening

Overuse of structure-based virtual screening has been seen to comprehend binding interactions and key molecular characteristics (Hassan et al., 2019). A computational methodology that was designed for the GOLD docking software was used to perform a virtual screening of about 1 million chemicals from the ZINC and internal databases. *N*-(4-hydroxyphenyl) icosa-5,8,11,14-tetraenamide was identified in this computational assessment as an inhibitor against tau aggregation and the GSK-3β enzyme (Lin et al., 2016). The ZINC database's VS approach for adding novel anti-amyloid-beta aggregation drugs was described by Rigi et al. Web servers were utilized to evaluate the pharmacological and toxicological characteristics of recently developed ligands (Rigi et al., 2017).

The variety of natural compounds' molecular structures make it difficult to employ them as active ingredients in natural therapies. Consequently, the recognition of bioactive natural substances is done using molecular dockings (Lin et al., 2016). For instance, around 4500 buildings' BACE1, GSK-3β, and other libraries have been tested against. To generate random forest models, use MAO-A. The natural alkaloid ligiodenine is active against MAO-A, while reaimine and simpodialine-β-N-oxide are effective against beta-site amyloid precursor protein-cleaving enzyme 1 (Staroń et al., 2020).

Several studies demonstrated the effective use of ligand-based virtual screening (LB–VS) within the search for opposing AD drugs. Using 2D footprint screening, fundamental characteristic and ADMET screening, 3D pharmacophore-founded inspection, as well as adaptable docking, multiplex and selective 5-HT6/5-HT2A ligands are presented (Staroń et al., 2020). Additional virtual screening techniques using optimized fragments were conducted with multipurpose activity against BACE1 and GSK-3β (Bottegoni et al., 2015). A structure screening protocol was used to identify hesperidin as a BACE1 inhibitor. Drug candidates for the treatment of schizophrenia were suggested to be found in two different 2D quantitative structure and relationship models: one for counter-amyloidogenic activity and the other for antioxidant activity. These models also revealed multi-potent phytochemicals (Chakraborty et al., 2016). Virtual screening pipelines have also been shown to be successful in using the 2D-QSAR framework for BACE1 blocking and the 2D quantitative structure and relationship method for antioxidant and anti-amyloidogenic activities (Chakraborty & Basu, 2017).

Serially, parallel, and hybrid approaches for data incorporation have been successfully used when ligand- and structure-based methodologies are coupled for the utilization of VS pipelines and platforms (Kim & Han, 2018). To investigate AD treatment, the cardiovascular system interactions, molecular docking, ADMET characterization, and simulated screening packages were utilized to find natural compounds that selectively inhibited metallothionein (MT)-III. Findings indicate the existence of strong, selective MT-III inhibitors that interact with significant amino acids through hydrogen bonding (Roy et al., 2015). To explore new AChE inhibitors, a mix of docking and ligand-based pharmacophore modeling was used. Two compounds were presented as selective inhibitors, and an innovative pharmacophore model of AChE was created based on current medications and included substantial interactions with the site of inhibition (Dhanjal et al., 2015).

Using virtual screening from the ZINC database, dibenzyl imidazolidine and triazole acetamide were added to target amyloidβ clumping and neural toxicity in PC12 cells (Das & Smid, 2017). In the realm of multi-target drug design, computational biology has garnered a lot of interest lately. The positive "hit" instances of biological activity are increased, and the time and expense to discovery are decreased thanks to these simulations (Olğaç et al., 2019). Future directions in drug design and discovery may benefit from the use of key chemical characteristics, binding interactions, morphological resemblance scaffold hopping, pharmacophore modeling, molecular interaction simulations, and 3D quantitative structure and relationship (Lu et al., 2018).

9.9 FUTURE-FOCUSED STRATEGIES

It is believed that the need for novel therapies to address the administration of psychiatry neurology and NDD conditions has increased due to the notable increase in affected populations, the impact on social and economic standings, and the ensuing economic implications on countries due to lost productivity. Scientists are always assessing the potential of emerging therapeutic targets to develop new particular and nonselective ligands.

However, because successful preclinical outcomes are rarely obtained in clinical studies, scientists are frequently quite apprehensive about engaging in this type of research due to the demanding and significant in vitro and preclinical research procedures associated with establishing new drugs, especially multifunctional drugs. It is essential to identify a lead chemical that exhibits the appropriate physiological function against numerous targets and to optimize it into an authorized medicine that possesses the desirable polypharmacology and a dependable, drug-like pharmacological persona (Hopkins, 2008). These are the two key steps in the drug-designing process.

Nevertheless, Hopkins noted that by integrating in silico methods with wide-ligand pharmacological sequencing as opposed to protein assays and gene expression arrays, drug design specialists will have another tool to assess the efficacy of preclinical polypharmacology (American Psychiatric Association Diagnostic and Statistical Manual of Mental Disorders, n.d.). Network pharmacology is a drug development methodology that acknowledges attempts to improve clinical efficacy while also understanding toxicity and side effects, which are two of the primary reasons for clinical trial failure. This approach incorporates systems biology, network analysis, connectivity, redundancy, and pleiotropy. It will take more work to improve pharmacological network search algorithms and methodologies, much as it took nearly 20 years for the concept of structure-based drug design to reach its current status in medicinal chemistry.

9.10 DISCUSSION AND CONCLUSIONS

This chapter presents a comprehensive overview of the "hot topic" of possible pharmacological applications for small molecules for use in the therapeutic management of psychiatry neurology and NDDs, drawn from the previous years' scientific writings. Our research technique includes terms connected to neurological illnesses, which is one of the most significant components of our study. These disorders include PD, AD attack, damage to the spinal cord, and traumatic brain damage, and they are principal worries on the rising mortality rate of today's world because they are closely linked to neuronal loss and degradation. A crucial examination of the papers included in this chapter may lead to particular discussion points.

One of the most important, in our estimation, is that the vast majority of the tiny compounds that have been found have primarily been studied for their possibility as treatments, particularly in disease-modeling creatures and, often, in rats. A dangerous process might result from a "too direct" transport of research findings from mammal models of NDDs to human pathology, which could lead to the creation and spread of inaccurate and biased information. This is true despite the fact that animal models have been essential in advancing our understanding of the pathogenetic processes, leading to a particular neurological illness, and have been critical in identifying specific pharmaceutical targets.

Additionally, it is important to retain critical aptitude, particularly when it comes to information on particular tiny chemicals which have been shown to enhance cognitive signs in some mouse methods. This is due to the fact that the deep and complex emotional components that may have a different effect on the cognitive changes linked to psychiatry neurology and NDDs in humans are entirely absent from rodent models.

The absence of accurate and thorough knowledge on the true capabilities of small molecules that demonstrate effectiveness in therapy on animal models of psychiatry neurology and NDDs to pass through the BBB and then effectively being transferred to the CNS is another crucial factor that makes it possible to translate data from rodent models to humans more directly. Further restrictions about the actual utility of the current in vitro BBB models further compound this. All of these worries may help to explain why certain small compounds, like clioquinol and PBT2, which were ultimately chosen for scientific studies and discovered to alleviate cognitive dysfunctions in humans, also came with quite substantial side effects. The conclusion is that, given the numerous "unexplored" areas of psychiatry neurology and NDDs, there is an urgent required to intensify efforts in designing thorough studies on the potential therapeutic application of tiny molecules in these conditions.

REFERENCES

Adlard, P. A., Cherny, R. A., Finkelstein, D. I., Gautier, E., Robb, E., Cortes, M., Volitakis, I., Liu, X., Smith, J. P., Perez, K., Laughton, K., Li, Q.-X., Charman, S. A., Nicolazzo, J. A., Wilkins, S., Deleva, K., Lynch, T., Kok, G., Ritchie, C. W., … Bush, A. I. (2008). Rapid restoration of cognition in Alzheimer's transgenic mice with 8-hydroxy quinoline analogs is associated with decreased interstitial Aβ. *Neuron*, *59*(1), pp. 43–55. https://doi.org/10.1016/j.neuron.2008.06.018

Albanese, F., Novello, S., & Morari, M. (2019). Autophagy and LRRK2 in the aging brain. *Frontiers in Neuroscience*, *13*, p. 1352. https://doi.org/10.3389/fnins.2019.01352

Altarescu, G., Hill, S., Wiggs, E., Jeffries, N., Kreps, C., Parker, C. C., Brady, R. O., Barton, N. W., & Schiffmann, R. (2001). The efficacy of enzyme replacement therapy in patients with chronic neuronopathic Gaucher's disease. *The Journal of Pediatrics*, *138*(4), pp. 539–547. https://doi.org/10.1067/mpd.2001.112171

Amenta, F., & Tayebati, S. (2008). Pathways of acetylcholine synthesis, transport and release as targets for treatment of adult-onset cognitive dysfunction. *Current Medicinal Chemistry*, *15*(5), pp. 488–498. https://doi.org/10.2174/092986708783503203

American Psychiatric Association Diagnostic and Statistical Manual of Mental Disorders (DSM-IV). (n.d.). *SpringerReference*. Springer-Verlag. https://doi.org/10.1007/springerreference_179660

Arena, G., Gelmetti, V., Torosantucci, L., Vignone, D., Lamorte, G., De Rosa, P., Cilia, E., Jonas, E. A., & Valente, E. M. (2013). PINK1 protects against cell death induced by mitochondrial depolarization, by phosphorylating Bcl-xL and impairing its pro-apoptotic cleavage. *Cell Death and Differentiation*, *20*(7), pp. 920–930. https://doi.org/10.1038/cdd.2013.19

Barbiero, J. K., Santiago, R. M., Lima, M. M. S., Ariza, D., Morais, L. H., Andreatini, R., & Vital, M. A. B. F. (2011). Acute but not chronic administration of pioglitazone promoted behavioral and neurochemical protective effects in the MPTP model of Parkinson's disease. *Behavioural Brain Research*, *216*(1), pp. 186–192. https://doi.org/10.1016/j.bbr.2010.07.033

Barnham, K. J., & Bush, A. I. (2008). Metals in Alzheimer's and Parkinson's diseases. *Current Opinion in Chemical Biology*, *12*(2), pp. 222–228. https://doi.org/10.1016/j.cbpa.2008.02.019

Blanz, J., & Saftig, P. (2016). Parkinson's disease: acid-glucocerebrosidase activity and alpha-synuclein clearance. *Journal of Neurochemistry*, *139*(S1), pp. 198–215. https://doi.org/10.1111/jnc.13517

Bonello, F., Hassoun, S.-M., Mouton-Liger, F., Shin, Y. S., Muscat, A., Tesson, C., Lesage, S., Beart, P. M., Brice, A., Krupp, J., Corvol, J.-C., & Corti, O. (2019). LRRK2 impairs PINK1/Parkin-dependent mitophagy via its kinase activity: pathologic insights into Parkinson's disease. *Human Molecular Genetics*, *28*(10), pp. 1645–1660. https://doi.org/10.1093/hmg/ddz004

Bordet, R., Ouk, T., Petrault, O., Gelé, P., Gautier, S., Laprais, M., Deplanque, D., Duriez, P., Staels, B., Fruchart, J. C., & Bastide, M. (2006). PPAR: a new pharmacological target for neuroprotection in stroke and neurodegenerative diseases. *Biochemical Society Transactions*, *34*(6), pp. 1341–1346. https://doi.org/10.1042/BST0341341

Bottegoni, G., Veronesi, M., Bisignano, P., Kacker, P., Favia, A. D., & Cavalli, A. (2015). Development and application of a virtual screening protocol for the identification of multitarget fragments. *ChemMedChem*, *11*(12), pp. 1259–1263. https://doi.org/10.1002/cmdc.201500521

Buchhave, P. (2012). Cerebrospinal fluid levels of β-amyloid 1-42, but not of tau, are fully changed already 5 to 10 years before the onset of Alzheimer dementia. *Archives of General Psychiatry*, *69*(1), p. 98. https://doi.org/10.1001/archgenpsychiatry.2011.155

Bush, A. I., & Tanzi, R. E. (2008). Therapeutics for Alzheimer 's disease based on the metal hypothesis. *Neurotherapeutics*, *5*(3), pp. 421–432. https://doi.org/10.1016/j.nurt.2008.05.001

Chakraborty, S., Bandyopadhyay, J., Chakraborty, S., & Basu, S. (2016). Multi-target screening mines hesperidin as a multi-potent inhibitor: implication in Alzheimer's disease therapeutics. *European Journal of Medicinal Chemistry*, *121*, pp. 810–822. https://doi.org/10.1016/j.ejmech.2016.03.057

Chakraborty, S., & Basu, S. (2017). Multi-functional activities of citrus flavonoid narirutin in Alzheimer's disease therapeutics: an integrated screening approach and in vitro validation. *International Journal of Biological Macromolecules*, *103*, pp. 733–743. https://doi.org/10.1016/j.ijbiomac.2017.05.110

Chen, G., Xie, Y., Zhou, D., Yang, L., Zou, L., Cheng, M., Liu, Y., & Li, N. (2020). In vivo evaluation and atom-based 3D-QSAR studies on saponins from shells of Xanthoceras sorbifolium Bunge as anti-AD agents. *Bioorganic Chemistry*, *94*, p. 103412. https://doi.org/10.1016/j.bioorg.2019.103412

Chen, J., Swope, D., & Dashtipour, K. (2007). Comprehensive review of rasagiline, a second-generation monoamine oxidase inhibitor, for the treatment of Parkinson's disease. *Clinical Therapeutics*, *29*(9), pp. 1825–1849. https://doi.org/10.1016/j.clinthera.2007.09.021

Chen, Y.-F., Kao, C.-H., Kirby, R., & Tsai, T.-F. (2009). Cisd2 mediates mitochondrial integrity and life span in mammals. *Autophagy*, *5*(7), pp. 1043–1045. https://doi.org/10.4161/auto.5.7.9351

Chimenti, F., Bolasco, A., Manna, F., Secci, D., Chimenti, P., Granese, A., Befani, O., Turini, P., Cirilli, R., Torre, F., Alcaro, S., Ortuso, F., & Langer, T. (2006). Synthesis, Biological evaluation and 3D-QSAR of 1,3,5-Trisubstituted-4,5- Dihydro-(1H)-Pyrazole derivatives as potent and highly selective monoamine oxidase A inhibitors. *Current Medicinal Chemistry*, *13*(12), pp. 1411–1428. https://doi.org/10.2174/092986706776872907

Choi, D.-H., Cristóvão, A. C., Guhathakurta, S., Lee, J., Joh, T. H., Beal, M. F., & Kim, Y.-S. (2012). NADPH oxidase 1-mediated oxidative stress leads to dopamine neuron death in Parkinson's disease. *Antioxidants & Redox Signaling*, *16*(10), pp. 1033–1045. https://doi.org/10.1089/ars.2011.3960

Cogo, S., Manzoni, C., Lewis, P. A., & Greggio, E. (2019). Leucine-rich repeat kinase 2 and lysosomal dyshomeostasis in Parkinson disease. *Journal of Neurochemistry*, *152*(3), pp. 273–283. https://doi.org/10.1111/jnc.14908

Cohen, S. I. A., Arosio, P., Presto, J., Kurudenkandy, F. R., Biverstål, H., Dolfe, L., Dunning, C., Yang, X., Frohm, B., Vendruscolo, M., Johansson, J., Dobson, C. M., Fisahn, A., Knowles, T. P. J., & Linse, S. (2015). A molecular chaperone breaks the catalytic cycle that generates toxic Aβ oligomers. *Nature Structural & Molecular Biology*, *22*(3), pp. 207–213. https://doi.org/10.1038/nsmb.2971

Cohen, S. I. A., Linse, S., Luheshi, L. M., Hellstrand, E., White, D. A., Rajah, L., Otzen, D. E., Vendruscolo, M., Dobson, C. M., & Knowles, T. P. J. (2013). Proliferation of amyloid-β42 aggregates occurs through a secondary nucleation mechanism. *Proceedings of the National Academy of Sciences*, *110*(24), pp. 9758–9763. https://doi.org/10.1073/pnas.1218402110

Collino, M., Aragno, M., Mastrocola, R., Gallicchio, M., Rosa, A. C., Dianzani, C., Danni, O., Thiemermann, C., & Fantozzi, R. (2006). Modulation of the oxidative stress and inflammatory response by PPAR-γ agonists in the hippocampus of rats exposed to cerebral ischemia/reperfusion. *European Journal of Pharmacology*, *530*(1–2), pp. 70–80. https://doi.org/10.1016/j.ejphar.2005.11.049

Dall'lgna, O. P., Porciúncula, L. O., Souza, D. O., Cunha, R. A., & Lara, D. R. (2003). Neuroprotection by caffeine and adenosine A 2A receptor blockade of β -amyloid neurotoxicity. *British Journal of Pharmacology*, *138*(7), pp. 1207–1209. https://doi.org/10.1038/sj.bjp.0705185

Dar, N. J., & Glazner, G. W. (2020). Deciphering the neuroprotective and neurogenic potential of soluble amyloid precursor protein alpha (sAPPα). *Cellular and Molecular Life Sciences*, *77*(12), pp. 2315–2330. https://doi.org/10.1007/s00018-019-03404-x

Dar, N. J., Na, R., & Ran, Q. (2022). Functional deficits of 5×FAD neural stem cells are ameliorated by glutathione peroxidase 4. *Cells*, *11*(11), p. 1770. https://doi.org/10.3390/cells11111770

Das, S., & Smid, S. D. (2017). Identification of dibenzyl imidazolidine and triazole acetamide derivatives through virtual screening targeting amyloid beta aggregation and neurotoxicity in PC12 cells. *European Journal of Medicinal Chemistry*, *130*, pp. 354–364. https://doi.org/10.1016/j.ejmech.2017.02.057

De Strooper, B., & Karran, E. (2016). The cellular phase of Alzheimer's disease. *Cell*, *164*(4), pp. 603–615. https://doi.org/10.1016/j.cell.2015.12.056

Dhanjal, J. K., Sharma, S., Grover, A., & Das, A. (2015). Use of ligand-based pharmacophore modeling and docking approach to find novel acetylcholinesterase inhibitors for treating Alzheimer's. *Biomedicine & Pharmacotherapy*, *71*, pp. 146–152. https://doi.org/10.1016/j.biopha.2015.02.010

Dinda, B., Dinda, M., Kulsi, G., Chakraborty, A., & Dinda, S. (2019). Therapeutic potentials of plant iridoids in Alzheimer's and Parkinson's diseases: a review. *European Journal of Medicinal Chemistry*, *169*, pp. 185–199. https://doi.org/10.1016/j.ejmech.2019.03.009

Dorababu, A. (2019). Critical evaluation of current Alzheimer's drug discovery (2018–19) & futuristic Alzheimer drug model approach. *Bioorganic Chemistry*, *93*, p. 103299. https://doi.org/10.1016/j.bioorg.2019.103299

Fernandez, H. H., Greeley, D. R., Zweig, R. M., Wojcieszek, J., Mori, A., & Sussman, N. M. (2010). Istradefylline as monotherapy for Parkinson disease: results of the 6002-US-051 trial. *Parkinsonism & Related Disorders*, *16*(1), pp. 16–20. https://doi.org/10.1016/j.parkreldis.2009.06.008

Foley, A., Hirst, W., Gallagher, H., Barry, C., Hagan, J., Upton, N., Walsh, F., Hunter, A., & Regan, C. (2008). The selective 5-HT6 receptor antagonists SB-271046 and SB-399885 potentiate NCAM PSA immunolabeling of dentate granule cells, but not neurogenesis, in the hippocampal formation of mature Wistar rats. *Neuropharmacology*, *54*(8), pp. 1166–1174. https://doi.org/10.1016/j.neuropharm.2008.03.012

Gaeta, A., & Hider, R. C. (2005). The crucial role of metal ions in neurodegeneration: the basis for a promising therapeutic strategy. *British Journal of Pharmacology*, *146*(8), pp. 1041–1059. https://doi.org/10.1038/sj.bjp.0706416

Geldenhuys, W. J., Darvesh, A. S., Funk, M. O., Van der Schyf, C. J., & Carroll, R. T. (2010). Identification of novel monoamine oxidase B inhibitors by structure-based virtual screening. *Bioorganic & Medicinal Chemistry Letters*, *20*(17), pp. 5295–5298. https://doi.org/10.1016/j.bmcl.2010.06.128

Gervasio, F. L., & Spiwok, V. (Eds.). (2018). Biomolecular simulations in structure-based drug discovery. Wiley. https://doi.org/10.1002/9783527806836

Giguère, V. (2008). Transcriptional control of energy homeostasis by the estrogen-related receptors. *Endocrine Reviews*, *29*(6), pp. 677–696. https://doi.org/10.1210/er.2008-0017

Glennon, R. A., Siripurapu, U., Roth, B. L., Kolanos, R., Bondarev, M. L., Sikazwe, D., Lee, M., & Dukat, M. (2010). The medicinal chemistry of 5-HT6 receptor ligands with a focus on arylsulfonyltryptamine analogs. *Current Topics in Medicinal Chemistry*, *10*(5), pp. 579–595. https://doi.org/10.2174/156802610791111542

Gurevich, E. V., & Gurevich, V. V. (2014). Therapeutic potential of small molecules and engineered proteins (pp. 1–12). DOI: 10.1007/978-3-642-41199-1_1

Haass, C., & Selkoe, D. J. (2007). Soluble protein oligomers in neurodegeneration: lessons from the Alzheimer's amyloid β-peptide. *Nature Reviews Molecular Cell Biology*, *8*(2), pp. 101–112. https://doi.org/10.1038/nrm2101

Habchi, J., Chia, S., Limbocker, R., Mannini, B., Ahn, M., Perni, M., Hansson, O., Arosio, P., Kumita, J. R., Challa, P. K., Cohen, S. I. A., Linse, S., Dobson, C. M., Knowles, T. P. J., & Vendruscolo, M. (2017). Systematic development of small molecules to inhibit specific microscopic steps of Aβ42 aggregation in Alzheimer's disease. *Proceedings of the National Academy of Sciences*, *114*(2). https://doi.org/10.1073/pnas.1615613114

Hardy, J., & De Strooper, B. (2017). Alzheimer's disease: where next for anti-amyloid therapies? *Brain*, *140*(4), pp. 853–855. https://doi.org/10.1093/brain/awx059

Hassan, M., Abbasi, M. A., Aziz-ur-Rehman, Siddiqui, S. Z., Shahzadi, S., Raza, H., Hussain, G., Shah, S. A. A., Ashraf, M., Shahid, M., Seo, S.-Y., & Malik, A. (2019). Designing of promising medicinal scaffolds for Alzheimer's disease through enzyme inhibition, lead optimization, molecular docking and dynamic simulation approaches. *Bioorganic Chemistry*, *91*, p. 103138. https://doi.org/10.1016/j.bioorg.2019.103138

Haycox, A., Armand, C., Murteira, S., Cochran, J., & François, C. (2009). Cost effectiveness of rasagiline and pramipexole as treatment strategies in early Parkinson's disease in the UK setting. *Drugs & Aging*, *26*(9), pp. 791–801. https://doi.org/10.2165/11316770-000000000-00000

Heneka, M. T., Sastre, M., Dumitrescu-Ozimek, L., Hanke, A., Dewachter, I., Kuiperi, C., O'Banion, K., Klockgether, T., Van Leuven, F., & Landreth, G. E. (2005). Acute treatment with the PPARγ agonist pioglitazone and ibuprofen reduces glial inflammation and Aβ1–42 levels in APPV717I transgenic mice. *Brain*, *128*(6), pp. 1442–1453. https://doi.org/10.1093/brain/awh452

Hertz, N. T., Berthet, A., Sos, M. L., Thorn, K. S., Burlingame, A. L., Nakamura, K., & Shokat, K. M. (2013). A neo-substrate that amplifies catalytic activity of parkinson's-disease-related kinase PINK1. *Cell*, *154*(4), pp. 737–747. https://doi.org/10.1016/j.cell.2013.07.030

Hindo, S. S., Mancino, A. M., Braymer, J. J., Liu, Y., Vivekanandan, S., Ramamoorthy, A., & Lim, M. H. (2009). small molecule modulators of copper-induced Aβ aggregation. *Journal of the American Chemical Society*, *131*(46), pp. 16663–16665. https://doi.org/10.1021/ja907045h

Honda, T., Kaikita, K., Tsujita, K., Hayasaki, T., Matsukawa, M., Fuchigami, S., Sugiyama, S., Sakashita, N., Ogawa, H., & Takeya, M. (2008). Pioglitazone, a peroxisome proliferator-activated receptor-γ agonist, attenuates myocardial ischemia–reperfusion injury in mice with metabolic disorders. *Journal of Molecular and Cellular Cardiology*, *44*(5), pp. 915–926. https://doi.org/10.1016/j.yjmcc.2008.03.004

Hong, L., & Sklar, L. A. (2014). Targeting GTPases in Parkinson's disease: comparison to the historic path of kinase drug discovery and perspectives. *Frontiers in Molecular Neuroscience*, *7*, p. 52. https://doi.org/10.3389/fnmol.2014.00052

Hopkins, A. L. (2008). Network pharmacology: the next paradigm in drug discovery. *Nature Chemical Biology*, *4*(11), pp. 682–690. https://doi.org/10.1038/nchembio.118

Hu, Z.-Y., Chen, B., Zhang, J.-P., & Ma, Y.-Y. (2017). Up-regulation of autophagy-related gene 5 (ATG5) protects dopaminergic neurons in a zebrafish model of Parkinson's disease. *The Journal of Biological Chemistry*, *292*(44), pp. 18062–18074. https://doi.org/10.1074/jbc.M116.764795

Hureau, C., Sasaki, I., Gras, E., & Faller, P. (2010). Two functions, one molecule: a metal-binding and a targeting moiety to combat Alzheimer's disease. *ChemBioChem*, *11*(7), pp. 950–953. https://doi.org/10.1002/cbic.201000102

Jiang, T.-F., Zhang, Y.-J., Zhou, H.-Y., Wang, H.-M., Tian, L.-P., Liu, J., Ding, J.-Q., & Chen, S.-D. (2013). Curcumin ameliorates the neurodegenerative pathology in A53T α-synuclein cell model of Parkinson's disease through the downregulation of mTOR/p70S6K signaling and the recovery of macroautophagy. *Journal of Neuroimmune Pharmacology*, *8*(1), pp. 356–369. https://doi.org/10.1007/s11481-012-9431-7

Joghataie, M. T., Roghani, M., Negahdar, F., & Hashemi, L. (2004). Protective effect of caffeine against neurodegeneration in a model of Parkinson's disease in rat: behavioral and histochemical evidence. *Parkinsonism & Related Disorders*, *10*(8), pp. 465–468. https://doi.org/10.1016/j.parkreldis.2004.06.004

Karran, E., & Hardy, J. (2014). A critique of the drug discovery and phase 3 clinical programs targeting the amyloid hypothesis for Alzheimer disease. *Annals of Neurology*, *76*(2), pp. 185–205. https://doi.org/10.1002/ana.24188

Kaundal, R. K., Iyer, S., Kumar, A., & Sharma, S. S. (2009). Protective effects of pioglitazone against global cerebral ischemic-reperfusion injury in gerbils. *Journal of Pharmacological Sciences*, *109*(3), pp. 361–367. https://doi.org/10.1254/jphs.08246FP

Kehler, J., & Nielsen, J. (2011). PDE10A inhibitors: novel therapeutic drugs for Schizophrenia. *Current Pharmaceutical Design*, *17*(2), pp. 137–150. https://doi.org/10.2174/138161211795049624

Khan, S., Bano, N., Ahamad, S., John, U., Dar, N. J., & Bhat, S. A. (2024). Excitotoxicity, oxytosis/ferroptosis, and neurodegeneration: emerging insights into mitochondrial mechanisms. *Aging and Disease*. https://doi.org/10.14336/AD.2024.0125-1

Kim, H., & Han, H. (2018). Computer-aided multi-target management of emergent Alzheimer's disease. *Bioinformation*, *14*(4), pp. 167–180. https://doi.org/10.6026/97320630014167

Klusa, V. Z., Isajevs, S., Svirina, D., Pupure, J., Beitnere, U., Rumaks, J., Svirskis, S., Jansone, B., Dzirkale, Z., Muceniece, R., Kalvinsh, I., & Vinters, H. V. (2010). Neuroprotective properties of mildronate, a small molecule, in a rat model of Parkinson's disease. *International Journal of Molecular Sciences*, *11*(11), pp. 4465–4487. https://doi.org/10.3390/ijms11114465

Kondapalli, C., Kazlauskaite, A., Zhang, N., Woodroof, H. I., Campbell, D. G., Gourlay, R., Burchell, L., Walden, H., Macartney, T. J., Deak, M., Knebel, A., Alessi, D. R., & Muqit, M. M. K. (2012). PINK1 is activated by mitochondrial membrane potential depolarization and stimulates Parkin E3 ligase activity by phosphorylating Serine 65. *Open Biology*, *2*(5), p. 120080. https://doi.org/10.1098/rsob.120080

Lamonaca, G., & Volta, M. (2020). Alpha-Synuclein and LRRK2 in synaptic autophagy: linking early dysfunction to late-stage pathology in Parkinson's disease. *Cells*, *9*(5), p. 1115. https://doi.org/10.3390/cells9051115

Law, A. J., Wang, Y., Sei, Y., O'Donnell, P., Piantadosi, P., Papaleo, F., Straub, R. E., Huang, W., Thomas, C. J., Vakkalanka, R., Besterman, A. D., Lipska, B. K., Hyde, T. M., Harrison, P. J., Kleinman, J. E., & Weinberger, D. R. (2012). Neuregulin 1-ErbB4-PI3K signaling in schizophrenia and phosphoinositide 3-kinase-p110δ inhibition as a potential therapeutic strategy. *Proceedings of the National Academy of Sciences*, *109*(30), pp. 12165–12170. https://doi.org/10.1073/pnas.1206118109

Leng, Y., Liang, M.-H., Ren, M., Marinova, Z., Leeds, P., & Chuang, D.-M. (2008). Synergistic neuroprotective effects of lithium and valproic acid or other histone deacetylase inhibitors in neurons: roles of glycogen synthase kinase-3 inhibition. *The Journal of Neuroscience*, *28*(10), pp. 2576–2588. https://doi.org/10.1523/JNEUROSCI.5467-07.2008

Levine, B., & Klionsky, D. J. (2017). Autophagy wins the 2016 Nobel prize in physiology or medicine: breakthroughs in baker's yeast fuel advances in biomedical research. *Proceedings of the National Academy of Sciences of the United States of America*, *114*(2), pp. 201–205. https://doi.org/10.1073/pnas.1619876114

LeWitt, P. A., Guttman, M., Tetrud, J. W., Tuite, P. J., Mori, A., Chaikin, P., & Sussman, N. M. (2008). Adenosine A 2A receptor antagonist istradefylline (KW-6002) reduces "off" time in Parkinson's disease: a double-blind, randomized, multicenter clinical trial (6002-US-005). *Annals of Neurology*, *63*(3), pp. 295–302. https://doi.org/10.1002/ana.21315

Li, J., Lang, M.-J., Mao, X.-B., Tian, L., & Feng, Y.-B. (2008). Antiapoptosis and mitochondrial effect of pioglitazone preconditioning in the ischemic/reperfused heart of rat. *Cardiovascular Drugs and Therapy*, *22*(4), pp. 283–291. https://doi.org/10.1007/s10557-008-6115-x

Lin, C.-H., Hsieh, Y.-S., Wu, Y.-R., Hsu, C.-J., Chen, H.-C., Huang, W.-H., Chang, K.-H., Hsieh-Li, H. M., Su, M.-T., Sun, Y.-C., Lee, G.-C., & Lee-Chen, G.-J. (2016). Identifying GSK-3β kinase inhibitors of Alzheimer's disease: virtual screening, enzyme, and cell assays. *European Journal of Pharmaceutical Sciences*, *89*, pp. 11–19. https://doi.org/10.1016/j.ejps.2016.04.012

Liu, G. Y., & Sabatini, D. M. (2020). mTOR at the nexus of nutrition, growth, ageing and disease. *Nature Reviews. Molecular Cell Biology*, *21*(4), pp. 183–203. https://doi.org/10.1038/s41580-019-0199-y

Liu, H., Yang, J., Wang, L., Xu, Y., Zhang, S., Lv, J., Ran, C., & Li, Y. (2017). Targeting β-amyloid plaques and oligomers: development of near-IR fluorescence imaging probes. *Future Medicinal Chemistry*, *9*(2), pp. 179–198. https://doi.org/10.4155/fmc-2016-0185

Lu, W., Zhang, R., Jiang, H., Zhang, H., & Luo, C. (2018). Computer-Aided Drug Design in Epigenetics. *Frontiers in Chemistry*, *6*, p. 57. https://doi.org/10.3389/fchem.2018.00057

Madav, Y., Wairkar, S., & Prabhakar, B. (2019). Recent therapeutic strategies targeting beta amyloid and tauopathies in Alzheimer's disease. *Brain Research Bulletin*, *146*, pp. 171–184. https://doi.org/10.1016/j.brainresbull.2019.01.004

Maia, L., & De Mendonça, A. (2002). Does caffeine intake protect from Alzheimer's disease? *European Journal of Neurology*, *9*(4), pp. 377–382. https://doi.org/10.1046/j.1468-1331.2002.00421.x

Manzine, P. R., Ettcheto, M., Cano, A., Busquets, O., Marcello, E., Pelucchi, S., Di Luca, M., Endres, K., Olloquequi, J., Camins, A., & Cominetti, M. R. (2019). ADAM10 in Alzheimer's disease: pharmacological modulation by natural compounds and its role as a peripheral marker. *Biomedicine & Pharmacotherapy*, *113*, p. 108661. https://doi.org/10.1016/j.biopha.2019.108661

Maria Fimia, G., Stoykova, A., Romagnoli, A., Giunta, L., Di Bartolomeo, S., Nardacci, R., Corazzari, M., Fuoco, C., Ucar, A., Schwartz, P., Gruss, P., Piacentini, M., Chowdhury, K., & Cecconi, F. (2007). Ambra1 regulates autophagy and development of the nervous system. *Nature*, *447*(7148), pp. 1121–1125. https://doi.org/10.1038/nature05925

Mitchinson, A., & Finkelstein, J. (2008). Small-molecule catalysis. *Nature*, *455*(7211), pp. 303–303. https://doi.org/10.1038/455303a

Molina-Holgado, F., Hider, R. C., Gaeta, A., Williams, R., & Francis, P. (2007). Metals ions and neurodegeneration. *BioMetals*, *20*(3–4), pp. 639–654. https://doi.org/10.1007/s10534-006-9033-z

Monte, D. A. Di. (2003). The environment and Parkinson's disease: is the nigrostriatal system preferentially targeted by neurotoxins? *The Lancet Neurology*, *2*(9), pp. 531–538. https://doi.org/10.1016/S1474-4422(03)00501-5

Morelli, M., Carta, A. R., & Jenner, P. (2009). Adenosine A2A receptors and Parkinson's disease (pp. 589–615). DOI: 10.1007/978-3-540-89615-9_18

Morphy, R., Kay, C., & Rankovic, Z. (2004). From magic bullets to designed multiple ligands. *Drug Discovery Today*, *9*(15), pp. 641–651. https://doi.org/10.1016/S1359-6446(04)03163-0

Morphy, R., & Rankovic, Z. (2005). Designed multiple ligands. An emerging drug discovery paradigm. *Journal of Medicinal Chemistry*, *48*(21), pp. 6523–6543. https://doi.org/10.1021/jm058225d

Mullin, S., Smith, L., Lee, K., D'Souza, G., Woodgate, P., Elflein, J., Hällqvist, J., Toffoli, M., Streeter, A., Hosking, J., Heywood, W. E., Khengar, R., Campbell, P., Hehir, J., Cable, S., Mills, K., Zetterberg, H., Limousin, P., Libri, V., … Schapira, A. H. V. (2020). Ambroxol for the treatment of patients with parkinson disease with and without glucocerebrosidase gene mutations: a nonrandomized, noncontrolled trial. *JAMA Neurology*, *77*(4), pp. 427–434. https://doi.org/10.1001/jamaneurol.2019.4611

Neumann, M., Müller, V., Kretzschmar, H. A., Haass, C., & Kahle, P. J. (2004). Regional Distribution of Proteinase K-Resistant α-Synuclein Correlates with Lewy Body Disease Stage. *Journal of Neuropathology & Experimental Neurology*, *63*(12), pp. 1225–1235. https://doi.org/10.1093/jnen/63.12.1225

Nunnari, J., & Suomalainen, A. (2012). Mitochondria: in sickness and in health. *Cell*, *148*(6), pp. 1145–1159. https://doi.org/10.1016/j.cell.2012.02.035

Nuytemans, K., Theuns, J., Cruts, M., & Van Broeckhoven, C. (2010). Genetic etiology of Parkinson disease associated with mutations in the SNCA, PARK2, PINK1, PARK7, and LRRK2 genes: a mutation update. *Human Mutation*, *31*(7), pp. 763–780. https://doi.org/10.1002/humu.21277

O'Hara, D. M., Pawar, G., Kalia, S. K., & Kalia, L. V. (2020). LRRK2 and α-synuclein: distinct or synergistic players in Parkinson's disease? *Frontiers in Neuroscience*, *14*, p. 577. https://doi.org/10.3389/fnins.2020.00577

Oh, M., Lee, J. H., Wang, W., Lee, H. S., Lee, W. S., Burlak, C., Im, W., Hoang, Q. Q., & Lim, H.-S. (2014). Potential pharmacological chaperones targeting cancer-associated MCL-1 and Parkinson disease-associated α-synuclein. *Proceedings of the National Academy of Sciences of the United States of America*, *111*(30), pp. 11007–11012. https://doi.org/10.1073/pnas.1320556111

Olğaç, A., Türe, A., Olğaç, S., Möller, S. (2019). Cloud-Based High Throughput Virtual Screening in Novel Drug Discovery. In: Kołodziej, J., González-Vélez, H. (eds) High-Performance Modelling and Simulation for Big Data Applications. Lecture Notes in Computer Science, vol 11400. Springer, Cham. https://doi.org/10.1007/978-3-030-16272-6_9

Palmieri, M., Impey, S., Kang, H., di Ronza, A., Pelz, C., Sardiello, M., & Ballabio, A. (2011). Characterization of the CLEAR network reveals an integrated control of cellular clearance pathways. *Human Molecular Genetics*, *20*(19), pp. 3852–3866. https://doi.org/10.1093/hmg/ddr306

Parsons, C. G., & Rammes, G. (2017). Preclinical to phase II amyloid beta (A_β) peptide modulators under investigation for Alzheimer's disease. *Expert Opinion on Investigational Drugs*, *26*(5), pp. 579–592. https://doi.org/10.1080/13543784.2017.1313832

Rauk, A. (2009). The chemistry of Alzheimer's disease. *Chemical Society Reviews*, *38*(9), pp. 2698. https://doi.org/10.1039/b807980n

Richter, F., Fleming, S. M., Watson, M., Lemesre, V., Pellegrino, L., Ranes, B., Zhu, C., Mortazavi, F., Mulligan, C. K., Sioshansi, P. C., Hean, S., De La Rosa, K., Khanna, R., Flanagan, J., Lockhart, D. J., Wustman, B. A., Clark, S. W., & Chesselet, M.-F. (2014). A GCase chaperone improves motor function in a mouse model of synucleinopathy. *Neurotherapeutics: The Journal of the American Society for Experimental NeuroTherapeutics*, *11*(4), pp. 840–856. https://doi.org/10.1007/s13311-014-0294-x

Rigi, G., Nakhaei, M. V. A., Eidipour, H., Najimi, A., Tajik, F., Taher, N., & Yarahmadi, K. (2017). Virtual screening following rational drug design based approach for introducing new anti amyloid beta aggregation agent. *Bioinformation*, *13*(2), pp. 42–45. https://doi.org/10.6026/97320630013042

Roy, S., Kumar, A., Baig, M. H., Masařík, M., & Provazník, I. (2015). Virtual screening, ADMET profiling, molecular docking and dynamics approaches to search for potent selective natural molecules based inhibitors against metallothionein-III to study Alzheimer's disease. *Methods*, *83*, pp. 105–110. https://doi.org/10.1016/j.ymeth.2015.04.021

Rugarli, E. I., & Langer, T. (2012). Mitochondrial quality control: a matter of life and death for neurons. *The EMBO Journal*, *31*(6), pp. 1336–1349. https://doi.org/10.1038/emboj.2012.38

Sagi, Y., Weinstock, M., & Youdim, M. B. H. (2003). Attenuation of MPTP-induced dopaminergic neurotoxicity by TV3326, a cholinesterase-monoamine oxidase inhibitor. *Journal of Neurochemistry*, *86*(2), pp. 290–297. https://doi.org/10.1046/j.1471-4159.2003.01801.x

Savolainen, M. H., Richie, C. T., Harvey, B. K., Männistö, P. T., Maguire-Zeiss, K. A., & Myöhänen, T. T. (2014). The beneficial effect of a prolyl oligopeptidase inhibitor, KYP-2047, on alpha-synuclein clearance and autophagy in A30P transgenic mouse. *Neurobiology of Disease*, *68*, pp. 1–15. https://doi.org/10.1016/j.nbd.2014.04.003

Savolainen, M. H., Yan, X., Myöhänen, T. T., & Huttunen, H. J. (2015). Prolyl oligopeptidase enhances α-synuclein dimerization via direct protein-protein interaction. *The Journal of Biological Chemistry*, *290*(8), pp. 5117–5126. https://doi.org/10.1074/jbc.M114.592931

Sawkar, A. R., Schmitz, M., Zimmer, K.-P., Reczek, D., Edmunds, T., Balch, W. E., & Kelly, J. W. (2006). Chemical chaperones and permissive temperatures alter the cellular localization of gaucher disease associated glucocerebrosidase variants. *ACS Chemical Biology*, *1*(4), pp. 235–251. https://doi.org/10.1021/cb600187q

Sayre, L. M., Perry, G., & Smith, M. A. (2008). Oxidative stress and neurotoxicity. *Chemical Research in Toxicology*, *21*(1), pp. 172–188. https://doi.org/10.1021/tx700210j

Schapira, A. H. V., & Olanow, C. W. (2009). Drug selection and timing of initiation of treatment in early Parkinson's disease. *Annals of Neurology*, *64*(S2), pp. S47–S55. https://doi.org/10.1002/ana.21460

Scott, L. E., & Orvig, C. (2009). Medicinal inorganic chemistry approaches to passivation and removal of aberrant metal ions in disease. *Chemical Reviews*, *109*(10), pp. 4885–4910. https://doi.org/10.1021/cr9000176

Shadrina MI, Slominsky PA, Limborska SA. Molecular mechanisms of pathogenesis of Parkinson's disease. Int Rev Cell Mol Biol. 2010;281:229-66. doi: 10.1016/S1937-6448(10)81006-8. PMID: 20460187.

Shastry, B. S. (2003). Neurodegenerative disorders of protein aggregation. *Neurochemistry International*, *43*(1), pp. 1–7. https://doi.org/10.1016/S0197-0186(02)00196-1

Shimazu, T., Inoue, I., Araki, N., Asano, Y., Sawada, M., Furuya, D., Nagoya, H., & Greenberg, J. H. (2005). A peroxisome proliferator-activated receptor-γ agonist reduces infarct size in transient but not in permanent ischemia. *Stroke*, *36*(2), pp. 353–359. https://doi.org/10.1161/01.STR.0000152271.21943.a2

Shugrue, C. R., & Miller, S. J. (2017). Applications of nonenzymatic catalysts to the alteration of natural products. *Chemical Reviews*, *117*(18), pp. 11894–11951. https://doi.org/10.1021/acs.chemrev.7b00022

Son, J. H., Shim, J. H., Kim, K. H., Ha, J. Y., & Han, J. Y. (2012). Neuronal autophagy and neurodegenerative diseases. *Experimental & Molecular Medicine*, *44*(2), pp. 89–98. https://doi.org/10.3858/emm.2012.44.2.031

Song, J.-X., Lu, J.-H., Liu, L.-F., Chen, L.-L., Durairajan, S. S. K., Yue, Z., Zhang, H.-Q., & Li, M. (2014). HMGB1 is involved in autophagy inhibition caused by SNCA/α-synuclein overexpression: a process modulated by the natural autophagy inducer corynoxine B. *Autophagy*, *10*(1), pp. 144–154. https://doi.org/10.4161/auto.26751

Srour, A. M., Dawood, D. H., Khalil, M. N. A., & Nofal, Z. M. (2019). Synthesis and 2D-QSAR study of dispiropyrrolodinyl-oxindole based alkaloids as cholinesterase inhibitors. *Bioorganic Chemistry*, *83*, pp. 226–234. https://doi.org/10.1016/j.bioorg.2018.10.030

Staroń, J., Kurczab, R., Warszycki, D., Satała, G., Krawczyk, M., Bugno, R., Lenda, T., Popik, P., Hogendorf, A. S., Hogendorf, A., Dubiel, K., Matłoka, M., Moszczyński-Pętkowski, R., Pieczykolan, J., Wieczorek, M., Zajdel, P., & Bojarski, A. J. (2020). Virtual screening-driven discovery of dual 5-HT6/5-HT2A receptor ligands with pro-cognitive properties. *European Journal of Medicinal Chemistry*, *185*, p. 111857. https://doi.org/10.1016/j.ejmech.2019.111857

Van der Schyf, C., Mandel, S., Geldenhuys, W., Amit, T., Avramovich, Y., Zheng, H., Fridkin, M., Gal, S., Weinreb, O., Am, O., Sagi, Y., & Youdim, M. (2007). Novel multifunctional anti-Alzheimer drugs with various CNS neurotransmitter targets and neuroprotective moieties. *Current Alzheimer Research*, *4*(5), pp. 522–536. https://doi.org/10.2174/156720507783018226

Watt, A. D., Jenkins, N. L., McColl, G., Collins, S., & Desmond, P. M. (2019). "To treat or not to treat": informing the decision for disease-modifying therapy in late-stage Alzheimer's disease. *Journal of Alzheimer's Disease*, *68*(4), pp. 1321–1323. https://doi.org/10.3233/jad-190033

Wauters, F., Cornelissen, T., Imberechts, D., Martin, S., Koentjoro, B., Sue, C., Vangheluwe, P., & Vandenberghe, W. (2020). LRRK2 mutations impair depolarization-induced mitophagy through inhibition of mitochondrial accumulation of RAB10. *Autophagy*, *16*(2), pp. 203–222. https://doi.org/10.1080/15548627.2019.1603548

Weikum, E. R., Liu, X., & Ortlund, E. A. (2018). The nuclear receptor superfamily: a structural perspective. *Protein Science: A Publication of the Protein Society*, *27*(11), pp. 1876–1892. https://doi.org/10.1002/pro.3496

Weinreb, O. (2008). The neuroprotective mechanism of action of the multimodal drug ladostigil. *Frontiers in Bioscience*, *13*, p. 5131. https://doi.org/10.2741/3069

Weinreb, O., Mandel, S., Bar-Am, O., Yogev-Falach, M., Avramovich-Tirosh, Y., Amit, T., & Youdim, M. B. H. (2009). Multifunctional neuroprotective derivatives of rasagiline as anti-alzheimer's disease drugs. *Neurotherapeutics*, *6*(1), pp. 163–174. https://doi.org/10.1016/j.nurt.2008.10.030

Weïwer, M., Lewis, M. C., Wagner, F. F., & Holson, E. B. (2013). Therapeutic Potential of Isoform Selective Hdac Inhibitors for the Treatment of Schizophrenia. *Future Medicinal Chemistry*, *5*(13), pp. 1491–1508. https://doi.org/10.4155/fmc.13.141

Wirawan, E., Lippens, S., Vanden Berghe, T., Romagnoli, A., Fimia, G. M., Piacentini, M., & Vandenabeele, P. (2012). Beclin1: a role in membrane dynamics and beyond. *Autophagy*, *8*(1), pp. 6–17. https://doi.org/10.4161/auto.8.1.16645

Wu, W., Lei, P., Liu, Q., Hu, J., Gunn, A. P., Chen, M., Rui, Y., Su, X., Xie, Z., Zhao, Y.-F., Bush, A. I., & Li, Y. (2008). Sequestration of copper from β-amyloid promotes selective lysis by cyclen-hybrid cleavage agents. *Journal of Biological Chemistry*, *283*(46), pp. 31657–31664. https://doi.org/10.1074/jbc.M804722200

Yang, G., Li, J., Cai, Y., Yang, Z., Li, R., & Fu, W. (2018). Glycyrrhizic acid alleviates 6-hydroxydopamine and corticosterone-induced neurotoxicity in SH-SY5Y cells through modulating autophagy. *Neurochemical Research*, *43*(10), pp. 1914–1926. https://doi.org/10.1007/s11064-018-2609-5

Youdim, M. B. H., & Buccafusco, J. J. (2005). CNS targets for multi-functional drugs in the treatment of Alzheimer's and Parkinson's diseases. *Journal of Neural Transmission*, *112*(4), pp. 519–537. https://doi.org/10.1007/s00702-004-0214-z

Zatta, P., Drago, D., Bolognin, S., & Sensi, S. L. (2009). Alzheimer's disease, metal ions and metal homeostatic therapy. *Trends in Pharmacological Sciences*, *30*(7), pp. 346–355. https://doi.org/10.1016/j.tips.2009.05.002

Zheng, H., Youdim, M. B. H., & Fridkin, M. (2009). Site-activated multifunctional chelator with acetylcholinesterase and neuroprotective–neurorestorative moieties for Alzheimer's therapy. *Journal of Medicinal Chemistry*, *52*(14), pp. 4095–4098. https://doi.org/10.1021/jm900504c

Zimmermann, G. R., Lehár, J., & Keith, C. T. (2007). Multi-target therapeutics: when the whole is greater than the sum of the parts. *Drug Discovery Today*, *12*(1–2), pp. 34–42. https://doi.org/10.1016/j.drudis.2006.11.008

10 Advances in Small Molecule Delivery Systems for the Treatment of Neurodegenerative Disease

Nargis Bano, Sameera Khan, Shakir Ahamad, Faaizah Fazal, Nawab John Dar, and Shahnawaz Ali Bhat

ABBREVIATIONS

AD	Alzheimer's Disease
AgNPs	Silver Nanoparticles
AuNPs	Gold Nanoparticles
BBB	Blood–Brain Barrier
BCSFB	Blood–Cerebrospinal fluid barrier
CNS	Central Nervous System
COMT	Catechol-*O*-methyltransferase
CSF	Cerebrospinal Fluid
HD	Huntington's Disease
IV	Intravenous Administration
MAO-B	Monoamine Oxidase-B
MNPs	Magnetic Nanoparticles
MRI	Magnetic Resonance Imaging
MSN-LEP-PIO	Leptin and Pioglitazone Cocktail-Infused Mesoporous Silica NPs
NMDA	*N*-methyl-D-aspartate
NPs	Nanoparticles
PD	Parkinson's Disease
PEG	Polyethylene Glycol
PLGA	Poly (lactide) (PLA), and poly (D, L-lactide-co-glycolic acid)
RES	Reticuloendothelial System
ROS	Reactive Oxygen Species
SA NPs	Serum Albumin Nanoparticles
SLNs	Solid–Lipid NPs
SPIONs	Superparamagnetic Iron Oxide Nanoparticles
TEER	Trans-endothelial Electrical Resistance
VIN	Versatile Immunosuppressive Nanoparticle

10.1 INTRODUCTION

The blood–brain barrier (BBB) is a crucial protective membrane that shields the central nervous system (CNS) from harmful toxins and pathogens circulating in the bloodstream (Wu et al., 2023). While this barrier is essential for maintaining brain health, it also presents a significant challenge for the treatment of CNS disorders (Wu et al., 2023). The BBB restricts the entry of most chemical

 DOI: 10.1201/9781003520610-10

drugs and biopharmaceuticals, complicating effective pharmacotherapy and limiting the ability to deliver therapeutic agents to the brain. As a result, treating CNS diseases has been notoriously difficult, often leading to low therapeutic efficacy and increased side effects due to drug accumulation in non-target tissues. The BBB is a semi-permeable barrier that encases the microvasculature of the CNS, formed by tightly connected endothelial cells that create extensive junctions (Abbott, 2013; Abbott et al., 2010; Ballabh et al., 2004). These junctions, together with various receptors, transporters, and efflux pumps, control the movement of molecules between the bloodstream and the brain (Abbott, 2013; Abbott et al., 2010; Ballabh et al., 2004). Due to the lack of paracellular or transcellular channels, molecules in the blood can access the interstitial fluid of the brain through only two mechanisms: lipid-mediated free diffusion or receptor-mediated transport and carrier-mediated transport (CMT). Although the BBB is successful in keeping harmful substances out of the brain, it also prevents nearly all large-molecule therapeutics and over 98% of small molecule drugs from reaching their target site (Banks, 2016; Pandit et al., 2020). Improving drug delivery across the BBB is a significant challenge in treating neurological diseases. The BBB's highly selective permeability is one of the main reasons why fewer than 10% of therapeutic agents for CNS disorders advance to clinical trials, as most drugs struggle to achieve sufficient brain penetration (Pangalos et al., 2007). Currently, drug development for the brain is largely limited to a small fraction of molecules, less than 2% of small molecules, that can cross the BBB via lipid-mediated diffusion (Ahamad et al., 2024b; Ahamad & Bhat, 2022b; Bhat et al., 2023; Khan et al., 2024; Pardridge, 2005). The BBB's tight junctions (TJs) allow only passive diffusion of lipid-soluble drugs with molecular weights under 400–600 Da, making it essential to enhance drug lipophilicity to improve BBB permeability. Consequently, many CNS drugs are small molecules capable of penetrating the BBB's phospholipid membrane through passive or carrier-mediated mechanisms. Over the past four decades, various brain drug delivery technologies have been developed to address the limitations of both small and large molecules. Typically, small molecules with molecular weights less than 400 Da and the ability to form less than eight hydrogen bonds can cross the BBB (Pardridge, 2012). However, most small molecules and virtually all large molecules lack these characteristics, presenting a significant obstacle. To overcome this, understanding the physicochemical and biological factors that influence BBB permeability is crucial. By leveraging the brain's natural transport mechanisms, it is possible to modify drugs to enhance their ability to cross the BBB. This knowledge can guide the design of small molecules and other therapeutics with improved delivery profiles, potentially leading to more effective treatments for brain diseases. In recent years, advancements in materials science and nanotechnology have opened new avenues for brain-targeted drug delivery (Allen & Cullis, 2004). Novel drug delivery systems are being designed to enhance the pharmacological and therapeutic profiles of drug molecules, particularly for targeting the brain (Kostarelos, 2003). These systems often utilize functional materials with specific sizes, architectures, and therapeutic properties to facilitate targeted drug delivery. They include a wide range of substrates—organic (Hwang et al., 2020), inorganic, and biomass-derived—that offer advantages such as high drug loading, biocompatibility, non-invasive delivery, prolonged blood circulation, and effective brain targeting (Le et al., 2021; Mitchell et al., 2021).

The history of brain drug delivery has evolved significantly since the early 20th century when it was first discovered that certain drugs could not cross the BBB (Pardridge, 2022). Over the decades, various brain drug delivery technologies have emerged, including BBB disruption, cerebrospinal fluid (CSF) delivery, trans-cranial delivery, lipid carriers, exosomes, nanoparticles (NPs), prodrugs, stem cells, gene therapy, and biologics (Pardridge, 2022). These technologies have expanded the possibilities for delivering drugs to the brain, allowing for the re-engineering of drugs that would otherwise be unable to penetrate the BBB (Pardridge, 2022). Small molecule drugs, owing to their size, can rapidly diffuse through biological fluids, cross biological barriers, and interact with various tissues and cell types. However, their therapeutic potential is often hampered by poor solubility (Savjani et al., 2012). With approximately 90% of preclinical drug candidates being low-solubility compounds (Kalepu & Nekkanti, 2015), addressing this challenge has become a critical focus in drug development (Abdulla et al., 2024; Ahamad et al., 2021, 2022, 2024a; Ahamad & Bhat, 2022a; Hussain et al., 2023; Kumar et al., 2022; Nisha et al., 2024). Delivery systems are increasingly employed to control solubility, dosing, and other delivery parameters for small molecules, and these strategies are being adapted for other classes of therapeutics as well. This chapter aims to provide a comprehensive overview of the BBB and explores the potential of brain-targeted drug delivery systems, particularly for small molecules, to inform and guide future research and development in this field.

10.2 TRADITIONAL DRUG DELIVERY SYSTEMS

To date, some of the invasive as well as non-invasive approaches of drug delivery to the brain have been adopted for successful treatments for neurodegenerative diseases. These approaches include oral drug delivery, intravenous drug delivery, intranasal (IN) drug delivery, and topical drug delivery approaches (Dong, 2018). For the treatment of CNS diseases, these conventional drug delivery routes are still being adopted based on the drug's pharmacokinetic and pharmacodynamic behaviors (Lu et al., 2014). For instance, some drugs are better absorbed orally, whereas some require intravenous administration (IV) for rapid and complete absorption. Topical delivery can be used for localized treatment, and IN delivery improves patient compliance (Lu et al., 2014). In this section we will delve into the traditional/conventional drug delivery approaches and their advantage and limitations in the context of neurodegenerative diseases.

Oral delivery refers to the administration of the drug by introducing it into the mouth. The systemic absorption of such medication can occur at any point as the drug transits through the GI tract (Mastropietro et al., 2017). This route of administration confers many advantages including a steady, consistent, and controllable delivery; the use of cost-effective oral formulations; and better patient compliance (Araújo et al., 2017; Banerjee et al., 2016; Choi et al., 2013, 2014; Hu & Luo, 2018). In addition to providing a large surface area for attachment and absorption, the presence of a viscous mucosal layer protects from the effect of various gastric juices as the drug moves along the GI tract (Araújo et al., 2017; Ensign et al., 2012; Leal et al., 2017). Owing to the presence of a large number of enterocytes in the intestine, the absorption is further enhanced (des Rieux et al., 2006; Kwon & Daniell, 2016; Ma et al., 2018; Maharjan et al., 2016). Currently, levodopa, a metabolic precursor of dopamine, is considered the most efficacious orally administered drug against Parkinson's disease (PD) (Ahlskog & Muenter, 2001; Goldenberg, 2008). Other approved oral medications include dopamine-receptor agonists (bromocriptine, ropinirole, and pramipexole) and COMT inhibitors (tolcapone and entacapone) (Ahlskog, 2001; Entacapone Improves Motor Fluctuations in Levodopa-Treated Parkinson's Disease Patients, 1997; Goldenberg, 2008; Kurth et al., 1997). Similarly, many FDA-approved drugs are also administered orally including donepezil, galantamine, rivastigmine (AChE inhibitors), and memantine (NMDAR receptor antagonist) (Bellotti et al., 2021; Homayun et al., 2019; Hyde et al., 2013; Tonda-Turo et al., 2019; Wen et al., 2017). These are available in the form of tablets, capsules, and solutions for daily oral administration (Bellotti et al., 2021; Homayun et al., 2019; Hyde et al., 2013; Tonda-Turo et al., 2019; Wen et al., 2017). Despite its many advantages the oral mode of delivery is not without its limitations (Bellotti et al., 2021; Homayun et al., 2019; Hyde et al., 2013; Tonda-Turo et al., 2019; Wen et al., 2017). The first-pass metabolic action by the liver and impedance provided by multiple barriers results in decreased concentration at the target site as well as slow absorption (Bellotti et al., 2021; Homayun et al., 2019; Hyde et al., 2013; Tonda-Turo et al., 2019; Wen et al., 2017). To overcome these limitations, frequent administration of the drugs may be required, which may lead to unwanted side effects including dizziness, nausea, and vomiting (Bellotti et al., 2021; Homayun et al., 2019; Hyde et al., 2013; Tonda-Turo et al., 2019; Wen et al., 2017).

IV refers to the direct administration of drugs into the veins using needles. The drug thus directly enters circulation, bypassing any physical barriers and guaranteeing absorption. This method has the added advantage of stringent dosage control and fastest delivery. Owing to this reduction in delivery time, the IV route is most preferred when emergency and acute responses are required (Homayun et al., 2019). FDA has approved the use of monoclonal amyloid targeting antibodies, aducanumab and lecanemab, for the treatment of Alzheimer's disease (AD) (Leinenga et al., 2021; Swanson et al., 2021). Both are administered intravenously over the course of weeks (Leinenga et al., 2021; Swanson et al., 2021). While aducanumab promotes clearance of insoluble and soluble Aβ, lecanemab selectively targets protofibrils (Leinenga et al., 2021; Swanson et al., 2021). Intravenous routes for drug delivery have disadvantages including infection by the needle at the site of injection, circulatory overload, phlebitis, and thrombosis (Korttila & Aromaa, 1980; Maxwell & Wilson, 2006).

Transdermal drug delivery is aimed at introducing drugs into the circulatory system by crossing the skin barriers (Brown et al., 2006). Intracellular, transcellular, and transappendageal pathways are the main pathways through which transdermal delivery is achieved (Ranade, 1991). This delivery route is preferential over both oral and IV drug administration routes in that the drugs are shielded against the first-pass effect of the liver compared to the oral route (Homayun et al., 2019). Being a non-invasive method, it is preferred over the intravenous system when there is a requirement for a sustained and chronic delivery of the drugs (Homayun et al., 2019). Rivastigmine, owing to its small size (400 Da) and solubility (lipophilic and hydrophobic), is a suitable candidate for transdermal drug delivery. FDA has approved the use of rivastigmine in the form of a transdermal patch for the management of AD (Cummings et al., 2007; Nunes et al., 2022). The drug is slowly released through the patch, and a stable, consistent plasma concentration is achieved (Lefèvre et al., 2008). Similarly, rotigotine, for the treatment of PD, can be administered through silicone-based transdermal patches ranging from 10 to 40 cm^2 in size with drug concentrations ranging from 4.5 to 18 mg (Antonini et al., 2015; Frampton, 2019; Giladi et al., 2007). The patch can be applied to healthy, intact skin of the abdomen, thigh, hip, fank, shoulder, or upper arm (Frampton, 2019). This method of drug delivery is particularly convenient for patients who suffer from symptoms like heartburn and bloating with oral medications (Woitalla et al., 2015). One of the challenges associated with this route of administration is that the molecules need to be soluble and to be less than 500 Da in size, thus limiting the variety of drugs available for transdermal administration (Winblad & Machado, 2008).

IN drug delivery refers to the administration of a drug into the nasal cavity. The drug enters the systemic circulation due to the highly vascularized nature of the respiratory epithelium (Homayun et al., 2019). Drugs can also directly enter the brain via the olfactory and trigeminal nerve cells present in the olfactory and respiratory epithelium (Dhuria et al., 2010; Illum, 2000). Drugs may opt for intracellular or extracellular mechanisms of transport as they travel through these nerves, passing the cribriform plate, to enter olfactory bulb and CSF. Subsequently, the drug is distributed as CSF travels through the brain (Selvaraj et al., 2017). IN drug delivery offers the same bioavailability profiles as IV with the added benefit of being non-invasive (Agarwal & Mishra, 1999; Jones et al., 1997; Misra & Kher, 2012; Vyas et al., 2005). The drugs can therefore be self-administered, making this approach more patient-friendly. Further drugs administered intranasally can avoid the first-pass metabolic action of the liver and offer quick absorption and rapid drug action (Misra & Kher, 2012). FDA is yet to approve any drug for IN administration for AD treatment. However, non-FDA-approved drugs like insulin, rivastigmine, and APH-1105 are currently being explored as potential candidates for IN delivery (Taléns-Visconti et al., 2023). Because of the individual differences in the nasal anatomy, the therapeutic effects of the drug may not be consistent among individuals. Drugs administered intranasally are rapidly cleared from the nasal mucosa and brain, thus necessitating frequent dosing to achieve and maintain therapeutic effects. Nasal irritation, dryness, and mucosal damage are other disadvantages of IN administration (Agrawal et al., 2018; Chatterjee et al., 2019; Illum, 2003; Kashyap & Shukla, 2019; Vyas et al., 2005).

10.3 CHALLENGES IN DRUG DELIVERY FOR SMALL MOLECULES

The BBB is a complicated cellular-based barrier that spans the brain, strictly regulating the uptake of brain nutrients, including glucose, nucleosides, amino acids, and fatty acids, and also shields the brain from potentially dangerous chemicals (Kaya & Ahishali, 2021). It is witnessed that under physiological and pathological conditions of the CNS, the maintenance of BBB integrity significantly contributes to normal neuronal functions and brain homeostasis (Andreone et al., 2017; O'Brown et al., 2018). Neuroglial units like astrocytes, pericytes, and non-fenestrated brain endothelial cells as well as the basement membrane are involved in the formation of BBB, controlling the permeability and the stability of brain microvasculature (Biddlestone-Thorpe et al., 2012). TJs between endothelial cells of the BBB are made up of a complex mixture of proteins, such as occludins, claudins, junctional adhesion molecules, and endothelial selective adhesion molecules (Cecchelli et al., 2007; Gao et al., 2022; Oller-Salvia et al., 2016). These limit the entrance of blood-borne molecules via paracellular pathways and high trans-endothelial electrical resistance of BBB and interact with the scaffolding proteins (cingulin, Par-3, Af-6, and ZO-1), avoiding the inflow of molecules from the blood into the brain (Crone & Olesen, 1982; Lipinski, 2000). It is noted that only lipophilic small molecules (<500 Da) can appreciably cross the BBB (Lapin et al., 2020).

An additional hurdle that needs to be overcome by the small molecules for their entry into the brain is crossing the blood-CSF barrier (BCSFB), mainly built by the epithelial cells of the choroid plexus (Johanson et al., 2011). This barrier functions in a similar way compared to BBB, with neighboring cells harboring tight connections between themselves that separate the blood from the CNS, sealing the paracellular cleft (Solár et al., 2020). A large group of multi-specific efflux transport proteins and detoxifying enzymes are found on BCSFB that work in synchrony to prevent harmful chemicals from entering the CNS (Ueno et al., 2016). The ATP-binding cassette transporters spanning the endothelial cells also provide resistance to the entry of drugs as well as small molecules into the brain. These membrane protein transporters require ATP to transport materials through cell membranes (Girardin, 2006). Some of the members of this family include multiple drug resistance protein 1 (MDR1) or ATP-binding cassette subfamily C member 4 (ABCC4), ATP-binding cassette subfamily G member 2(ABCG2), and breast cancer resistance protein (Khalin et al., 2014). These transporters use the energy derived from ATP hydrolysis to pump out a wide range of substances, including drugs and toxins, back into the bloodstream (Kooij et al., 2012).

Several strategies are adopted by small molecules to gain entry into the brain such as utilizing endogenous protein CMT mechanisms expressed on the BBB, optimizing physicochemical properties to facilitate passive paracellular and transcellular diffusion and employing BBB disruption strategies to allow paracellular drug transport (Mikitsh & Chacko, 2014). Most small molecule therapeutics reach the brain via passive diffusion across the BBB. For a small molecule to cross BBB via diffusion, it must possess certain characteristics such as it should follow Lipinski's rule of five in terms of low molecular weight (400–500 Da) and the molecule must be lipophilic with a range of lipophilicity (log P) lying between 1.5 and 2.5, to traverse the hydrophobic phospholipid bilayer of a cell membrane by passive diffusion (Levin et al., 1980; Misra et al., 2003). Apart from this, other chemical properties of small molecules such as polar surface area, charge, hydrogen bonding, and molecular flexibility greatly influence BBB permeability. For example, molecules having a large polar surface area with an upper limit estimated between 60 and 90 Å do not readily diffuse through the BBB (Begley, 2004; Pajouhesh & Lenz, 2005). Moreover, molecules with significant electrostatic charge will not passively diffuse across the BBB. It is seen that the net movement of a compound having the same anionic charge as that of a hydrophobic brain capillary endothelial cell membrane is thermodynamically unfavorable, because of the mutual repulsion of negative electrostatic charge (Mikitsh & Chacko, 2014). Also, an increase in the number of hydrogen bonds (H-bond) harms the passive diffusion of small molecules across the BBB (with <5 H-bond donors and <10 H-bond acceptors being acceptable for a CNS drug candidate) (Pajouhesh & Lenz, 2005). Moreover, it is seen that the solvation energy of the drugs is related to its diffusion across the BBB; the higher the solvation energy of the drug in water, the more hydrophilic the drug is, thereby resulting in less interaction with the lipophilic cell membrane and thus limited drug transport by passive diffusion (Pajouhesh & Lenz, 2005). Nowadays, molecular flexibility is the currently used filter to forecast a drug's BBB permeability prediction. It is the measurement of rotatable single bonds that are not engaged in a ring structure or attached to a non-terminal heavy atom (Leeson & Davis, 2004).

When compared to other drug classes, CNS drugs comprise five or fewer rotatable bonds, which suggests that the conformational range for BBB permeability is constrained (Leeson & Davis, 2004). It is well known that many small molecules do not meet the criteria of being both highly lipid soluble and having a molecular weight under 400–600 Da (Bellettato & Scarpa, 2018; Bourasset et al., 2020). Even if the molecule is small and lipid soluble, its permeability is limited due to TJs as well as efflux transporters present on BBB as already stated above, which pump many substances back into the bloodstream (Guan, 2024). Even if it enters the brain, the pharmacokinetics and distribution of small molecules are influenced by several factors such as tissue perfusion, binding to plasma protein, and the unique anatomical and physiological properties of the brain (Bourasset et al., 2020). Moreover, even if small molecules possess high lipid solubility, which offers them the advantage to cross BBB efficiently, its non-specific distribution in other tissues causing potential toxicity cannot be neglected (Bellettato & Scarpa, 2018).

Due to these challenges, alternative methods such as NP-based drug delivery systems have emerged, which have revolutionized the era of drug-delivering strategies, particularly in the context of neurodegenerative diseases. NPs reduce the frequency of administration by releasing drugs in a sustained manner, prolong in vivo drug circulation half-life, improve the solubility of poorly water-soluble drugs, and minimize adverse systemic effects by delivering drugs preferentially to the target tissues (Wagner et al., 2006; Zhang et al., 2008). As a result, numerous NP platforms have been developed or proposed for drug delivery applications, which are discussed below.

10.4 ADVANCES IN DRUG DELIVERY SYSTEMS FOR SMALL MOLECULES

The term "NPs" commonly refers to materials having diameters below 200 nm. These materials provide appropriate nanoplatforms on which the medication can be loaded and subsequently transported, favoring its selective release to the target location and minimizing systemic exposure and adverse effects. Some of the plethora of NPs used to transport drugs efficiently into the target organs are discussed below.

10.4.1 Micelles

Micelles are aggregated colloid vesicles composed of either amphiphilic copolymers or amphiphilic surfactants (Aliabadi & Lavasanifar, 2006). They feature a core-shell structural design with an exterior hydrophilic environment composed of PEG, which protects the micelles from the reticuloendothelial system (RES) (a group of reticuloendothelial cells that actively clears the particles via phagocytosis after their systemic introduction) and gives them stability in an aqueous environment by prolonging their circulation period in the bloodstream (Ghezzi et al., 2021). The inner hydrophobic core of the micelles is synthesized using chemicals such as polycaprolactone, phospholipids, polypropylene glycols, and fatty acids which allows the loading of hydrophobic drugs (Ghezzi et al., 2021).

10.4.2 Liposomes

Liposomes are spherical vesicles with diameters ranging from 50 to 1,000 nm and are made of lipid bilayers with a unique shape that allows them to entrap hydrophobic compounds within the lipid bilayers and encapsulate hydrophilic molecules inside the central aqueous compartment, protecting them from degradation (Pastorino et al., 2019). Liposomes are advantageous to use over many other carriers in clinical investigations particularly due to better pharmacokinetics and biodistribution of drugs, which is mainly due to their innate properties such as high stability, high drug-loading efficiency, and high biocompatibility (Shah et al., 2020).

10.4.3 Polymeric NPs

Polymeric NPs are solid colloidal dispersions with diameters ranging from 10 to 100 nm composed of biocompatible and biodegradable polymers such as poly (alkyl cyanoacrylate), polyesters such as poly (lactide) (PLA), and poly (D, L-lactide-co-glycolic acid) (PLGA) (Hadavi & Poot, 2016). They possess a hydrophilic corona that gives NPs steric stability and a core of dense polymer matrix to enclose the lipophilic drugs (Lu et al., 2011). The drug that has to be delivered may be adsorbed, encapsulated, or chemically linked to the surface of polymeric NPs (Lu et al., 2011).

10.4.4 Exosomes

Exosomes are endosomal products that encapsulate drugs and target CNS disorders. After extracting exosomes and performing post-modifications like liposome procedures, exosomes serve as potential candidates for drug delivery (Sarko & McKinney, 2017). Having the same membrane composition as host cells, exosomes can target diseased cells by invading biological barriers like the BBB and specific cellular membranes (Sarko & McKinney, 2017). Exosomes also express antiphagocytic markers (CD47), which provide them the ability to evade macrophages and monocytes of the RES, lowering immunological responses and subsequently their rate of blood clearance (Li et al., 2020). Exosomes are ornamented with active-targeting ligands such as PEG and dextran sulfate to improve targeting efficiency, cellular absorption, and circulation half-life (Morishita et al., 2017).

10.4.5 Solid–Lipid NPs

Solid–lipid NPs (SLNs) serve as one of the most common modern drug delivery arsenal. These are aqueous colloidal nanocarrier systems composed of physiological lipids such as waxes, steroids, fatty acids, and triglycerides, which solidify upon cooling when dispersed in water or an aqueous surfactant solution (Duan et al., 2020). SLNs, like any other NPs, must be modified according to the need to maximize the drug-loading capacity and long-term stability (Paliwal et al., 2020).

10.4.6 Nanoemulsions

Nanoemulsions are either water-in-oil (W/O) or oil-in-water (O/W) colloidal particulate systems with a size range of 100–500 nm made of edible oils, surface-active substances (surfactants), and water (Zhi et al., 2022). The combination of nanoemulsions with oils high in omega-3 polyunsaturated fatty acids provides the nanocarriers with specialized qualities such as breaking biological barriers, including the BBB, and so aids in the fast delivery of medications to peripheral areas, including the brain (Barkat et al., 2020).

10.4.7 Dendrimers

Dendrimers have a 3-D symmetrical design with an inner core from which a variety of hyperbranched called "generations" with diverse chemical functional groups at the outer terminal surface emerge, allowing them to be readily functionalized with different ligands (Dias et al., 2020). Dendrimers of various generations, complexity, and building block materials have been produced, among which polypropylene imine, polyamidoamine, and polylysine dendrimers are of particular interest for the delivery of both hydrophilic and hydrophobic drug molecules (Madaan et al., 2014).

10.4.8 Metal NPs

Metal NPs have come into the limelight due to their ability to cross the BBB and brain accumulation (Sintov et al., 2016). Due to certain features like surface changes, size, and stability, metal NPs may be easily modified for effective brain targeting (Vio et al., 2017). For efficient drug delivery to the CNS, they are ornamented with various brain-targeted ligands, such as antibodies, proteins, and small molecules such as mannose (Yim et al., 2012). Among various metallic NPs, gold and silver NPs have been the most exploited and extensively used for CNS delivery.

Gold NPs (AuNPs) have been used in CNS imaging and targeting (Kumar et al., 2013). It is seen that when exosome-derived membranes are combined with synthetic AuNPs, it provides unique features for brain transport (Kumar et al., 2013). Covalently coated AuNPs with an organic layer improve biocompatibility, biophysical characteristics, and targeting (Khongkow et al., 2019). AuNPs' core includes plasmonic capabilities (the capability to interfere with electromagnetic radiation particularly due to free electrons), making them suitable for micro-CT scanning or X-ray imaging applications (Sarfraz & Khan, 2021). Silver NPs (SNPs) translocate to the brain through the blood stream (Ravindran et al., 2013). The distribution of silver NPs (AgNPs) was witnessed across the BBB in a rat brain micro-vessel vascular endothelial cell in vitro model, demonstrating that AgNPs may cross the BBB quite efficiently (Ravindran et al., 2013). Transcytosis of capillary endothelial cells may be the primary mechanism by which AgNPs cross the BBB (Tang et al., 2010).

10.4.9 Magnetic NPs

Magnetic NPs (MNPs) used in imaging and clinical investigations possess unique features as they are attracted to the magnetic field that can be used in drug targeting and cell sorting (Wu et al., 2019). These NPs can be guided to specific locations using an externally induced magnetic field gradient. Once at the desired spot, they can be held until the therapy is completed and removed when needed (Stueber et al., 2021). Magnetic drug carriers can carry a large drug dosage, resulting in high local concentrations while avoiding toxicity and other unfavorable side effects associated with high drug concentrations in other body areas (El-Boubbou, 2018).

10.4.10 Carbon Nanotubes

Carbon nanotubes (CNTs) often fall into a new age of nanovehicles for drug delivery. CNTs are divided into single-walled carbon nanotubes and multi-walled carbon nanotubes (MWCNTs) (Dineshkumar et al., 2015). This classification depends on the number of inner layers viewed as rolled-up graphene sheets (Dineshkumar et al., 2015). Because of their small diameter (0.4–100 nm), the surface area of CNTs is very large, thereby facilitating effective transit via tight capillaries (Venkataraman et al., 2019).

10.5 BENEFITS OF ADVANCED DRUG DELIVERY SYSTEMS IN NEURODEGENERATIVE DISEASES

Neurodegenerative diseases are prevalent, incurable neurological disorders that lead to memory, movement, language, and intelligence impairments, posing a severe threat to the lives and health of patients globally (Duan et al., 2023). As already mentioned, the presence of BBB strictly regulates the transport of substances between the blood and the brain, which includes drugs as well as small molecule therapeutics (Kaya & Ahishali, 2021). NPs can be used as drug transport vehicles and are also known as NP-based drug delivery systems (Duan et al., 2023). These systems not only improve therapeutic efficacy and stability but also make it easier for drugs and pharmaceuticals to get through the BBB (Duan et al., 2023). In the following section we will elaborate on the use of various small molecule therapeutics in the context of neurodegenerative diseases such as AD, PD, and Huntington's disease (HD) and how the use of NPs to transport small molecules has led to the increase in drug bioavailability, their efficacy, and curbing of the side effects.

AD and PD are the top two neurodegenerative disorders with very little or no treatment at all. R-flurbiprofen, a nonsteroidal anti-inflammatory drug, is known to confer neuroprotective potential in AD against Aβ-peptide toxicity, particularly by preventing the mitochondrial calcium overload and depolarizing the mitochondria (Sanz-Blasco et al., 2008). A phase 2 trial found preliminary evidence for R-flurbiprofen's possible benefits in patients with moderate AD (Wilcock et al., 2008). However, phase 3 trials in patients with mild AD failed to show any therapeutic effects (Green, 2009; Wilcock et al., 2009). It was reported that administration of R-flurbiprofen orally, ultimately leads to its poor penetration in BBB, and serves as one key reason for the gap observed between preclinical and clinical studies (Imbimbo, 2009). It was observed that when R-flurbiprofen was delivered via serum albumin (SA) NPs intranasally into C57BL/6 mice, its accumulation in the brain significantly increased and also enhanced the mitochondrial respiration (both basal and maximal), proving that this strategy could be used to mitigate mitochondrial dysfunction in the AD (Table 10.1) (Wong & Ho, 2017). Similarly, tacrine, which is the first licensed inhibitor of acetylcholinesterase for AD treatment, also has some drawbacks which include its short elimination time, poor bioavailability, and hepatotoxicity (de los Ríos & Marco-Contelles, 2019). It was reported that the nasal delivery of tacrine hydrochloride in SA NPs carrying β-cyclodextrin derivatives improved the loading efficiency and drug permeability into porcine mucin, thereby increasing its uptake into the brain. Furthermore, these NPs have proven advantageous in modulating mucoadhesiveness (Luppi et al., 2011). In an attempt to improve the delivery of the drug to the brain, rivastigmine, a reversible cholinesterase inhibitor approved by the FDA to treat AD, was also administered by its binding to poly (n-butyl cyanoacrylate) NPs both alone and in combination with the NPs coated with polysorbate 80 (non-ionic surfactant) (Wilson et al., 2008). In the case of the NPs with 1% surfactant, the drug concentration in the brain increased 3.2-fold and was found beneficial in mitigating the disease (Table 10.1) (Wilson et al., 2008). Donepezil, which is also an acetylcholinesterase (AChE) inhibitor, is also an FDA-approved drug that is advised to be taken daily to maintain efficacy and treat moderate AD (Benjamin & Burns, 2007). It is witnessed that the daily intake of donepezil leads to side effects such as nausea and diarrhea along with a high risk of gastrointestinal bleeding and stomach ulcers (Campbell et al., 2017; Valis et al., 2017). In studies of donepezil along with NPs, it was seen that donepezil-loaded cholesterol-modified pullulan NPs showed sustained drug release and enhanced brain targeting, which was far superior to that of administration of free donepezil (Table 10.1) (Tao et al., 2021). Another study has reported the use of PLGA-b-PEG NPs conjugated with donepezil. This nanoformulation has shown a controlled release profile with high BBB penetration to induce the destabilization of Aβ fibrils (Baysal et al., 2017). The use of these NPs maintained a steady concentration of donepezil in the brain, reducing the frequency of dosing and potentially minimizing side effects (Figueiredo et al., 2023).

A new genre of oligopeptide drugs known as the β-sheet breakers are designed to target Aβ-sheets in AD, particularly by disrupting the beta-sheet conformation via their proline amino acid (Wisniewski & Sadowski, 2008). These have proven to hold good potential against the treatment of AD (Bölükbaşı Hatip & Hatip-Al-Khatib, 2013). They, however, have the disadvantage of rapid clearance as well as poor entry through the BBB with intravenous injection with a short plasma half-life. In their delivery mechanisms via NPs it was seen that the PEG–PLA NPs functionalized with TGN (a peptide that acts as a shuttle for NP entry into the brain) and QSH (a peptide with high affinity for Aβ plaques) were used for the delivery of β-sheet breaker H102 (HKQLPFFEED) to AD-induced ICR mice (Table 10.1) (Zhang et al., 2014). These NPs enhanced the efficient targeted delivery of the drug through the BBB and as a consequence showed improved results in AD therapy (Zheng et al., 2017). Another β-sheet breaker peptide drug, iAβ5, has been shown to inhibit aggregate formation in AD brains but has faced restricted entry through the BBB. PLGA NPs decorated with anti-transferrin receptor antibodies (OX26) as well as anti-amyloid antibodies (DE2B4) have been shown to improve the cellular uptake of the β-sheet breaker peptide iAβ5 (Loureiro et al., 2016). Moreover, the attachment of BBB permeability peptide (B6) at the PEG–PLA NP surface is now being used as a more targeted approach to invading the BBB. This facilitated better transport

TABLE 10.1
Overview of Small Molecules, Their Mechanisms, NP Delivery Systems, Targeted Brain Diseases, and Therapeutic Effects

NPs	Drug	Category	Disease	Therapeutic Effects	References
Liposome	Rivastigmine	Cholinesterase inhibitor	AD	This formulation resulted in faster memory regain and amelioration of metabolic disturbances as well as abolished the amyloidogenic effect of $AlCl_3$ administration.	Ismail et al. (2013)
Serum albumin (SA) NPs	R-flurbiprofen	Nonsteroidal anti-inflammatory drugs (NSAIDs)	AD	Exhibited increased brain-to-plasma ratio profile and attenuated mitochondrial dysfunction in AD.	Wong and Ho (2017)
Cholesterol-modified pullulan NPs	Donepezil	Acetylcholinesterase inhibitor	AD	Showed sustained drug release profile and enhanced brain targeting compared to free donepezil. Destabilizes Aβ fibrils.	Tao et al. (2021)
Microparticles of polyanhydride sebacic acid copolymer	Bethanechol	Acetylcholinesterase-resistant choli nomimetic	AD	Significant improvement in the radial maze test within 10 days of microparticle induction.	Howard et al. (1989)
Polymeric NPs	H102 (HKQLPFFEED)	β-sheet breaker	AD	Disrupt the beta-sheet conformation of amyloid beta fibrils.	Zhang et al. (2014)
Liposomes	Curcumin	Antioxidant phytochemical	AD	Enhanced BBB permeation, delayed $A\beta_{1-42}$ peptide aggregation.	Lajoie and Shusta (2015); Mourtas et al. (2014)
Single-walled carbon nanotubes	Acetylcholine	Neurotransmitter	AD	These NPs target the lysosomes of the cells and are distributed specifically in neurons and neurites. Learning and memory ability were enhanced in kainic acid-induced AD mice.	Yang et al. (2010)
Borneol and lactoferrin polymeric NPs	Levodopa	Dopamine precursor	PD	Increased the permeability and striatum-specificity of dopamine transport and improved motor functions.	Tang et al. (2019)
Polymeric NPs	Selegiline	MAO-B inhibitor	PD	Sustained brain uptake of the drug, lowered cataleptic response in PD mice and decreased MAO-B activity.	Bali and Salve (2019)
Solid–lipid NPs	Apomorphine	Dopamine agonist	PD	The bioavailability was increased up to 12–13 fold from control and targeted the cerebellum, brainstem, and striatum. Increased neuroprotection by increasing the number of dopamine and its metabolites in 6-OHDA rats.	Tsai et al. (2011)
Lipid nanocarriers	Engeletin	Flavanol glycoside	HD	Reduce mitochondrial dysfunction and suppress inflammation via the Keap1/Nrf2 pathway.	Smriti et al. (2023)
Nanoemulsion in capmul MCM oil	Tetrabenazine	Reversible VMAT2 inhibitor	HD	Enhanced permeability of nasal mucosa, exhibited greater C-max and AUC, and reduced the hyperkinetic movement.	Arora et al. (2020)

along with promoting the accumulation of larger neuropeptides as well as peptide drugs through the BBB via clathrin-mediated and lipid raft-mediated endocytosis (Liu et al., 2013a).

Some of the small molecule phytochemicals are also gaining limelight to treat neurodegenerative disorders linked with oxidative stress, due to their antioxidant and anti-inflammatory properties (Olufunmilayo et al., 2023). Curcumin, a natural polyphenol found in turmeric, is a strong antioxidant that possesses anti-inflammatory properties and is widely used in the treatment of neurodegenerative diseases (Esmaealzadeh et al., 2024). Irrespective of these, curcumin possesses low bioavailability, with most of it rapidly being metabolized and eliminated from the body (Dei Cas & Ghidoni, 2019; Stohs et al., 2020). Curcumin has low bioavailability, meaning only a small fraction of the consumed amount is absorbed into the bloodstream. Most of it is rapidly metabolized and eliminated from the body (Dei Cas & Ghidoni, 2019; Stohs et al., 2020). Also, the chemical structure of curcumin makes it less likely to cross the BBB, rendering it water-insoluble, thus limiting its ability to pass through tightly packed cells of BBB (Assi et al., 2023). According to a study, curcumin-loaded liposomes functionalized with anti-transferrin antibodies significantly increased their transcytosis into BBB via transferrin receptor and retarded the aggregation of Aβ1–42 peptides (Table 10.1) (Lajoie & Shusta, 2015; Mourtas et al., 2014).

PD is a complex progressive neurodegenerative disease that is characterized by tremor, bradykinesia, and rigidity, ultimately leading to postural instability as the disease progresses (Tarakad & Jankovic, 2020). Many drugs have been developed to treat PD with dopamine agonists as well as dopamine precursors being on the frontline to treat PD. However, the results have corroborated that it is because 3,4-dihydroxyphenylalanine decarboxylase, an enzyme that converts levodopa to dopamine, gets decreased significantly in PD patients, they become less responsive to levodopa (Krishna et al., 2014; Salat & Tolosa, 2013). Moreover, rapid degradation of levodopa occurs mainly either by the enzymes present in the gastrointestinal tract or by the decarboxylation performed by L-DOPA decarboxylase during the hepatic first-pass metabolism as well as in the systemic circulation, and hence the drug has a half-life of around 50 minutes (Ahmad et al., 2022; LeWitt, 2015). This results in low oral L-DOPA bioavailability (about 30%) and low percentages of L-DOPA reaching the brain (roughly 1%), coupled with peripheral side effects such as headaches, dizziness, vomiting, and sleeplessness (Palmer, 2011). To improve the effectiveness of treatment and minimize side effects, a unique method involves dopamine delivery directly to the brain via nose-to-brain administration in the form of biodegradable polymeric NPs encapsulating dopamine. It was found that borneol, which is a bicyclic monoterpene and can enhance brain penetration, and lactoferrin (Lf)-co-modified, NP (Lf-BNPs)-encapsulated dopamine increased the permeability and striatum-specificity of dopamine transport, conferring neuroprotection in 6-OHDA-treated Sprague–Dawley rats following IN administration (Table 10.1) (Qi et al., 2013; Tang et al., 2019). The pharmacodynamics of co-modified NPs have revealed that, although borneol monoterpene is effective in promoting drug transport to the brain, drugs were still found to be metabolized in non-targeting tissues, causing serious adverse effects (Ren et al., 2013). It is via Lf that the drug delivery to the striatum is very precise as the Lf receptor is abundantly expressed on the neurons of the brain related to neurodegenerative diseases such as PD and AD and also on the blood capillaries as well as the apical surface of respiratory epithelial cells (Elfinger et al., 2007; Liu et al., 2013b; Meng et al., 2018; Suzuki et al., 2005).

The use of monoamine oxidase-B (MAO-B) inhibitors is another current strategy adopted to treat PD. MAO-B is an enzyme responsible for the degradation of dopamine. MAO-B inhibitors function by inhibiting the degradation of dopamine thereby increasing dopamine levels. One of the MAO-B inhibitors that are used in the treatment of PD are selegiline, rasagiline, and safinamide (Tan et al., 2022). Studies have demonstrated these inhibitors can also act as neuroprotectants from oxidative stress (Ebadi et al., 2002; Siddiqui & Plosker, 2005). Selegiline is sold under the brand name Emsam as transdermal patches to treat PD (Jessen et al., 2008). It has been observed that the transdermal route of selegiline administration has varied absorption rates among individuals, mainly depending on the patient's skin condition. Additionally, clinical trial data have revealed that individuals administered with selegiline transdermal patches experienced skin reactions such as rash, redness, and discomfort (Silva et al., 2021). So the use of nanoemulsions for transdermal permeation enhancement of selegiline has come into play. The research found that selegiline encapsulated in PLGA NPs and applied to transdermal films made of ethylene vinyl acetate showed C-max values that were similar to those of selegiline given intravenously (Table 10.1) (Bali & Salve, 2019). Studies on the in vivo brain distribution showed an increase in the brain uptake of intravenously administered selegiline solution, although the uptake was for a short duration. The NP formulation in contrast has demonstrated a sustained brain uptake of the drug and a long-lasting pharmacological action. Additionally, this strategy lowered the cataleptic response in mice, decreased MAO-B activity, and increased dopamine levels, and a skin test demonstrated its safety for up to 24 hours of usage (Bali & Salve, 2019).

The use of NP-based delivery of small molecule therapeutics is also witnessed in the case of HD. The hallmarks of HD, which is a hereditary neurodegenerative disease that progresses over time, are the continuous degradation of cognitive functions and memory as well as the progressive development of involuntary muscle movements affecting the hands, feet, face, and torso (Armstrong & Miyasaki, 2012). Even though the pathogenesis of HD has still not been resolved and there is no known treatment, there are several therapeutic alternatives such as antipsychotics, mood stabilizers, and antidepressants ranging from managing its symptoms and signs to enhancing the patient's quality of life. Tetrabenazine is the only drug for HD that has received approval from the FDA for the treatment of chorea associated with HD (Kenney et al., 2007; Mestre & Ferreira, 2012) Tetrabenazine is a reversible inhibitor of the central vesicular monoamine transporter transporter type 2, which regulates the brain's various monoaminergic systems including dopamine, serotonin, and noradrenaline (Thibaut et al., 1995). Although tetrabenazine is successfully absorbed from the gastrointestinal tract, it is exposed to extensive first-pass metabolism, which means that a very small amount of it is available for systemic circulation than that absorbed from the gastrointestinal tract. This results in a low oral bioavailability of nearly 5%, which in turn causes a low brain bioavailability (Arora et al., 2020). Therefore, nanoemulsions are gaining wide importance as lipid-based systems to deliver lipophilic drugs to the brain using the IN route. It was observed that tetrabenazine nanoemulsions were prepared in capmul MCM oil, a lipophilic vehicle that can function as a solubilizer for poorly soluble compounds particularly due to its medium chain glycerides, showed significant enhancement in the permeability of nasal mucosa, nearly 1.68 times compared to tetrabenazine solution administered intravenously in Wistar rats, thereby reducing the hyperkinetic movement associated with HD (Table 10.1) (Sindhu et al., 2018). According to pharmacokinetic studies, tetrabenazine nanoemulsion exhibited greater C-max and area under curve (AUC) in Wistar rats compared to tetrabenazine solution given intravenously (Arora et al., 2020). The use of NP drug delivery systems is also paving the way for imparting protective responses in brain injury and cerebral ischemia models. It is reported that riluzole, a sodium channel inhibitor and a non-competitive *N*-methyl-D-aspartate receptor (NMDA) receptor antagonist, inhibits the release of glutamate neurotransmitters, prevents nerve damage, and confers neuroprotection by treatment of severe motor neuron disorders in ALS (Carbone et al., 2012). It is also found to be neuroprotective in cerebral ischemia, AD, and spinal cord ischemia (Lips et al., 2000; Matthews et al., 2021; Wahl et al., 1993). It was seen that riluzole possesses certain properties such as poor water solubility, short half-life, and some side effects such as dizziness, nausea, and stomach pain at higher concentrations, thus limiting its use as a therapeutic agent (Verma et al., 2016). To overcome these limitations, riluzole was transported to the brain in the form of NPs. A study has found that chitosan-conjugated NIPAAM (*N*-isopropylacrylamide) NPs coated with Tween80 carried large amounts of riluzole across BBB, which produced significant neuroprotective effects even at very low concentrations, in cerebral ischemic Wistar rats (Verma et al., 2016). By these studies it is concluded that NPs mediated the delivery of therapeutic small molecules in neurodegenerative diseases and have proven far more effective than the conventional drug delivery paradigm in terms of enhanced BBB penetration, enhanced stability and bioavailability of drugs, and targeted delivery to the diseased brain regions.

10.6 FUTURE DIRECTIONS IN DRUG DELIVERY SYSTEMS FOR SMALL MOLECULES

10.6.1 Personalized Medicine and Targeted Drug Delivery

The tendency to fine-tune and tailor an NP formulation by flexibly controlling drug dosage, formulation, and their physicochemical properties has a great influence on the healthcare sector, particularly for pediatric and geriatric patients. This concept is crucial, especially in light of the drug response, physiology, and genetic makeup of the patients (Ahmed et al., 2021; Jamróz et al., 2018; Vaz & Kumar, 2021). Based on the idea of "one size fits all," the current medical treatment approach employs the administration of the same drugs at the same dosages and frequency (Litman, 2019). Treatment of different patients using this "one-size-fits-all" paradigm has resulted in varying responses. Therefore, an effective strategy that treats patients particularly based on their physiology, age, drug response, and the affected protein needs to be formulated (Ahmed et al., 2021). Many neurodegenerative disorders involve a multitude of factors like genetic defects, environmental factors, and epigenetic modifications (miRNA dysregulation, DNA methylation, and histone/chromatin remodeling) in the onset and progression of pathology; therefore, they are unlikely to benefit from "one-size-fits-all" strategy (Griñán-Ferré et al., 2024). In this regard, the development of the precise/personalized medicine approach may put forward an excellent possibility for differential diagnosis, to identify preclinical stages of the disease, and provide adequate and timely treatments alternative to the conventional treatments, which are normally prescribed at later stages of the disease (Strafella et al., 2018).

Neurodegenerative disorders are characterized by the death of neurons, for instance, cholinergic neurons in AD, dopaminergic neurons in PD, and MSN neurons in HD, and NPs relieve the symptoms of the disease by selectively targeting these neurons (Hu et al., 2019; Qian et al., 2022; Yu et al., 2022). For instance, NPs loaded with acetylcholine esterase inhibitors and NMDA antagonists (galantamine, donepezil, rivastigmine, and memantine) reach the target region, releasing the cargo and inhibiting enzymes that increase cholinergic transmission, and prevent current flow via NMDA channels, alleviating AD symptoms (Poudel & Park, 2022). Dopamine replacement therapy is currently in use to curve the decreased levels of dopamine; treatment with dopamine-loaded PLGA NPs mitigates physical endurance, anxiety, and coordination in PD animal models (Monge-Fuentes et al., 2021). Progressive deposition of insoluble protein aggregates like Aβ, Tau, α-syn, and mHTT is reported in the brain (Shahnawaz et al., 2020; Song et al., 2018; Zhong et al., 2019). Cyclodextrin-based NPs conjugated with antisense oligonucleotide were reported to successfully decrease the mHTT protein and HD-causing allele SNP1 in striatal neurons and HD patient-derived fibroblasts (Mendonça et al., 2023). Similarly, p-tau was targeted in the tauP301L transgenic mice model of AD using Au–PEG conjugates; they moderately restored the cognition and decreased p-tau burden, by mitigating the malfunctioned proteostasis and protein quality control system by acting as nanochaperons (Bhattacharyya et al., 2017; Vimal et al., 2020).

Additionally, the production of reactive oxygen species (ROS), inflammation, mitochondrial dysfunction, and altered autophagy are observed in these neurodegenerative disorders (Rivera et al., 2023; Roca-Agujetas et al., 2021; Simpson & Oliver, 2020). Oxidative stress is a result of imbalanced ROS production and their removal by antioxidants; therefore, NPs are conjugated with drugs that act as ROS scavengers and increase the production of antioxidants (Bao et al., 2018; Ciofani et al., 2014; Yan et al., 2024). Resveratrol combined with SLNs has shown improved efficacy by reducing oxidative stress and improving cognitive impairments in vascular dementia by triggering the Nrf2/HO-1 pathway, which is linked to the production of antioxidants (Loureiro et al., 2017; Poudel & Park, 2022; Yadav et al., 2018). Nanoliposomes loaded with polyphenols were able to restore the ROS level and the expression of genes related to antioxidant responses, including glutathione synthetase, superoxide dismutase 1 and 2, and catalase, to physiological levels in an in vitro model of PD (Marino et al., 2021; Mistretta et al., 2023).

Similarly, an overactive brain immune milieu can be managed with a versatile immunosuppressive NP (VIN) that incorporates an engineering CXCL12 biomimetic decoy (Shi et al., 2022). The CXCR4-overexpressing mesenchymal stem cell (MSC) membrane shell of VIN, among other things, improves the homing of NPs to cerebral ischemic foci; additionally, the loaded A151 (a cyclic GMP-AMP synthase [cGAS] inhibitor, telomerase repeat sequence) blocks the cGAS–STING axis in microglia, causing the microglia to polarize into an M2-like phenotype involved in anti-inflammation (Shi et al., 2022). Moreover, NPs-based delivery of multifunctional hybrid peptide consisting of neuroprotective peptide S14G-humanin and antioxidant peptide SS31 successfully restored mitochondrial function through STAT3 and PGC-1α pathways, prevented tau hyperphosphorylation and Aβ deposition, and alleviated memory dysfunction and cholinergic neuronal loss in 3xTg-AD mice (Qian et al., 2022). Autophagy can destroy aggregate-prone proteins such as α-synuclein, huntingtin, and misfolded peripheral myelin protein 22 (Nixon & Rubinsztein, 2024). This is why encouraging autophagy in neurons could be a useful treatment strategy. NPs synthesized from zinc oxide/iron oxide/carbon core and a surface terminated with amine/hemin/arginine/trehalose, respectively, induced cellular autophagy, promoted clearance of toxic huntingtin aggregates from neuron cells, and increased cell viability (Sarkar et al., 2024). All over, the studies have suggested that personalized medicine that specifically targets and treats the patient in terms of their physiological responses toward the drug and type of pathology like aggregation of protein, oxidative stress, mitochondrial dysfunction, and autophagy could be a better approach to successfully treat the patient and improve the effect of the drug together with increasing the chances of survival of the patient.

10.6.2 Combination Therapy Using Multiple Drug Delivery Systems

Nanoscale drug-loading particles have been effectively used to deliver a range of imaging and therapeutic compounds for diagnostic purposes and treatments throughout the past 20 years (Grosso et al., 2023; Liu et al., 2016). Still, the development of therapeutic NPs for brain disorders like AD, PD, and HD remains challenging. Myriads of pathway dysfunctions are typically involved in the complicated pathophysiology of brain illnesses, necessitating the co-delivery of multiple medications to produce synergistic effects (Liu et al., 2016). And the presence of BBB, which is considered to be one of the body's most compact structures, makes it very challenging to administer therapeutic medicines systemically by non-invasive means to the brain. The multifunctional NPs, containing more than one drug delivered using multiple drug delivery systems, offer a practical approach to resolving the aforementioned problems. Incorporation of many therapeutic compounds into a single formulation can increase patient compliance and streamline dosage regimens, and the different drug-loading modes effectively reduce the interaction among drugs (Liu et al., 2016). For instance, in AD Aβ deposition and tau hyperphosphorylation are the two primary pathogenic features, and researchers successfully targeted Aβ and tau fibrils simultaneously using a multidrug delivery system (Liu et al., 2020; 2016). Scientists have developed a multifunctional polymeric NP using PEGylated dendrigraft poly-L-lysines (DGLs) for co-delivering BACE 1-AS shRNA and D-peptides (D-TLKIVW) against Aβ and tau fibrils, respectively (Liu et al., 2016; Muolokwu et al., 2024). In this strategy, both the D-peptide and brain-targeted RVG29 (rabies virus glycoprotein composed of 29 amino acid peptides) that bind with specific n-acetylcholine receptors, which are widely expressed in BBB and brain parenchyma cells, were modified to DGLs through bi-functional PEG linker (Liu et al., 2016; Zhu et al., 2021). The resultant DGLs-PEG-RVG29-D-peptide was subsequently complexed with plasmid DNA encoding BACE1-AS shRNA, yielding DGLs-PEG-RVG29-D-peptide/pshBACE1-AS NPs. Intravascular administration of the NPs has revealed that the therapeutic gene BACE 1-AS shRNA reduced the expression of BACE1, which in turn decreased Aβ plaques; furthermore, D-peptide inhibited tau-fibril formation (Table 10.2) (Liu et al., 2016). W20/XD4-superparamagnetic iron oxide nanoparticle (SPION) multifunctional NPs were developed by the conjugation of class A scavenger receptor (SR-A) activator XD4 and oligomer-specific scFv antibody W20 onto PEGylated superparamagnetic iron oxide NPs (SPIONs).

TABLE 10.2
NPs, Their Modifications, Co-Delivery of Conjugated Drugs, and Their Impact on Disease Pathology

Nanoparticle	Modification	Conjugate	Disease	Therapeutic Effects	References
Dendrigraft	D-Peptide and brain-targeted RVG29 were modified to dendrigraft poly-L-lysines (DGLs) through a bi-functional PEG linker. The resultant DGLs-PEG-RVG29-D-peptide was further conjugated with plasmid DNA encoding BACE1-AS shRNA, the therapeutic gene, yielding DGLs-PEG-RVG29-D-peptide/pshBACE1-AS NPs	D-Peptide (all-D amino acid inhibitor, sequence: D-TLKIVW) prevents the p-tau-associated fibril formation and BACE1-AS shRNA encoding plasmids	AD	Decreased the levels of BACE1, Aβ deposition, disrupted p-tau related fibril formation, and improved cognitive function	Liu et al. (2016)
ssSPIONS	W20/XD4-SPIONs were constructed by the conjugation of oligomer-specific scFv antibody W20 and class A scavenger receptor (SR-A) activator XD4 onto PEG-coated superparamagnetic iron oxide nanoparticles (SPIONs)	scFv antibody W20 and microglial class A scavenger receptor (SR-A) activator XD4	AD	Increase uptake of Aβ oligomers by microglia	Liu et al. (2020)
Human serum albumin NPs	Clioquinol and donepezil co-encapsulated in human serum albumin NPs (dcHGT NPs) are designed, which are modified with monosialotetrahexosylganglioside (GM1) incorporated into protein structure and transcriptional activator protein (TAT) peptides conjugated on the surface	Clioquinol (metal-ion chelating agent) and donepezil (acetylcholinesterase (AChE) inhibitor)	AD	Significantly inhibited and eliminated *Aβ* aggregation, alleviated acetylcholine-related inflammation in microglial cells, decreased neuron morphological changes, protected primary neurons from *Aβ* oligomer-induced neurotoxicity, and restored memory deficits	Yang et al. (2020)
Lipid NPs	Curcumin and piperine coloaded glyceryl monooleate (GMO) based lipid NPs coated with surfactants 140 F68 and vitamin E TPGS.	Curcumin (antioxidant, anti-inflammatory) and piperine (antioxidant, anti-inflammatory).	PD	Decreased αS aggregation, oxidative stress, apoptosis, promoted autophagic pathway, rescued motor coordination, decreased dopaminergic neuronal degeneration.	Kundu et al. (2016)
Diblock copolymer	NH_2–PEO–PCL diblock copolymer encapsulated with curcumin and L-DOPA, these NPs were then modified with GSH	Curcumin and L-DOPA (dopamine precursor)	PD	Increased the viability of cells in vitro	Mogharbel et al. (2022)
Lipid NPs	Curcumin and DHA encapsulated NPs prepared by self-assembly of lyotropic lipids and PEGylated amphiphilic derivatives	Curcumin and DHA		Promoted neuronal survival	Guerzoni et al. (2017)
Mesoporous silica NPs	Mesoporous silica NPs functionalized with (3-Aminopropyl) triethoxysilane (AP), which is followed by covalent incorporation of leptin and encapsulation with pioglitazone	Leptin (antioxidant, anti-inflammatory) and pioglitazone (anti-inflammatory)	ALS	Slowed down the disease progression and improved motor performance	Díaz-García et al. (2022)

W20/XD4-SPIONs have shown higher efficacy compared to W20-SPIONs and XD4-SPIONs alone. W20/XD4-SPIONs selectively bind with the Aβ oligomers and significantly promote their clearance by scavenger receptor activation and microglial phagocytosis (Table 10.2) (Liu et al., 2020).

Aberrant alpha-synuclein (α-Syn) aggregation and oxidative stress are two important factors leading to dopaminergic neuronal cell death (Mehta et al., 2019; Won et al., 2022). Thus, combined therapy against PD is required, which has the potential to prevent α-Syn aggregation and modify the equilibrium of ROS. Due to poor pharmacokinetic properties of curcumin NPs like curcumin encapsulated mesoporous silica NPs in inhibiting α-Syn fibrillation, ROS, and restoration of tyrosine hydroxylase (Taebnia et al., 2016). Curcumin was encapsulated with piperine in a liposomal formulation coated with glycerol monooleate to enhance brain delivery (Kundu et al., 2016). Due to the antioxidant properties of curcumin and piperine, the combined therapy greatly increased protection against α-Syn aggregation-driven Parkinson's both in vivo and in vitro by enhancing motor coordination and recovering autophagy-lysosome pathway and mitochondrial function (Table 10.2) (Kundu et al., 2016). Furthermore, ROS-responsive multifunctional NPs were constructed by combining polydopamine (PDA)-based curcumin-loaded NPs with RVG coating (Lei et al., 2023). PDA-based nanomaterials containing many reductive functional groups, including imine and catechol, are more efficient in scavenging ROS both in vivo and in vitro (Lei et al., 2023). These NPs successfully alleviated ROS and inhibited α-Syn aggregation and neurobehavioral impairment in rotenone-induced PD models (Table 10.2) (Lei et al., 2023). A study has synthesized leptin and pioglitazone cocktail-infused mesoporous silica NPs (MSN-LEP-PIO). Treatment of TDP-43^{A315T} mice with MSN-LEP-PIO slowed down the progression of ALS. Also, it alleviated motor dysfunction and decreased energy demand at the later stages of the disease symptoms (Table 10.2) (Díaz-García et al., 2022). rPZDCu having ROS scavenging, thrombolytic, and neuroprotective function was developed based on an ultrasmall $Cu_{4.6}O$ NP, docosahexanoic acid, and thrombolytic agent rt-PA (Kong et al., 2024). In the carotid artery clot model and the embolic middle cerebral artery occlusion (MCAO) ischemic stroke model of rats, rPZDCu demonstrated good thrombolytic efficacy and strong thrombus-targeting ability, which were both made possible by the platelet cell membrane coating on the NP surface (Kong et al., 2024). Additionally, in the rat embolic MCAO model, rPZDCu demonstrated a successful escape from macrophage phagocytosis, efficient activation of microglia polarization, and a successful recovery of neurobiological and behavioral functioning (Kong et al., 2024). Thus, the use of NPs for the co-delivery of drugs is a better approach to tackling multiple pathologies simultaneously and effectively in neurodegenerative disorders. This approach also increases the efficacy of the drugs, decreases the required dosage, and effectively alleviates the pathologies associated with the respective disease.

10.6.3 Integration of Technology for Real-Time Drug Monitoring

NPs have transformed the field of drug delivery and monitoring. Drug delivery to particular cells or tissues can be facilitated by engineering NPs, which reduces side effects and increases therapeutic efficacy. They can be constructed to deliver medications in a controlled manner, providing a consistent and long-lasting therapeutic impact (Agogo-Mawuli & Siderovski, 2022). NPs can also be employed to identify drug levels in the body in real time, providing useful information for modifying dosages and optimizing treatment outcomes (Agogo-Mawuli & Siderovski, 2022). Compared to conventional techniques like immunoassays or chromatographic procedures, NP-based sensors may be more affordable and quicker, making drug monitoring more readily accessible (Agogo-Mawuli & Siderovski, 2022). Imaging and diagnosis are the two primary foundations of personalized medicine (Cash et al., 2015). But to progress in these areas, we must create new methods and tools for in vivo monitoring and imaging of physiologically significant analytes (Cash et al., 2015).

Magnetic resonance imaging (MRI) is non-invasive and repeatable over time; it is one of the most widely used techniques for tracking medical disorders (Grover et al., 2015). Additionally, it is a promising method for tracking the precise targeting and biodistribution of nanomedicines (Ju & Cho, 2023; Liu et al., 2019). Image contrast in MRI is based on the T1 or T2 relaxation periods of protons from various biomolecular structures (Grover et al., 2015). T1, or the spin-lattice relaxation time, indicates the length of time it takes an excited proton to return to its initial state. While T2, or the spin–spin relaxation time, refers to the rate at which the MRI signal fades following excitation (Grover et al., 2015). Whether or not T1 or T2 is weighted determines the contrast of the MR picture. For example, fat appears brighter in a T1-weighted image than in a T2-weighted one because fat loses transverse magnetization more quickly than water (Long & Bulte, 2009). Therefore, MRI monitoring can be used to track the release of drugs from nanocarriers, and its effectiveness is dependent on the contrast-generating principle, which provides distinct signals depending on whether a drug is within or outside of the nanocarrier (Ju & Cho, 2023). However, the conventional MRI technique did not allow for the image of individual Aβ in the human brain (Tafoya et al., 2017). It is evident that new approaches are required to improve the contrast between individual plaques and surrounding brain tissue as well as the plaques' apparent dimensions to successfully use MRI methods to measure individual plaques inside the brains of living humans (Tafoya et al., 2017). This will enable the plaques to be measured directly using the 1.5–3T clinical MRI equipment currently in use (Tafoya et al., 2017). The two main categories of contrast agents that have been investigated for MRI of Aβ plaques both in vivo and ex vivo are ionic gadolinium chelates and SPIONs (Avasthi et al., 2020; Neilio & Ginat, 2024; Neuwelt et al., 2015; Tafoya et al., 2017). Due to the poor MR molar relaxivity of gadolinium preparations (rl ~ r2 ~5 Hz/mM28), patients have to receive relatively large contrast agent injections (~10 g) (Tanifum et al., 2016). Conversely, SPIONs exhibit exceptionally high MR relaxivities of r2*~950 Hz/mM36, and one preparation, ferumoxytol, is being administered to patients. Because SPIONs have much higher MR relaxivity, much lower dosages of contrast agents (about 100–500 mg) can be adopted (Finn et al., 2016; Hawkett et al., 2015; Neuwelt et al., 2015, 2017). Researchers have synthesized an anti-amyloid-SPION NP conjugated with resveratrol that attached with Aβ plaques of ~50 μm diameter, which resulted in large local magnetic susceptibility gradients. These gradients dephased the tissue proton magnetization and created areas of MR signal loss, which showed the plaques as hypointense areas in contrast to the brighter brain background (Tafoya et al., 2017). The presence of the local magnetic susceptibility gradients and the use of MRI sequences having more sensitivity to these gradients like FLASH/SWI and MGE enhanced the detection of individual plaques (Tafoya et al., 2017). Moreover, the real-time monitoring of the drug was enabled by MRI because the release of the drug decreases the levels of Aβ plaque deposition and neuroinflammation (Tafoya et al., 2017). Intravenous treatment with W20-SPIONs that are sensitive, specific, and non-toxic and can reach the brain of transgenic PD and HD mice provides MRI signals to differentiate between healthy and diseased mice. W20-SPIONs have the potential for diagnostics in HD and PD at an early stage and for evaluating the efficacy of new drugs (Liu et al., 2019).

Another approach to directly and continuously measure the real-time drug release is through the electrochemical method. Electrochemical devices have been widely employed for continuous (industrial and environmental) monitoring purposes because they provide a quick return on chemical information (Wang, 2002). For instance, real-time measurements have been made for many years using electrochemical sensors, such as pH and oxygen electrodes (Mora et al., 2009). Electrochemical sensors detect analytes of interest by oxidizing or reducing them on the surface of a working electrode, resulting in a signal proportional to the analyte's concentration (Bedioui et al., 2018; Tabish et al., 2023). Signals can be detected using either amperometric or voltammetric methods (Gill et al., 2019; Jensen et al., 2013). In amperometry, the voltage between the working and reference electrodes is utilized to measure the current produced using a two-electrode mode system (Tabish et al., 2023). This measures the current created by analyte oxidation using a fixed working electrode voltage against the reference electrode (Brown & Schoenfisch, 2019; Tabish et al., 2023). In voltammetric techniques, the current generated is measured using a three-electrode system and a variable voltage that is delivered as a function of time (Tabish et al., 2023). To reduce the stray capacitance effect and as they provide rapid testing, sensitivity, quick response time, the magnitude of the signal, and efficiency in terms of lower detection limits, amperometric methods are preferred over voltammetric approaches (Cha et al., 2010; Tajik et al., 2021). In a study, PBS buffer was mixed with the doxorubicin-loaded liposomes at either pH 7.4 (pH of blood) or pH 5.2 (pH of cytoplasm) (Mora et al., 2009). The voltammetric sensor was then used to measure the amount of free doxorubicin that was released from the liposomes. Square-wave voltammograms of doxorubicin at various concentrations were generated, and then the release kinetic from the liposome was intrapolated based on the standard curve (Mora et al., 2009).

Traditional optical imaging methods are hampered by light's shallow penetration depth and scattering, which reduces spatial resolution in deep tissue (Cash et al., 2015; Kircher et al., 2012). Many of these limitations are mitigated by photoacoustic imaging, which achieves acoustically defined spatial resolution and imaging depths of up to several centimeters (Kircher et al., 2012). This approach is non-invasive and is used to monitor the drug that has vibrations by the photoacoustic effect (Tang et al., 2023; Wang & Yao, 2016). As a result of the cation being extracted into the polymer's core, the internal pH of the nanosensor changes, changing a chromoionophore's protonation state and the nanosensor's optical characteristics (Cash et al., 2015). The photoacoustic spectrum varies in tandem with changes in the chromoionophore's absorption spectrum (Cash et al., 2015). Both the absorbance and photoacoustic intensity at 515 nm and the absorbance and photoacoustic intensity at 660 nm decrease with increasing lithium concentration (Cash et al., 2015). The ratiometric index from the observations at these two-peak wavelengths that rise linearly with lithium concentration is calculated (Cash et al., 2015). However, this imaging technique still has not been employed for real-time drug monitoring in neurodegenerative disorders, but its use in bipolar disorder revealed its successful use in the brain (Cash et al., 2015).

10.7 CONCLUSION

Due to the multifactorial nature of neurodegenerative disease, the exact cause of these diseases is still not fully understood. Neurodegenerative diseases involve several arrays of pathophysiological processes, including protein misfolding, oxidative stress, and inflammation (Emerit et al., 2004). This poses challenges in developing targeted treatments for neurodegenerative diseases. Rapid innovations in nanotechnology could lead to the production of nanocarriers that help to overcome the BBB; target specific brain regions or cell types to improve the precision of drug delivery; and release drugs in a controlled and sustained manner to ensure a consistent therapeutic effect over time and tailoring them for easy surface modification to carry different types of drugs, including small molecules, proteins and nucleic acids. Thus, these NPs with conjugated drugs have proven advantageous over conventional drug formulations, which possess great difficulty in traversing BBB. It is witnessed that while nanotherapeutic approaches offer a wide alternative for treating neurodegenerative diseases by meeting contemporary therapeutic inadequacies and shortcomings, additional research is required before its clinical application to fully understand the long-term effects, pharmacokinetics, toxicity, and biodistribution. Subsequently, because of their minute size, inventing and adopting novel methodologies to precisely isolate and quantify NPs is especially important for the development of nanodrugs for neurodegenerative diseases. Moreover, standardized protocols are to be adopted for synthesis, evaluation, and characterization, which are of utmost importance to ensure the safety and efficacy of NPs. With the emergence of nano informatics in drug delivery, which is centered on the use of informatics tools to better understand and control the behavior of NPs in biological systems, drug delivery approaches are poised for a revolution in the context of disease management.

REFERENCES

Abbott, N. J. (2013). Blood–brain barrier structure and function and the challenges for CNS drug delivery. *Journal of Inherited Metabolic Disease*, *36*(3), pp. 437–449. https://doi.org/10.1007/s10545-013-9608-0

Abbott, N. J., Patabendige, A. A. K., Dolman, D. E. M., Yusof, S. R., & Begley, D. J. (2010). Structure and function of the blood–brain barrier. *Neurobiology of Disease*, *37*(1), pp. 13–25. https://doi.org/10.1016/j.nbd.2009.07.030

Abdulla, M., Hussain, M. K., & Ahamad, S. (2024). Trifluoromethylnitrone: a versatile building block for synthesizing trifluoromethyl-containing heterocyclic compounds. *Organic & Biomolecular Chemistry*, *22*(26), pp. 5242–5256. https://doi.org/10.1039/D4OB00849A

Agarwal, V., & Mishra, B. (1999). Recent trends in drug delivery systems: intranasal drug delivery. *Indian Journal of Experimental Biology*, *37*(1), pp. 6–16.

Agogo-Mawuli, P. S., & Siderovski, D. P. (2022). Recent advances in nanosensors for therapeutic drug monitoring (TDM). In *Recent Advances in Therapeutic Drug Monitoring and Clinical Toxicology* (pp. 233–253). Springer International Publishing. https://doi.org/10.1007/978-3-031-12398-6_14

Agrawal, M., Saraf, S., Saraf, S., Antimisiaris, S. G., Chougule, M. B., Shoyele, S. A., & Alexander, A. (2018). Nose-to-brain drug delivery: an update on clinical challenges and progress towards approval of anti-Alzheimer drugs. *Journal of Controlled Release*, *281*, pp. 139–177. https://doi.org/10.1016/j.jconrel.2018.05.011

Ahamad, S., Abdulla, M., Saquib, M., & Kamil Hussain, M. (2024a). Pseudo-natural products: expanding chemical and biological space by surpassing natural constraints. *Bioorganic Chemistry*, p. 107525. https://doi.org/10.1016/j.bioorg.2024.107525

Ahamad, S., Bano, N., Khan, S., Hussain, M. K., & Bhat, S. A. (2024b). Unraveling the puzzle of therapeutic peptides: a promising frontier in Huntington 's disease treatment. *Journal of Medicinal Chemistry*, *67*(2), pp. 783–815. https://doi.org/10.1021/acs.jmedchem.3c01131

Ahamad, S., & Bhat, S. A. (2022a). Recent update on the development of PCSK9 inhibitors for hypercholesterolemia treatment. *Journal of Medicinal Chemistry*, *65*(23), pp. 15513–15539. https://doi.org/10.1021/acs.jmedchem.2c01290

Ahamad, S., & Bhat, S. A. (2022b). The emerging landscape of small-molecule therapeutics for the treatment of Huntington's disease. *Journal of Medicinal Chemistry*, *65*(24), pp. 15993–16032. https://doi.org/10.1021/acs.jmedchem.2c00799

Ahamad, S., Branch, S., Harrelson, S., Hussain, M. K., Saquib, M., & Khan, S. (2021). Primed for global coronavirus pandemic: emerging research and clinical outcome. *European Journal of Medicinal Chemistry*, *209*, p. 112862. https://doi.org/10.1016/j.ejmech.2020.112862

Ahamad, S., Mathew, S., Khan, W. A., & Mohanan, K. (2022). Development of small-molecule PCSK9 inhibitors for the treatment of hypercholesterolemia. *Drug Discovery Today*, *27*(5), pp. 1332–1349. https://doi.org/10.1016/j.drudis.2022.01.014

Ahlskog, J. E. (2001). Parkinson's disease: medical and surgical treatment. *Neurologic Clinics*, *19*(3), pp. 579–605. https://doi.org/10.1016/S0733-8619(05)70036-0

Ahlskog, J. E., & Muenter, M. D. (2001). Frequency of levodopa-related dyskinesias and motor fluctuations as estimated from the cumulative literature. *Movement Disorders*, *16*(3), pp. 448–458. https://doi.org/10.1002/mds.1090

Ahmad, M. Z., Sabri, A. H. Bin, Anjani, Q. K., Domínguez-Robles, J., Abdul Latip, N., & Hamid, K. A. (2022). Design and development of levodopa loaded polymeric nanoparticles for intranasal delivery. *Pharmaceuticals*, *15*(3), p. 370. https://doi.org/10.3390/ph15030370

Ahmed, T. A., Felimban, R. I., Tayeb, H. H., Rizg, W. Y., Alnadwi, F. H., Alotaibi, H. A., Alhakamy, N. A., Abd-Allah, F. I., Mohamed, G. A., Zidan, A. S., & El-Say, K. M. (2021). Development of multi-compartment 3D-printed tablets loaded with self-nanoemulsified formulations of various drugs: a new strategy for personalized medicine. *Pharmaceutics*, *13*(10), p. 1733. https://doi.org/10.3390/pharmaceutics13101733

Aliabadi, H. M., & Lavasanifar, A. (2006). Polymeric micelles for drug delivery. *Expert Opinion on Drug Delivery*, *3*(1), pp. 139–162. https://doi.org/10.1517/17425247.3.1.139

Allen, T. M., & Cullis, P. R. (2004). Drug delivery systems: entering the mainstream. *Science*, *303*(5665), pp. 1818–1822. https://doi.org/10.1126/science.1095833

Andreone, B. J., Chow, B. W., Tata, A., Lacoste, B., Ben-Zvi, A., Bullock, K., Deik, A. A., Ginty, D. D., Clish, C. B., & Gu, C. (2017). Blood-brain barrier permeability is regulated by lipid transport-dependent suppression of caveolae-mediated transcytosis. *Neuron*, *94*(3), pp. 581–594.e5. https://doi.org/10.1016/j.neuron.2017.03.043

Antonini, A., Bauer, L., Dohin, E., Oertel, W. H., Rascol, O., Reichmann, H., Schmid, M., Singh, P., Tolosa, E., & Chaudhuri, K. R. (2015). Effects of rotigotine transdermal patch in patients with Parkinson's disease presenting with non-motor symptoms – results of a double-blind, randomized, placebo-controlled trial. *European Journal of Neurology*, *22*(10), pp. 1400–1407. https://doi.org/10.1111/ene.12757

Araújo, F., das Neves, J., Martins, J. P., Granja, P. L., Santos, H. A., & Sarmento, B. (2017). Functionalized materials for multistage platforms in the oral delivery of biopharmaceuticals. *Progress in Materials Science*, *89*, pp. 306–344. https://doi.org/10.1016/j.pmatsci.2017.05.001

Armstrong, M. J., & Miyasaki, J. M. (2012). Evidence-based guideline: pharmacologic treatment of chorea in Huntington disease [RETIRED]. *Neurology*, *79*(6), pp. 597–603. https://doi.org/10.1212/WNL.0b013e318263c443

Arora, A., Kumar, S., Ali, J., & Baboota, S. (2020). Intranasal delivery of tetrabenazine nanoemulsion via olfactory region for better treatment of hyperkinetic movement associated with Huntington's disease: pharmacokinetic and brain delivery study. *Chemistry and Physics of Lipids*, *230*, p. 104917. https://doi.org/10.1016/j.chemphyslip.2020.104917

Assi, A.-A., Farrag, M. M. Y., Badary, D. M., Allam, E. A. H., & Nicola, M. A. (2023). Protective effects of curcumin and Ginkgo biloba extract combination on a new model of Alzheimer's disease. *Inflammopharmacology*, *31*(3), pp. 1449–1464. https://doi.org/10.1007/s10787-023-01164-6

Avasthi, A., Caro, C., Pozo-Torres, E., Leal, M. P., & García-Martín, M. L. (2020). Magnetic nanoparticles as MRI contrast agents. *Topics in Current Chemistry*, *378*(3), p. 40. https://doi.org/10.1007/s41061-020-00302-w

Bali, N. R., & Salve, P. S. (2019). Selegiline nanoparticle embedded transdermal film: an alternative approach for brain targeting in Parkinson's disease. *Journal of Drug Delivery Science and Technology*, *54*, p. 101299. https://doi.org/10.1016/j.jddst.2019.101299

Ballabh, P., Braun, A., & Nedergaard, M. (2004). The blood–brain barrier: an overview. *Neurobiology of Disease*, *16*(1), pp. 1–13. https://doi.org/10.1016/j.nbd.2003.12.016

Banerjee, A., Qi, J., Gogoi, R., Wong, J., & Mitragotri, S. (2016). Role of nanoparticle size, shape and surface chemistry in oral drug delivery. *Journal of Controlled Release*, *238*, pp. 176–185. https://doi.org/10.1016/j.jconrel.2016.07.051

Banks, W. A. (2016). From blood–brain barrier to blood–brain interface: new opportunities for CNS drug delivery. *Nature Reviews Drug Discovery*, *15*(4), pp. 275–292. https://doi.org/10.1038/nrd.2015.21

Bao, X., Zhao, J., Sun, J., Hu, M., & Yang, X. (2018). Polydopamine nanoparticles as efficient scavengers for reactive oxygen species in periodontal disease. *ACS Nano*, *12*(9), pp. 8882–8892. https://doi.org/10.1021/acsnano.8b04022

Barkat, M. A., Harshita, Rizwanullah, M., Pottoo, F. H., Beg, S., Akhter, S., & Ahmad, F. J. (2020). Therapeutic nanoemulsion: concept to delivery. *Current Pharmaceutical Design*, *26*(11), pp. 1145–1166. https://doi.org/10.2174/1381612826666200317140600

Baysal, I., Ucar, G., Gultekinoglu, M., Ulubayram, K., & Yabanoglu-Ciftci, S. (2017). Donepezil loaded PLGA-b-PEG nanoparticles: their ability to induce destabilization of amyloid fibrils and to cross blood brain barrier in vitro. *Journal of Neural Transmission*, *124*(1), pp. 33–45. https://doi.org/10.1007/s00702-016-1527-4

Bedioui, F., Ismail, A., & Griveau, S. (2018). Electrochemical detection of nitric oxide and S-nitrosothiols in biological systems: past, present & future. *Current Opinion in Electrochemistry*, *12*, pp. 42–50. https://doi.org/10.1016/j.coelec.2018.04.014

Begley, D. J. (2004). Delivery of therapeutic agents to the central nervous system: the problems and the possibilities. *Pharmacology & Therapeutics*, *104*(1), pp. 29–45. https://doi.org/10.1016/j.pharmthera.2004.08.001

Bellettato, C. M., & Scarpa, M. (2018). Possible strategies to cross the blood–brain barrier. *Italian Journal of Pediatrics*, *44*(S2), p. 131. https://doi.org/10.1186/s13052-018-0563-0

Bellotti, E., Schilling, A. L., Little, S. R., & Decuzzi, P. (2021). Injectable thermoresponsive hydrogels as drug delivery system for the treatment of central nervous system disorders: a review. *Journal of Controlled Release*, *329*, pp. 16–35. https://doi.org/10.1016/j.jconrel.2020.11.049

Benjamin, B., & Burns, A. (2007). Donepezil for Alzheimer's disease. *Expert Review of Neurotherapeutics*, *7*(10), pp. 1243–1249. https://doi.org/10.1586/14737175.7.10.1243

Bhat, S. A., Ahamad, S., Dar, N. J., Siddique, Y. H., & Nazir, A. (2023). The emerging landscape of natural small-molecule therapeutics for Huntington's disease. *Current Neuropharmacology*, *21*(4), pp. 867–889. https://doi.org/10.2174/1570159X21666230216104621

Bhattacharyya, S., Kim, K., & Teizer, W. (2017). Remodeling tau and prion proteins using nanochaperons. *Advanced Biosystems*, *1*(10). https://doi.org/10.1002/adbi.201700108

Biddlestone-Thorpe, L., Marchi, N., Guo, K., Ghosh, C., Janigro, D., Valerie, K., & Yang, H. (2012). Nanomaterial-mediated CNS delivery of diagnostic and therapeutic agents. *Advanced Drug Delivery Reviews*, *64*(7), pp. 605–613. https://doi.org/10.1016/j.addr.2011.11.014

Bölükbaşı Hatip, F. F., & Hatip-Al-Khatib, I. (2013). Effects of β-sheet breaker peptides on altered responses of thoracic aorta in rats' Alzheimer's disease model induced by intraamygdaloid Aβ40. *Life Sciences*, *92*(3), pp. 228–236. https://doi.org/10.1016/j.lfs.2012.12.004

Bourasset, F., Auvity, S., Thorne, R. G., & Scherrmann, J.-M. (2020). In *Brain Distribution of Drugs: Brain Morphology, Delivery Routes, and Species Differences* (pp. 97–120). https://doi.org/10.1007/164_2020_402

Brown, M. B., Martin, G. P., Jones, S. A., & Akomeah, F. K. (2006). Dermal and transdermal drug delivery systems: current and future prospects. *Drug Delivery*, *13*(3), pp. 175–187. https://doi.org/10.1080/10717540500455975

Brown, M. D., & Schoenfisch, M. H. (2019). Selective and sensocompatible electrochemical nitric oxide sensor with a bilaminar design. *ACS Sensors*, *4*(7), pp. 1766–1773. https://doi.org/10.1021/acssensors.9b00170

Campbell, N. L., Perkins, A. J., Gao, S., Skaar, T. C., Li, L., Hendrie, H. C., Fowler, N., Callahan, C. M., & Boustani, M. A. (2017). Adherence and tolerability of Alzheimer's disease medications: a pragmatic randomized trial. *Journal of the American Geriatrics Society*, *65*(7), pp. 1497–1504. https://doi.org/10.1111/jgs.14827

Carbone, M., Duty, S., & Rattray, M. (2012). Riluzole neuroprotection in a parkinson's disease model involves suppression of reactive astrocytosis but not GLT-1 regulation. *BMC Neuroscience*, *13*(1), p. 38. https://doi.org/10.1186/1471-2202-13-38

Cash, K. J., Li, C., Xia, J., Wang, L. V., & Clark, H. A. (2015). Optical drug monitoring: photoacoustic imaging of nanosensors to monitor therapeutic lithium in vivo. *ACS Nano*, *9*(2), pp. 1692–1698. https://doi.org/10.1021/nn5064858

Cecchelli, R., Berezowski, V., Lundquist, S., Culot, M., Renftel, M., Dehouck, M.-P., & Fenart, L. (2007). Modelling of the blood–brain barrier in drug discovery and development. *Nature Reviews Drug Discovery*, *6*(8), pp. 650–661. https://doi.org/10.1038/nrd2368

Cha, W., Tung, Y.-C., Meyerhoff, M. E., & Takayama, S. (2010). Patterned electrode-based amperometric gas sensor for direct nitric oxide detection within microfluidic devices. *Analytical Chemistry*, *82*(8), pp. 3300–3305. https://doi.org/10.1021/ac100085w

Chatterjee, B., Gorain, B., Mohananaidu, K., Sengupta, P., Mandal, U. K., & Choudhury, H. (2019). Targeted drug delivery to the brain via intranasal nanoemulsion: available proof of concept and existing challenges. *International Journal of Pharmaceutics*, *565*, pp. 258–268. https://doi.org/10.1016/j.ijpharm.2019.05.032

Choi, H.-J., Ebersbacher, C. F., Kim, M.-C., Kang, S.-M., & Montemagno, C. D. (2013). A mechanistic study on the destabilization of whole inactivated influenza virus vaccine in gastric environment. *PLoS One*, *8*(6), p. e66316. https://doi.org/10.1371/journal.pone.0066316

Choi, H.-J., Kim, M.-C., Kang, S.-M., & Montemagno, C. D. (2014). The osmotic stress response of split influenza vaccine particles in an acidic environment. *Archives of Pharmacal Research*, *37*(12), pp. 1607–1616. https://doi.org/10.1007/s12272-013-0257-5

Ciofani, G., Genchi, G. G., Mazzolai, B., & Mattoli, V. (2014). Transcriptional profile of genes involved in oxidative stress and antioxidant defense in PC12 cells following treatment with cerium oxide nanoparticles. *Biochimica et Biophysica Acta (BBA) - General Subjects*, *1840*(1), pp. 495–506. https://doi.org/10.1016/j.bbagen.2013.10.009

Crone, C., & Olesen, S. P. (1982). Electrical resistance of brain microvascular endothelium. *Brain Research*, *241*(1), pp. 49–55. https://doi.org/10.1016/0006-8993(82)91227-6

Cummings, J., Lefèvre, G., Small, G., & Appel-Dingemanse, S. (2007). Pharmacokinetic rationale for the rivastigmine patch. *Neurology*, *69*(4_suppl_1). https://doi.org/10.1212/01.wnl.0000281846.40390.50

de los Ríos, C., & Marco-Contelles, J. (2019). Tacrines for Alzheimer's disease therapy. III. The PyridoTacrines. *European Journal of Medicinal Chemistry*, *166*, pp. 381–389. https://doi.org/10.1016/j.ejmech.2019.02.005

Dei Cas, M., & Ghidoni, R. (2019). Dietary curcumin: correlation between bioavailability and health potential. *Nutrients*, *11*(9), p. 2147. https://doi.org/10.3390/nu11092147

des Rieux, A., Fievez, V., Garinot, M., Schneider, Y.-J., & Préat, V. (2006). Nanoparticles as potential oral delivery systems of proteins and vaccines: a mechanistic approach. *Journal of Controlled Release*, *116*(1), pp. 1–27. https://doi.org/10.1016/j.jconrel.2006.08.013

Dhuria, S. V., Hanson, L. R., & Frey, W. H. (2010). Intranasal delivery to the central nervous system: mechanisms and experimental considerations. *Journal of Pharmaceutical Sciences*, *99*(4), pp. 1654–1673. https://doi.org/10.1002/jps.21924

Dias, A. P., da Silva Santos, S., da Silva, J. V., Parise-Filho, R., Igne Ferreira, E., Seoud, O. El, & Giarolla, J. (2020). Dendrimers in the context of nanomedicine. *International Journal of Pharmaceutics*, *573*, p. 118814. https://doi.org/10.1016/j.ijpharm.2019.118814

Díaz-García, D., Ferrer-Donato, Á., Méndez-Arriaga, J. M., Cabrera-Pinto, M., Díaz-Sánchez, M., Prashar, S., Fernandez-Martos, C. M., & Gómez-Ruiz, S. (2022). Design of mesoporous silica nanoparticles for the treatment of amyotrophic lateral sclerosis (ALS) with a therapeutic cocktail based on leptin and pioglitazone. *ACS Biomaterials Science & Engineering*, *8*(11), pp. 4838–4849. https://doi.org/10.1021/acsbiomaterials.2c00865

Dineshkumar, B., Krishnakumar, K., Bhatt, A., Paul, D., Cherian, J., John, A., & Suresh, S. (2015). Single-walled and multi-walled carbon nanotubes based drug delivery system: cancer therapy: a review. *Indian Journal of Cancer*, *52*(3), p. 262. https://doi.org/10.4103/0019-509X.176720

Dong, X. (2018). Current strategies for brain drug delivery. *Theranostics*, *8*(6), pp. 1481–1493. https://doi.org/10.7150/thno.21254

Duan, L., Li, X., Ji, R., Hao, Z., Kong, M., Wen, X., Guan, F., & Ma, S. (2023). Nanoparticle-based drug delivery systems: an inspiring therapeutic strategy for neurodegenerative diseases. *Polymers*, *15*(9), p. 2196. https://doi.org/10.3390/polym15092196

Duan, Y., Dhar, A., Patel, C., Khimani, M., Neogi, S., Sharma, P., Siva Kumar, N., & Vekariya, R. L. (2020). A brief review on solid lipid nanoparticles: part and parcel of contemporary drug delivery systems. *RSC Advances*, *10*(45), pp. 26777–26791. https://doi.org/10.1039/D0RA03491F

Ebadi, M., Sharma, S., Shavali, S., & El Refaey, H. (2002). Neuroprotective actions of selegiline. *Journal of Neuroscience Research*, *67*(3), pp. 285–289. https://doi.org/10.1002/jnr.10148

El-Boubbou, K. (2018). Magnetic iron oxide nanoparticles as drug carriers: clinical relevance. *Nanomedicine*, *13*(8), pp. 953–971. https://doi.org/10.2217/nnm-2017-0336

Elfinger, M., Maucksch, C., & Rudolph, C. (2007). Characterization of lactoferrin as a targeting ligand for nonviral gene delivery to airway epithelial cells. *Biomaterials*, *28*(23), pp. 3448–3455. https://doi.org/10.1016/j.biomaterials.2007.04.011

Emerit, J., Edeas, M., & Bricaire, F. (2004). Neurodegenerative diseases and oxidative stress. *Biomedicine & Pharmacotherapy*, *58*(1), pp. 39–46. https://doi.org/10.1016/j.biopha.2003.11.004

Ensign, L. M., Cone, R., & Hanes, J. (2012). Oral drug delivery with polymeric nanoparticles: the gastrointestinal mucus barriers. *Advanced Drug Delivery Reviews*, *64*(6), pp. 557–570. https://doi.org/10.1016/j.addr.2011.12.009

Entacapone Improves Motor Fluctuations in Levodopa- Treated Parkinson's Disease Patients. (1997). *Annals of Neurology*, *42*(5), pp. 747–755. https://doi.org/10.1002/ana.410420511

Esmaealzadeh, N., Miri, M. S., Mavaddat, H., Peyrovinasab, A., Ghasemi Zargar, S., Sirous Kabiri, S., Razavi, S. M., & Abdolghaffari, A. H. (2024). The regulating effect of curcumin on NF-κB pathway in neurodegenerative diseases: a review of the underlying mechanisms. *Inflammopharmacology*, *32*(4), pp. 2125–2151. https://doi.org/10.1007/s10787-024-01492-1

Figueiredo, E. C., Seabra, C. L., Mendes, T. V., Rosa, M. A., Pinho, C. D. L. N., de Oliveira Figueiredo, D. M., Boralli, V. B., Junior, M. K. A., de Lima, M. E. T., & Reis, S. (2023). Molecularly imprinted nanoparticles as drug carriers to the brain. *Journal of Materials Science*, *58*(46), pp. 17578–17593. https://doi.org/10.1007/s10853-023-09053-7

Finn, J. P., Nguyen, K.-L., Han, F., Zhou, Z., Salusky, I., Ayad, I., & Hu, P. (2016). Cardiovascular MRI with ferumoxytol. *Clinical Radiology*, *71*(8), pp. 796–806. https://doi.org/10.1016/j.crad.2016.03.020

Frampton, J. E. (2019). Rotigotine transdermal patch: a review in Parkinson's disease. *CNS Drugs*, *33*(7), pp. 707–718. https://doi.org/10.1007/s40263-019-00646-y

Gao, X., Xu, J., Yao, T., Liu, X., Zhang, H., & Zhan, C. (2022). Peptide-decorated nanocarriers penetrating the blood-brain barrier for imaging and therapy of brain diseases. *Advanced Drug Delivery Reviews*, *187*, p. 114362. https://doi.org/10.1016/j.addr.2022.114362

Ghezzi, M., Pescina, S., Padula, C., Santi, P., Del Favero, E., Cantù, L., & Nicoli, S. (2021). Polymeric micelles in drug delivery: an insight of the techniques for their characterization and assessment in biorelevant conditions. *Journal of Controlled Release*, *332*, pp. 312–336. https://doi.org/10.1016/j.jconrel.2021.02.031

Giladi, N., Boroojerdi, B., Korczyn, A. D., Burn, D. J., Clarke, C. E., & Schapira, A. H. V. (2007). Rotigotine transdermal patch in early Parkinson's disease: a randomized, double-blind, controlled study versus placebo and ropinirole. *Movement Disorders*, *22*(16), pp. 2398–2404. https://doi.org/10.1002/mds.21741

Gill, A., Zajda, J., & Meyerhoff, M. E. (2019). Comparison of electrochemical nitric oxide detection methods with chemiluminescence for measuring nitrite concentration in food samples. *Analytica Chimica Acta*, 1077, pp. 167–173. https://doi.org/10.1016/j.aca.2019.05.065

Girardin, F. (2006). Membrane transporter proteins: a challenge for CNS drug development. *Dialogues in Clinical Neuroscience*, *8*(3), pp. 311–321. https://doi.org/10.31887/DCNS.2006.8.3/fgirardin

Goldenberg, M. M. (2008). Medical management of Parkinson's disease. *P & T: A Peer-Reviewed Journal for Formulary Management*, *33*(10), pp. 590–606.

Green, R. C. (2009). Effect of tarenflurbil on cognitive decline and activities of daily living in patients with mild Alzheimer disease: a randomized controlled trial. *JAMA*, *302*(23), p. 2557. https://doi.org/10.1001/jama.2009.1866

Griñán-Ferré, C., Bellver-Sanchis, A., Guerrero, A., & Pallàs, M. (2024). Advancing personalized medicine in neurodegenerative diseases: the role of epigenetics and pharmacoepigenomics in pharmacotherapy. *Pharmacological Research*, *205*, p. 107247. https://doi.org/10.1016/j.phrs.2024.107247

Grosso, C., Silva, A., Delerue-Matos, C., & Barroso, M. F. (2023). Single and multitarget systems for drug delivery and detection: up-to-date strategies for brain disorders. *Pharmaceuticals*, *16*(12), p. 1721. https://doi.org/10.3390/ph16121721

Grover, V. P. B., Tognarelli, J. M., Crossey, M. M. E., Cox, I. J., Taylor-Robinson, S. D., & McPhail, M. J. W. (2015). Magnetic resonance imaging: principles and techniques: lessons for clinicians. *Journal of Clinical and Experimental Hepatology*, *5*(3), pp. 246–255. https://doi.org/10.1016/j.jceh.2015.08.001

Guan, X. (2024). Glutathione transporter as a target for brain drug delivery. *Medicinal Chemistry Research*. https://doi.org/10.1007/s00044-024-03225-z

Guerzoni, L. P. B., Nicolas, V., & Angelova, A. (2017). In vitro modulation of TrkB receptor signaling upon sequential delivery of curcumin-DHA loaded carriers towards promoting neuronal survival. *Pharmaceutical Research*, *34*(2), pp. 492–505. https://doi.org/10.1007/s11095-016-2080-4

Hadavi, D., & Poot, A. A. (2016). Biomaterials for the treatment of Alzheimer's disease. *Frontiers in Bioengineering and Biotechnology*, *4*. https://doi.org/10.3389/fbioe.2016.00049

Hawkett, B., Pham, B., Jain, N., Kuchel, P., Chapman, B., Bickley, S., & Jones, S. (2015). The interaction of sterically stabilized magnetic nanoparticles with fresh human red blood cells. *International Journal of Nanomedicine*, 6645. https://doi.org/10.2147/IJN.S93225

Homayun, B., Lin, X., & Choi, H.-J. (2019). Challenges and recent progress in oral drug delivery systems for biopharmaceuticals. *Pharmaceutics*, *11*(3), p. 129. https://doi.org/10.3390/pharmaceutics11030129

Howard, M. A., Gross, A., Grady, M. S., Langer, R. S., Mathiowitz, E., Winn, H. R., & Mayberg, M. R. (1989). Intracerebral drug delivery in rats with lesion-induced memory deficits. *Journal of Neurosurgery*, *71*(1), pp. 105–112. https://doi.org/10.3171/jns.1989.71.1.0105

Hu, Q., & Luo, Y. (2018). Recent advances of polysaccharide-based nanoparticles for oral insulin delivery. *International Journal of Biological Macromolecules*, *120*, pp. 775–782. https://doi.org/10.1016/j.ijbiomac.2018.08.152

Hu, Y.-B., Zhang, Y.-F., Wang, H., Ren, R.-J., Cui, H.-L., Huang, W.-Y., Cheng, Q., Chen, H.-Z., & Wang, G. (2019). miR-425 deficiency promotes necroptosis and dopaminergic neurodegeneration in Parkinson's disease. *Cell Death & Disease*, *10*(8), p. 589. https://doi.org/10.1038/s41419-019-1809-5

Hussain, M. K., Ali, R., Ahamad, S., Khan, M. F., & Saquib, M. (2023). Ivermectin. In *Chemistry and Biological Activities of Ivermectin* (pp. 121–139). Wiley. https://doi.org/10.1002/9781394168033.ch6

Hwang, D., Ramsey, J. D., & Kabanov, A. V. (2020). Polymeric micelles for the delivery of poorly soluble drugs: rom nanoformulation to clinical approval. *Advanced Drug Delivery Reviews*, *156*, pp. 80–118. https://doi.org/10.1016/j.addr.2020.09.009

Hyde, C., Peters, J., Bond, M., Rogers, G., Hoyle, M., Anderson, R., Jeffreys, M., Davis, S., Thokala, P., & Moxham, T. (2013). Evolution of the evidence on the effectiveness and cost-effectiveness of acetylcholinesterase inhibitors and memantine for Alzheimer's disease: systematic review and economic model. *Age and Ageing*, *42*(1), pp. 14–20. https://doi.org/10.1093/ageing/afs165

Illum, L. (2000). Transport of drugs from the nasal cavity to the central nervous system. *European Journal of Pharmaceutical Sciences*, *11*(1), pp. 1–18. https://doi.org/10.1016/S0928-0987(00)00087-7

Illum, L. (2003). Nasal drug delivery—possibilities, problems and solutions. *Journal of Controlled Release*, *87*(1–3), pp. 187–198. https://doi.org/10.1016/S0168-3659(02)00363-2

Imbimbo, B. P. (2009). Why did tarenflurbil fail in Alzheimer's disease? *Journal of Alzheimer's Disease*, *17*(4), pp. 757–760. https://doi.org/10.3233/JAD-2009-1092

Ismail, M. F., ElMeshad, & Salem. (2013). Potential therapeutic effect of nanobased formulation of rivastigmine on rat model of Alzheimer's disease. *International Journal of Nanomedicine*, *393*. https://doi.org/10.2147/IJN.S39232

Jamróz, W., Szafraniec, J., Kurek, M., & Jachowicz, R. (2018). 3D printing in pharmaceutical and medical applications – recent achievements and challenges. *Pharmaceutical Research*, *35*(9), p. 176. https://doi.org/10.1007/s11095-018-2454-x

Jensen, G. C., Zheng, Z., & Meyerhoff, M. E. (2013). Amperometric nitric oxide sensors with enhanced selectivity over carbon monoxide via platinum oxide formation under alkaline conditions. *Analytical Chemistry*, *85*(21), pp. 10057–10061. https://doi.org/10.1021/ac402633t

Jessen, L., Kovalick, L. J., & Azzaro, A. J. (2008). The selegiline transdermal system (emsam): a therapeutic option for the treatment of major depressive disorder. *P & T : A Peer-Reviewed Journal for Formulary Management*, *33*(4), pp. 212–246.

Johanson, C. E., Stopa, E. G., & McMillan, P. N. (2011). *The Blood–Cerebrospinal Fluid Barrier: Structure and Functional Significance* (pp. 101–131). https://doi.org/10.1007/978-1-60761-938-3_4

Jones, N. S., Quraishi, S., & Mason, J. D. (1997). The nasal delivery of systemic drugs. *International Journal of Clinical Practice*, *51*(5), pp. 308–311.

Ju, S., & Cho, H.-Y. (2023). Biohybrid nanoparticle-based in situ monitoring of in vivo drug delivery. *Biosensors*, *13*(12), p. 1017. https://doi.org/10.3390/bios13121017

Kalepu, S., & Nekkanti, V. (2015). Insoluble drug delivery strategies: review of recent advances and business prospects. *Acta Pharmaceutica Sinica B*, *5*(5), pp. 442–453. https://doi.org/10.1016/j.apsb.2015.07.003

Kashyap, K., & Shukla, R. (2019). Drug delivery and targeting to the brain through nasal route: mechanisms, applications and challenges. *Current Drug Delivery*, *16*(10), pp. 887–901. https://doi.org/10.2174/1567201816666191029122740

Kaya, M., & Ahishali, B. (2021). Basic physiology of the blood-brain barrier in health and disease: a brief overview. *Tissue Barriers*, *9*(1), p. 1840913. https://doi.org/10.1080/21688370.2020.1840913

Kenney, C., Hunter, C., Davidson, A., & Jankovic, J. (2007). Short-term effects of tetrabenazine on chorea associated with Huntington's disease. *Movement Disorders*, *22*(1), pp. 10–13. https://doi.org/10.1002/mds.21161

Khalin, I., Alyautdin, R., Ismail, N. M., Haron, M. H., & Kuznetsov, D. (2014). Nanoscale drug delivery systems and the blood–brain barrier. *International Journal of Nanomedicine*, *795*. https://doi.org/10.2147/IJN.S52236

Khan, S., Bano, N., Ahamad, S., John, U., Dar, N. J., & Bhat, S. A. (2024). Excitotoxicity, oxytosis/ferroptosis, and neurodegeneration: emerging insights into mitochondrial mechanisms. *Aging and Disease*. https://doi.org/10.14336/AD.2024.0125-1

Khongkow, M., Yata, T., Boonrungsiman, S., Ruktanonchai, U. R., Graham, D., & Namdee, K. (2019). Surface modification of gold nanoparticles with neuron-targeted exosome for enhanced blood–brain barrier penetration. *Scientific Reports*, *9*(1), p. 8278. https://doi.org/10.1038/s41598-019-44569-6

Kircher, M. F., de la Zerda, A., Jokerst, J. V, Zavaleta, C. L., Kempen, P. J., Mittra, E., Pitter, K., Huang, R., Campos, C., Habte, F., Sinclair, R., Brennan, C. W., Mellinghoff, I. K., Holland, E. C., & Gambhir, S. S. (2012). A brain tumor molecular imaging strategy using a new triple-modality MRI-photoacoustic-Raman nanoparticle. *Nature Medicine*, *18*(5), pp. 829–834. https://doi.org/10.1038/nm.2721

Kong, J., Zou, R., Chu, R., Hu, N., Liu, J., Sun, Y., Ge, X., Mao, M., Yu, H., & Wang, Y. (2024). An Ultrasmall Cu/Cu2O nanoparticle-based diselenide-bridged nanoplatform mediating reactive oxygen species scavenging and neuronal membrane enhancement for targeted therapy of ischemic stroke. *ACS Nano*, *18*(5), pp. 4140–4158. https://doi.org/10.1021/acsnano.3c08734

Kooij, G., van Horssen, J., Bandaru, V. V. R., Haughey, N. J., & de Vries, H. E. (2012). The role of ATP-binding cassette transporters in neuro-inflammation: relevance for bioactive lipids. *Frontiers in Pharmacology*, *3*. https://doi.org/10.3389/fphar.2012.00074

Korttila, K., & Aromaa, U. (1980). Venous complications after intravenous injection of Diazepam, Flunitrazepam, Thiopentone and Etomidate. *Acta Anaesthesiologica Scandinavica*, *24*(3), pp. 227–230. https://doi.org/10.1111/j.1399-6576.1980.tb01540.x

Kostarelos, K. (2003). Rational design and engineering of delivery systems for therapeutics: biomedical exercises in colloid and surface science. *Advances in Colloid and Interface Science*, *106*(1–3), pp. 147–168. https://doi.org/10.1016/S0001-8686(03)00109-X

Krishna, R., Ali, M., & Moustafa, A. A. (2014). Effects of combined MAO-B inhibitors and levodopa vs. monotherapy in Parkinson™s disease. *Frontiers in Aging Neuroscience*, *6*. https://doi.org/10.3389/fnagi.2014.00180

Kumar, A., Khan, W. A., Ahamad, S., & Mohanan, K. (2022). Trifluorodiazoethane: a versatile building block to access trifluoromethylated heterocycles. *Journal of Heterocyclic Chemistry*, *59*(4), pp. 607–632. https://doi.org/10.1002/jhet.4416

Kumar, A., Zhang, X., & Liang, X.-J. (2013). Gold nanoparticles: emerging paradigm for targeted drug delivery system. *Biotechnology Advances*, *31*(5), pp. 593–606. https://doi.org/10.1016/j.biotechadv.2012.10.002

Kundu, P., Das, M., Tripathy, K., & Sahoo, S. K. (2016). Delivery of dual drug loaded lipid based nanoparticles across the blood–brain barrier impart enhanced neuroprotection in a rotenone induced mouse model of Parkinson's disease. *ACS Chemical Neuroscience*, *7*(12), pp. 1658–1670. https://doi.org/10.1021/acschemneuro.6b00207

Kurth, M. C., Adler, C. H., St. Hilaire, M., Singer, C., Waters, C., LeWitt, P., Chernik, D. A., Dorflinger, E. E., & Yoo, K. (1997). Tolcapone improves motor function and reduces levodopa requirement in patients with Parkinson's disease experiencing motor fluctuations. *Neurology*, *48*(1), pp. 81–87. https://doi.org/10.1212/WNL.48.1.81

Kwon, K.-C., & Daniell, H. (2016). Oral delivery of protein drugs bioencapsulated in plant cells. *Molecular Therapy*, *24*(8), pp. 1342–1350. https://doi.org/10.1038/mt.2016.115

Lajoie, J. M., & Shusta, E. V. (2015). Targeting receptor-mediated transport for delivery of biologics across the blood-brain barrier. *Annual Review of Pharmacology and Toxicology*, *55*(1), pp. 613–631. https://doi.org/10.1146/annurev-pharmtox-010814-124852

Lapin, N. A., Gill, K., Shah, B. R., & Chopra, R. (2020). Consistent opening of the blood brain barrier using focused ultrasound with constant intravenous infusion of microbubble agent. *Scientific Reports*, *10*(1), p. 16546. https://doi.org/10.1038/s41598-020-73312-9

Le, Q.-V., Lee, J., Lee, H., Shim, G., & Oh, Y.-K. (2021). Cell membrane-derived vesicles for delivery of therapeutic agents. *Acta Pharmaceutica Sinica B*, *11*(8), pp. 2096–2113. https://doi.org/10.1016/j.apsb.2021.01.020

Leal, J., Smyth, H. D. C., & Ghosh, D. (2017). Physicochemical properties of mucus and their impact on transmucosal drug delivery. *International Journal of Pharmaceutics*, *532*(1), pp. 555–572. https://doi.org/10.1016/j.ijpharm.2017.09.018

Leeson, P. D., & Davis, A. M. (2004). Time-related differences in the physical property profiles of oral drugs. *Journal of Medicinal Chemistry*, *47*(25), pp. 6338–6348. https://doi.org/10.1021/jm049717d

Lefèvre, G., Sędek, G., Jhee, S. S., Leibowitz, M. T., Huang, H.-L., Enz, A., Maton, S., Ereshefsky, L., Pommier, F., Schmidli, H., & Appel-Dingemanse, S. (2008). Pharmacokinetics and pharmacodynamics of the novel daily rivastigmine transdermal patch compared with twice-daily capsules in Alzheimer's disease patients. *Clinical Pharmacology & Therapeutics*, *83*(1), pp. 106–114. https://doi.org/10.1038/sj.clpt.6100242

Lei, L., Tu, Q., Zhang, X., Xiang, S., Xiao, B., Zhai, S., Yu, H., Tang, L., Guo, B., Chen, X., & Zhang, C. (2023). Stimulus-responsive curcumin-based polydopamine nanoparticles for targeting Parkinson's disease by modulating α-synuclein aggregation and reactive oxygen species. *Chemical Engineering Journal*, *461*, p. 141606. https://doi.org/10.1016/j.cej.2023.141606

Leinenga, G., Koh, W. K., & Götz, J. (2021). A comparative study of the effects of Aducanumab and scanning ultrasound on amyloid plaques and behavior in the APP23 mouse model of Alzheimer disease. *Alzheimer's Research & Therapy*, *13*(1), p. 76. https://doi.org/10.1186/s13195-021-00809-4

Levin, V. A., Patlak, C. S., & Landahl, H. D. (1980). Heuristic modeling of drug delivery to malignant brain tumors. *Journal of Pharmacokinetics and Biopharmaceutics*, *8*(3), p. 257–296. https://doi.org/10.1007/BF01059646

LeWitt, P. A. (2015). Levodopa therapy for Parkinson's disease: pharmacokinetics and pharmacodynamics. *Movement Disorders*, *30*(1), pp. 64–72. https://doi.org/10.1002/mds.26082

Li, C., Hou, X., Zhang, P., Li, J., Liu, X., Wang, Y., Guan, Q., & Zhou, Y. (2020). Exosome-based tumor therapy: opportunities and challenges. *Current Drug Metabolism*, *21*(5), pp. 339–351. https://doi.org/10.2174/1389200221666200515103354

Lipinski, C. A. (2000). Drug-like properties and the causes of poor solubility and poor permeability. *Journal of Pharmacological and Toxicological Methods*, *44*(1), pp. 235–249. https://doi.org/10.1016/S1056-8719(00)00107-6

Lips, J., de Haan, P., Bodewits, P., Vanicky, I., Dzoljic, M., Jacobs, M. J., & Kalkman, C. J. (2000). Neuroprotective effects of Riluzole and Ketamine during transient spinal cord Ischemia in the rabbit. *Anesthesiology*, *93*(5), pp. 1303–1311. https://doi.org/10.1097/00000542-200011000-00025

Litman, T. (2019). Personalized medicine—concepts, technologies, and applications in inflammatory skin diseases. *APMIS*, *127*(5), pp. 386–424. https://doi.org/10.1111/apm.12934

Liu, X.-G., Zhang, L., Lu, S., Liu, D.-Q., Zhang, L.-X., Yu, X.-L., & Liu, R.-T. (2020). Multifunctional superparamagnetic iron oxide nanoparticles conjugated with Aβ oligomer-specific scFv antibody and Class A scavenger receptor activator show early diagnostic potentials for Alzheimer's disease. *International Journal of Nanomedicine*, *15*, pp. 4919–4932. https://doi.org/10.2147/IJN.S240953

Liu, X., Lu, S., Liu, D., Zhang, L., Zhang, L., Yu, X., & Liu, R. (2019). ScFv-conjugated superparamagnetic iron oxide nanoparticles for MRI-based diagnosis in transgenic mouse models of Parkinson's and Huntington's diseases. *Brain Research*, 1707, pp. 141–153. https://doi.org/10.1016/j.brainres.2018.11.034

Liu, Y., An, S., Li, J., Kuang, Y., He, X., Guo, Y., Ma, H., Zhang, Y., Ji, B., & Jiang, C. (2016). Brain-targeted co-delivery of therapeutic gene and peptide by multifunctional nanoparticles in Alzheimer's disease mice. *Biomaterials*, *80*, pp. 33–45. https://doi.org/10.1016/j.biomaterials.2015.11.060

Liu, Z., Gao, X., Kang, T., Jiang, M., Miao, D., Gu, G., Hu, Q., Song, Q., Yao, L., Tu, Y., Chen, H., Jiang, X., & Chen, J. (2013a). B6 peptide-modified PEG-PLA nanoparticles for enhanced brain delivery of neuroprotective peptide. *Bioconjugate Chemistry*, *24*(6), pp. 997–1007. https://doi.org/10.1021/bc400055h

Liu, Z., Jiang, M., Kang, T., Miao, D., Gu, G., Song, Q., Yao, L., Hu, Q., Tu, Y., Pang, Z., Chen, H., Jiang, X., Gao, X., & Chen, J. (2013b). Lactoferrin-modified PEG-co-PCL nanoparticles for enhanced brain delivery of NAP peptide following intranasal administration. *Biomaterials*, *34*(15), pp. 3870–3881. https://doi.org/10.1016/j.biomaterials.2013.02.003

Long, C. M., & Bulte, J. W. (2009). In vivo tracking of cellular therapeutics using magnetic resonance imaging. *Expert Opinion on Biological Therapy*, *9*(3), pp. 293–306. https://doi.org/10.1517/14712590802715723

Loureiro, J. A., Gomes, B., Fricker, G., Coelho, M. A. N., Rocha, S., & Pereira, M. C. (2016). Cellular uptake of PLGA nanoparticles targeted with anti-amyloid and anti-transferrin receptor antibodies for Alzheimer's disease treatment. *Colloids and Surfaces B: Biointerfaces*, *145*, pp. 8–13. https://doi.org/10.1016/j.colsurfb.2016.04.041

Loureiro, J., Andrade, S., Duarte, A., Neves, A., Queiroz, J., Nunes, C., Sevin, E., Fenart, L., Gosselet, F., Coelho, M., & Pereira, M. (2017). Resveratrol and grape extract-loaded solid lipid nanoparticles for the treatment of Alzheimer's fisease. *Molecules*, *22*(2), p. 277. https://doi.org/10.3390/molecules22020277

Lu, C.-T., Zhao, Y.-Z., Wong, H. L., Cai, J., Peng, L., & Tian, X.-Q. (2014). Current approaches to enhance CNS delivery of drugs across the brain barriers. *International Journal of Nanomedicine*, *2241*. https://doi.org/10.2147/IJN.S61288

Lu, X.-Y., Wu, D.-C., Li, Z.-J., & Chen, G.-Q. (2011). In Polymer *N*anoparticles (pp. 299–323). https://doi.org/10.1016/B978-0-12-416020-0.00007-3

Luppi, B., Bigucci, F., Corace, G., Delucca, A., Cerchiara, T., Sorrenti, M., Catenacci, L., Di Pietra, A. M., & Zecchi, V. (2011). Albumin nanoparticles carrying cyclodextrins for nasal delivery of the anti-Alzheimer drug tacrine. *European Journal of Pharmaceutical Sciences*, *44*(4), pp. 559–565. https://doi.org/10.1016/j.ejps.2011.10.002

Ma, S., Wang, L., Huang, X., Wang, X., Chen, S., Shi, W., Qiao, X., Jiang, Y., Tang, L., Xu, Y., & Li, Y. (2018). Oral recombinant Lactobacillus vaccine targeting the intestinal microfold cells and dendritic cells for delivering the core neutralizing epitope of porcine epidemic diarrhea virus. *Microbial Cell Factories*, *17*(1), p. 20. https://doi.org/10.1186/s12934-018-0861-7

Madaan, K., Kumar, S., Poonia, N., Lather, V., & Pandita, D. (2014). Dendrimers in drug delivery and targeting: drug-dendrimer interactions and toxicity issues. *Journal of Pharmacy and Bioallied Sciences*, *6*(3), p. 139. https://doi.org/10.4103/0975-7406.130965

Maharjan, S., Singh, B., Jiang, T., Yoon, S.-Y., Li, H.-S., Kim, G., Gu, M. J., Kim, S. J., Park, O.-J., Han, S. H., Kang, S.-K., Yun, C.-H., Choi, Y.-J., & Cho, C.-S. (2016). Systemic administration of RANKL overcomes the bottleneck of oral vaccine delivery through microfold cells in ileum. *Biomaterials*, *84*, pp. 286–300. https://doi.org/10.1016/j.biomaterials.2016.01.043

Marino, A., Battaglini, M., Desii, A., Lavarello, C., Genchi, G., Petretto, A., & Ciofani, G. (2021). Liposomes loaded with polyphenol-rich grape pomace extracts protect from neurodegeneration in a rotenone-based in vitro model of Parkinson's disease. *Biomaterials Science*, *9*(24), pp. 8171–8188. https://doi.org/10.1039/D1BM01202A

Mastropietro, D., Park, K., & Omidian, H. (2017). 4.23 Polymers in oral drug delivery. In *Comprehensive Biomaterials II* (pp. 430–444). Elsevier. https://doi.org/10.1016/B978-0-12-803581-8.09291-2

Matthews, D. C., Mao, X., Dowd, K., Tsakanikas, D., Jiang, C. S., Meuser, C., Andrews, R. D., Lukic, A. S., Lee, J., Hampilos, N., Shafiian, N., Sano, M., David Mozley, P., Fillit, H., McEwen, B. S., Shungu, D. C., & Pereira, A. C. (2021). Riluzole, a glutamate modulator, slows cerebral glucose metabolism decline in patients with Alzheimer's disease. *Brain*, *144*(12), pp. 3742–3755. https://doi.org/10.1093/brain/awab222

Maxwell, M. J., & Wilson, M. J. A. (2006). Complications of blood transfusion. *Continuing Education in Anaesthesia Critical Care & Pain*, *6*(6), pp. 225–229. https://doi.org/10.1093/bjaceaccp/mkl053

Mehta, N. J., Marwah, P. K., & Njus, D. (2019). Are proteinopathy and oxidative stress two sides of the same coin? *Cells*, *8*(1), p. 59. https://doi.org/10.3390/cells8010059

Mendonça, M. C. P., Sun, Y., Cronin, M. F., Lindsay, A. J., Cryan, J. F., & O'Driscoll, C. M. (2023). Cyclodextrin-based nanoparticles for delivery of antisense oligonucleotides targeting Huntingtin. *Pharmaceutics*, *15*(2), p. 520. https://doi.org/10.3390/pharmaceutics15020520

Meng, Q., Wang, A., Hua, H., Jiang, Y., Wang, Y., Mu, H., Wu, Z., & Sun, K. (2018). Intranasal delivery of Huperzine A to the brain using lactoferrin-conjugated N-trimethylated chitosan surface-modified PLGA nanoparticles for treatment of Alzheimer's disease. *International Journal of Nanomedicine*, *13*, pp. 705–718. https://doi.org/10.2147/IJN.S151474

Mestre, T. A., & Ferreira, J. J. (2012). An evidence-based approach in the treatment of Huntington's disease. *Parkinsonism & Related Disorders*, *18*(4), pp. 316–320. https://doi.org/10.1016/j.parkreldis.2011.10.021

Mikitsh, J. L., & Chacko, A.-M. (2014). Pathways for small molecule delivery to the central nervous system across the blood-brain barrier. *Perspectives in Medicinal Chemistry*, *6*, p. PMC.S13384. https://doi.org/10.4137/PMC.S13384

Misra, A., Ganesh, S., Shahiwala, A., & Shah, S. P. (2003). Drug delivery to the central nervous system: a review. *Journal of Pharmacy & Pharmaceutical Sciences : A Publication of the Canadian Society for Pharmaceutical Sciences, Societe Canadienne Des Sciences Pharmaceutiques*, *6*(2), pp. 252–273.

Misra, A., & Kher, G. (2012). Drug delivery systems from nose to brain. *Current Pharmaceutical Biotechnology*, *13*(12), pp. 2355–2379. https://doi.org/10.2174/138920112803341752

Mistretta, M., Farini, A., Torrente, Y., & Villa, C. (2023). Multifaceted nanoparticles: emerging mechanisms and therapies in neurodegenerative diseases. *Brain*, *146*(6), pp. 2227–2240. https://doi.org/10.1093/brain/awad014

Mitchell, M. J., Billingsley, M. M., Haley, R. M., Wechsler, M. E., Peppas, N. A., & Langer, R. (2021). Engineering precision nanoparticles for drug delivery. *Nature Reviews Drug Discovery*, *20*(2), pp. 101–124. https://doi.org/10.1038/s41573-020-0090-8

Mogharbel, B. F., Cardoso, M. A., Irioda, A. C., Stricker, P. E. F., Slompo, R. C., Appel, J. M., de Oliveira, N. B., Perussolo, M. C., Saçaki, C. S., da Rosa, N. N., Dziedzic, D. S. M., Travelet, C., Halila, S., Borsali, R., & de Carvalho, K. A. T. (2022). Biodegradable nanoparticles loaded with levodopa and curcumin for treatment of Parkinson's disease. *Molecules*, *27*(9), p. 2811. https://doi.org/10.3390/molecules27092811

Monge-Fuentes, V., Biolchi Mayer, A., Lima, M. R., Geraldes, L. R., Zanotto, L. N., Moreira, K. G., Martins, O. P., Piva, H. L., Felipe, M. S. S., Amaral, A. C., Bocca, A. L., Tedesco, A. C., & Mortari, M. R. (2021). Dopamine-loaded nanoparticle systems circumvent the blood–brain barrier restoring motor function in mouse model for Parkinson's disease. *Scientific Reports*, *11*(1), p. 15185. https://doi.org/10.1038/s41598-021-94175-8

Mora, L., Chumbimuni-Torres, K. Y., Clawson, C., Hernandez, L., Zhang, L., & Wang, J. (2009). Real-time electrochemical monitoring of drug release from therapeutic nanoparticles. *Journal of Controlled Release*, *140*(1), pp. 69–73. https://doi.org/10.1016/j.jconrel.2009.08.002

Morishita, M., Takahashi, Y., Nishikawa, M., & Takakura, Y. (2017). Pharmacokinetics of exosomes—an important factor for elucidating the biological roles of exosomes and for the development of exosome-based therapeutics. *Journal of Pharmaceutical Sciences*, *106*(9), pp. 2265–2269. https://doi.org/10.1016/j.xphs.2017.02.030

Mourtas, S., Lazar, A. N., Markoutsa, E., Duyckaerts, C., & Antimisiaris, S. G. (2014). Multifunctional nanoliposomes with curcumin–lipid derivative and brain targeting functionality with potential applications for Alzheimer disease. *European Journal of Medicinal Chemistry*, *80*, pp. 175–183. https://doi.org/10.1016/j.ejmech.2014.04.050

Muolokwu, C. E., Chaulagain, B., Gothwal, A., Mahanta, A. K., Tagoe, B., Lamsal, B., & Singh, J. (2024). Functionalized nanoparticles to deliver nucleic acids to the brain for the treatment of Alzheimer's disease. *Frontiers in Pharmacology*, *15*. https://doi.org/10.3389/fphar.2024.1405423

Neilio, J. M., & Ginat, D. T. (2024). Emerging head and neck tumor targeting contrast agents for the purpose of CT, MRI, and multimodal diagnostic imaging: a molecular review. *Diagnostics*, *14*(15), p. 1666. https://doi.org/10.3390/diagnostics14151666

Neuwelt, A., Langsjoen, J., Byrd, T., Eberhardt, S., Mlady, G., Baca, J., Bocklage, T., Martinez, C., Orner, J., Rivera, M., & Sillerud, L. O. (2017). Ferumoxytol negatively enhances T 2 -weighted MRI of pedal osteomyelitis in vivo. *Journal of Magnetic Resonance Imaging*, *45*(4), pp. 1241–1245. https://doi.org/10.1002/jmri.25437

Neuwelt, A., Sidhu, N., Hu, C.-A. A., Mlady, G., Eberhardt, S. C., & Sillerud, L. O. (2015). Iron-based superparamagnetic nanoparticle contrast agents for MRI of infection and inflammation. *American Journal of Roentgenology*, *204*(3), pp. W302–W313. https://doi.org/10.2214/AJR.14.12733

Nisha, R., Kumar, P., Mishra, N., Maurya, P., Ahmad, S., Singh, N., & Saraf, S. A. (2024). Appraisal of folate functionalized bosutinib cubosomes against hepatic cancer cells: in-vitro, in-silico, and in-vivo pharmacokinetic study. *International Journal of Pharmaceutics*, *654*, p. 123975. https://doi.org/10.1016/j.ijpharm.2024.123975

Nixon, R. A., & Rubinsztein, D. C. (2024). Mechanisms of autophagy–lysosome dysfunction in neurodegenerative diseases. *Nature Reviews Molecular Cell Biology*. https://doi.org/10.1038/s41580-024-00757-5

Nunes, D., Loureiro, J. A., & Pereira, M. C. (2022). Drug delivery systems as a strategy to improve the efficacy of FDA-approved Alzheimer's drugs. *Pharmaceutics*, *14*(11), p. 2296. https://doi.org/10.3390/pharmaceutics14112296

O'Brown, N. M., Pfau, S. J., & Gu, C. (2018). Bridging barriers: a comparative look at the blood–brain barrier across organisms. *Genes & Development*, *32*(7–8), pp. 466–478. https://doi.org/10.1101/gad.309823.117

Oller-Salvia, B., Sánchez-Navarro, M., Giralt, E., & Teixidó, M. (2016). Blood–brain barrier shuttle peptides: an emerging paradigm for brain delivery. *Chemical Society Reviews*, *45*(17), pp. 4690–4707. https://doi.org/10.1039/C6CS00076B

Olufunmilayo, E. O., Gerke-Duncan, M. B., & Holsinger, R. M. D. (2023). Oxidative stress and antioxidants in neurodegenerative disorders. *Antioxidants*, *12*(2), p. 517. https://doi.org/10.3390/antiox12020517

Pajouhesh, H., & Lenz, G. R. (2005). Medicinal chemical properties of successful central nervous system drugs. *NeuroRX*, *2*(4), pp. 541–553. https://doi.org/10.1602/neurorx.2.4.541

Paliwal, R., Paliwal, S. R., Kenwat, R., Kurmi, B. Das, & Sahu, M. K. (2020). Solid lipid nanoparticles: a review on recent perspectives and patents. *Expert Opinion on Therapeutic Patents*, *30*(3), pp. 179–194. https://doi.org/10.1080/13543776.2020.1720649

Palmer, A. M. (2011). The role of the blood brain barrier in neurodegenerative disorders and their treatment. *Journal of Alzheimer's Disease*, *24*(4), pp. 643–656. https://doi.org/10.3233/JAD-2011-110368

Pandit, R., Chen, L., & Götz, J. (2020). The blood-brain barrier: physiology and strategies for drug delivery. *Advanced Drug Delivery Reviews*, *165–166*, pp. 1–14. https://doi.org/10.1016/j.addr.2019.11.009

Pangalos, M. N., Schechter, L. E., & Hurko, O. (2007). Drug development for CNS disorders: strategies for balancing risk and reducing attrition. *Nature Reviews Drug Discovery*, *6*(7), pp. 521–532. https://doi.org/10.1038/nrd2094

Pardridge, W. M. (2005). The blood-brain barrier: bottleneck in brain drug development. *NeuroRX*, *2*(1), pp. 3–14. https://doi.org/10.1602/neurorx.2.1.3

Pardridge, W. M. (2012). Drug transport across the blood–brain barrier. *Journal of Cerebral Blood Flow & Metabolism*, *32*(11), pp. 1959–1972. https://doi.org/10.1038/jcbfm.2012.126

Pardridge, W. M. (2022). A Historical review of brain drug delivery. *Pharmaceutics*, *14*(6), p. 1283. https://doi.org/10.3390/pharmaceutics14061283

Pastorino, F., Brignole, C., Di Paolo, D., Perri, P., Curnis, F., Corti, A., & Ponzoni, M. (2019). Overcoming biological barriers in neuroblastoma therapy: the vascular targeting approach with liposomal drug nanocarriers. *Small*, *15*(10), p. 1804591. https://doi.org/10.1002/smll.201804591

Poudel, P., & Park, S. (2022). Recent advances in the treatment of Alzheimer's disease using nanoparticle-based drug delivery systems. *Pharmaceutics*, *14*(4), p. 835. https://doi.org/10.3390/pharmaceutics14040835

Qi, H.-P., Gao, X.-C., Zhang, L.-Q., Wei, S.-Q., Bi, S., Yang, Z.-C., & Cui, H. (2013). In vitro evaluation of enhancing effect of borneol on transcorneal permeation of compounds with different hydrophilicities and molecular sizes. *European Journal of Pharmacology*, *705*(1–3), pp. 20–25. https://doi.org/10.1016/j.ejphar.2013.02.031

Qian, K., Bao, X., Li, Y., Wang, P., Guo, Q., Yang, P., Xu, S., Yu, F., Meng, R., Cheng, Y., Sheng, D., Cao, J., Xu, M., Wu, J., Wang, T., Wang, Y., Xie, Q., Lu, W., & Zhang, Q. (2022). Cholinergic neuron targeting nanosystem delivering hybrid peptide for combinatorial mitochondrial therapy in Alzheimer's disease. *ACS Nano*, *16*(7), pp. 11455–11472. https://doi.org/10.1021/acsnano.2c05795

Ranade, V. V. (1991). Drug delivery systems. 6. Transdermal drug delivery. *The Journal of Clinical Pharmacology*, *31*(5), pp. 401–418. https://doi.org/10.1002/j.1552-4604.1991.tb01895.x

Ravindran, A., Chandran, P., & Khan, S. S. (2013). Biofunctionalized silver nanoparticles: advances and prospects. *Colloids and Surfaces B: Biointerfaces*, *105*, pp. 342–352. https://doi.org/10.1016/j.colsurfb.2012.07.036

Ren, J., Zou, M., Gao, P., Wang, Y., & Cheng, G. (2013). Tissue distribution of borneol-modified ganciclovir-loaded solid lipid nanoparticles in mice after intravenous administration. *European Journal of Pharmaceutics and Biopharmaceutics*, *83*(2), pp. 141–148. https://doi.org/10.1016/j.ejpb.2012.10.018

Rivera, J., Gangwani, L., & Kumar, S. (2023). Mitochondria localized microRNAs: an unexplored miRNA niche in Alzheimer's disease and aging. *Cells*, *12*(5), p. 742. https://doi.org/10.3390/cells12050742

Roca-Agujetas, V., Barbero-Camps, E., de Dios, C., Podlesniy, P., Abadin, X., Morales, A., Marí, M., Trullàs, R., & Colell, A. (2021). Cholesterol alters mitophagy by impairing optineurin recruitment and lysosomal clearance in Alzheimer's disease. *Molecular Neurodegeneration*, *16*(1), p. 15. https://doi.org/10.1186/s13024-021-00435-6

Salat, D., & Tolosa, E. (2013). Levodopa in the treatment of Parkinson's disease: current status and new developments. *Journal of Parkinson's Disease*, *3*(3), pp. 255–269. https://doi.org/10.3233/JPD-130186

Sanz-Blasco, S., Valero, R. A., Rodríguez-Crespo, I., Villalobos, C., & Núñez, L. (2008). Mitochondrial Ca2+ overload underlies Aβ oligomers neurotoxicity providing an unexpected mechanism of neuroprotection by NSAIDs. *PLoS One*, *3*(7), p. e2718. https://doi.org/10.1371/journal.pone.0002718

Sarfraz, N., & Khan, I. (2021). Plasmonic gold nanoparticles (AuNPs): properties, synthesis and their advanced energy, environmental and biomedical applications. *Chemistry – An Asian Journal*, *16*(7), pp. 720–742. https://doi.org/10.1002/asia.202001202

Sarkar, A. K., Kaur, G., Seth, P., Jana, N. R., & Jana, N. R. (2024). Autophagy-induced nanoparticle-based clearing of toxic Huntingtin protein aggregates from neuron cells. *ACS Applied Nano Materials*, *7*(3), pp. 3468–3478. https://doi.org/10.1021/acsanm.3c06292

Sarko, D. K., & McKinney, C. E. (2017). Exosomes: origins and therapeutic potential for neurodegenerative disease. *Frontiers in Neuroscience*, *11*. https://doi.org/10.3389/fnins.2017.00082

Savjani, K. T., Gajjar, A. K., & Savjani, J. K. (2012). Drug solubility: importance and enhancement techniques. *ISRN Pharmaceutics*, 2012, pp. 1–10. https://doi.org/10.5402/2012/195727

Selvaraj, K., Gowthamarajan, K., & Karri, V. V. S. R. (2017). Nose to brain transport pathways an overview: potential of nanostructured lipid carriers in nose to brain targeting. *Artificial Cells, Nanomedicine, and Biotechnology*, pp. 1–8. https://doi.org/10.1080/21691401.2017.1420073

Shah, S., Dhawan, V., Holm, R., Nagarsenker, M. S., & Perrie, Y. (2020). Liposomes: advancements and innovation in the manufacturing process. *Advanced Drug Delivery Reviews*, *154–155*, pp. 102–122. https://doi.org/10.1016/j.addr.2020.07.002

Shahnawaz, M., Mukherjee, A., Pritzkow, S., Mendez, N., Rabadia, P., Liu, X., Hu, B., Schmeichel, A., Singer, W., Wu, G., Tsai, A.-L., Shirani, H., Nilsson, K. P. R., Low, P. A., & Soto, C. (2020). Discriminating α-synuclein strains in Parkinson's disease and multiple system atrophy. *Nature*, *578*(7794), pp. 273–277. https://doi.org/10.1038/s41586-020-1984-7

Shi, J., Yang, Y., Yin, N., Liu, C., Zhao, Y., Cheng, H., Zhou, T., Zhang, Z., & Zhang, K. (2022). Engineering CXCL12 biomimetic decoy-integrated versatile immunosuppressive nanoparticle for ischemic stroke therapy with management of overactivated brain immune microenvironment. *Small Methods*, *6*(1). https://doi.org/10.1002/smtd.202101158

Siddiqui, M. A. A., & Plosker, G. L. (2005). Rasagiline. *Drugs & Aging*, *22*(1), pp. 83–91. https://doi.org/10.2165/00002512-200522010-00006

Silva, S., Almeida, A., & Vale, N. (2021). Importance of nanoparticles for the delivery of antiparkinsonian drugs. *Pharmaceutics*, *13*(4), p. 508. https://doi.org/10.3390/pharmaceutics13040508

Simpson, D. S. A., & Oliver, P. L. (2020). ROS generation in microglia: understanding oxidative stress and inflammation in neurodegenerative disease. *Antioxidants*, *9*(8), p. 743. https://doi.org/10.3390/antiox9080743

Sindhu, P., Kumar, S., Iqbal, B., Ali, J., & Baboota, S. (2018). Duloxetine loaded-microemulsion system to improve behavioral activities by upregulating serotonin and norepinephrine in brain for the treatment of depression. *Journal of Psychiatric Research*, *99*, pp. 83–95. https://doi.org/10.1016/j.jpsychires.2018.01.015

Sintov, A. C., Velasco-Aguirre, C., Gallardo-Toledo, E., Araya, E., & Kogan, M. J. (2016). Metal nanoparticles as targeted carriers circumventing the blood–brain barrier (pp. 199–227). https://doi.org/10.1016/bs.irn.2016.06.007

Smriti, Singla, M., Gupta, S., Porwal, O., Nasser Binjawhar, D., Sayed, A. A., Mittal, P., El-Demerdash, F. M., Algahtani, M., Singh, S. K., Dua, K., Gupta, G., Bawa, P., Altyar, A. E., & Abdel-Daim, M. M. (2023). Theoretical design for covering Engeletin with functionalized nanostructure-lipid carriers as neuroprotective agents against Huntington's disease via the nasal-brain route. *Frontiers in Pharmacology*, *14*. https://doi.org/10.3389/fphar.2023.1218625

Solár, P., Zamani, A., Kubíčková, L., Dubový, P., & Joukal, M. (2020). Choroid plexus and the blood–cerebrospinal fluid barrier in disease. *Fluids and Barriers of the CNS*, *17*(1), p. 35. https://doi.org/10.1186/s12987-020-00196-2

Song, H., Li, H., Guo, S., Pan, Y., Fu, Y., Zhou, Z., Li, Z., Wen, X., Sun, X., He, B., Gu, H., Zhao, Q., Wang, C., An, P., Luo, S., Hu, Y., Xie, X., & Lu, B. (2018). Targeting Gpr52 lowers mutant HTT levels and rescues Huntington's disease-associated phenotypes. *Brain*, *141*(6), pp. 1782–1798. https://doi.org/10.1093/brain/awy081

Stohs, S. J., Chen, O., Ray, S. D., Ji, J., Bucci, L. R., & Preuss, H. G. (2020). Highly bioavailable forms of curcumin and promising avenues for curcumin-based research and application: a review. *Molecules*, *25*(6), p. 1397. https://doi.org/10.3390/molecules25061397

Strafella, C., Caputo, V., Galota, M. R., Zampatti, S., Marella, G., Mauriello, S., Cascella, R., & Giardina, E. (2018). Application of precision medicine in neurodegenerative diseases. *Frontiers in Neurology*, *9*. https://doi.org/10.3389/fneur.2018.00701

Stueber, D. D., Villanova, J., Aponte, I., Xiao, Z., & Colvin, V. L. (2021). Magnetic nanoparticles in biology and medicine: past, present, and future trends. *Pharmaceutics*, *13*(7), p. 943. https://doi.org/10.3390/pharmaceutics13070943

Suzuki, Y. A., Lopez, V., & Lönnerdal, B. (2005). Lactoferrin. *Cellular and Molecular Life Sciences*, *62*(22), pp. 2560–2575. https://doi.org/10.1007/s00018-005-5371-1

Swanson, C. J., Zhang, Y., Dhadda, S., Wang, J., Kaplow, J., Lai, R. Y. K., Lannfelt, L., Bradley, H., Rabe, M., Koyama, A., Reyderman, L., Berry, D. A., Berry, S., Gordon, R., Kramer, L. D., & Cummings, J. L. (2021). A randomized, double-blind, phase 2b proof-of-concept clinical trial in early Alzheimer's disease with lecanemab, an anti-Aβ protofibril antibody. *Alzheimer's Research & Therapy*, *13*(1), p. 80. https://doi.org/10.1186/s13195-021-00813-8

Tabish, T. A., Zhu, Y., Shukla, S., Kadian, S., Sangha, G. S., Lygate, C. A., & Narayan, R. J. (2023). Graphene nanocomposites for real-time electrochemical sensing of nitric oxide in biological systems. *Applied Physics Reviews*, *10*(4). https://doi.org/10.1063/5.0162640

Taebnia, N., Morshedi, D., Yaghmaei, S., Aliakbari, F., Rahimi, F., & Arpanaei, A. (2016). Curcumin-loaded amine-functionalized mesoporous silica nanoparticles inhibit α-synuclein fibrillation and reduce its cytotoxicity-associated effects. *Langmuir*, *32*(50), pp. 13394–13402. https://doi.org/10.1021/acs.langmuir.6b02935

Tafoya, M. A., Madi, S., & Sillerud, L. O. (2017). Superparamagnetic nanoparticle-enhanced MRI of Alzheimer's disease plaques and activated microglia in 3X transgenic mouse brains: contrast optimization. *Journal of Magnetic Resonance Imaging*, *46*(2), pp. 574–588. https://doi.org/10.1002/jmri.25563

Tajik, S., Dourandish, Z., Jahani, P. M., Sheikhshoaie, I., Beitollahi, H., Shahedi Asl, M., Jang, H. W., & Shokouhimehr, M. (2021). Recent developments in voltammetric and amperometric sensors for cysteine detection. *RSC Advances*, *11*(10), pp. 5411–5425. https://doi.org/10.1039/D0RA07614G

Taléns-Visconti, R., de Julián-Ortiz, J. V., Vila-Busó, O., Diez-Sales, O., & Nácher, A. (2023). Intranasal drug administration in Alzheimer-type dementia: towards clinical applications. *Pharmaceutics*, *15*(5), p. 1399. https://doi.org/10.3390/pharmaceutics15051399

Tan, Y.-Y., Jenner, P., & Chen, S.-D. (2022). Monoamine oxidase-B inhibitors for the treatment of Parkinson's disease: past, present, and future. *Journal of Parkinson's Disease*, *12*(2), pp. 477–493. https://doi.org/10.3233/JPD-212976

Tang, J., Xiong, L., Zhou, G., Wang, S., Wang, J., Liu, L., Li, J., Yuan, F., Lu, S., Wan, Z., Chou, L., & Xi, T. (2010). Silver nanoparticles crossing through and distribution in the blood-brain barrier in vitro. *Journal of Nanoscience and Nanotechnology*, *10*(10), pp. 6313–6317. https://doi.org/10.1166/jnn.2010.2625

Tang, S.-J., Zhang, M., Sun, J., Meng, J.-W., Xiong, X., Gong, Q., Jin, D., Yang, Q.-F., & Xiao, Y.-F. (2023). Single-particle photoacoustic vibrational spectroscopy using optical microresonators. *Nature Photonics*, *17*(11), pp. 951–956. https://doi.org/10.1038/s41566-023-01264-3

Tang, S., Wang, A., Yan, X., Chu, L., Yang, X., Song, Y., Sun, K., Yu, X., Liu, R., Wu, Z., & Xue, P. (2019). Brain-targeted intranasal delivery of dopamine with borneol and lactoferrin co-modified nanoparticles for treating Parkinson's disease. *Drug Delivery*, *26*(1), pp. 700–707. https://doi.org/10.1080/10717544.2019.1636420

Tanifum, E. A., Ghaghada, K., Vollert, C., Head, E., Eriksen, J. L., & Annapragada, A. (2016). A novel liposomal nanoparticle for the imaging of amyloid plaque by magnetic resonance imaging. *Journal of Alzheimer's Disease*, *52*(2), pp. 731–745. https://doi.org/10.3233/JAD-151124

Tao, X., Mao, S., Zhang, Q., Yu, H., Li, Y., He, X., Yang, S., Zhang, Z., Yi, Z., Song, Y., & Feng, X. (2021). Brain-targeted polysorbate 80-emulsified donepezil drug-loaded nanoparticles for neuroprotection. *Nanoscale Research Letters*, *16*(1), p. 132. https://doi.org/10.1186/s11671-021-03584-1

Tarakad, A., & Jankovic, J. (2020). Recent advances in understanding and treatment of Parkinson's disease. *Faculty Reviews*, *9*. https://doi.org/10.12703/b/9-6

Thibaut, F., Faucheux, B. A., Marquez, J., Villares, J., Menard, J. F., Agid, Y., & Hirsch, E. C. (1995). Regional distribution of monoamine vesicular uptake sites in the mesencephalon of control subjects and patients with Parkinson's disease: a postmortem study using tritiated tetrabenazine. *Brain Research*, *692*(1–2), pp. 233–243. https://doi.org/10.1016/0006-8993(95)00674-F

Tonda-Turo, C., Origlia, N., Mattu, C., Accorroni, A., & Chiono, V. (2019). Current Limitations in the treatment of Parkinson's and Alzheimer's diseases: state-of-the-art and future perspective of polymeric carriers. *Current Medicinal Chemistry*, *25*(41), pp. 5755–5771. https://doi.org/10.2174/0929867325666180221125759

Tsai, M.-J., Huang, Y.-B., Wu, P.-C., Fu, Y.-S., Kao, Y.-R., Fang, J.-Y., & Tsai, Y.-H. (2011). Oral apomorphine delivery from solid lipid nanoparticles with different monostearate emulsifiers: pharmacokinetic and behavioral evaluations. *Journal of Pharmaceutical Sciences*, *100*(2), pp. 547–557. https://doi.org/10.1002/jps.22285

Ueno, M., Chiba, Y., Murakami, R., Matsumoto, K., Kawauchi, M., & Fujihara, R. (2016). Blood–brain barrier and blood–cerebrospinal fluid barrier in normal and pathological conditions. *Brain Tumor Pathology*, *33*(2), pp. 89–96. https://doi.org/10.1007/s10014-016-0255-7

Valis, M., Masopust, J., Vysata, O., Hort, J., Dolezal, R., Tomek, J., Misik, J., Kuca, K., & Karasova, J. Z. (2017). Concentration of donepezil in the cerebrospinal fluid of AD patients: evaluation of dosage sufficiency in standard treatment strategy. *Neurotoxicity Research*, *31*(1), pp. 162–168. https://doi.org/10.1007/s12640-016-9672-y

Vaz, V. M., & Kumar, L. (2021). 3D printing as a promising tool in personalized medicine. *AAPS PharmSciTech*, *22*(1), p. 49. https://doi.org/10.1208/s12249-020-01905-8

Venkataraman, A., Amadi, E. V., Chen, Y., & Papadopoulos, C. (2019). Carbon nanotube assembly and integration for applications. *Nanoscale Research Letters*, *14*(1), p. 220. https://doi.org/10.1186/s11671-019-3046-3

Verma, S. K., Arora, I., Javed, K., Akhtar, M., & Samim, M. (2016). Enhancement in the neuroprotective power of riluzole against cerebral ischemia using a brain targeted drug delivery vehicle. *ACS Applied Materials & Interfaces*, *8*(30), pp. 19716–19723. https://doi.org/10.1021/acsami.6b01776

Vimal, S. K., Zuo, H., Wang, Z., Wang, H., Long, Z., & Bhattacharyya, S. (2020). Self-therapeutic nanoparticle that alters tau protein and ameliorates tauopathy toward a functional nanomedicine to tackle Alzheimer's. *Small*, *16*(16). https://doi.org/10.1002/smll.201906861

Vio, V., Marchant, M. J., Araya, E., & Kogan, M. K. (2017). Metal nanoparticles for the treatment and diagnosis of neurodegenerative brain diseases. *Current Pharmaceutical Design*, *23*(13), pp. 1916–1926. https://doi.org/10.2174/1381612823666170105152948

Vyas, T. K., Babbar, A. K., Sharma, R. K., & Misra, A. (2005). Intranasal mucoadhesive microemulsions of zolmitriptan: preliminary studies on brain-targeting. *Journal of Drug Targeting*, *13*(5), pp. 317–324. https://doi.org/10.1080/10611860500246217

Vyas, T., Shahiwala, A., Marathe, S., & Misra, A. (2005). Intranasal drug delivery for brain targeting. *Current Drug Delivery*, *2*(2), pp. 165–175. https://doi.org/10.2174/1567201053586047

Wagner, V., Dullaart, A., Bock, A.-K., & Zweck, A. (2006). The emerging nanomedicine landscape. *Nature Biotechnology*, *24*(10), pp. 1211–1217. https://doi.org/10.1038/nbt1006-1211

Wahl, F., Allix, M., Plotkine, M., & Boulu, R. G. (1993). Effect of riluzole on focal cerebral ischemia in rats. *European Journal of Pharmacology*, *230*(2), pp. 209–214. https://doi.org/10.1016/0014-2999(93)90804-Q

Wang, J. (2002). Real-time electrochemical monitoring: toward green analytical chemistry. *Accounts of Chemical Research*, *35*(9), pp. 811–816. https://doi.org/10.1021/ar010066e

Wang, L. V, & Yao, J. (2016). A practical guide to photoacoustic tomography in the life sciences. *Nature Methods*, *13*(8), pp. 627–638. https://doi.org/10.1038/nmeth.3925

Wen, M. M., El-Salamouni, N. S., El-Refaie, W. M., Hazzah, H. A., Ali, M. M., Tosi, G., Farid, R. M., Blanco-Prieto, M. J., Billa, N., & Hanafy, A. S. (2017). Nanotechnology-based drug delivery systems for Alzheimer's disease management: technical, industrial, and clinical challenges. *Journal of Controlled Release*, *245*, pp. 95–107. https://doi.org/10.1016/j.jconrel.2016.11.025

Wilcock, G. K., Black, S. E., Balch, A. H., Amato, D. A., Beelen, A. P., Schneider, L. S., Green, R. C., Swabb, E. A., & Zavitz, K. H. (2009). O1-04-07: safety and efficacy of tarenflurbil in subjects with mild Alzheimer's disease: results from an 18-month international multi-center phase 3 trial. *Alzheimer's & Dementia*, *5*(4S_Part_3). https://doi.org/10.1016/j.jalz.2009.05.220

Wilcock, G. K., Black, S. E., Hendrix, S. B., Zavitz, K. H., Swabb, E. A., & Laughlin, M. A. (2008). Efficacy and safety of tarenflurbil in mild to moderate Alzheimer's disease: a randomised phase II trial. *The Lancet Neurology*, *7*(6), pp. 483–493. https://doi.org/10.1016/S1474-4422(08)70090-5

Wilson, B., Samanta, M. K., Santhi, K., Kumar, K. P. S., Paramakrishnan, N., & Suresh, B. (2008). Poly(n-butylcyanoacrylate) nanoparticles coated with polysorbate 80 for the targeted delivery of rivastigmine into the brain to treat Alzheimer's disease. *Brain Research*, 1200, pp. 159–168. https://doi.org/10.1016/j.brainres.2008.01.039

Winblad, B., & Machado, J. C. (2008). Use of rivastigmine transdermal patch in the treatment of Alzheimer's disease. *Expert Opinion on Drug Delivery*, *5*(12), pp. 1377–1386. https://doi.org/10.1517/17425240802542690

Wisniewski, T., & Sadowski, M. (2008). Preventing β-amyloid fibrillization and deposition: β-sheet breakers and pathological chaperone inhibitors. *BMC Neuroscience*, *9*(S2), p. S5. https://doi.org/10.1186/1471-2202-9-S2-S5

Woitalla, D., Kassubek, J., Timmermann, L., Lauterbach, T., Berkels, R., Grieger, F., & Müller, T. (2015). Reduction of gastrointestinal symptoms in Parkinson's disease after a switch from oral therapy to rotigotine transdermal patch: a non-interventional prospective multicenter trial. *Parkinsonism & Related Disorders*, *21*(3), pp. 199–204. https://doi.org/10.1016/j.parkreldis.2014.11.024

Won, S. J., Fong, R., Butler, N., Sanchez, J., Zhang, Y., Wong, C., Tambou Nzoutchoum, O., Huynh, A., Pan, J., & Swanson, R. A. (2022). Neuronal oxidative stress promotes α-synuclein aggregation in vivo. *Antioxidants*, *11*(12), p. 2466. https://doi.org/10.3390/antiox11122466

Wong, L. R., & Ho, P. C. (2017). Role of serum albumin as a nanoparticulate carrier for nose-to-brain delivery of R-flurbiprofen: implications for the treatment of Alzheimer's disease. *Journal of Pharmacy and Pharmacology*, *70*(1), pp. 59–69. https://doi.org/10.1111/jphp.12836

Wu, D., Chen, Q., Chen, X., Han, F., Chen, Z., & Wang, Y. (2023). The blood–brain barrier: structure, regulation, and drug delivery. *Signal Transduction and Targeted Therapy*, *8*(1), p. 217. https://doi.org/10.1038/s41392-023-01481-w

Wu, K., Su, D., Liu, J., Saha, R., & Wang, J.-P. (2019). Magnetic nanoparticles in nanomedicine: a review of recent advances. *Nanotechnology*, *30*(50), p. 502003. https://doi.org/10.1088/1361-6528/ab4241

Yadav, A., Sunkaria, A., Singhal, N., & Sandhir, R. (2018). Resveratrol loaded solid lipid nanoparticles attenuate mitochondrial oxidative stress in vascular dementia by activating Nrf2/HO-1 pathway. *Neurochemistry International*, *112*, pp. 239–254. https://doi.org/10.1016/j.neuint.2017.08.001

Yan, Z., Xiao, P., Ji, P., Su, R., Ren, Z., Xu, L., Qiu, X., & Li, D. (2024). Enhanced breast cancer therapy using multifunctional lipid-coated nanoparticles combining curcumin chemotherapy and nitric oxide gas delivery. *Scientific Reports*, *14*(1), p. 18107. https://doi.org/10.1038/s41598-024-69229-2

Yang, H., Mu, W., Wei, D., Zhang, Y., Duan, Y., Gao, J., Gong, X., Wang, H., Wu, X., Tao, H., & Chang, J. (2020). A novel targeted and high-efficiency nanosystem for combinational therapy for Alzheimer's disease. *Advanced Science*, *7*(19). https://doi.org/10.1002/advs.201902906

Yang, Z., Zhang, Y., Yang, Y., Sun, L., Han, D., Li, H., & Wang, C. (2010). Pharmacological and toxicological target organelles and safe use of single-walled carbon nanotubes as drug carriers in treating Alzheimer disease. *Nanomedicine: Nanotechnology, Biology and Medicine*, *6*(3), pp. 427–441. https://doi.org/10.1016/j.nano.2009.11.007

Yim, Y. S., Choi, J., Kim, G. T., Kim, C. H., Shin, T.-H., Kim, D. G., & Cheon, J. (2012). A facile approach for the delivery of inorganic nanoparticles into the brain by passing through the blood–brain barrier (BBB). *Chemical Communications*, *48*(1), pp. 61–63. https://doi.org/10.1039/C1CC15113D

Yu, D., Zarate, N., White, A., Coates, D., Tsai, W., Nanclares, C., Cuccu, F., Yue, J. S., Brown, T. G., Mansky, R. H., Jiang, K., Kim, H., Nichols-Meade, T., Larson, S. N., Gundry, K., Zhang, Y., Tomas-Zapico, C., Lucas, J. J., Benneyworth, M., … Gomez-Pastor, R. (2022). CK2 alpha prime and alpha-synuclein pathogenic functional interaction mediates synaptic dysregulation in Huntington's disease. *Acta Neuropathologica Communications*, *10*(1), p. 83. https://doi.org/10.1186/s40478-022-01379-8

Zhang, C., Zheng, X., Wan, X., Shao, X., Liu, Q., Zhang, Z., & Zhang, Q. (2014). The potential use of H102 peptide-loaded dual-functional nanoparticles in the treatment of Alzheimer's disease. *Journal of Controlled Release*, *192*, pp. 317–324. https://doi.org/10.1016/j.jconrel.2014.07.050

Zhang, L., Gu, F., Chan, J., Wang, A., Langer, R., & Farokhzad, O. (2008). Nanoparticles in medicine: therapeutic applications and developments. *Clinical Pharmacology & Therapeutics*, *83*(5), pp. 761–769. https://doi.org/10.1038/sj.clpt.6100400

Zheng, X., Zhang, C., Guo, Q., Wan, X., Shao, X., Liu, Q., & Zhang, Q. (2017). Dual-functional nanoparticles for precise drug delivery to Alzheimer's disease lesions: targeting mechanisms, pharmacodynamics and safety. *International Journal of Pharmaceutics*, *525*(1), pp. 237–248. https://doi.org/10.1016/j.ijpharm.2017.04.033

Zhi, Z., Liu, R., Wang, W., Dewettinck, K., & Van Bockstaele, F. (2022). Recent progress in oil-in-water-in-oil (O/W/O) double emulsions. *Critical Reviews in Food Science and Nutrition*, pp. 1–12. https://doi.org/10.1080/10408398.2022.2029346

Zhong, L., Xu, Y., Zhuo, R., Wang, T., Wang, K., Huang, R., Wang, D., Gao, Y., Zhu, Y., Sheng, X., Chen, K., Wang, N., Zhu, L., Can, D., Marten, Y., Shinohara, M., Liu, C.-C., Du, D., Sun, H., … Chen, X.-F. (2019). Soluble TREM2 ameliorates pathological phenotypes by modulating microglial functions in an Alzheimer's disease model. *Nature Communications*, *10*(1), pp. 1365. https://doi.org/10.1038/s41467-019-09118-9

Zhu, L., Xu, L., Wu, X., Deng, F., Ma, R., Liu, Y., Huang, F., & Shi, L. (2021). Tau-targeted multifunctional nanoinhibitor for Alzheimer's Disease. *ACS Applied Materials & Interfaces*, *13*(20), pp. 23328–23338. https://doi.org/10.1021/acsami.1c00257

11 An Enigmatic Role of Small Molecule Therapeutics in Neurodegeneration

Rukhsana Hassan, Feroze Ahmad Mir, Umeair Kazime, Aarizo Tabasum, Sajad Ahmad Bhat, Mir Mohammad Hussain, Mushtaq Ahmad, Gh Nabi Bhat, and Gh Rasool Bhat

ABBREVIATIONS

AchEI	Acetylcholine Esterase Inhibitor
AD	Alzheimer's Disease
ADC	Antibody–Drug Conjugate
AI	Artificial Intelligence
ALS	Amyotrophic Lateral Sclerosis
BBB	Blood–Brain Barrier
CHMP	Committee for Medicinal Products for Human Use
CNS	Central Nervous System
DA	Dopamine Agonists
DLB	Dementia with Lewy Body
EMEA	Evaluation of Medical Products
FDA	Food and Drug Administration
FTD	Frontotemporal Dementia
HD	Huntington's Disease
IV	Intravenous
LBD	Lewy Body Dementia
LD/CD	Levodopa/Carbidopa
L-Dopa	Levodopa
MAA	Marketing Authorization Application
MAO	Monoamine oxidase
MAPT	Microtubule-Associated Protein Tau
MCI	Mild Cognitive Impairment
MEM	Memantine
MMSE	Mini Mental State Examination
MRS	Magnetic Resonance Spectroscopy
NAA	N-Acetyl Aspartate
NDD	Neurodegenerative Disease
NMDA	N-Methyl D-Aspartate
NMDAR	N-Methyl D-Aspartate Receptor
PD	Parkinson's Disease
RSM	RNA-targeting Small Molecules

DOI: 10.1201/9781003520610-11

SMA	Spinal Muscular Atrophy
SOD1	Superoxide Dismutase 1
UPDRS	Unified Parkinson's Disease Rating Scales

11.1 INTRODUCTION

Neurodegenerative disorders (NDDs) typically present a progressive clinical course and are distinguished by specific regional neuronal loss in the central nervous system (CNS). The predominant characteristic of most neurodegenerative disorders is the progressive deterioration of neurons, their structure, and function due to impaired synapses, neuronal networks, and the accumulation of abnormal protein variants in the brain. Neurodegeneration represents the fundamental pathophysiological variation observed in most neurodegenerative disorders. The process of neurodegeneration involves intricate etiology, and one of its key hallmarks is detrimental protein homeostasis, which results in the accumulation of inclusion bodies and insoluble deposits in different parts of the brain (Rapp et al., 2015). The exact mechanisms associated with neurodegeneration in NDDs are still unknown. However, several pathological features that may contribute to the process of neurodegeneration include neuroinflammation, mitochondrial dysfunction, and disequilibrium in protein homeostasis (Jellinger, 2010). The perseverance of all these pathological features slows down the process of the self-defense mechanism, which ultimately induces programmed cell death through various pathways including apoptosis, autophagy, necrosis, and parthanatos (Fan et al., 2017).

Despite the well-established Mendelian inheritance in neurodegenerative disorders, most cases are sporadic (Matilla-Dueñas et al., 2017). The clinical manifestations and neuropathological features of both sporadic and hereditary neurodegenerative disorders are indistinguishable, demonstrating that the underlying pathological mechanisms of both these conditions are the same. The likelihood of developing these disorders increases with age, and their prevalence and incidence are expected to increase dramatically as life expectancies rise globally. These neurodegenerative disorders affect millions of people globally. Recent studies have strongly advocated that both age and a combination of genetic and environmental factors contribute to the development of these disorders (Longhena et al., 2017). Moreover, neurodegeneration is highly influenced by environmental factors, notwithstanding the expression of specific genes associated with neurodegenerative disorders. Furthermore, recent studies have revealed that a single neurodegenerative disorder may be associated with multiple pathologies. Therefore, these disorders may become life-threatening, depending on their type and stage. Since most bodily functions are controlled by the brain, neurodegenerative disorders primarily affect the brain, limiting individuals' basic physiological activities including speaking, motion, and maintaining balance, along with other complicated tasks like bowel and urine movements and other mental abilities. The majority of neurodegenerative disorders progress without any period of improvement; however, certain therapies aim to alleviate symptoms, alleviate existing pain, and/or restore stability and movement. Common neurodegenerative disorders with neurodegeneration as a pathological hallmark include amyotrophic lateral sclerosis, Huntington's disease, Parkinson's disease (PD), Alzheimer's disease (AD), and motor neuron disease (spinal muscular atrophy [SMA]), as shown in Table 11.1 (Bhat et al., 2023; Harding et al., 2015; Jain & Chen-Plotkin, 2018; Martin, 1999). Even though major advances have been made to unravel the underlying molecular mechanism of NDDs, to date there is no such treatment available that can prevent, cure, or delay the progression of these NDDs. In this context, several innovative techniques including CRISPR-Cas gene editing technique and cytotherapy have garnered significant interest due to their ability to commendably address the challenge of toxic protein accumulations in different NDD models. Nonetheless, these tactics, being extortionate and arduous, are challenging to administer into the nervous system due to the highly complex structure of BBB. Consequently, the quest for a small molecule that can efficiently decrease the accumulation of toxic protein aggregates is of utmost importance. For the treatments of these NDDs, several drug discovery efforts have investigated extensive collections of artificial compounds to address these NDDs, and a small

TABLE 11.1
Showing Different Aspects of Important Neurological Disorders

Neurological Disorder	NeuroPathologic Hallmark	Affected Protein	Main Anatomic Vulnerability	Inheritance Pattern	Predominantly Affected Site
AD (Alzheimer's disease)	Accumulation of β-amyloid plaques, neurofibrillary tangles, and nerve cell damage	APP, PSEN1, PSEN2, APOE	Basal forebrain Frontal and temporal lobes Limbic structures Locus coeruleus Olfactory bulb	Autosomal dominant	Cerebral cortex
PD (Parkinson's disease)	Neuronal degeneration in substantia nigra, accumulation of Lewy bodies	Alpha-synuclein	Cerebral cortex Substantia nigra	Autosomal dominant and autosomal recessive	Basal ganglia
HD (Huntington's disease)	Neuronal degeneration in the striatum, basal ganglia, and cerebral cortex	HTT	Cerebral cortex, basal ganglia	Autosomal dominant	Basal ganglia
ALS (amyotrophic lateral sclerosis)	Degeneration of pyramidal Betz cells in anterior horn cells of the spinal cord and lower motor neurons of the brainstem	TDP43, SOD1, FUS	Motor cortex Brainstem motor neurons Spinal cord motor neurons	Varies depending on the gene involved	Motor system
SMA (spinal muscular atrophy)	Loss of lower motor neurons	Survival motor neuron	Brainstem motor neurons Spinal cord motor neurons	Autosomal recessive	Motor system

number of promising options have already progressed to clinical trials (Pérez-Arancibia et al., 2023). Furthermore, the identification of potential new targets for NDDs has led to the unearthing of novel small molecules that have shown their effectiveness in preclinical trials. Recent studies have highlighted the important approach in emphasizing the crucial role of small molecules in addressing NDD. The main advantage of using these small molecules is that they offer cost-effectiveness, an easy mode of delivery compared to other therapeutic strategies, highlighting the pragmatic utility of small molecules, potentially enhancing patient compliance (Pérez-Arancibia et al., 2023).

11.2 CHALLENGES IN TREATING NEURODEGENERATIVE DISEASES

NDDs are treated only for symptoms, and currently there are no disease-modifying therapies available that can completely cure these NDDs. However, significant progress has been made in understanding the causes of NDDs over the past few decades, and effective disease-modifying strategies will likely be discovered. Regardless of this progress, there are still many challenges that need to be addressed to make significant advancements in the proper treatment of these NDDs. The three chief challenges that need to be addressed to make significant improvements in the detection and treatment of NDDs include the following (Angeloni et al., 2022).

i. **Making progress in understanding misfolded proteins**

   Despite significant advances in the field, there is still unreliability about the exact mechanism of toxic misfolded proteins that occur during the progression of NDDs. In healthy cells, misfolded proteins are either degraded or refolded with the help of chaperone proteins. Disturbances caused by factors such as stress, mutations, or aging can disrupt this process, resulting in the escape of proteins from quality control and leading to aggregate formation. Although studies in model organisms have provided valuable insights, more research is needed to replicate the biochemical and behavioral phenotypes seen in humans with NDDs (Calabrese et al., 2022; Iadanza et al., 2018).

ii. **Advancing early diagnostic tools**

From a clinical standpoint, the ability to diagnose NDDs in their initial stages could enhance patients' responsiveness to treatment and deepen our understanding of disease progression. The development of reliable and specific biomarkers holds great potential. A combined approach involving disease marker identification, neuroimaging, and diagnostic testing promised to uncover more sensitive and specific NDD biomarkers, thus paving the way for personalized treatment (Noble & Burns, 2010).

iii. **Accessing more effective therapies**

To slow down the progression of neurodegenerative processes, there is a need for more effective therapies. While stem cell therapies show potential, accessibility and effectiveness across different genetic backgrounds remain challenging. Identifying the specific disease-associated proteins responsible for the condition is a key step in understanding the most effective therapeutic targets. Phenotypic screening presents an attractive method for identifying novel targets and therapeutic molecules for NDD, considering the shared mechanisms and phenotypes across different conditions (Fan et al., 2017). Advancements in technologies such as robotic microscopy, CRISPR screening, and machine learning methods (AI/ML) are aiding in handling the vast amount of data generated by screening processes. Nevertheless, there is still a gap between the accumulation of data and the development of diagnostic and therapeutic solutions. Further studies are critical to unravel the complex molecular mechanisms involved in these pathways (Espay et al., 2016). While considerable progress has been made in understanding NDDs, ongoing efforts to integrate data related to the molecules involved in NDD progression will play a key role in developing novel therapeutic strategies.

11.3 SMALL MOLECULE THERAPEUTICS: AN OVERVIEW

Small molecules are organic compounds intended to imitate, enhance, or reduce the behavior of natural products in the body. With a molecular weight below 500 Da, these compounds can easily pass through cells (Lipinski et al., 1997). They boast relatively simple structures that can be tailored to meet specific therapeutic purposes. Small molecule drugs are usually stable and hardly require special storage conditions. The anticipated behavior of these small molecules within the body, or in vivo, typically follows a predictable pattern, enabling the development of uncomplicated dosing regimens, often administered orally, which patients find convenient to follow (Beck et al., 2018). Due to their ability to move through the body easily, small molecule drugs can treat a wide array of diseases by moving from the gut through the bloodstream to the site of action, permeating cell membranes to reach intracellular targets (Beck et al., 2018). They can be administered in various forms such as pills, inhalers, suppositories, or injectables, making them highly adjustable. Several advantages make small molecules crucial in therapeutics, including the ability for oral administration and cell membrane permeability, enabling them to reach specific tissues and targets. In this regard, small molecule drugs can be designed to obtain specific affinity and selectivity (Poduval et al., 2023) and are capable of influencing the function of diverse proteins, including protein–protein interactions, by making complexes with their targets (Scott et al., 2016; Southey & Brunavs, 2023). The chemical structure of small molecules can be tailored to selectively interact with specific biological targets. By modifying the atomic composition of small molecules, their overall properties can be finely modified to induce only the desired response. This pliability thus allows searching of the entire "chemical space," which gives these small molecule approaches a distinct advantage over other methods. The process of designing a small molecule drug and ensuring that these small molecule drugs perform as expected while minimizing unwanted side effects demands meticulous design and synthetic expertise from researchers, often spanning several years (Southey & Brunavs, 2023).

11.4 ADVANTAGES AND DISADVANTAGES OF SMALL MOLECULE THERAPEUTICS

Small molecules offer numerous advantages that make them highly intriguing to the biopharmaceutical community. They are relatively simple stable structures that can be specifically tailored to converge therapeutic objectives and hardly need specific storage conditions. Moreover, their anticipated behavior in the body follows easy dosing protocols which patients find convenient to follow. The main advantage of small molecule drugs is their oral route of administration, providing convenience for patients and potentially improving treatment adherence compared to injectable or intravenous (IV) drugs. This is due to the cost-effectiveness and convenience associated with small molecules. Furthermore, small molecules are easier to manufacture at scale compared to biologics and tend to be more chemically stable. By modifying the atomic composition of small molecules, their properties can be finely tuned to elicit only the desired response, giving them remarkable benefits above other treatment modalities. Small molecule drugs have the ability to selectively target specific receptors or enzymes, resulting in inaccurate therapeutic effects while minimizing their off-target outcomes, and thus the patients may experience reduced side effects when using these small molecule drugs. However, it is crucial to recognize that the chief limitation of traditional small molecule therapeutics lies in their inherent specificity for a single function, independent of the patient's physiological state, as they do not have mechanisms for responding to bodily feedback (Gurevich & Gurevich, 2014).

11.5 THERAPEUTIC APPROACHES IN MANAGEMENT OF COMMON NDDS

Several therapeutic approaches currently used in the management of neurodegenerative disorders are often disease-specific and can either improve the symptoms or target the disease pathogenesis

11.5.1 Alzheimer's Disease

AD is an advancing and irreparable NDD among elderly people that alters expressed cognition (Lane et al., 2017). The symptoms of AD usually begin at the age of 20 years or above with trivial abnormalities affecting mental capacity that are imperceptible in the affected individuals, which leads to severe depression in elderly patients (>65 years old) that ultimately results in death. As the world population ages (> 65 years old), dementia becomes one of the most important health concerns globally. AD is one of the most common forms of dementia and accounts for nearly about 60% of all cases of dementia with its incidence increasing twofold after 65 years of age (Mendez, 2017). The age-related incidence of AD notably rises from < 1% at 65 years of age to 6% every year after 85 years of age, and its prevalence rises from 10% after 65 years of age to 40% after the age of 85 years. After cardiovascular disease and cancer, AD is considered the third major cause of death in developed countries (Huang et al., 2016). The main characteristic symptoms of AD include progressive memory loss, reduced cognitive function, language skills, and some brain alterations that imperceptibly worsen over time and cause the forfeiture of even basic daily activities (Hu et al., 2016). The prime pathological features of AD are phosphorylated Tau-induced neurofibrillary tangles and beta-amyloid plaques (AB) (Congdon & Sigurdsson, 2018). The beta-amyloid plaque is a chief instigator, which triggers the aggregation and development of AD through cumulation, and patients with such conditions reveal a deficiency of acetylcholine, accountable for transmitting the nerve impulses inside the brain. In the pathogenesis of AD, the two significant pathways of neurotoxicity include the formation of Aβ oligomer and the generation of free radicals. The formation of

Aβ and tau proteins results in redox imbalance and the generation of excessive reactive oxygen species that occurs because of unusual aggregation of transition metals and mitochondrial dysfunction that induces neurotoxicity and the progression of AD (Zhao & Zhao, 2013). The pathology of AD and its development is influenced by several risk factors, such as vascular conditions, environment, genetics, and lifestyle. Mutations in the PSEN1 and PSEN2 genes, as well as apolipoprotein, are implicated in the development of AD (Zhao & Zhao, 2013). Owing to the complicated interactions of AD with genes and proteins and its complex pathophysiological features, no effective drugs were available, which can cure AD or provide long-term treatment benefits to patients (Liu et al., 2017). The current therapeutics approved by the Food and Drug Administration (FDA) for the treatment of AD (Guzior et al., 2015) include the inhibitors of acetylcholine esterase inhibitors (AChEIs) such as donepezil and *N*-methyl-D-aspartate (NMDA) receptor antagonist memantine (MEM) (Reitz et al., 2011).

11.5.1.1 Donepezil (Unique Molecule) (Discovery and Development)

Donepezil hydrochloride, a piperidine derivative, with the formula of $C_{24}H_{29}NO_3.HCl$ and molecular weight of 416 Da is a centrally acting, rapid, and reversible AChEIs predominantly used for the treatment of mild-to-moderate AD (Information NCfB: **PubChem compound summary** in: National Center for Biotechnology Information Bethesda, MD, USA; 2021). It is an AChEI that was first produced by the Eisai Company (Information NCfB: **PubChem compound summary** in: National Center for Biotechnology Information Bethesda, MD, USA; 2021). Acetylcholine esterase (AChE) disintegrates acetylcholine following its release from the presynapse, thereby increasing acetylcholine availability at the synapses and enhancing its cholinergic transmission. It is the second drug approved by the U.S. FDA for the treatment of AD in mild, moderate, and severe dementia and is now marketed not only in the United States but globally in nearly 90 countries for the treatment of mild-to-moderate AD conditions (Levine et al., 2022). In 1996, donepezil was first traded in the United States, followed by its registration in 1997 in many other countries including the European Union and Canada (Pooladgar et al., 2022). Donepezil was approved in Japan in 2014 for Lewy body dementia. The benefit of donepezil has a small effect size, although a patient-level investigation has demonstrated a larger benefit in early responders compared to non-responders (Levine et al., 2022). Even though donepezil does not alter the root cause of the disease, it has shown some promising results in palliating some specific symptoms like improving cognition and behavior in AD patients.

11.5.1.1.1 Mechanism of Action

Donepezil hydrochloride is an AChEI that plays a pivotal role in the degradation of acetylcholine upon its subsequent release from the presynaptic membrane. It binds reversibly with AChEIs, disrupting the hydrolysis of acetylcholine, which results in an increase in the presence of acetylcholine at the synaptic junctions, thereby augmenting cholinergic transmission (Lieberman et al., 2023). Various in vitro studies indicate that donepezil's anticholinesterase activity is relatively specific to AChEI in the brain and is structurally different from other anticholinesterase molecules like tacrine and physostigmine. Along with its cholinergic outcomes, donepezil may have some noncholinergic mechanisms. It is known to upregulate nicotinic receptors in cortical neurons, potentially contributing to its neuroprotective properties, and effectively downregulate neuroinflammatory responses, including microglial and astrocytic activation associated with AD (Seltzer, 2005).

11.5.1.1.2 Clinical Trials and Outcomes

More than 250 clinical trials have been conducted to investigate the efficacy of donepezil, with a significant focus on its effects on AD patients. Studies of early-stage AD patients classified by MMSE between 21 and 26 showed that 10 mg/day of donepezil resulted in moderate cognitive improvement, particularly in memory, as measured by ADAS-cog (7th Congress of the European Federation of Neurological Societies (EFNS 2003), 2002; Seltzer, 2004). Clinical trials including mild-to-moderate AD patients exhibit that 5–10 mg/day doses of donepezil led to a modest increase in cognition and functional measures, as evaluated by ADAS-cog, CIBIC-plus Clinician's Interview-Based Assessment of Change-Plus, and CDR-SB scales (CDR-SB) (Burns et al., 1999; Rogers et al., 1998). The AWARE study, Aricept Washout, and REchallenge which included 817 patients in phase 3 clinical trial, evaluated the effects of continued treatment in patients who did not respond to 5 or 10 mg/day of donepezil during the first 6 months of treatment. During this clinical trial, donepezil was generally well-tolerated and demonstrated a positive impact on memory. Moreover, it provides a small benefit on daily activities of living and behavioral symptoms for the majority of patients at 9 months, even those who did not initially experience these benefits (Johannsen et al., 2006). Long-term studies demonstrated that donepezil manifested moderate effectiveness in treating those AD patients who were on donepezil for 3 years, showing less cognitive decline than patients who took placebo for 1 year. Pharmacoeconomic studies investigated in the United States and Europe generally portrayed that treating AD patients with donepezil reduces the cost of care. In the context of mild cognitive impairment (MCI), studies conducted on more than 1800 patients on the efficacy of 5–10 mg/day of donepezil specify different effects on cognitive and functional endpoints (Doody et al., 2009; Salloway et al., 2004). Furthermore, donepezil has regulatory approval for the treatment of severe AD in the United States, Canada, and Japan and has been approved in Japan for dementia with Lewy bodies (DLB) based on the results of clinical trials (Ikeda et al., 2013, 2015). Additional clinical trials conducted between 2016 and 2018 evaluated pharmacokinetics, placement on distinct areas of the body, and outcomes of heat on drug delivery, for differentiating between donepezil patches to pills in a total of 599 healthy adults. Based on the results of these studies, the FDA permitted marketing approval in March 2022. The patches continue for 1 week, giving 5 or 10 mg of the drug daily. The main side effects were itching or irritation at the application site, muscle spasms, headache, constipation, diarrhea, and dizziness. Results from one study demonstrated that the patch caused fewer gastrointestinal and nervous system side effects than pills (Tariot et al., 2022). A different patch formulation was approved in Korea in 2021 (Han et al., 2022) and by a Japanese company. Ongoing trials are scrutinizing donepezil for traumatic brain damage, healing of wounds in Type 2 diabetes, and cognitive functions attributed to chemotherapy, surgery, or radiation therapy. Additionally, it is usually used as a standard of care in trials assessing many investigational drugs as add-on therapy, as shown in Table 11.2.

11.5.1.1.3 Current Status and Future Directions of Donepezil

Although donepezil is a key cholinesterase inhibitor for treating AD, its therapeutic efficacy greatly differs among patients ranging from 20% to 60%. Studies have indicated that various genetic factors play a significant role in the diverse pharmacodynamics and pharmacokinetics of donepezil, resulting in different clinical outcomes. Additionally, various studies have demonstrated that specific liver enzyme genetic polymorphisms like CYP2D6 and CYP3A4(7) and APOE alleles like APOE

TABLE 11.2
Showing the Detailed Clinical Trials of Memantine

Serial No.	Aim of Clinical Trial	Duration of Study	Total No. of Participants	Current Status
01	To evaluate the efficacy of donepezil in AD patients at the early stage of the disease	24 weeks	153	Completed
02	A multicenter double-blind study aimed to evaluate the safety and efficacy of donepezil in AD patients with mild-to-moderate symptoms	24 weeks	473	Completed
03	A transdermal donepezil patch treatment in mild-to-moderate AD patients to evaluate its safety and efficacy	24 weeks	340	Completed
04	To investigate continued donepezil treatment in patients with AD for whom clinical benefit was initially judged to be undetermined	24 weeks	619	Completed

E3 also influence the efficacy of donepezil (Lu et al., 2020). Another challenge in treating AD is due to its complex nature, characterized by the involvement of diverse pathways and symptoms that demand the use of a combination of various therapies for effective treatment of AD. Efforts have been made to develop a new derivative of donepezil targeting different pathways and receptors, which involve exploring the components of the donepezil, benzylpiperidine and dimethoxy indanone, that specifically bind to or potentially replace them with compounds such as flavonoids and curcumin, pyridyl methyl piperidine, benzylpiperazine, pyrimidyl piperazine, heterocyclic rings, ferulic acid (Mohsin & Ahmad, 2020). These modifications aim to create derivatives with novel properties, including anti-oxidative, anti-dementia, and monoamine oxidase inhibition-like effects. In light of these complexities, further research is needed to improve therapeutic responses and minimize potential side effects in the treatment of AD.

11.5.1.2 Memantine (Wonderful Molecule)

MEM is a subclass of glutamate receptor which acts as an antagonist for the NMDA receptor and decelerates the neurotoxicity implicated in AD and other NDDs (Kuns et al., 2024). MEM is classified as an uncompetitive, low-affinity, open-channel blocker that selectively targets the extrasynaptic NMDA receptor (NMDAR) in its open state. By doing so, the drug effectively hampers the disruption of normal synaptic transmission, thereby guarding against further impairment attributed to excitotoxicity-induced neuronal cell death. It is most frequently utilized in conjunction with AChEIs for the efficacious management of Alzheimer's dementia and has garnered approval from the U.S. FDA. This drug's adverse effects prevent an extremely active glutaminergic system in the CNS, resulting in neurotoxicity noticed in AD (Cacabelos et al., 1999; Wong et al., 2019). This examination delves into the pharmacology, indications, adverse effects, drug interactions, and contraindications of MEM. It underscores the collaborative work of interprofessional healthcare teams in the treatment of patients with dementia (Rogawski & Wenk, 2003).

11.5.1.2.1 Discovery and Development of MEM

In 1972, Merz Pharma applied for a German patent for MEM as a promising treatment for various NDDs (Lipton, 2007, 2006). In the early 1990s, the company started research on the efficacy and safety of MEM in nursing patients with severe dementia. In 1999, Forest Laboratories acquired the license for using MEM and commenced a series of clinical trials for the treatment of moderate to severe AD. MEM was approved for the treatment of moderately severe to severe AD in 2002 by the European Agency for the Evaluation of Medical Products (EMEA). MEM's efficacy was assessed as monotherapy or combined with AChEIs (Tu et al., 2014). The dosing regimen of 10 mg bid was based on the results of earlier clinical trials. Currently, MEM has been accepted widely for the treatment of AD, and in 2018 it was the second-most commonly used drug for the treatment of dementia globally, accounting for 32.7% of total use (Ju et al., 2021). Up to 2022, 114 MEM AD drug brands have been registered, and in the United States, MEM is well known as Namenda (Dembitsky et al., 2020). Nonetheless, MEM may instigate a number of side effects. Common adverse reactions include hallucinations, confusion, headache, dizziness, high blood pressure, sleepiness, restlessness and the chief issue with MEM and other approved treatments for AD is that they do not slow down the disease's progression but only have a symptomatic effect. However, efforts are being aimed at developing new molecules that can modify the course of AD progression (Shafiei-Irannejad et al., 2021).

11.5.1.2.2 Mechanism of Action

MEM is an uncompetitive inimical of the NMDA subtype of glutamate receptors in the CNS. AD is caused due to excessive stimulation of glutamate, the chief excitatory amino acid in the CNS, resulting in neuronal damage and excitotoxicity. The NMDAR is a type of ion channel that is voltage-gated, and its activity is regulated by changes in the electrical potential across the cell membrane. In its unstimulated state, the NMDAR is blocked by magnesium ions. However, upon stimulation, by binding of neurotransmitters, the magnesium ions are displaced, allowing the invasion of calcium ions upon subsequent receptor activation. Excessive stimulation of the NMDAR can result in a chronically active state of the receptor, contributing to the pathological process. Thus MEM works by blocking excessive stimulation of the NMDAR, helping to restore a more balanced and controlled level of activity (Rogawski & Wenk, 2003).

11.5.1.2.3 Clinical Trials and Outcomes

It is evident that while preclinical trials for AD using MEM show promise, the clinical results have been ineffective in providing a cure, at best only managing to slow cognitive decline (Alley et al., 2010). For instance, in one clinical trial study where 28 mg of MEM was orally administered on a daily basis for 24 weeks (Grossberg et al., 2013), the assessment of cognitive function was conducted using the Severe Impairment Battery, a neuropsychological test for severely demented patients. The study concluded that high doses of MEM, when used in combination with an AChEI, provided behavioral benefits for patients with moderate to severe AD. Similarly, Howard and colleagues obtained comparable outcomes using donepezil in conjunction with MEM (Howard et al., 2015). Consequently, this conjunction therapy of AChEIs with MEM holds the promise to provide advantages that are implausible when the drug is used in isolation. A clinical multicenter trial including mild-to-moderate AD evaluating the combined effect of galantamine with MEM over a duration of 12 months is underway (Clinical trial: NCT01921972) (Peters et al., 2015); however, the results of this study are yet to be delineated. If the expected outcomes are positive, the drug combination is anticipated to increase synaptic activity and halt excitotoxicity. Another combination therapy of masitinib with MEM has exhibited the potential to markedly ameliorate cognitive behavior (Piette et al., 2011). Consequently, it is evident that further studies are warranted to examine more promising combinatory treatments for delaying or preventing AD, given that it is an intricate disease which needs to be approached through numerous targets. Furthermore, the studies demonstrated that the combination was more effective than consuming either substance individually. Additionally, a recent clinical trial (AD-IDEA) is currently under analysis; however, currently no findings are delineated from this clinical trial regarding improving cognitive function. Lindquist et al. documented a familial case in which a patient with FTD, a neurodegenerative disorder in adults characterized by dementia and behavioral disturbances, received MEM at a maintenance dose of 10 mg twice daily. In this particular case, the patient had a pathogenic mutation (R406W) in the microtubule-associated protein tau (MAPT) gene, and MEM treatment appeared to halt the progression of the disease symptoms. However, a clinical trial (NCT00545974) conducted at multiple centers, utilizing a randomized, double-blind, placebo-controlled design, assessed the effectiveness of MEM in individuals with mild-to-moderate FTD and concluded that MEM treatment did not ameliorate these dementia symptoms, as shown in Table 11.3 (Lemire et al., 2018; Vercelletto et al., 2011).

11.5.1.2.4 Current Status of MEM and Future Directions

MEM is approved for treating moderate to severe AD patients and is commonly used as an off-label drug for the management of mild AD in the United States, which involves using MEM for purposes beyond its officially approved indications, doses, regimes, routes, or patient populations (McShane

TABLE 11.3
Showing the Detailed Clinical Trials of Memantine

S. No.	Clinical Trial No.	Duration of Study	Total Number of Participants	Current Status
01	NCT00322153	24 weeks	676	Completed
02	NCT01921972	12 months	232	Completed
03	NCT00976118	24 weeks	34	Completed
04	NCT00545974	26 weeks	100	Completed

et al., 2019). Nonetheless, it is the only drug approved for the treatment of AD since 2003. Thus, research on MEM for the treatment of AD holds remarkable importance for the elderly population (Liu and Howard, 2021; Vaz and Silvestre, 2020). Studies indicate that the effectiveness of MEM in treating mild AD varies greatly from its efficacy in moderate to severe AD and MEM does not exhibit benefits for mild AD, while showing improvement in patients with moderate to severe AD symptoms with a favorable safety and efficacy profile. By ameliorating symptoms like impaired memory and cognitive decline, MEM may increase the performance of everyday activities and decrease agitation in AD patients with moderate to severe symptoms. The clinical diversity of AD renders it impractical for one drug to manifest significant efficiency, signifying that the optimal drug therapy may involve the utilization of numerous drugs (McShane et al., 2019). Regarding the FDA-approved drugs, aducanumab is the only compound authorized by the FDA to prevent brain amyloid aggregation or clear existing amyloid plaques. However, the basic therapeutic mechanism of aducanumab has not been thoroughly studied, and every single new drug has failed in clinical studies (Mullane & Williams, 2019). Thus, it is crucial to raise attention to how we can maximize the impact of currently approved drugs. MEM's ability to maintain normal cellular metabolism and membrane function by inhibiting glutamate hyperactivation makes it a promising avenue for further exploration in AD treatment.

11.5.2 Parkinson's Disease

PD, a movement disorder, is the second-most frequent NDD after AD, which gets worse overtime. As neurons in some parts of the brain debilitate, impair, or die, patients experience problems with movement like tremors, muscle stiffness, Bradykinesia, and impaired balance. While the disease progresses, patients have difficulty completing simple tasks such as talking and walking (Bordelon & Keener, 2016). Although the exact cause of PD is not known, both genetic and environmental factors may play a role in the development of PD, and age is considered as prime risk factor in disease development (Cova & Priori, 2018). Apart from genetic and environmental factors, several other factors that promote the progression of PD include excess intake of caffeine, smoking, and other environmental factors. The four key symptoms of PD include posture instability, rigidity, resting tremor, and bradykinesia termed parkinsonism, which is clinically and pathologically differentiated by the accumulation of ubiquitinated protein in neuron's cytoplasm (known as Lewy bodies) and loss of neurons (Kouli et al., 2018). The age-related frequency of dementia in patients with PD varies from 12.4% in 50–59 years of age and 68.7% in more than 80 years of age (Tanner & Goldman, 1996). Almost 15% of PD patients have a family history that results from mutations in genes PARK7, PINK1 or SNCA, LRRK2, and PARK2. PD mostly occurs when neurons in the brain are damaged, and many areas of the brain are affected, which results in the loss of neurons in the substantia nigra. The neurons of the substantia nigra make dopamine, which is accountable for sending signals between the substantia nigra and corpus striatum, essential for normal movement. Numerous studies have investigated that virtually 60–80% of dopamine-making cells in substantia nigra disappear in PD patients before the onset of symptoms (Poewe et al., 2017). The Lewy bodies are also affected in brain cells of most patents with PD. The exact reason that these Lewy bodies are formed and what role they play is not known. Some studies propose that protein disposal systems are defective in PD, inducing these defective proteins to cause detrimental effects and to activate cell apoptosis (Balestrino & Schapira, 2019). Furthermore, it has been observed in many studies that due to the presence of environmental toxins, the genes and proteins affected in both sporadic and inherited PD conditions are similar. Apart from this, the neuron dysfunction started by inflammatory responses is related to cognitive and motor damage, which promotes the breakdown of the BBB. Cognitive decline and motor dysfunction are distinctive properties of DLBs caused by the aggregation of α synuclein protein in neurons; low level of dopamine and synaptic dysfunction are usually observed in PD patients (Armstrong & Okun, 2020).

11.5.2.1 Pathophysiology of PD

The pathophysiology of PD remains largely idiopathic, indicating an unknown cause. It likely involves the interplay of host vulnerability and environmental factors. While a small percentage of cases are genetically linked, extensive research is being conducted on genetic factors. Physiologically, the symptoms related to PD are due to the loss of various neurotransmitters, specifically dopamine. As the disease advances, more affected cells are lost, resulting in the worsening of symptoms over time. The disease follows a highly distinct course, with some patients demonstrating minimal symptoms as they age, while others encounter a quick progression of symptoms (Bonnet & Houeto, 1999). PD is a highly complex NDD with a sequential progression. Evidence from various studies suggests that it initially affects the dorsal motor nucleus of the vagus nerve and the olfactory bulbs and nucleus, followed by the locus coeruleus, and eventually progresses to the substantia nigra. Cortical areas of the brain are affected in later stages. Damage to various neuronal systems underlies complex pathophysiologic changes, resulting in damage not only in the motor system but also in the cognitive and neuropsychological systems (Kwan & Whitehill, 2011). Regardless of extensive efforts to better understand the cellular mechanisms underlying the development of PD, there are currently no treatments available that can prevent, cure, or remarkably delay the progression of the disease (Troncoso-Escudero et al., 2020). Current therapeutics for PD treatment involve small molecules aimed at correcting the loss of fine motor control. These approaches involve the use of drugs to enhance dopamine availability with dopamine precursors (levodopa) and those that inhibit dopamine degeneration by monoamine oxidase (Pérez-Arancibia et al., 2023).

11.5.2.2 Levodopa (L-Dopa) (Inimitable Molecule)

Levodopa serves as the precursor to dopamine and is commonly utilized as a replacement for dopamine in the treatment of PD. It is particularly effective in treating the slow movements associated with PD and is the most effective medication for enhancing the quality of life in individuals with idiopathic PD. Levodopa is generally prescribed to PD patients when their symptoms become more challenging to manage with other antiparkinsonian drugs. Moreover, recent findings indicate that levodopa may potentially slow down the progression of PD or provide increased benefits even after the discontinuation of the drug (Ebada et al., 2019; Hampel et al., 2017; Romagnolo et al., 2018).

11.5.2.2.1 Mechanism of Action

The degeneration of the substantia nigra in PD disturbs the nigrostriatal pathway, leading to reduced levels of dopamine in the striatum. Unlike dopamine, levodopa is able to penetrate the blood–brain barrier (BBB). Levodopa is converted to dopamine in both the CNS and periphery to inflate its effectiveness and reduce potential side effects (Wagner et al., 2023). Levodopa is commonly administered in conjunction with peripheral decarboxylase inhibitors (such as carbidopa and benserazide). Dopamine decarboxylase inhibitors prevent the transformation of levodopa to dopamine in the periphery, allowing more levodopa to penetrate the BBB. Once converted to dopamine, postsynaptic dopaminergic receptors are activated, compensating for the reduced endogenous dopamine (Ogungbenro et al., 2015).

11.5.2.2.2 Clinical Trials and Outcomes

In 1975, the U.S. FDA approved the levodopa/carbidopa (LD/CD) pill, marketed under the name Sinemet. This marked a significant milestone in the treatment of PD. The formulations have since evolved to include disintegrating tablets, immediate or extended-release tablets, and generics, with a dosing frequency of three or four times daily. Fast forward to 2015, when the FDA approved two new LD/CD formulations aimed at addressing fluctuations in plasma concentration to prolong drug action and reduce off-time (Hauser et al., 2013). Another innovative development was the LD/CD intestinal gel, Duopa, which significantly decreased daily off-time compared to immediate-release oral tablets. However, it's worth noting that Duopa requires surgical implantation of a feeding tube and is tailored for advanced PD patients with severe motor fluctuations (Olanow et al., 2014).

TABLE 11.4
Showing the Detailed Clinical Trials of Levodopa (L-Dopa)

S. No.	Aim of Clinical Trial	Duration of Study	Total Number of Participants	Current Status
01	To evaluate the safety and efficacy of IPX066 in advanced Parkinson's disease (PD) (NCT00974974)	13 weeks	471	Completed
02	To investigate the safety, efficacy, and suitability of levodopa–carbidopa intestinal gel in levodopa-responsive Parkinson's patients (NCT00660387)	12 weeks	35	Completed
03	Pharmacokinetic investigation of CVT-301 (levodopa inhalation powder) (NCT03887884)	3 months	23	Completed
04	To differentiate the continuous subcutaneous infusion of ABBV-951 with oral levodopa/carbidopa tablets for treatment of motor fluctuations in adult subjects with PD	12 months	174	Completed

In 2018, an inhaled powder formulation of levodopa, Inbrija, was approved by the FDA as a rescue treatment for OFF episodes occurring between oral levodopa doses. Inbrija delivers LD/CD to the bloodstream within minutes, outperforming the conventional pill form in a phase 3 trial (Safirstein et al., 2020). Studies have shown that the initial trials in healthy volunteers and people with PD have manifested promising results, with the potential to revolutionize treatment for PD. Research and development are ongoing, with phase 3 and phase 1 studies continuously building on the progress made so far. Despite experiencing some regulatory hurdles, the advancements in LD/CD formulations offer hope for improving the lives of individuals living with PDs, as shown in Table 11.4.

11.5.2.2.3 Future Directions

Various new levodopa formulations are being explored, including oral long-acting pills, transdermal formulations, continuous infusion, and different approaches to overcome technical challenges. Studies have shown that a LD/CD gel formulation for enteral infusion can give more compatible clinical effects and decrease motor problems in an advanced stage of PD. Levodopa prodrugs are also in development, with some showing promising results in maintaining continuous levodopa levels. Additionally, a transdermal patch is being developed with the aim of improving the solubility and stability of levodopa (Hauser, 2008; Nyholm, 2006).

11.5.2.3 Rasagiline (Extraordinary Molecule)

Rasagiline, marketed as Azilect, used in managing the early symptoms of PD is an irreversible inhibitor of MAO. It can be used as an initial monotherapy or as an adjunct therapy to levodopa. Rasagiline mesylate is known to be a highly potent selective MAO-B inhibitor that helps to increase the availability of neurotransmitters at the synaptic level. Compared with selegiline, another approved treatment for PD, rasagiline has shown greater potency in inhibiting MAO-B and does not exhibit undesirable amphetamine-related effects. Rasagiline is categorized into two vital molecular species, A and B, and is also found in membranes of mitochondria in various parts of the body (Naoi & Maruyama, 2009). It primarily targets MAO-B, which is mainly found in the human brain and regulates the metabolic degradation of dopamine and phenylethylamine. Moreover, ex vivo animal studies have demonstrated rasagiline's potency, selectivity, and irreversible inhibition of MAO-B (Leegwater-Kim & Bortan, 2010).

11.5.2.3.1 Discovery and Development

Azilect (rasagiline) is a medication used for the treatment of PD. It can be prescribed as a standalone treatment during the early stages of the disease and also as a supplement to levodopa for patients in the moderate to advanced stages of the disease. The drug comes in the form of circular tablets available in doses of 0.5–1 mg for oral administration. The development of Azilect was a collaborative effort between Technion—the Israel Institute of Technology and Teva Pharmaceutical, a pharmaceutical company based in Israel. Teva later licensed the marketing of rasagiline in Japan to Takeda, a pharmaceutical company in April 2014, with approval in Japan granted in March 2018. It was first approved in Israel in January 2005 for marketing, followed by its approval by EMA (European Medicines Agency) and the U.S. FDA in February 2005 and May 2006, respectively. In June 2014, the FDA expanded its label to include its use as an additive to dopamine agonists (DAs). Currently, rasagiline is available in nearly 50 countries worldwide including the United States, Canada, Israel, Mexico, Asian countries, and all European Union countries (Gouda et al., 2022).

11.5.2.3.2 Mechanism of Action

The exact way in which rasagiline works is not fully understood. However, it is thought to act by inhibiting the activity of MAO-B, an enzyme involved in the breakdown of dopamine. By doing so, it is believed to increase the levels of dopamine in the striatum, a region of the brain, thus helping to restore dopamine levels to normal. This elevation in dopamine levels and the resulting increase in dopaminergic activity are believed to be responsible for rasagiline's positive effects, such as supporting normal mobility, balance, and coordination, particularly in models of dopaminergic motor dysfunction (Gouda et al., 2022).

11.5.2.3.3 Clinical Trials and Outcomes

The safety and effectiveness of rasagiline was investigated in several clinical trials. In the TEMPO clinical trial, which lasted 26 weeks, rasagiline was compared to a placebo in 404 patients with early PD symptoms (Gouda et al., 2022). The study showed that rasagiline was significantly more effective than the placebo in reducing the Unified Parkinson's Disease Rating Scale (UPDRS) score from baseline to endpoint. In another 12-month phase of the TEMPO trial, patients who received rasagiline (1 and 2 mg doses) showed less decline in total UPDRS compared to those who had delayed treatment with rasagiline. The results suggest that starting rasagiline early may slow the progression of PD-related impairment. Additionally, when rasagiline was administered in combination with levodopa, it significantly reduced "off-time" and improved motor function in PD patients. The LARGO trial, involving 697 PD patients, also demonstrated that rasagiline reduced off-time, similar to the effects of entacapone, and was observed for its long duration of action. The ADAGIO clinical trial proposed that treatment with rasagiline during the early stages of PD with a dose of 1 mg daily has a neuroprotective effect and decelerates the development of PD. However, a 24-week study involving 170 people with PD and MCI did not find any improvement in cognitive function with 1 mg of rasagiline daily (Weintraub et al., 2016). While some small studies reported improved sleep quality with rasagiline, a recent randomized controlled trial found no evidence that it slows disease progression in PD. Apart from PD, rasagiline has been evaluated for use in amyotrophic lateral sclerosis (ALS), schizophrenia, multiple-system atrophy, progressive supranuclear palsy (PSP), serotonin syndrome, and retinal detachment. It's worth noting that sembragiline, another MAO-B inhibitor, yielded negative results in an AD trial in July 2015. Finally, a study of 11 participants with AD, treated for 1.7 years with 1 mg of rasagiline per day along with donepezil, showed improved blood flow in specific brain regions, as measured by single-photon emission computed tomography (SPECT). This study did not measure cognitive function, as shown in Table 11.5 (Song et al., 2020).

11.5.2.3.4 Current Status and Future Direction

Current treatments for PD effectively manage the symptoms but do not slow down the disease's progression. These treatments mainly work by increasing dopamine levels using different drugs. However, there is a need for new treatments that can protect the neurons and modify the course of the disease by preventing or delaying the death of dopaminergic and other types of neuronal cells, especially in the early stages of PD. Some drugs, such as rasagiline, have shown potential for modifying the disease,

TABLE 11.5
Showing the Detailed Clinical Trials of Rasagiline

S. No.	Aim of Clinical Trial	Duration of Study	Total Number of Participants	Current Status
01	A placebo-controlled trial of rasagiline for mild cognitive disability in Parkinson's disease	24 weeks	170	Completed
02	To evaluate the outcomes of monoamine oxidase B (MAO-B) inhibitor (rasagiline) approved for Parkinson's disease	24 weeks	150	Completed
03	To study the efficacy and safety of rasagiline in levodopa-treated patients with Parkinson's disease and motor neuron changes	18 weeks	88	Completed
04	To investigate the results of rasagiline on regional cerebral flow (rCBF) in patients with AD using single-photon emission computed tomography (SPECT)	1.6 years	22	Completed

and researchers are investigating their neuroprotective effects. It is important to further understand how these drugs work to develop new medications for PD that can both increase dopamine levels to protect degenerating neurons. Additionally, identifying the best way to administer these drugs to maximize their effectiveness is crucial (Bonuccelli & Del Dotto, 2006; Chaudhuri et al., 2006).

11.5.3 Amyotrophic Lateral Sclerosis

ALS, also known as Lou Gehrig's disease, is a progressive paralytic NDD that affects the upper and lower motor neurons and is the most common motor neuron disease with both sporadic and familial forms (Brown & Al-Chalabi, 2017). The precise cause of ALS is currently unknown, but it is believed to result from a combination of both genetic and environmental factors. More than 120 genes have been associated with ALS, and it is thought that the accumulation of abnormal proteins due to aberrant RNA processing may play a role in the development of the disease (Wang et al., 2023). A number of genes such as TDP-43, FUS, and C9ORF72 have been implicated in familial ALS, and mutations in the C9ORF72 gene are the most common genetic mutations associated with ALS. Superoxide dismutase 1 (SOD1), responsible for converting superoxide radicals to oxygen and hydrogen peroxide, was the first gene discovered in relation to ALS (Brown & Al-Chalabi, 2017; Quinn & Elman, 2020).

Furthermore, the cytoskeleton variables and functional deficits at the neuromuscular junction have also been observed in ALS patients. ALS primarily presents as sporadic cases, with approximately 90–95% of cases being sporadic. The remaining 5–10% of cases are familial. It has been observed that the majority of patients with sporadic ALS have an underlying genetic susceptibility, while the rest are attributed to environmental influences. The lifetime risk of developing ALS is approximately 1:350 for men and 1:400 for women. The mean age of onset is 64 years, with the risk being highest at around 75 years of age. The global prevalence of ALS is estimated to be four to five cases per 100,000 people, with 1–2.6 new cases per 100,000 people annually. It is projected that there will be 380,000 cases of ALS worldwide by 2040 (Brown & Al-Chalabi, 2017; Tzeplaeff et al., 2023; Wolfson et al., 2023).

11.5.3.1 Pathophysiology

The pathological hallmarks of ALS include deterioration of the pyramidal Betz cells in the motor cortex, anterior horn cells of the spinal cord, and lower cranial motor nuclei of the brainstem. The degeneration of the spinal cord atrophies, along with the affected muscles, results in a variety of intracellular inclusions, common in degenerating neurons (Brown & Al-Chalabi, 2017; Saberi et al., 2015). The accretion of TDP-43 is present in most cases of sporadic ALS and some cases of familial

ALS (Piao et al., 2003). The interaction of epigenetics, a propagating channel (oxidative stress, excitotoxicity, defective RNA processing, intraneuronal protein aggregates, mitochondrial dysfunction, neuroinflammation, or protein misfolding), and final clinical manifestations have also been advocated. The precise underlying pathological pathway of ALS remains elusive and potentially multifaceted, posing significant challenges in finding effective therapies to retard or potentially halt disease progression. Currently, riluzole and edaravone are the approved drugs in ALS treatment, demonstrating moderate efficacy in disease progression.

11.5.3.2 Riluzole (Remarkable Molecule)

Riluzole, a glutamate antagonist, is used as an anticonvulsant (anticonvulsants) to perpetuate the survivance of ALS patients. Riluzole is marketed under the trade name Rilutek and with other names, which marginally extends life expectancy by 2–3 months without remarkable improvements in quality of life (Song & Narahashi, 1996). It is a lipophilic compound, with a chemical formula of $C_8H_5F_3N_2OS$ and a molecular mass of 234.2 g/mol; belongs to the benzothiazole drug class; and was approved by the FDA in 1995 and has been currently marketed in dozens of countries around the world since the late 1990s. Studies have reported that it reduces tau pathology and improves memory performance in TauP301L-expressing mice (Hunsberger et al., 2015). Moreover, studies on animal models demonstrated that riluzole prevents age-related changes in gene expression similar to those seen in AD (Pereira et al., 2017).

11.5.3.2.1 Mechanism of Action

Riluzole, a member of the benzothiazole class of drugs, is used to treat patients with ALS, a progressive and fatal neurodegenerative disorder. Riluzole is believed to have neuroprotective properties, as it has been shown to be effective in various experimental models of neuronal injury related to excitotoxic mechanisms. Even though the exact mechanism of how riluzole works is not known yet, it is widely accepted that it exerts an inhibitory influence on glutamate release, a key excitatory neurotransmitter associated with brain signaling and function (Martin et al., 1993). Riluzole's role as a glutamate inhibitor works by activating the G-protein-dependent signaling cascade and inactivating the voltage-gated sodium channels and interfering with intracellular events, following the binding of transmitters at excitatory amino acid receptors (Hubert et al., 1994).

11.5.3.2.2 Clinical Trials and Outcomes

Riluzole, a member of class benzothiazole, has been approved for the treatment of ALS patients due to its modest extension of survival or delay in the insertion of a breathing tube in several clinical trials. However, initial reports of symptomatic or slowdown in muscle degeneration were not consistently supported in subsequent studies (Group et al., 2013; Lacomblez et al., 1996). Moreover, Riluzole has accomplished clinical trials of phase 3 study for the treatment of PD and Huntington's disease (HD) but yielded negative outcomes, leading to the discontinuation of both programs (Bensimon et al., 2009; Landwehrmeyer et al., 2007). A phase 2 6-month clinical trial including 50 participants was conducted at the University of Rockefeller in April 2013 to evaluate the effect of riluzole in those AD patients who were previously treated with donepezil. This clinical trial studied the alterations in cognitive function and brain concentration of the neuronal viability marker *N*-acetyl aspartate (NAA) using magnetic resonance spectroscopy (MRS) and FDG-PET. Outcomes of this clinical showed that out of the 50 patients, 26 received riluzole 50 mg twice daily, and 24 received placebo tablets. The study concluded that glucose metabolism declined significantly less in the posterior cingulate of riluzole-treated patients compared to those on placebo. Additionally, FDG-PET exhibited a positive correlation with cognitive function. Though no variations were observed in NAA levels, glutamate was enhanced after 3 and 6 months of treatment (Matthews et al., 2021). Furthermore, riluzole is being evaluated for other neurologic and neuropsychiatric conditions in various trials, including ataxia, Fragile X, spinal cord injury, autism, bipolar disorder, obsessive-compulsive disorder (OCD), and SMA, as shown in Table 11.6.

TABLE 11.6
Showing the Detailed Clinical Trials of Riluzole

S. No.	Aim of Clinical Trial	Duration of Study	Total Number of Participants	Current Status
01	To evaluate the efficacy and safety of the antiglutamate agent riluzole	12 months	155	Completed
02	Double-blind, placebo-controlled, multicenter study to confirm findings and assess drug efficacy at different doses of riluzole	Less than 5 years	959	Completed
03	Clinical trial of riluzole as a promising disease-modifying agent in Parkinson plus disorders (NNIPPS: Neuroprotection and Natural History in Parkinson Plus Syndromes)	36 months	767	Completed
04	Double-blind phase 2 randomized clinical trial of riluzole in patients with a diagnosis of mild Alzheimer's disease	6 months	97	Completed

11.5.3.3 Edaravone (Outstanding Molecule)

Edaravone, commercially known as Radicava, is a crystalline solid with a chemical formula of $C_{10}H_{10}N_2O$ and a molecular mass of 174.2 g/mol. Initially authorized in Japan for acute ischemic stroke treatment in 2001, the FDA sanctioned its IV infusion for ALS treatment in 2017. Edaravone functions as a free radical scavenger, offering antioxidant effects to alleviate oxidative stress-related damage. It is believed to delay disease progression by inhibiting lipid peroxidation, providing neuroprotective effects against the impact of free hydroxyl and peroxynitrite radicals on neuronal health in conditions such as ischemic strokes and neurodegenerative disorders (Berry et al., 2021). Edaravone is a neuroprotective agent with antioxidant properties and operates as a free radical scavenger. Edaravone works to rummage ROS, involved in in ALS and cerebral ischemia. Edaravone received its first approval in Japan in 2001 for the treatment of ischemic stroke. Later, it acquired approval for the treatment of ALS in other countries including Japan, South Korea, and Canada. The oral suspension formulation of edaravone was approved in May 2022 by the FDA and Health Canada in November 2022. Edaravone had initially been granted orphan designation by the EMA on June 19, 2015, and was under regulatory review in Europe. Nonetheless, the drug manufacturer, Mitsubishi Tanabe Pharma, opted to withdraw the Marketing Authorization Application (MAA) for edaravone from the European market on May 24, 2019. This decision was made based on a feedback request by the Committee for Medicinal Products for Human Use (CHMP) for a long-term efficacy and safety study. Edaravone has also undergone investigation for potential use in other conditions such as AD, neuropathic pain, and ischemia nerve injury. In the United States and Canada, edaravone is indicated for the treatment of ALS, and it is also approved for treating acute ischemic stroke in Japan.

11.5.3.3.1 Mechanism of Action

Edaravone is known for its ability to slow down the progression of neurological disorders like ischemic stroke and ALS by reducing neuronal damage or death. It functions as a free radical scavenger, targeting hydroxyl radicals and peroxynitrite radicals to limit their generation (Watanabe et al., 2008). Although the exact mechanism of ALS is not fully understood, it is believed to deliver curative effects via its antioxidant properties and oxidative stress (Brooks et al., 2000). Considering that oxidative damage plays a role in the development of ALS and brain infraction, edaravone's ability to inhibit lipid peroxidation, suppress damage to endothelial cells induced by lipid peroxides, and scavenge free radicals may offer neuroprotective benefits (Lapin et al., 2020). Furthermore, studies have suggested that edaravone could have anti-inflammatory properties, as it has been found to inhibit neutrophil activation, suppress the expression of iNOS (inducible nitric oxide synthase) and nNOS (neuronal nitric oxide synthase), and ameliorate ROS-induced inflammatory oxidative stress after ischemic brain reperfusion and protect neurons from the damage caused by abnormal buildup of these molecules as cells consume energy (Lapin et al., 2020).

11.5.3.3.2 Clinical Trials and Outcomes

The clinical trial data seems to show mixed results, suggesting that the treatment may be more effective for certain patients due to the diverse nature of the disease. Although the initial phase III study did not demonstrate significant differences from placebo in overall efficacy, further investigation into specific subgroups of ALS patients revealed potential benefits for those with greater baseline functionality (Yoshino & Kimura, 2006). Additionally, a post-hoc analysis of the subsequent phase III trial suggested a statistically significant improvement in the ALS Functional Rating Scale Revised (ALSFRS-R) for patients with forced vital capacity of >80% and disease duration less than 2 years (Abe et al., 2014). In 2020, a study was conducted to examine the lasting effectiveness of administering edaravone intravenously during an open-label extension period. The results indicated a sustained significant improvement in ALSFRS-R scores for the edaravone treatment group compared to the placebo group over the 48-week period (Berry et al., 2021). However, a separate cohort study conducted by the German Motor Neuron Disease Network did not find any disease-modifying benefits associated with edaravone treatment, highlighting the need for further research. On the oral suspension front, a phase 3 clinical trial presented at the 32nd International Symposium of ALS/MND in December 2021 reported encouraging safety and tolerability results for a 24-week period, with no serious adverse events related to the treatment. Looking ahead, there's an ongoing long-term safety extension study of oral edaravone, which will provide insights into the treatment's effects over a 96-week period. This study aims to expand our understanding of the long-term efficacy and safety of oral edaravone, as shown in Table 11.7 (Berry et al., 2021).

11.5.3.3.3 Current Status and Future Directions

The antioxidant therapy with edaravone shows promising results in the treatment of NDDs such as ALS and AD. Various studies suggest that edaravone may have therapeutic effects on a broad spectrum of diseases related to oxidative stress. With the number of ALS cases expected to rise globally, efforts to optimize current treatments and develop new options are crucial. Personalizing treatment through genetic testing and early intervention is seen as essential for future therapies. Additionally, more studies are needed to understand the different mechanisms affecting the progression of ALS and to determine the efficacy and safety of edaravone in different stages, early, middle, and late stages, of the disease. Moreover, to firmly establish the efficacy and safety of edaravone as a treatment option, conducting more extensive studies and clinical trials outside of Japan, involving a larger global population, would provide highly valuable insights (Yamashita & Abe, 2024).

TABLE 11.7
Showing the Detailed Clinical Trials of Edaravone

S. No.	Aim of Clinical Trial	Duration of Study	Total Number of Participants	Current Status
01	Phase 2 clinical trial aimed to investigate the therapeutic effects of edaravone on ALS	6 months	29	Completed
02	To evaluate the safety and efficacy of edaravone in the treatment of amyotrophic lateral sclerosis (NCT00330681)	36 weeks	246	Completed
03	To investigate the effects of MCI-186 on the treatment of ALS, a phase 3 clinical trial	24 weeks	137	Completed
04	Radicava (edaravone) findings in biomarkers from ALS (REFINE-ALS) (NCT04259255)	24 weeks	300	Active

11.6 NEW AND EMERGING SMALL MOLECULE THERAPEUTIC APPROACHES

The development of small molecule drugs has played a key role in shaping medical advancements and increasing patient care. Small molecule therapeutics offer vital properties such as the ability to cross biological barriers, modulate various biological targets, and manifest oral bioavailability, which makes them suitable for standard oral delivery. Additionally, the modular structure of small molecules and their availability through chemical synthesis allow for rapid variations in their chemical structure and systematic improvement of their properties. These features continuously expand the potential for advanced approaches that can overcome the latest target spaces, contributing to the classical toolbox for various innovative approaches in the drug development process (Lamptey et al., 2022).

11.7 INNOVATIVE APPROACHES

The various innovative approaches in drug discovery and development, including RNA-targeting small molecules and antibody–drug conjugates (ADCs), have unlocked the doors to novel treatment modalities and paved the way for addressing previously undruggable biological targets. The emergence of RNA-targeting small molecules (RSMs) as a new treatment modality has shifted the perception of targeting human RNA, offering promising treatment options previously considered undruggable. The development of Evrysdi (risdiplam), the first human RSM drug, approved to treat SMA marked a significant milestone in the field (Costales et al., 2020; *Discovery of Risdiplam, a Selective Survival of Motor Neuron2 (SMN2) Gene Splicing Modifier for the Treatment of Spinal Muscular Atrophy (SMA)*, n.d.; Hargrove, 2020).

ADCs have further evolved as prominent examples of small molecule drug conjugates, involving the manipulation of highly selective antibodies to deliver small molecule drugs to their specific target sites and employ pharmacological activity with high specificity (Khongorzul et al., 2020). Furthermore, small molecules play a significant role in facilitating cell and gene therapy, with growing significance in improving the safety, efficacy, and manufacturing of these novel therapeutic modalities. Small molecule additives in cell therapy and gene therapy contribute to enhanced product quality, improved toxicological profiles, and therapeutic efficacy, promising advancements in the next generation of therapies (Schott et al., 2019).

11.8 ARTIFICIAL INTELLIGENCE (AI/ML)

The integration of digitalization and artificial intelligence (AI) has brought about a stupendous transformation in the field of small molecule drug discovery. Leveraging ML approaches has proven mainly effective in the context of small molecules because of their modular chemical structure, which facilitates translation into machine-readable formats (Savage, 2021; Venkatasubramanian & Mann, 2022). Furthermore, the availability of extensive historical data sets containing information on synthetic methods, physicochemical characteristics, and protein–target interactions further supports the application of ML in the development of small molecule therapeutics. The utilization of ML and computational methods aims to expedite this process, ultimately enabling the provision of optimized candidate compounds in a significantly reduced timeframe. The first three small molecules created with the assistance of AI have advanced to clinical trials, showcasing the new era of AI in small molecule drug discovery (Beck et al., 2022; Yingngam & Sethabouppha, 2024).

11.9 FUTURE DIRECTIONS

The chief attribute of small molecules is their flexible nature and virtually unlimited number of probabilities to produce biologically active molecules from the latest combinations of their components, addressing the vast virtual chemical space using computational methods, and AI will be important for sustained success in small molecule drug discovery (Hoffmann & Gastreich, 2019). Moreover, the influence of these small molecule drugs on medical treatment, lifestyle, and society as a whole is becoming progressively evident in addressing prevalent health concerns in NDDs in aging societies. Thus, there is hope for the establishment of promising future therapies. While there is currently no cure for NDDs, many candidate disease-modifying drugs have already reached clinical trials, with a significant proportion encompassing small molecules. This offers cautious optimism that better therapies for these diseases will become available in the future (Abbott, 2011; Cummings et al., 2019).

11.10 CHALLENGES IN DEVELOPING SMALL MOLECULE THERAPEUTICS FOR NEURODEGENERATION

Neurodegeneration has been identified as the key pathophysiological feature in neurodegenerative disorders (Merelli et al., 2013). Despite the unremitting efforts by contemporary science to develop surgical or medical remedies, the results have not been encouraging yet. The main challenges of the current small molecule therapeutics, which have led to the existing needs for advanced therapeutic approaches for the treatment of these NDDs, include the BBB, the main hurdle toward the effective management of NDDs (Daneman & Prat, 2015). The BBB is a tightly modulated barrier that controls the passage of substances from the bloodstream into the brain and consists of brain microvascular endothelial cells, tight junctions, neurons, astrocytes, and basal membranes, which all work together to limit the entry of molecules and proteins (Daneman & Prat, 2015). The BBB presents a formidable challenge for drug delivery due to the absence of fenestrations in brain capillary endothelial cells, which restricts the diffusion of substances (Hawkins & Davis, 2005; Serlin et al., 2015). Furthermore, inter-endothelial junctions and efflux transporters can limit the passage of molecules into the brain. Molecules that do cross the BBB must navigate through either the paracellular or transcellular pathway, and their physicochemical properties, such as size, molecular weight, lipid solubility, and charge, greatly influence their ability to traverse the barrier (Hladky & Barrand, 2018). The effectiveness of systematically administered drugs is heavily influenced by their pharmacokinetic properties (Dong, 2018). Once the drug is administered, it faces various challenges on its journey to the target site, predominantly the brain. Another prime challenge is the presence of various plasma proteins that can bind to drugs, reducing the amount of free drug accessible for carrying to the brain (Ghuman et al., 2005). Certain drugs are also eliminated rapidly by the body's clearance organs, leading to only a limited amount remaining in the bloodstream. Furthermore, the interaction between drugs and target cells can impact drug absorption. For example, drug molecules can affect cells in ways that hinder absorption, such as by blocking channels, altering membrane potential, or changing cell conformation (Krol, 2012). Thus, there is an urgent need to explore the pharmacokinetic framework of small molecules in relevant animal models that can effectively cross the BBB to reach the target site and provide medical breakthroughs for NDDs.

11.11 CONCLUSION

NDDs such as AD, PD, and ALS are highly intricate and heterogeneous. Owing to the nature of neurons in the human CNS, as well as the BBB that protects the brain, treating these diseases became arduous. Although new small molecule therapeutics have demonstrated their efficacy in preclinical studies, these treatment options are limited and symptomatic; thus, there is a dire need to discover new drugs to target the complexity of neurodegeneration. Researchers are exploring various innovative approaches to address this challenge. Moreover, medicinal chemists are focusing on tailoring existing molecules and developing new compounds to target NDDs. The prediction of a drug's ability to cross the BBB is crucial for drug development for the CNS. However, computational tools and in vitro screening are being used to evaluate the drug-like properties and permeability of potential CNS drug candidates. Furthermore, identifying new molecules that can potently target the biochemical pathways of NDDs will provide new opportunities for drug development and offer potential solutions for targeting complex NDDs and improving the pharmacokinetic properties of treatments. Thus, there is an optimism that, in the future, advanced disease-modifying tactics will be discovered, which will be proven to be successful.

CONFLICT OF INTEREST

None.

REFERENCES

7th Congress of the European Federation of Neurological Societies (EFNS 2003). (2002). *Autonomic Neuroscience*, *101*(1–2), p. III. https://doi.org/10.1016/s1566-0702(02)00221-7

Abbott, A. (2011). Dementia: a problem for our age. *Nature*, *475*(7355), pp. S2–S4. https://doi.org/10.1038/475s2a

Abe, K., Itoyama, Y., Sobue, G., Tsuji, S., Aoki, M., Doyu, M., Hamada, C., Kondo, K., Yoneoka, T., Akimoto, M., Yoshino, H., & Group, E. A. L. S. S. (2014). Confirmatory double-blind, parallel-group, placebo-controlled study of efficacy and safety of edaravone (MCI-186) in amyotrophic lateral sclerosis patients. *Amyotrophic Lateral Sclerosis & Frontotemporal Degeneration*, *15*(7–8), pp. 610–617. https://doi.org/10.3109/21678421.2014.959024

Alley, G. M., Bailey, J. A., Chen, D., Ray, B., Puli, L. K., Tanila, H., Banerjee, P. K., & Lahiri, D. K. (2010). Memantine lowers amyloid-beta peptide levels in neuronal cultures and in APP/PS1 transgenic mice. *Journal of Neuroscience Research*, *88*(1), pp. 143–154. https://doi.org/10.1002/jnr.22172

Angeloni, C., Malaguti, M., Prata, C., Freschi, M., Barbalace, M. C., & Hrelia, S. (2022). Mechanisms underlying neurodegenerative disorders and potential neuroprotective activity of agrifood by-products. *Antioxidants (Basel, Switzerland)*, *12*(1), p. 94. https://doi.org/10.3390/antiox12010094

Armstrong, M. J., & Okun, M. S. (2020). Diagnosis and treatment of Parkinson disease. *JAMA*, *323*(6), p. 548. https://doi.org/10.1001/jama.2019.22360

Balestrino, R., & Schapira, A. H. V. (2019). Parkinson disease. *European Journal of Neurology*, *27*(1), pp. 27–42. https://doi.org/10.1111/ene.14108

Beck, H., Härter, M., Haß, B., Schmeck, C., & Baerfacker, L. (2022). Small molecules and their impact in drug discovery: a perspective on the occasion of the 125th anniversary of the Bayer chemical research laboratory. *Drug Discovery Today*, *27*(6), pp. 1560–1574. https://doi.org/10.1016/j.drudis.2022.02.015

Beck, H., Jeske, M., Thede, K., Stoll, F., Flamme, I., Akbaba, M., Ergüden, J.-K., Karig, G., Keldenich, J., Oehme, F., Militzer, H.-C., Hartung, I. V, & Thuss, U. (2018). Discovery of molidustat (BAY 85–3934): a small-molecule oral HIF-prolyl hydroxylase (HIF-PH) inhibitor for the treatment of renal anemia. *ChemMedChem*, *13*(10), pp. 988–1003. https://doi.org/10.1002/cmdc.201700783

Bensimon, G., Ludolph, A., Agid, Y., Vidailhet, M., Payan, C., Leigh, P. N., & Group, N. S. (2009). Riluzole treatment, survival and diagnostic criteria in Parkinson plus disorders: the NNIPPS study. *Brain: A Journal of Neurology*, *132*(Pt 1), pp. 156–171. https://doi.org/10.1093/brain/awn291

Berry, J., Brooks, B., Genge, A., Heiman-Patterson, T., Appel, S., Benatar, M., Bowser, R., Cudkowicz, M., Gooch, C., Shefner, J., Westra, J., Agnese, W., Merrill, C., Nelson, S., & Apple, S. (2021). Radicava/Edaravone findings in biomarkers from amyotrophic lateral sclerosis (REFINE-ALS): protocol and study design. *Neurology. Clinical Practice*, *11*(4), pp. e472–e479. https://doi.org/10.1212/CPJ.0000000000000968

Bhat, S. A., Ahamad, S., Dar, N. J., Siddique, Y. H., & Nazir, A. (2023). The emerging landscape of natural small-molecule therapeutics for Huntington's disease. *Current Neuropharmacology*, *21*(4), pp. 867–889. https://doi.org/10.2174/1570159X21666230216104621

Bonnet, A. M., & Houeto, J. L. (1999). Pathophysiology of Parkinson's disease. *Biomedicine & Pharmacotherapy*, *53*(3), pp. 117–121. https://doi.org/10.1016/s0753-3322(99)80076-6

Bonuccelli, U., & Del Dotto, P. (2006). New pharmacologic horizons in the treatment of Parkinson disease. *Neurology*, *67*(7_suppl_2). https://doi.org/10.1212/wnl.67.7_suppl_2.s30

Bordelon, Y., & Keener, A. (2016). Parkinsonism. *Seminars in Neurology*, *36*(4), pp. 330–334. https://doi.org/10.1055/s-0036-1585097

Brooks, B. R., Miller, R. G., Swash, M., & Munsat, T. L. (2000). El Escorial revisited: revised criteria for the diagnosis of amyotrophic lateral sclerosis. *Amyotrophic Lateral Sclerosis and Other Motor Neuron Disorders*, *1*(5), pp. 293–299. https://doi.org/10.1080/146608200300079536

Brown, R. H., & Al-Chalabi, A. (2017). Amyotrophic lateral sclerosis. *New England Journal of Medicine*, *377*(2), pp. 162–172. https://doi.org/10.1056/nejmra1603471

Burns, A., Rossor, M., Hecker, J., Gauthier, S., Petit, H., Möller, H.-J., Rogers, S. L., & Friedhoff, L. T. (1999). The effects of Donepezil in Alzheimer's disease – results from a multinational trial *Dementia and Geriatric Cognitive Disorders*, *10*(3), pp. 237–244. https://doi.org/10.1159/000017126

Cacabelos, R., Takeda, M., & Winblad, B. (1999). The glutamatergic system and neurodegeneration in dementia: preventive strategies in Alzheimer's disease. *International Journal of Geriatric Psychiatry*, *14*(1), pp. 3–47. https://doi.org/10.1002/(sici)1099-1166(199901)14:1<3::aid-gps897>3.0.co;2–7

Calabrese, G., Molzahn, C., & Mayor, T. (2022). Protein interaction networks in neurodegenerative diseases: from physiological function to aggregation. *Journal of Biological Chemistry*, *298*(7), p. 102062. https://doi.org/10.1016/j.jbc.2022.102062

Chaudhuri, K. R., Martinez-Martin, P., Schapira, A. H. V, Stocchi, F., Sethi, K., Odin, P., Brown, R. G., Koller, W., Barone, P., MacPhee, G., Kelly, L., Rabey, M., MacMahon, D., Thomas, S., Ondo, W., Rye, D., Forbes, A., Tluk, S., Dhawan, V., … Olanow, C. W. (2006). International multicenter pilot study of the first comprehensive self-completed nonmotor symptoms questionnaire for Parkinson's disease: The NMSQuest study. *Movement Disorders*, *21*(7), pp. 916–923. https://doi.org/10.1002/mds.20844

Congdon, E. E., & Sigurdsson, E. M. (2018). Tau-targeting therapies for Alzheimer disease. *Nature Reviews. Neurology*, *14*(7), 399–415. https://doi.org/10.1038/s41582-018-0013-z

Costales, M. G., Childs-Disney, J. L., Haniff, H. S., & Disney, M. D. (2020). How we think about targeting RNA with small molecules. *Journal of Medicinal Chemistry*, *63*(17), pp. 8880–8900. https://doi.org/10.1021/acs.jmedchem.9b01927

Cova, I., & Priori, A. (2018). Diagnostic biomarkers for Parkinson's disease at a glance: where are we? *Journal of Neural Transmission (Vienna, Austria : 1996)*, *125*(10), pp. 1417–1432. https://doi.org/10.1007/s00702-018-1910-4

Cummings, J., Lee, G., Ritter, A., Sabbagh, M., & Zhong, K. (2019). Alzheimer's disease drug development pipeline: 2019. *Alzheimer's & Dementia (New York, N. Y.)*, *5*, pp. 272–293. https://doi.org/10.1016/j.trci.2019.05.008

Daneman, R., & Prat, A. (2015). The blood–brain barrier. *Cold Spring Harbor Perspectives in Biology*, *7*(1), p. a020412. https://doi.org/10.1101/cshperspect.a020412

Dembitsky, V. M., Dzhemileva, L., Gloriozova, T., & D'yakonov, V. (2020). Natural and synthetic drugs used for the treatment of the dementia. *Biochemical and Biophysical Research Communications*, *524*(3), pp. 772–783. https://doi.org/10.1016/j.bbrc.2020.01.123

Discovery of Risdiplam, a Selective Survival of Motor Neuron2 (SMN2) Gene Splicing Modifier for the Treatment of Spinal Muscular Atrophy (SMA). (n.d.). American Chemical Society (ACS). https://doi.org/10.1021/acs.jmedchem.8b00741.s001

Dong, X. (2018). Current strategies for brain drug delivery. *Theranostics*, *8*(6), pp. 1481–1493. https://doi.org/10.7150/thno.21254

Doody, R. S., Ferris, S. H., Salloway, S., Sun, Y., Goldman, R., Watkins, W. E., Xu, Y., & Murthy, A. K. (2009). Donepezil treatment of patients with MCI. *Neurology*, *72*(18), pp. 1555–1561. https://doi.org/10.1212/01.wnl.0000344650.95823.03

Ebada, M. A., Alkanj, S., Ebada, M., Abdelkarim, A. H., Diab, A., Aziz, M. A. E., Soliman, A. M., Fayed, N., Bahbah, E. I., & Negida, A. (2019). Safety and efficacy of Levetiracetam for the management of Levodopa-induced Dyskinesia in patients with Parkinson's disease: a systematic review. *CNS & Neurological Disorders - Drug Targets*, *18*(4), pp. 317–325. https://doi.org/10.2174/1871527318666190314101314

Espay, A. J., Bonato, P., Nahab, F. B., Maetzler, W., Dean, J. M., Klucken, J., Eskofier, B. M., Merola, A., Horak, F., Lang, A. E., Reilmann, R., Giuffrida, J., Nieuwboer, A., Horne, M., Little, M. A., Litvan, I., Simuni, T., Dorsey, E. R., Burack, M. A., … Technology, M. D. S. T. F. on. (2016). Technology in Parkinson's disease: challenges and opportunities. *Movement Disorders: Official Journal of the Movement Disorder Society*, *31*(9), pp. 1272–1282. https://doi.org/10.1002/mds.26642

Fan, J., Dawson, T. M., & Dawson, V. L. (2017). Cell death mechanisms of neurodegeneration. In *Advances in Neurobiology* (pp. 403–425). Springer International Publishing. https://doi.org/10.1007/978-3-319-57193-5_16

Ghuman, J., Zunszain, P. A., Petitpas, I., Bhattacharya, A. A., Otagiri, M., & Curry, S. (2005). Structural basis of the drug-binding specificity of human serum albumin. *Journal of Molecular Biology*, *353*(1), pp. 38–52. https://doi.org/10.1016/j.jmb.2005.07.075

Gouda, N. A., Elkamhawy, A., & Cho, J. (2022). Emerging therapeutic strategies for Parkinson's disease and future prospects: a 2021 update. *Biomedicines, 10*(2), p. 371. https://doi.org/10.3390/biomedicines10020371

Grossberg, G. T., Manes, F., Allegri, R. F., Gutiérrez-Robledo, L. M., Gloger, S., Xie, L., Jia, X. D., Pejović, V., Miller, M. L., Perhach, J. L., & Graham, S. M. (2013). The safety, tolerability, and efficacy of once-daily memantine (28 mg): a multinational, randomized, double-blind, placebo-controlled trial in patients with moderate-to-severe Alzheimer's disease taking cholinesterase inhibitors. *CNS Drugs, 27*(6), pp. 469–478. https://doi.org/10.1007/s40263-013-0077-7

Group, U.-L. S., Morrison, K. E., Dhariwal, S., Hornabrook, R., Savage, L., Burn, D. J., Khoo, T. K., Kelly, J., Murphy, C. L., Al-Chalabi, A., Dougherty, A., Leigh, P. N., Wijesekera, L., Thornhill, M., Ellis, C. M., O'Hanlon, K., Panicker, J., Pate, L., Ray, P., … Majid, T. (2013). Lithium in patients with amyotrophic lateral sclerosis (LiCALS): a phase 3 multicentre, randomised, double-blind, placebo-controlled trial. *The Lancet. Neurology, 12*(4), pp. 339–345. https://doi.org/10.1016/S1474-4422(13)70037-1

Gurevich, E. V, & Gurevich, V. V. (2014). Therapeutic potential of small molecules and engineered proteins. *Handbook of Experimental Pharmacology, 219*, pp. 1–12. https://doi.org/10.1007/978-3-642-41199-1_1

Guzior, N., Wieckowska, A., Panek, D., & Malawska, B. (2015). Recent development of multifunctional agents as potential drug candidates for the treatment of Alzheimer's disease. *Current Medicinal Chemistry, 22*(3), pp. 373–404. https://doi.org/10.2174/0929867321666141106122628

Hampel, H., O'Bryant, S. E., Durrleman, S., Younesi, E., Rojkova, K., Escott-Price, V., Corvol, J.-C., Broich, K., Dubois, B., & Lista, S. (2017). A precision medicine initiative for Alzheimer's disease: the road ahead to biomarker-guided integrative disease modeling. *Climacteric, 20*(2), pp. 107–118. https://doi.org/10.1080/13697137.2017.1287866

Han, H. J., Park, M. Y., Park, K. W., Park, K. H., Choi, S. H., Kim, H.-J., Yang, D. W., Ebenezer, E. G. A. M., Yang, Y.-H., Kewalram, G. M., Han, S.-H., & Study, on B. the I.-301. (2022). A multinational, multicenter, randomized, double-blind, active comparator, Phase III clinical trial to evaluate the efficacy and safety of Donepezil transdermal patch in patients with Alzheimer's disease. *Journal of Clinical Neurology (Seoul, Korea), 18*(4), pp. 428–436. https://doi.org/10.3988/jcn.2022.18.4.428

Harding, B. N., Kariya, S., Monani, U. R., Chung, W. K., Benton, M., Yum, S. W., Tennekoon, G., & Finkel, R. S. (2015). Spectrum of neuropathophysiology in spinal muscular atrophy type I. *Journal of Neuropathology and Experimental Neurology, 74*(1), pp. 15–24. https://doi.org/10.1097/NEN.0000000000000144

Hargrove, A. E. (2020). Small molecule-RNA targeting: starting with the fundamentals. *Chemical Communications (Cambridge, England), 56*(94), pp. 14744–14756. https://doi.org/10.1039/d0cc06796b

Hauser, R. A. (2008). Levodopa: past, present, and future. *European Neurology, 62*(1), pp. 1–8. https://doi.org/10.1159/000215875

Hauser, R. A., Hsu, A., Kell, S., Espay, A. J., Sethi, K., Stacy, M., Ondo, W., O'Connell, M., & Gupta, S. (2013). Extended-release carbidopa-levodopa (IPX066) compared with immediate-release carbidopa-levodopa in patients with Parkinson's disease and motor fluctuations: a phase 3 randomised, double-blind trial. *The Lancet Neurology, 12*(4), pp. 346–356. https://doi.org/10.1016/s1474-4422(13)70025-5

Hawkins, B. T., & Davis, T. P. (2005). The blood-brain barrier/neurovascular unit in health and disease. *Pharmacological Reviews, 57*(2), pp. 173–185. https://doi.org/10.1124/pr.57.2.4

Hladky, S. B., & Barrand, M. A. (2018). Elimination of substances from the brain parenchyma: efflux via perivascular pathways and via the blood-brain barrier. *Fluids and Barriers of the CNS, 15*(1), p. 30. https://doi.org/10.1186/s12987-018-0113-6

Hoffmann, T., & Gastreich, M. (2019). The next level in chemical space navigation: going far beyond enumerable compound libraries. *Drug Discovery Today, 24*(5), pp. 1148–1156. https://doi.org/10.1016/j.drudis.2019.02.013

Howard, R., McShane, R., Lindesay, J., Ritchie, C., Baldwin, A., Barber, R., Burns, A., Dening, T., Findlay, D., Holmes, C., Jones, R., Jones, R., McKeith, I., Macharouthu, A., O'Brien, J., Sheehan, B., Juszczak, E., Katona, C., Hills, R., … Phillips, P. P. J. (2015). Nursing home placement in the Donepezil and Memantine in moderate to severe Alzheimer's disease (DOMINO-AD) trial: secondary and post-hoc analyses. *The Lancet Neurology, 14*(12), pp. 1171–1181. https://doi.org/10.1016/s1474-4422(15)00258-6

Hu, H., Tan, C.-C., Tan, L., & Yu, J.-T. (2016). A Mitocentric view of Alzheimer's disease. *Molecular Neurobiology, 54*(8), pp. 6046–6060. https://doi.org/10.1007/s12035-016-0117-7

Huang, S.-W., Wang, W., Zhang, M.-Y., Liu, Q.-B., Luo, S.-Y., Peng, Y., Sun, B., Wu, D.-L., & Song, S.-J. (2016). The effect of ethyl acetate extract from persimmon leaves on Alzheimer's disease and its underlying mechanism. *Phytomedicine, 23*(7), pp. 694–704. https://doi.org/10.1016/j.phymed.2016.03.009

Hubert, J. P., Delumeau, J. C., Glowinski, J., Prémont, J., & Doble, A. (1994). Antagonism by riluzole of entry of calcium evoked by NMDA and veratridine in rat cultured granule cells: evidence for a dual mechanism of action. *British Journal of Pharmacology, 113*(1), pp. 261–267. https://doi.org/10.1111/j.1476-5381.1994.tb16203.x

Hunsberger, H. C., Weitzner, D. S., Rudy, C. C., Hickman, J. E., Libell, E. M., Speer, R. R., Gerhardt, G. A., & Reed, M. N. (2015). Riluzole rescues glutamate alterations, cognitive deficits, and tau pathology associated with P301L tau expression. *Journal of Neurochemistry*, *135*(2), pp. 381–394. https://doi.org/10.1111/jnc.13230

Iadanza, M. G., Jackson, M. P., Hewitt, E. W., Ranson, N. A., & Radford, S. E. (2018). A new era for understanding amyloid structures and disease. *Nature Reviews Molecular Cell Biology*, *19*(12), pp. 755–773. https://doi.org/10.1038/s41580-018-0060-8

Ikeda, M., Mori, E., Kosaka, K., Iseki, E., Hashimoto, M., Matsukawa, N., Matsuo, K., & Nakagawa, M. (2013). Long-term safety and efficacy of Donepezil in patients with Dementia with Lewy bodies: results from a 52-week, open-label, multicenter extension study. *Dementia and Geriatric Cognitive Disorders*, *36*(3–4), pp. 229–241. https://doi.org/10.1159/000351672

Ikeda, M., Mori, E., Matsuo, K., Nakagawa, M., & Kosaka, K. (2015). Donepezil for dementia with Lewy bodies: a randomized, placebo-controlled, confirmatory phase III trial. *Alzheimer's Research & Therapy*, *7*(1), p. 4. https://doi.org/10.1186/s13195-014-0083-0

Jain, N., & Chen-Plotkin, A. S. (2018). Genetic modifiers in neurodegeneration. *Current Genetic Medicine Reports*, *6*(1), pp. 11–19. https://doi.org/10.1007/s40142-018-0133-1

Jellinger, K. A. (2010). Basic mechanisms of neurodegeneration: a critical update. *Journal of Cellular and Molecular Medicine*, *14*(3), pp. 457–487. https://doi.org/10.1111/j.1582-4934.2010.01010.x

Johannsen, P., Salmon, E., Hampel, H., Xu, Y., Richardson, S., Qvitzau, S., & Schindler, R. (2006). Assessing therapeutic efficacy in a progressive disease. *CNS Drugs*, *20*(4), pp. 311–325. https://doi.org/10.2165/00023210-200620040-00005

Ju, C., Wong, I. C. K., Lau, W. C. Y., Man, K. K. C., Brauer, R., Ma, T., Alsharif, A., Alwafi, H., Lau, K. K., Chan, E. W., Chui, C. S. L., Li, X., & Wei, L. (2021). Global trends in symptomatic medication use against dementia in 66 countries/regions from 2008 to 2018. *European Journal of Neurology*, *28*(12), pp. 3979–3989. https://doi.org/10.1111/ene.15053

Khongorzul, P., Ling, C. J., Khan, F. U., Ihsan, A. U., & Zhang, J. (2020). Antibody–drug conjugates: a comprehensive review. *Molecular Cancer Research*, *18*(1), pp. 3–19. https://doi.org/10.1158/1541-7786.mcr-19-0582

Kouli, A., Torsney, K. M., & Kuan, W.-L. (2018). Parkinson's disease: etiology, neuropathology, and pathogenesis. In *Parkinson's Disease: Pathogenesis and Clinical Aspects* (pp. 3–26). Codon Publications. https://doi.org/10.15586/codonpublications.parkinsonsdisease.2018.ch1

Krol, S. (2012). Challenges in drug delivery to the brain: nature is against us. *Journal of Controlled Release*, *164*(2), pp. 145–155. https://doi.org/10.1016/j.jconrel.2012.04.044

Kuns, B., Rosani, A., Patel, P., & Varghese, D. (2024). Memantine. In *StatPearls*. https://www.ncbi.nlm.nih.gov/pubmed/14530799

Kwan, L. C., & Whitehill, T. L. (2011). Perception of speech by individuals with Parkinson's disease: a review. *Parkinson's Disease*, 2011, p. 389767. https://doi.org/10.4061/2011/389767

Lacomblez, L., Bensimon, G., Meininger, V., Leigh, P. N., & Guillet, P. (1996). Dose-ranging study of riluzole in amyotrophic lateral sclerosis. *The Lancet*, *347*(9013), pp. 1425–1431. https://doi.org/10.1016/s0140-6736(96)91680-3

Lamptey, R. N. L., Chaulagain, B., Trivedi, R., Gothwal, A., Layek, B., & Singh, J. (2022). A Review of the common neurodegenerative disorders: current therapeutic approaches and the potential role of nanotherapeutics. *International Journal of Molecular Sciences*, *23*(3), p. 1851. https://doi.org/10.3390/ijms23031851

Landwehrmeyer, G. B., Dubois, B., de Yébenes, J. G., Kremer, B., Gaus, W., Kraus, P. H., Przuntek, H., Dib, M., Doble, A., Fischer, W., & Ludolph, A. C. (2007). Riluzole in Huntington's disease: a 3-year, randomized controlled study. *Annals of Neurology*, *62*(3), pp. 262–272. https://doi.org/10.1002/ana.21181

Lane, C. A., Hardy, J., & Schott, J. M. (2017). Alzheimer's disease. *European Journal of Neurology*, *25*(1), pp. 59–70. https://doi.org/10.1111/ene.13439

Lapin, N. A., Gill, K., Shah, B. R., & Chopra, R. (2020). Consistent opening of the blood brain barrier using focused ultrasound with constant intravenous infusion of microbubble agent. *Scientific Reports*, *10*(1), p. 16546. https://doi.org/10.1038/s41598-020-73312-9

Leegwater-Kim, J., & Bortan, E. (2010). The role of rasagiline in the treatment of Parkinson's disease. *Clinical Interventions in Aging*, *5*, pp. 149–156. https://doi.org/10.2147/cia.s4145

Lemire, P., Brangier, A., Beaudenon, M., Duval, G. T., & Annweiler, C. (2018). Cognitive changes under memantine according to vitamin D status in Alzheimer patients: an exposed/unexposed cohort pilot study. *The Journal of Steroid Biochemistry and Molecular Biology*, *175*, pp. 151–156. https://doi.org/10.1016/j.jsbmb.2016.12.019

Levine, S. Z., Goldberg, Y., Yoshida, K., Samara, M., Cipriani, A., Iwatsubo, T., Leucht, S., & Furukawa, T. A. (2022). Early- and subsequent- response of cognitive functioning in Alzheimer's disease: individual-participant data from five pivotal randomized clinical trials of donepezil. *Journal of Psychiatric Research, 148*, pp. 159–164. https://doi.org/10.1016/j.jpsychires.2022.01.055

Lieberman, O. J., Lee, S., & Zabinski, J. (2023). Donepezil treatment is associated with improved outcomes in critically ill dementia patients via a reduction in delirium. *Alzheimer's & Dementia : The Journal of the Alzheimer's Association, 19*(5), pp. 1742–1751. https://doi.org/10.1002/alz.12807

Lipinski, C. A., Lombardo, F., Dominy, B. W., & Feeney, P. J. (1997). Experimental and computational approaches to estimate solubility and permeability in drug discovery and development settings. *Advanced Drug Delivery Reviews, 23*(1–3), pp. 3–25. https://doi.org/10.1016/s0169-409x(96)00423-1

Lipton, S. (2007). Pathologically-Activated Therapeutics for Neuroprotection: Mechanism of NMDA Receptor Block by Memantine and S-Nitrosylation. *Current Drug Targets, 8*(5), pp. 621–632. https://doi.org/10.2174/138945007780618472

Lipton, S. A. (2006). Paradigm shift in neuroprotection by NMDA receptor blockade: Memantine and beyond. *Nature Reviews Drug Discovery, 5*(2), pp. 160–170. https://doi.org/10.1038/nrd1958

Liu, K. Y., & Howard, R. (2021). Can we learn lessons from the FDA's approval of aducanumab? *Nature Reviews Neurology, 17*(11), pp. 715–722. https://doi.org/10.1038/s41582-021-00557-x

Liu, Z., Zhang, A., Sun, H., Han, Y., Kong, L., & Wang, X. (2017). Two decades of new drug discovery and development for Alzheimer's disease. *RSC Advances, 7*(10), pp. 6046–6058. https://doi.org/10.1039/c6ra26737h

Longhena, F., Spano, P., & Bellucci, A. (2017). Targeting of disordered proteins by small molecules in neurodegenerative diseases. In *Handbook of Experimental Pharmacology* (pp. 85–110). Springer International Publishing. https://doi.org/10.1007/164_2017_60

Lu, J., Wang, X., Wan, L., Fu, J., Huo, Y., Zhao, Y., & Guo, C. (2020). Gene polymorphisms affecting the pharmacokinetics and pharmacodynamics of Donepezil efficacy. *Frontiers in Pharmacology, 11*, p. 934. https://doi.org/10.3389/fphar.2020.00934

Martin, D., Thompson, M. A., & Nadler, J. V. (1993). The neuroprotective agent riluzole inhibits release of glutamate and aspartate from slices of hippocampal area CA1. *European Journal of Pharmacology, 250*(3), pp. 473–476. https://doi.org/10.1016/0014-2999(93)90037-i

Martin, J. B. (1999). Molecular basis of the neurodegenerative disorders. *New England Journal of Medicine, 340*(25), pp. 1970–1980. https://doi.org/10.1056/nejm199906243402507

Matilla-Dueñas, A., Corral-Juan, M., Rodríguez-Palmero Seuma, A., Vilas, D., Ispierto, L., Morais, S., Sequeiros, J., Alonso, I., Volpini, V., Serrano-Munuera, C., Pintos-Morell, G., Álvarez, R., & Sánchez, I. (2017). Rare neurodegenerative diseases: clinical and genetic update. In *Advances in Experimental Medicine and Biology* (pp. 443–496). Springer International Publishing. https://doi.org/10.1007/978-3-319-67144-4_25

Matthews, D. C., Mao, X., Dowd, K., Tsakanikas, D., Jiang, C. S., Meuser, C., Andrews, R. D., Lukic, A. S., Lee, J., Hampilos, N., Shafiian, N., Sano, M., David Mozley, P., Fillit, H., McEwen, B. S., Shungu, D. C., & Pereira, A. C. (2021). Riluzole, a glutamate modulator, slows cerebral glucose metabolism decline in patients with Alzheimer's disease. *Brain, 144*(12), pp. 3742–3755. https://doi.org/10.1093/brain/awab222

McShane, R., Westby, M. J., Roberts, E., Minakaran, N., Schneider, L., Farrimond, L. E., Maayan, N., Ware, J., & Debarros, J. (2019). Memantine for dementia. *The Cochrane Database of Systematic Reviews, 3*(3), p. CD003154. https://doi.org/10.1002/14651858.CD003154.pub6

Mendez, M. F. (2017). Early-onset Alzheimer disease. *Neurologic Clinics, 35*(2), pp. 263–281. https://doi.org/10.1016/j.ncl.2017.01.005

Merelli, A., Czornyj, L., & Lazarowski, A. (2013). Erythropoietin: a neuroprotective agent in Cerebral Hypoxia, neurodegeneration, and epilepsy. *Current Pharmaceutical Design, 19*(38), pp. 6791–6801. https://doi.org/10.2174/1381612811319380011

Mohsin, N. ul A., & Ahmad, M. (2020). Donepezil: a review of the recent structural modifications and their impact on anti-Alzheimer activity. *Brazilian Journal of Pharmaceutical Sciences, 56*. https://doi.org/10.1590/s2175-97902019000418325

Mullane, K., & Williams, M. (2019). Preclinical models of Alzheimer's disease: relevance and translational validity. *Current Protocols in Pharmacology, 84*(1). https://doi.org/10.1002/cpph.57

Naoi, M., & Maruyama, W. (2009). Functional mechanism of neuroprotection by inhibitors of type B monoamine oxidase in Parkinson's disease. *Expert Review of Neurotherapeutics, 9*(8), pp. 1233–1250. https://doi.org/10.1586/ern.09.68

Noble, W., & Burns, M. P. (2010). Challenges in neurodegeneration research. *Frontiers in Psychiatry*, *1*, p. 7. https://doi.org/10.3389/fpsyt.2010.00007

Nyholm, D. (2006). Enteral levodopa/carbidopa gel infusion for the treatment of motor fluctuations and dyskinesias in advanced Parkinson's disease. *Expert Review of Neurotherapeutics*, *6*(10), pp. 1403–1411. https://doi.org/10.1586/14737175.6.10.1403

Ogungbenro, K., Pertinez, H., & Aarons, L. (2015). Empirical and semi-mechanistic modelling of double-peaked pharmacokinetic profile phenomenon due to gastric emptying. *The AAPS Journal*, *17*(1), pp. 227–236. https://doi.org/10.1208/s12248-014-9693-5

Olanow, C. W., Kieburtz, K., Odin, P., Espay, A. J., Standaert, D. G., Fernandez, H. H., Vanagunas, A., Othman, A. A., Widnell, K. L., Robieson, W. Z., Pritchett, Y., Chatamra, K., Benesh, J., Lenz, R. A., Antonini, A., & Group, L. H. S. (2014). Continuous intrajejunal infusion of levodopa-carbidopa intestinal gel for patients with advanced Parkinson's disease: a randomised, controlled, double-blind, double-dummy study. *The Lancet. Neurology*, *13*(2), pp. 141–149. https://doi.org/10.1016/S1474-4422(13)70293-X

Pereira, A. C., Gray, J. D., Kogan, J. F., Davidson, R. L., Rubin, T. G., Okamoto, M., Morrison, J. H., & McEwen, B. S. (2017). Age and Alzheimer's disease gene expression profiles reversed by the glutamate modulator riluzole. *Molecular Psychiatry*, *22*(2), pp. 296–305. https://doi.org/10.1038/mp.2016.33

Pérez-Arancibia, R., Cisternas-Olmedo, M., Sepúlveda, D., Troncoso-Escudero, P., & Vidal, R. L. (2023). Small molecules to perform big roles: the search for Parkinson's and Huntington's disease therapeutics. *Frontiers in Neuroscience*, *16*. https://doi.org/10.3389/fnins.2022.1084493

Peters, O., Fuentes, M., Joachim, L. K., Jessen, F., Luckhaus, C., Kornhuber, J., Pantel, J., Hüll, M., Schmidtke, K., Rüther, E., Möller, H.-J., Kurz, A., Wiltfang, J., Maier, W., Wiese, B., Frölich, L., & Heuser, I. (2015). Combined treatment with memantine and galantamine-CR compared with galantamine-CR only in antidementia drug naïve patients with mild-to-moderate Alzheimer's disease. *Alzheimer's & Dementia (New York, N. Y.)*, *1*(3), pp. 198–204. https://doi.org/10.1016/j.trci.2015.10.001

Piao, Y.-S., Wakabayashi, K., Kakita, A., Yamada, M., Hayashi, S., Morita, T., Ikuta, F., Oyanagi, K., & Takahashi, H. (2003). Neuropathology with clinical correlations of sporadic amyotrophic lateral sclerosis: 102 autopsy cases examined between 1962 and 2000. *Brain Pathology (Zurich, Switzerland)*, *13*(1), pp. 10–22. https://doi.org/10.1111/j.1750-3639.2003.tb00002.x

Piette, F., Belmin, J., Vincent, H., Schmidt, N., Pariel, S., Verny, M., Marquis, C., Mely, J., Hugonot-Diener, L., Kinet, J.-P., Dubreuil, P., Moussy, A., & Hermine, O. (2011). Masitinib as an adjunct therapy for mild-to-moderate Alzheimer's disease: a randomised, placebo-controlled phase 2 trial. *Alzheimer's Research & Therapy*, *3*(2), p. 16. https://doi.org/10.1186/alzrt75

Poduval, P., Parsekar, S., & Meena, S. N. (2023). Small molecules vs biologics. In *New Horizons in Natural Compound Research* (pp. 179–199). Elsevier. https://doi.org/10.1016/b978-0-443-15232-0.00001-1

Poewe, W., Seppi, K., Tanner, C. M., Halliday, G. M., Brundin, P., Volkmann, J., Schrag, A.-E., & Lang, A. E. (2017). Parkinson disease. *Nature Reviews Disease Primers*, *3*(1). https://doi.org/10.1038/nrdp.2017.13

Pooladgar, P., Sakhabakhsh, M., Taghva, A., & Soleiman-Meigooni, S. (2022). Donepezil beyond Alzheimer's disease? a narrative review of therapeutic potentials of Donepezil in different diseases. *Iranian Journal of Pharmaceutical Research : IJPR*, *21*(1), p. e128408. https://doi.org/10.5812/ijpr-128408

Quinn, C., & Elman, L. (2020). Amyotrophic lateral sclerosis and other motor neuron diseases. *Continuum: Lifelong Learning in Neurology*, *26*(5), pp. 1323–1347. https://doi.org/10.1212/con.0000000000000911

Rapp, T., Chauvin, P., Costa, N., & Molinier, L. (2015). Health economic considerations in neurodegenerative disorders. In *Imaging in Neurodegenerative Disorders* (pp. 42–53). Oxford University Press. https://doi.org/10.1093/med/9780199671618.003.0004

Reitz, C., Brayne, C., & Mayeux, R. (2011). Epidemiology of Alzheimer disease. *Nature Reviews. Neurology*, *7*(3), pp. 137–152. https://doi.org/10.1038/nrneurol.2011.2

Rogawski, M. A., & Wenk, G. L. (2003). The neuropharmacological basis for the use of memantine in the treatment of Alzheimer's disease. *CNS Drug Reviews*, *9*(3), pp. 275–308. https://doi.org/10.1111/j.1527-3458.2003.tb00254.x

Rogers, S. L., Farlow, M. R., Doody, R. S., Mohs, R., & Friedhoff, L. T. (1998). A 24-week, double-blind, placebo-controlled trial of donepezil in patients with Alzheimer's disease. *Neurology*, *50*(1), pp. 136–145. https://doi.org/10.1212/wnl.50.1.136

Romagnolo, A., Merola, A., Artusi, C. A., Rizzone, M. G., Zibetti, M., & Lopiano, L. (2018). Levodopa-induced neuropathy: a systematic review. *Movement Disorders Clinical Practice*, *6*(2), pp. 96–103. https://doi.org/10.1002/mdc3.12688

Saberi, S., Stauffer, J. E., Schulte, D. J., & Ravits, J. (2015). Neuropathology of amyotrophic lateral sclerosis and its variants. *Neurologic Clinics*, *33*(4), pp. 855–876. https://doi.org/10.1016/j.ncl.2015.07.012

Safirstein, B. E., Ellenbogen, A., Zhao, P., Henney, H. R., Kegler-Ebo, D. M., & Oh, C. (2020). Pharmacokinetics of inhaled Levodopa administered with oral Carbidopa in the fed state in patients with Parkinson's disease. *Clinical Therapeutics*, *42*(6), pp. 1034–1046. https://doi.org/10.1016/j.clinthera.2020.04.004

Salloway, S., Ferris, S., Kluger, A., Goldman, R., Griesing, T., Kumar, D., & Richardson, S. (2004). Efficacy of donepezil in mild cognitive impairment. *Neurology*, *63*(4), pp. 651–657. https://doi.org/10.1212/01.wnl.0000134664.80320.92

Savage, N. (2021). Tapping into the drug discovery potential of AI. *Biopharma Dealmakers*. https://doi.org/10.1038/d43747-021-00045-7

Schott, J. W., León-Rico, D., Ferreira, C. B., Buckland, K. F., Santilli, G., Armant, M. A., Schambach, A., Cavazza, A., & Thrasher, A. J. (2019). Enhancing Lentiviral and Alpharetroviral transduction of human hematopoietic stem cells for clinical application. *Molecular Therapy. Methods & Clinical Development*, *14*, pp. 134–147. https://doi.org/10.1016/j.omtm.2019.05.015

Scott, D. E., Bayly, A. R., Abell, C., & Skidmore, J. (2016). Small molecules, big targets: drug discovery faces the protein–protein interaction challenge. *Nature Reviews Drug Discovery*, *15*(8), pp. 533–550. https://doi.org/10.1038/nrd.2016.29

Seltzer, B. (2004). Efficacy of Donepezil in early-stage Alzheimer disease. *Archives of Neurology*, *61*(12), p. 1852. https://doi.org/10.1001/archneur.61.12.1852

Seltzer, B. (2005). Donepezil: a review. *Expert Opinion on Drug Metabolism & Toxicology*, *1*(3), pp. 527–536. https://doi.org/10.1517/17425255.1.3.527

Serlin, Y., Shelef, I., Knyazer, B., & Friedman, A. (2015). Anatomy and physiology of the blood-brain barrier. *Seminars in Cell & Developmental Biology*, *38*, pp. 2–6. https://doi.org/10.1016/j.semcdb.2015.01.002

Shafiei-Irannejad, V., Abbaszadeh, S., Janssen, P. M. L., & Soraya, H. (2021). Memantine and its benefits for cancer, cardiovascular and neurological disorders. *European Journal of Pharmacology*, *910*, p. 174455. https://doi.org/10.1016/j.ejphar.2021.174455

Song, I.-U., Im, J. J., Jeong, H., Na, S.-H., & Chung, Y.-A. (2020). Possible neuroprotective effects of rasagiline in Alzheimer's disease: a SPECT study. *Acta Radiologica*, *62*(6), pp. 784–790. https://doi.org/10.1177/0284185120940264

Song, J.-H., & Narahashi, T. (1996). Differential effects of the pyrethroid tetramethrin on tetrodotoxin-sensitive and tetrodotoxin-resistant single sodium channels. *Brain Research*, *712*(2), pp. 258–264. https://doi.org/10.1016/0006-8993(95)01449-7

Southey, M. W. Y., & Brunavs, M. (2023). Introduction to small molecule drug discovery and preclinical development. *Frontiers in Drug Discovery*, *3*. https://doi.org/10.3389/fddsv.2023.1314077

Tanner, C. M., & Goldman, S. M. (1996). Epidemiology of Parkinson's disease. *Neurologic Clinics*, *14*(2), pp. 317–335. https://doi.org/10.1016/S0733-8619(05)70259-0

Tariot, P. N., Braeckman, R., & Oh, C. (2022). Comparison of steady-state pharmacokinetics of Donepezil transdermal delivery system with oral Donepezil. *Journal of Alzheimer's Disease : JAD*, *90*(1), pp. 161–172. https://doi.org/10.3233/JAD-220530

Troncoso-Escudero, P., Sepulveda, D., Pérez-Arancibia, R., Parra, A. V., Arcos, J., Grunenwald, F., & Vidal, R. L. (2020). On the right track to treat movement disorders: promising therapeutic approaches for Parkinson's and Huntington's disease. *Frontiers in Aging Neuroscience*, *12*. https://doi.org/10.3389/fnagi.2020.571185

Tu, S., Okamoto, S., Lipton, S. A., & Xu, H. (2014). Oligomeric Aβ-induced synaptic dysfunction in Alzheimer's disease. *Molecular Neurodegeneration*, *9*, p. 48. https://doi.org/10.1186/1750-1326-9-48

Tzeplaeff, L., Wilfling, S., Requardt, M. V., & Herdick, M. (2023). Current state and future directions in the therapy of ALS. *Cells*, *12*(11), p. 1523. https://doi.org/10.3390/cells12111523

Vaz, M., & Silvestre, S. (2020). Alzheimer's disease: tecent treatment strategies. *European Journal of Pharmacology*, *887*, p. 173554. https://doi.org/10.1016/j.ejphar.2020.173554

Venkatasubramanian, V., & Mann, V. (2022). Artificial intelligence in reaction prediction and chemical synthesis. *Current Opinion in Chemical Engineering*, *36*, p. 100749. https://doi.org/10.1016/j.coche.2021.100749

Vercelletto, M., Boutoleau-Bretonnière, C., Volteau, C., Puel, M., Auriacombe, S., Sarazin, M., Michel, B.-F., Couratier, P., Thomas-Antérion, C., Verpillat, P., Gabelle, A., Golfier, V., Cerato, E., & Lacomblez, L. (2011). Memantine in behavioral variant frontotemporal dementia: negative results. *Journal of Alzheimer's Disease*, *23*(4), pp. 749–759. https://doi.org/10.3233/jad-2010-101632

Wagner, M. J., Daniel, C. P., Plaisance, C. J., Borne, G. E., Ahmadzadeh, S., Shekoohi, S., & Kaye, A. D. (2023). Apomorphine for Parkinson's disease: pharmacologic and clinical considerations. *Expert Opinion on Emerging Drugs*, *28*(4), pp. 275–281. https://doi.org/10.1080/14728214.2023.2278677

Wang, H., Guan, L., & Deng, M. (2023). Recent progress of the genetics of amyotrophic lateral sclerosis and challenges of gene therapy. *Frontiers in Neuroscience*, *17*, p. 1170996. https://doi.org/10.3389/fnins.2023.1170996

Watanabe, T., Tahara, M., & Todo, S. (2008). The novel antioxidant edaravone: from bench to bedside. *Cardiovascular Therapeutics*, *26*(2), pp. 101–114. https://doi.org/10.1111/j.1527-3466.2008.00041.x

Weintraub, D., Hauser, R. A., Elm, J. J., Pagan, F., Davis, M. D., & Choudhry, A. (2016). Rasagiline for mild cognitive impairment in Parkinson's disease: a placebo-controlled trial. *Movement Disorders*, *31*(5), pp. 709–714. https://doi.org/10.1002/mds.26617

Wolfson, C., Gauvin, D. E., Ishola, F., & Oskoui, M. (2023). Global prevalence and incidence of amyotrophic lateral sclerosis: a systematic review. *Neurology*, *101*(6), pp. e613–e623. https://doi.org/10.1212/WNL.0000000000207474

Wong, K. H., Riaz, M. K., Xie, Y., Zhang, X., Liu, Q., Chen, H., Bian, Z., Chen, X., Lu, A., & Yang, Z. (2019). Review of current strategies for delivering Alzheimer's disease drugs across the blood-brain barrier. *International Journal of Molecular Sciences*, *20*(2), p. 381. https://doi.org/10.3390/ijms20020381

Yamashita, T., & Abe, K. (2024). Update on antioxidant therapy with edaravone: expanding applications in neurodegenerative diseases. *International Journal of Molecular Sciences*, *25*(5), p. 2945. https://doi.org/10.3390/ijms25052945

Yingngam, B., & Sethabouppha, B. (2024). Advanced AI applications for drug discovery. In *Advances in Medical Technologies and Clinical Practice* (pp. 42–86). IGI Global. https://doi.org/10.4018/979-8-3693-2333-5.ch003

Yoshino, H., & Kimura, A. (2006). Investigation of the therapeutic effects of edaravone, a free radical scavenger, on amyotrophic lateral sclerosis (Phase II study). *Amyotrophic Lateral Sclerosis*, *7*(4), pp. 247–251. https://doi.org/10.1080/17482960600881870

Zhao, Y., & Zhao, B. (2013). Oxidative stress and the pathogenesis of Alzheimer's disease. *Oxidative Medicine and Cellular Longevity*, 2013, p. 316523. https://doi.org/10.1155/2013/316523

12 Bridging the Gap in Neurodegenerative Disease Therapy (NDT)

Shikha Sharma, Showkat Dar, and Nawab John Dar

ABBREVIATIONS

ADRD	Alzheimer's Disease and Related Dementias
AI	Artificial Intelligence
CARD	Central for Alzheimer's and Related Dementias
CNS	Central Nervous System
DAO	D-Amino Acid Oxidase
ECM	Extracellular Matrix
EEG	Electroencephalograms
EMG	Electromyography
FUS	Fused-in Sarcoma
HMM	Hidden Markov Model
iPSC	Induced Pluripotent Stem Cell
MATR3	Matrin 3
MEG	Magnetoencephalogram
ML	Machine Learning
MRI	Magnetic Resonance Imaging
ND	Neurodegenerative Disease
NGS	Next-Generation Sequencing
NIA	National Institute of Aging
NMJ	Neuromuscular Junction
OPTN	Optineurin
PET	Positron Emission Tomography
PFN-1	Profilin 1
SETX	Senataxin
SIGMAR1	Sigma Receptor 1
SOD1	Superoxide Dismutase 1
SPET	Single Proton Emission Tomography
SQSTM1	Sequestosome 1
TBK1	TANK-Binding Kinase 1
TDP43	Tar DNA-binding Protein 43
UBQLN2	Ubiquilin 2
VAPB	VAMP-associated protein B
VCP	Valosin-Containing Protein

DOI: 10.1201/9781003520610-12

12.1 INTRODUCTION

Neurodegenerative diseases (NDs) present a significant global health challenge, with an increasing prevalence due to aging populations worldwide. Despite decades of research, effective treatments for most NDs remain elusive (Dar & Glazner, 2020; Palanisamy et al., 2023). This chapter explores the latest advances in bridging the gap between our understanding of these diseases and the development of successful therapies. The gaps in the NDs were more prominent during the COVID-19 pandemic, as it severely disrupted routine health services, revealing significant cracks in support for those patients. As per a report by University College London, an estimate of up to 75% mortality rates was quantified among people with dementia during the pandemic. Alzheimer's Society noted a 4% decline in dementia diagnoses (Overcoming Gaps in the Treatment of Neurodegenerative Disease, 2020).

The current approach (internationally) to dementia research is insufficient and urges for greater collaboration, regulatory flexibility, and financial incentives to support long-term investment in finding a cure (Vogel, 2014), highlighting the complexity of neurological diseases, requiring combined therapy approaches with the need for more research in targets for drug development. Additionally, experts suggest that earlier intervention could be key, as current treatments might be administered too late (Dar et al., 2017, 2018). The call to action included pooling data among companies to reduce inefficiencies and considering special regulatory categories for dementia drugs to expedite development (Vogel, 2014).

Mostly the NDs are focused on treatment at a very late stage especially due to their asymptomatic nature (Downing et al., 2015). For example, in the case of Alzheimer's disease (AD), the visible symptoms appear only in its dementia stage, overlooking the preclinical and prodromal phases (Gregory et al., 2022). However, with a push from the financial point of view as well as scientific and ethical aspects, the shift is being focused on earlier stages of the disease – known as the Triad approach. This shift aims to improve brain health outcomes and delay the onset of symptoms. The proposed approach for clinical trials involves three main strategies (Triad approach): using experimental medicine to better understand disease mechanisms, conducting phase 2 learning trials to guide phase 3 trials, and employing precision medicine to target subgroups of patients with similar characteristics. Additionally, it's crucial for clinical services around the world to adopt brain health clinic models to detect NDs early and improve overall prognosis (Gregory et al., 2022).

In 2013, the G8 Dementia Summit aimed to find a cure or disease-modifying therapy by 2025, but new evidence suggests that this goal is very unrealistic. Instead, researchers emphasize the importance of broad societal policies for dementia prevention throughout life. The Summit resulted in commitments to innovation, increased research funding, and international collaboration, but these lacked specificity and accountability. Emphasizing the need for public health approaches, the report suggests that reducing risk factors like hypertension, obesity, smoking, and social isolation could prevent up to 40% of dementia cases. The COVID-19 pandemic has highlighted gaps in dementia

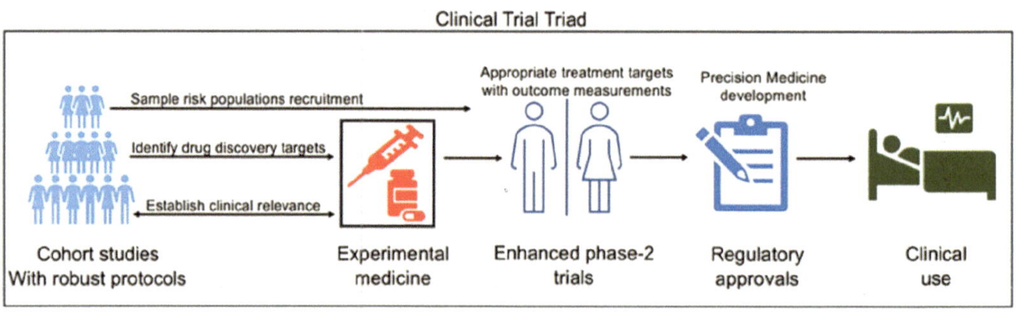

FIGURE 12.1 Clinical trial Triad overview showing the process of developing clinically useful drugs/applications. (Adopted from Gregory et al., 2022.)

care and the need for structural changes to promote dementia prevention. Moving forward, updated commitments and action plans should balance biomedical and public health responses to maximize benefits for the aging population (Fox & Petersen, 2013; Shah et al., 2016; Wallace et al., 2021). At the First WHO Ministerial Conference on Global Action Against Dementia in March 2015, delegates from 80 countries agreed to work together to reduce the global burden of dementia. They identified 59 key research areas, focusing on prevention, care quality, diagnosis, and treatment and emphasized the need for ongoing collaboration and monitoring to implement a global action plan effectively (Shah et al., 2016).

As per the 2017 Alzheimer's Association International Conference, only a fraction of the 3,000 abstracts dealt with drug treatments for neurological disorders and dementia with no new drugs for dementia. Researchers are shifting focus more to prevention, instead of developing new drugs. Furthermore, new findings emphasized the importance of reducing risk factors for dementia, e.g., the Lancet Commission suggested that up to 35% of dementia cases could be prevented by addressing issues like hypertension, obesity, smoking, and social isolation. Other studies highlighted how crucial factors like good sleep quality, managing hearing loss, and following a diet are for cognitive health. The significance of cognitive rehabilitation and careful treatment of conditions like anemia in dementia patients was also stressed, showcasing non-drug approaches as promising ways to prevent and manage the disease (Shah et al., 2016; Wise, 2017). Novel and modern innovations in neurodegenerative disorder like the "healthy home" system using sensors and artificial intelligence (AI) to monitor dementia patients are being developed to help them live independently (Overcoming Gaps in the Treatment of Neurodegenerative Disease, 2020).

To fill the gaps in the research in understating neurological disorders and therapy, various initiatives are being undertaken to fill this caveat. These include the initiatives at the molecular or cell level; tissue or organoid level; lab-on-chips, brain-on-chips, or microfluidic chips; bioprinting; or the animal model. Each of these models has some advantages and disadvantages (Table 12.1 adopted from Slanzi et al., 2020). We will go through some of such initiatives in the following text to have ideas for these programs.

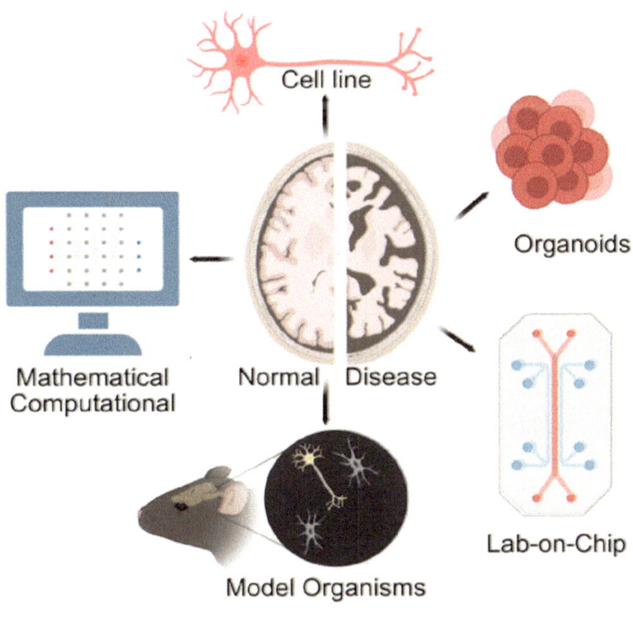

FIGURE 12.2 Different prototypes used for studying the neuronal diseases. In the center is the illustration of brain cross section with normal and disease halves, followed by five different types of procedures used to study neurodegenerative diseases (clockwise from top is representation of cell lines, organoids, lab on chip, model organism (like mouse) and mathematical/computational methods).

TABLE 12.1
In Vitro Systems Used for the Neuronal Studies

Cell Line	Type	Advantages	Disadvantages
H4	Human neuroglioma cells	Easy to culture and transfect	Lacks dopaminergic phenotype
HEK 293	Immortalized human embryonic kidney cells	Homogenous populations, suitable for large-scale experiments	High passage numbers can lead to genetic and epigenetic alterations. Non-neuronal cell type
SH-SY5Y	Human neuroblastoma cells	Useful to study α-synuclein aggregation and mutations. Differentiate into neuronal-like cells exhibiting cholinergic, dopaminergic, or noradrenergic phenotypes	Neuroblastoma origin may influence differentiation, viability, growth performance, metabolic properties and genomic stability. Multiple differentiation protocols lead to different outcomes
PC12	Pheochromocytoma-derived cell line from the rat adrenal medulla	Synthesizes, releases, and stores catecholamines. Easy to handle and homogeneous	Not human. Derived from a neural tumor, which may alter signaling pathways
LUHMES	Immortalized human embryonic mesencephalic cells	Already used in co-cultures with astrocytes and in a 3D culture system, prolonged survival after differentiation. More sensitive to toxins than other dopaminergic cell lines once differentiated	Low transfection efficiency
Primary neurons	Prepared from embryonic rodent brain tissue	Similar to human neurons in terms of morphology and physiology. Similar proliferation rate to human neurons. Suitable for the generation of genetic models	Ethical problems. Mixed culture. Variations among different culture preparations and difficult to maintain. Dissection procedure can introduce experimental variability. Species-dependent differences
iPSCs	Derived from patients	Capacity for self-renewal. Potential to differentiate into any cell type. Allow the generation of autologous pluripotent cells from any individual for disease modeling. Suitable for large-scale studies and personalized medicine	Challenging to identify disease-specific cell phenotypes that better represent pathogenesis. Do not mimic aging. Requirement of standardized protocols and quality controls to reduce technical variation. High costs
Organoids	Derived from patients	Provide a 3D environment of multiple cell types. Organized structure. Enhanced cellular maturity. Promising for screening compounds targeting the central nervous system. Possibility to study PD-related genes	Highly variable. Need for improved vascularization and optimization of differentiation protocols. Time-consuming and expensive. Ethical problems

12.2 CELL LINE MODELS

In vitro cell lines are readily available and invaluable tools for studying ND, allowing researchers to investigate the mechanisms of disease and test potential treatments in a controlled environment (Cetin et al., 2022). These models bridge the gap between traditional animal studies and human clinical trials, offering insights into human-specific pathophysiology that animal models often fail to replicate.

Primary culture cells isolated and cultivated dopaminergic neurons from post-mortem adult brains are used primarily, sometimes contaminated with glial cells, can be advantageous for studying interactions between neurons and glia (Slanzi et al., 2020). However, despite the insights gained from primary cultures, they do not fully replicate the cellular organization and extracellular matrix (ECM) found in the central nervous system (CNS). To overcome these limitations several cell lines are commonly used that include cell lines like SH-SY5Y cells, which are human neuroblastoma cell line neuropathy research due to its ability to differentiate into dopaminergic neurons or Neuro2a, HT22 which are mouse lines (Lone et al., 2016; Xicoy et al., 2017). Cells derived from human embryonic mesencephalic tissue (like LUHMES cells) are often used to model Parkinson's disease (PD) and differentiated into dopaminergic neurons (Zhang et al., 2014). Induced pluripotent stem cell (iPSC)-derived neurons from patients (or control subjects) with NDs, such as AD and PD, offer a patient-specific model for studying genetic influences on disease progression (Dräger et al., 2022).

Starting to understand NDs from the very genetic level, the National Institutes of Health (NIH – USA) launched a massive cross-disciplinary program under the Center for Alzheimer's and Related Dementia and the National Institute of Aging collaboration known as iPSC neurodegenerative disease initiative (iNDI) (Dräger et al., 2022). iNDI has two phases that involve generating and characterizing these isogenic iPSC lines (neurons and glia) and then conducting extensive analysis to understand how mutations disrupt cellular processes. The open-access nature of the iNDI datasets and cell lines is intended to facilitate collaborative research and accelerate the discovery of new therapeutic strategies for AD and related dementia (ADRD). The idea here was to generate the disease-associated genetic variation from the isogenic iPSCs and follow them at different levels of genomic, epigenomic, transcriptomic, proteomic, and morphological levels. The initiative will enable detailed studies of how specific genetic variants affect cellular functions, using state-of-the-art tools such as CRISPR-Cas9 for genetic engineering and high-throughput robotics for cell culture and differentiation. This study will represent the largest genome engineering project, creating isogenic-induced iPSC lines to model over 134 genetic variants, associated with ADRD. This initiative addresses the important need for high-quality, standardized models that can be used to study the underlying mechanisms of ADRD, testing various conditions or hypotheses, and to develop new therapeutic targets. The iNDI project envisions generating a large-scale repository of isogenic-engineered iPSC lines to harbor ADRD-related mutations. These cells, along with comprehensive datasets, will be made widely available to the research community via the Jackson Laboratory for Genomic Medicine in Farmington, Connecticut (Ramos et al., 2021).

There have been advancements in the program, e.g., the development of "A fully automated FAIMS-DIA mass spectrometry-based proteomic pipeline" (Reilly et al., 2023). The team introduced a fully automated, optimized mass spectrometry-based proteomics pipeline that enhances reproducibility and adaptability for large-scale research. By using commercially available devices and advanced informatics, the platform ensures deep proteome coverage with high reproducibility, making it suitable for modern proteomic laboratories and contract research organizations (Reilly et al., 2023). There are other labs that are working on the other aspects as well including long-read sequencing pipelines and analysis including the discovery of novel isoforms in these cell lines.

12.3 MULTICELLULAR OR ORGANOID MODELS

The cell lines provide us with valuable options to study neurodegeneration at the cellular level, but they often fail to replicate the complex neuronal cell environment; therefore, other levels of mimicking the brain environment were developed (Peters, 2021). The next level in the hierarchy of such research areas is the co-culturing of the cells. For example, microglia, the brain's immune cells, play a key role in neurodegenerative and neurodevelopmental diseases by modulating neuro-inflammation and oxidative stress, enabling the study of microglial impact on neurons, both positive and negative (Roqué & Costa, 2017). Another such study on PD demonstrates that polyacrylonitrile-based nano-scaffolds effectively support the growth, differentiation, and survival of neuron- and astrocyte-like cells, offering a promising method for co-culturing cells for studying disease mechanisms and discovering neuroprotective agents (Chemmarappally et al., 2020).

Various co-culturing methods are used, which include growing two or more different cell types together by direct contact or separately: transwell system, microfluidic platforms, or hydrogel-based co-culture (Liu et al., 2022; Monteiro et al., 2024). As the name suggests, co-culturing methods involve growing two or more cell types together to study their interactions in a controlled environment. The transwell system separates cells on opposite sides of a permeable membrane to study paracrine signaling. Microfluidic platforms and hydrogel-based co-cultures offer additional precision and complexity, enabling spatial separation of cells and providing a three-dimensional (3D) environment that better mimics tissue architecture and supports the studies of tissue development and organ-specific functions (Liu et al., 2022). Other neuropathy models include organoids, developed to study these diseases in more detail (Monteiro et al., 2024; Oddo et al., 2019). Creating an accurate in vitro brain model is challenging due to the complexity of replicating the diverse cell types, cell–cell interactions, and the ECM found in the brain. For example, for the proper functioning of neurons, glial cells play vital roles in supporting, nourishing, and protecting neurons, neural stem cells and endothelial cells contribute to brain function, and the ECM has bioactive cues and specific mechanical properties (Pereira et al., 2023).

Organoids (derived mainly from stem cells) are advanced 3D cultural model systems that closely mimic the spatial structure and function of human organs, offering a more accurate in vitro representation than traditional animal models (Jgamadze et al., 2020; Pasko et al., 2022). These models make them invaluable for studying NDs, brain injuries, and potential therapeutic interventions. Organoids allow researchers to observe the intricate interactions within the brain's microenvironment and trace cellular changes in real time, providing deeper insights into disease mechanisms and potential treatments, including the use of organoid transplants for brain injury repair (Jgamadze et al., 2020).

12.4 LAB-ON-CHIP OR BRAIN-ON-CHIP MODELS

The latest advancements and combination of material science and cell biology have led to the creation of organ-on-a-chip devices, which replicate key functions of various organs, including the brain – brain on chip (Amirifar et al., 2022). Such platforms offer new insights into physiological processes and disease mechanisms, and they provide a more accurate means of testing drug effects compared to traditional methods (Maoz, 2021). Furthermore, these tools address the limitations of conventional cell culture and animal models, which often fail to accurately represent human physiology and raise ethical concerns (Amirifar et al., 2022). Although the development of brain-on-a-chip systems is still in its early stages, they hold great potential for advancing research in neurophysiology, pathology, and neuropharmacology (Amartumur et al., 2024). There are studies that show that the gut microbiome plays a more crucial role in influencing the CNS through the "microbiome–brain–gut" axis, where bacterial metabolic products directly impact brain development and function (Loh et al., 2024). Advanced in vitro models, such as SHIME, SIMGI, and organ-on-a-chip systems, are now used to study this relationship by accurately simulating the gastrointestinal tract and its interactions with the brain (Ceppa et al., 2020; Raimondi et al., 2019; Van de Wiele et al., 2015). These models help researchers understand how gut bacteria contribute to NDs by replicating the transfer of bacterial neurotoxins and other interactions within the body.

12.5 MODEL ORGANISMS

To reproduce the characteristics of the neurological disorders, the characteristic phenotypes were replicated in the animal models, e.g., non-human primates, rodents, *Caenorhabditis elegans*, and *Drosophila* (Dawson et al., 2018; Dovonou et al., 2023; Jeibmann & Paulus, 2009). Multiple techniques are employed for different neurological disorders although none of the methods can model the disease completely; therefore a combination of in vivo and in vitro and other methods are combined to complement them (Ma et al., 2022). For example, PD that involves the loss of

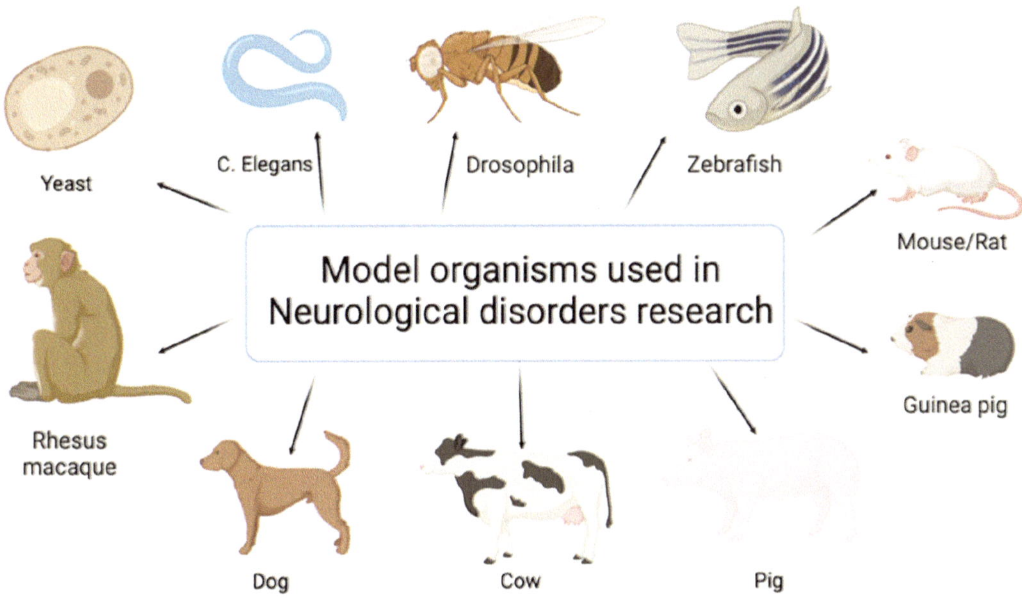

FIGURE 12.3 Various model organisms used as model organisms in neurological research ranging from: 1) simple single cellular organism Yeast (*Saccharomyces*), 2) multicellular invertebrate organisms like *C. elegans*, *Drosophila*, 3) Vertebrates like Zebrafish, Mouse/Rat, Guinea pig, Dog, Pig, Cow 4) Vertebrate-primates Rhesus macaque.

dopamine neurons forming Lewy bodies after the accumulation of α-synuclein can be induced by drugs, neurotoxins, agrochemicals, genetic mutations/models, or viral vectors (Dovonou et al., 2023). Similarly, there can be different phenotypes or characteristics that might be associated with the specific situation; e.g., for modeling amyotrophic lateral sclerosis (ALS) in rodents we might be interested in either ALS-like phenotypes that include locomotor impairment, coordination defects, early lethality, weakness, muscle atrophy, coordination defects, decrease in body weight, and late-stage paralysis. Otherwise, if we are looking for ALS-like pathologies, we will focus on upper and lower motor neuron loss, neuroinflammation, inclusions and other relevant pathology, axonopathy, neuromuscular junction (NMJ) abnormalities, and muscle denervation (Todd & Petrucelli, 2022). It is very difficult to study all the factors and how they impact each other in the disease conditions; however, genetic mutation of their genetic factors can help us uncover their roles, as shown in Table 12.2.

Genetic models, where we have a single gene mutation in the model organisms, provide more insights into the disease mechanism and pathways; however, they do not replicate the exact complex disease conditions like AD or PD (Tello et al., 2022).

Another domain that is emerging is stem cell-based human–animal brain chimeras that can provide a deeper understanding and more accurate mapping of human cellular physiology (Ji et al., 2024). These models are usually engineered by integrating induced pluripotent stem cells (iPSCs) into animal brains at various developmental stages where they mimic exact human-specific features and can be directed to various disease types or specific mutations of that disease (Chesselet & Carmichael, 2012; Ji et al., 2024). Furthermore, based on the cell types, these chimeras can be further sub-categorized into specific domains or used for specific purposes in understanding these disorders: neuronal chimeras – to study disorders such as AD, Down's syndrome, autism, Huntington's disease, and PD models; astroglial chimeras – to understand the role of astrocytes in the diseases; oligodendroglial chimeras – models that are crucial for studying demyelination disorders; microglial chimeras – models that allow the study of human microglia in a living brain environment, which is vital for understanding their role in NDs.

TABLE 12.2
Factors Responsible for Various AI Phenotypes and Their Impact

Protein Factor	Role	Associated ALS Phenotype	Mutation Impact
SOD1 (superoxide dismutase 1)	Antioxidant defense enzyme	Motor neuron loss, paralysis, early lethality, muscle denervation	Gain-of-function mutations lead to toxic aggregation
TDP43 (TAR DNA-binding protein 43)	RNA-binding protein involved in RNA splicing and stress response	Neuronal loss, cytoplasmic inclusions, motor dysfunction	Mutations or mislocalization result in toxic inclusions and motor neuron degeneration
FUS (fused-in sarcoma)	RNA-binding protein involved in RNA metabolism, DNA repair, stress response	Motor neuron degeneration, axonopathy, NMJ abnormalities	Mutations lead to cytoplasmic mislocalization and toxic aggregation
MATR3 (Matrin 3)	Nuclear matrix protein involved in RNA processing	Motor impairment, myopathic changes, motor neuron loss	Toxic gain-of-function mutations impact RNA processing, leading to motor neuron degeneration
OPTN (optineurin)	Autophagy adaptor and regulator of inflammatory signaling	NMJ defects, progressive axonal degeneration, TDP43 inclusions	Loss-of-function mutations impair autophagy and protein clearance, contributing to neurodegeneration
VCP (valosin-containing protein)	AAA-ATPase involved in protein degradation pathways	Muscle pathology, motor neuron degeneration, pTDP43 inclusions	Mutations impair protein clearance, leading to neurodegeneration and the inclusion of body formation
UBQLN2 (ubiquilin 2)	Shuttling factor in the ubiquitin-proteasome system	Cognitive impairment, locomotor defects, synaptic dysfunction, protein aggregation	Mutations impair the clearance of misfolded proteins, leading to protein inclusions and neurodegeneration
SETX (senataxin)	RNA/DNA helicase involved in DNA repair	Progressive motor abnormalities, TDP43 mislocalization, smaller motor neurons	Likely toxic gain-of-function mutations disrupt RNA processing and lead to motor neuron degeneration
SQSTM1 (sequestosome 1, p62)	Autophagy receptor involved in protein degradation	Protein inclusions, neurodegeneration	Mutations disrupt autophagy, leading to the accumulation of misfolded proteins and neurodegeneration
VAPB (VAMP-associated protein B)	ER-associated protein involved in protein trafficking and UPR	Locomotor phenotypes, ER stress, motor neuron loss	Mutations cause protein aggregation and ER stress, leading to neurodegeneration
C9ORF72	Involved in endosomal trafficking	RNA foci, DPR accumulation, TDP43 proteinopathy, neurodegeneration	Hexanucleotide repeat expansion leads to toxic gain-of-function, contributing to motor neuron degeneration
PFN1 (profilin 1)	Regulates actin polymerization	Motor neuron loss, muscle denervation, cytoplasmic aggregates	Gain-of-function mutations destabilize the protein, promoting toxic aggregation
SIGMAR1 (sigma 1 receptor)	ER chaperone involved in mitochondrial function	Locomotor defects, motor neuron loss, ER stress	Mutations mislocalize the receptor, disrupting ER–mitochondria interactions and contributing to neurodegeneration
DAO (D-amino acid oxidase)	Degrades D-serine, a co-agonist of NMDA receptors	Motor neuron loss, gait abnormalities, kyphosis	Mutations impair D-serine degradation, leading to excitotoxicity and motor neuron degeneration
TBK1 (TANK-binding kinase 1)	Regulates autophagy and inflammatory responses	Cognitive impairment, motor neuron degeneration, pTDP43 mislocalization	Loss-of-function mutations disrupt autophagy and inflammatory signaling, contributing to neurodegeneration

12.6 COMPUTATIONAL BIOLOGY AND DATA SCIENCE

There is a pressing demand for patient care and diagnostics, which are still waiting for a reliable and sure shot treatment; it becomes important to employ advanced AI, machine learning (ML), and big data management to help diagnose, monitor, manage, and treat them (Hassan et al., 2022). As we know that NDs are heterogeneous, we need to develop new methods and approaches that integrate the datasets from different methods to fill the gaps for developing therapeutic or diagnostic approaches (Hassan et al., 2022; Jedynak et al., 2012). With the advancement of high-throughput methods like next-generation sequencing and other methods (genome-wide associative studies), there is an increase in data generation related to various aspects of neurological research (Di Resta et al., 2021; Rexach et al., 2019). Additionally, various bioinformatic and computational methods as well as other methods have been developed to analyze and elucidate this large data. These methods, on the genomic level, have helped to decipher the genetic factors responsible for these disorders or the risk factors involved in these complications (Di Resta et al., 2021; Rexach et al., 2019).

Furthermore, with the advancement of magnetic resonance imaging (MRI), researchers can enhance the diagnosis and classification of neurological disorders (Ali, 2024; Carvalho Macruz et al., 2024; Ghasempour Dabbaghi et al., 2023). ML algorithms (like support vector machines) and neural networks (convolutional or deep) are used for classification tasks or the detection of complex pattern recognition tasks as in medical images (e.g., MRI, positron emission tomography (PET) scans) and electroencephalograms (EEGs) to detect disease-specific patterns. Random forests and hidden Markov models are used for scenarios to distinguish between different disease stages or types, especially for tracking the progression of diseases (Rao et al., 2024; Tăuţan et al., 2021).

Similarly, other computational methods are expended for imaging techniques like MRI, PET, single-photon emission tomography (SPET), electrophysiological data analysis (EEG), magnetoencephalogram, electromyography to build new models that help bridge the gaps in diagnostics or therapeutics development (Jones et al., 2022; Rao et al., 2024; Tăuţan et al., 2021). Other such data that is used to assist clinicians in diagnostics and managing patients include wearable sensors and movement analysis and sleep disorder analysis (polysomnography or actigraphy) (Tăuţan et al., 2021). All this data is integrated using advanced computational methods and mathematical models to fill the gap and improve the diagnostics and management of neurological disorder patients (Jones et al., 2022). Other computational methods that are engaged to fill the gap and push neurological research to the next level include simulations, mathematical and computational (deep learning) modeling neurodegeneration, computer-aided drug designs, and systems biology (Baig et al., 2018; Moore et al., 2023; Moravveji et al., 2024; Savelieff et al., 2022; Tuladhar et al., 2021). Each of these domains is vast in its own depth and is beyond the scope of this chapter.

12.7 AI AND ML

AI and ML are revolutionizing ND therapy (NDT) research by enabling the analysis of complex, high-dimensional data and accelerating the discovery of novel therapeutic targets and biomarkers. These advanced computational techniques are transforming various aspects of NDT research, from genomics to drug discovery and clinical applications.

One of the most significant contributions of AI in NDT research is its ability to analyze large-scale genomic, proteomic, and metabolomic datasets. Advanced ML algorithms, particularly deep learning models, can identify intricate patterns and relationships within these complex datasets that may not be apparent through traditional analytical methods. For instance, a study by Hampel et al. (2021) demonstrated the use of AI algorithms to integrate multi-omics data for identifying novel biomarkers and drug targets in AD (Hampel et al., 2021). The researchers employed ML techniques to analyze genomic, proteomic, and metabolomic data from AD patients, revealing new potential therapeutic targets and improving our understanding of the disease's underlying mechanisms.

AI-powered image analysis is another area where significant advancements have been made in NDT research. Neuroimaging plays a crucial role in the diagnosis and monitoring of NDs, and AI algorithms can enhance the interpretation of these complex imaging data. A study by Wen et al. (2020) showcased the application of deep learning models for analyzing brain MRI scans to detect early signs of AD with high accuracy (Wen et al., 2020). The AI model was able to identify subtle structural changes in the brain that are indicative of early-stage AD, potentially enabling earlier diagnosis and intervention. ML models are also being employed to predict drug efficacy and toxicity, potentially reducing the need for extensive animal testing in the early stages of drug development. A study by Vamathevan et al. (2019) highlighted the use of ML models in predicting drug–target interactions and potential side effects, demonstrating the potential of these approaches in accelerating the drug discovery process for NDs (Vamathevan et al., 2019). These AI-driven approaches can significantly reduce the time and cost associated with traditional drug discovery methods, allowing researchers to focus on the most promising candidates for further development.

12.8 MULTI-OMICS APPROACHES

The integration of multi-omics technologies provides a comprehensive view of disease processes in neurodegenerative disorders, offering unprecedented insights into the complex molecular mechanisms underlying these conditions. Multi-omics approaches combine data from various biological levels, including genomics, transcriptomics, proteomics, and metabolomics, to create a holistic understanding of disease pathology and progression.

A groundbreaking study by Nativio et al. (2020) demonstrated the power of multi-omics integration in revealing new insights into AD mechanisms (Nativio et al., 2020). The researchers combined genomic, transcriptomic, and proteomic data from post-mortem brain samples of AD patients and healthy controls. This integrated approach uncovered novel gene regulatory networks and protein interactions associated with AD pathology, providing potential new targets for therapeutic intervention.

Single-cell multi-omics approaches have emerged as a powerful tool for studying disease heterogeneity at an unprecedented resolution. These techniques allow researchers to analyze multiple molecular features simultaneously at the individual cell level, providing insights into cellular subpopulations and their roles in disease progression. A study by Mathys et al. (2019) utilized single-cell RNA sequencing to characterize cellular heterogeneity in AD, revealing distinct transcriptional signatures associated with disease progression (Mathys et al., 2019). This high-resolution approach identified specific cell populations that may be more vulnerable to AD pathology, offering new avenues for targeted therapeutic strategies.

12.9 IN SILICO DRUG DISCOVERY AND REPURPOSING

Computational approaches are accelerating drug discovery for NDs by enabling rapid screening of large compound libraries and identifying potential drug candidates more efficiently. Virtual screening and molecular docking studies have become invaluable tools in the early stages of drug discovery, allowing researchers to predict the binding affinity and interactions between potential drug molecules and their targets.

A comprehensive study by Ali et al. (2022) showcased the application of virtual screening and molecular docking techniques in identifying novel inhibitors of alpha-synuclein aggregation, a key pathological process in PD (Ali et al., 2022). The researchers screened a large library of compounds using sophisticated docking algorithms, identifying several promising candidates that showed potential in preventing alpha-synuclein aggregation. This in silico approach significantly expedited the drug discovery process, allowing researchers to focus on the most promising compounds for further experimental validation.

Network-based approaches have also gained prominence in identifying repurposing opportunities for existing drugs in treating NDs. A study by Wen et al. (2020) employed a network-based drug repositioning strategy to identify potential therapeutic candidates for AD (Wen et al., 2020). By analyzing the complex interactions between drugs, targets, and disease-associated genes, the researchers identified several FDA-approved drugs that could potentially be repurposed for AD treatment. This approach not only accelerates the drug discovery process but also offers a cost-effective strategy for identifying new therapeutic options.

12.10 BIOMARKER DEVELOPMENT AND LIQUID BIOPSIES

Advances in biomarker research are improving the diagnosis and treatment monitoring of NDs. The development of blood-based biomarkers for the early detection of NDTs has been a major focus of recent research, offering the potential for less invasive and more accessible diagnostic tools (Rehman et al., 2023).

A landmark study by Nakamura et al. (2018) demonstrated the development of a highly sensitive blood test for detecting amyloid-β, a key biomarker of AD (Nakamura et al., 2018). The researchers used a novel technique called immunoprecipitation coupled with mass spectrometry to detect minute amounts of amyloid-β in blood samples, achieving a level of sensitivity comparable to more invasive and expensive PET imaging methods. This blood-based biomarker approach could potentially revolutionize early detection and monitoring of AD progression.

Liquid biopsies using circulating nucleic acids or exosomes have emerged as promising non-invasive diagnostic and prognostic tools for NDs. A study by Fiandaca et al. (2015) showcased the potential of using exosomal biomarkers in blood samples for the early detection of AD (Fiandaca et al., 2015). The researchers identified a panel of exosomal proteins that could distinguish AD patients from healthy controls with high accuracy, even in the preclinical stages of the disease. This non-invasive approach offers the potential for early diagnosis and monitoring of disease progression without the need for more invasive procedures.

These advancements in biomarker development and liquid biopsy techniques are paving the way for more personalized and effective approaches to diagnosing and treating NDs, potentially enabling earlier intervention and improved patient outcomes.

12.11 SUMMARY AND FUTURE DIRECTIONS

NDs, particularly dementia, represent a significant and growing global health challenge. The increasing prevalence of these conditions, coupled with the aging population worldwide, underscores the urgent need for effective prevention strategies and treatments. Despite decades of research, progress in developing curative therapies has been limited. However, recent advancements in technology, computational methods, and interdisciplinary approaches are opening new avenues for understanding, diagnosing, and potentially treating these complex disorders.

Key areas of progress and future directions include:

1. **Early intervention and prevention**: There is a growing consensus on the importance of early intervention and prevention strategies. Future research should focus on identifying modifiable risk factors and developing interventions that can be implemented decades before the onset of clinical symptoms. This includes lifestyle modifications, management of vascular risk factors, and potentially preventive pharmacological treatments.
2. **Precision medicine**: The future of ND treatment lies in personalized approaches. Advanced genetic and biomarker profiling will enable tailored therapies based on individual disease mechanisms and risk factors. This approach may lead to more effective treatments and better patient outcomes.

3. **Advanced in vitro models**: The development of more sophisticated in vitro models, such as 3D organoids and brain-on-a-chip systems, will provide more accurate representations of human brain physiology. These models will be crucial for studying disease mechanisms and testing potential therapies before clinical trials.
4. **AI and ML**: The integration of AI and ML in ND research will continue to grow, enabling more efficient analysis of complex datasets, prediction of disease progression, and identification of novel therapeutic targets. These technologies will be particularly valuable in analyzing multi-omics data and neuroimaging results.
5. **Multi-omics approaches**: Integrating data from various omics platforms (genomics, proteomics, metabolomics, etc.) will provide a more comprehensive understanding of ND pathophysiology. This holistic approach may reveal new insights into disease mechanisms and potential intervention points.
6. **In silico drug discovery and repurposing**: Computational approaches will accelerate the drug discovery process, enabling rapid screening of potential compounds and repurposing of existing drugs. This could significantly reduce the time and cost associated with bringing new treatments to market.
7. **Biomarker development**: The development of reliable, non-invasive biomarkers for early diagnosis and disease monitoring is a critical area for future research. Blood-based biomarkers and liquid biopsies hold particular promise for enabling earlier intervention and more precise tracking of disease progression.
8. **Collaborative research initiatives**: Future progress will depend on increased collaboration between academic institutions, pharmaceutical companies, and regulatory bodies. Sharing of data and resources will be crucial for advancing our understanding of NDs and developing effective treatments.
9. **Patient-centered approaches**: Future research and care models should prioritize patient-reported outcomes and quality-of-life measures. This includes developing interventions that not only target disease pathology but also improve overall well-being and functional capacity.
10. **Digital health technologies**: The integration of wearable devices, smartphone applications, and remote monitoring technologies will play an increasing role in early detection, disease monitoring, and delivery of interventions.

In conclusion, the future of ND research and treatment lies in a multidisciplinary, technology-driven approach that combines biological insights with computational power and patient-centered care. By focusing on early intervention, personalized medicine, and innovative research methodologies, we can hope to make significant strides in preventing, treating, and potentially curing these devastating diseases. The challenges are substantial, but the potential impact on global health and quality of life for millions of people makes this one of the most important frontiers in medical research.

REFERENCES

Ali, R., Alam, A., K. Rajput, S., & Ahmad, R. (2022). Pharmacotherapeutics and molecular docking studies of alpha-synuclein modulators as promising therapeutics for Parkinson's disease. *Biocell*, *46*(12), pp. 2681–2694. https://doi.org/10.32604/biocell.2022.021224

Ali, S. S. A. (2024). Brain MRI sequence and view plane identification using deep learning. *Frontiers in Neuroinformatics*, *18*, p. 1373502. https://doi.org/10.3389/fninf.2024.1373502

Amartumur, S., Nguyen, H., Huynh, T., Kim, T. S., Woo, R.-S., Oh, E., Kim, K. K., Lee, L. P., & Heo, C. (2024). Neuropathogenesis-on-chips for neurodegenerative diseases. *Nature Communications*, *15*(1), p. 2219. https://doi.org/10.1038/s41467-024-46554-8

Amirifar, L., Shamloo, A., Nasiri, R., de Barros, N. R., Wang, Z. Z., Unluturk, B. D., Libanori, A., Ievglevskyi, O., Diltemiz, S. E., Sances, S., Balasingham, I., Seidlits, S. K., & Ashammakhi, N. (2022). Brain-on-a-chip: recent advances in design and techniques for microfluidic models of the brain in health and disease. *Biomaterials*, *285*, p. 121531. https://doi.org/10.1016/j.biomaterials.2022.121531

Baig, M. H., Ahmad, K., Rabbani, G., Danishuddin, M., & Choi, I. (2018). Computer aided drug design and its application to the development of potential drugs for neurodegenerative disorders. *Current Neuropharmacology*, *16*(6), pp. 740–748. https://doi.org/10.2174/1570159X15666171016163510

Carvalho Macruz, F. B. de, Dias, A. L. M. P., Andrade, C. S., Nucci, M. P., Rimkus, C. de M., Lucato, L. T., Rocha, A. J. da, & Kitamura, F. C. (2024). The new era of artificial intelligence in neuroradiology: current research and promising tools TT – A nova era da inteligência artificial em neurorradiologia: pesquisa atual e ferramentas promissoras. *Arquivos de Neuro-Psiquiatria*, *82*(6), pp. 1–12. https://doi.org/10.1055/s-0044-1779486

Ceppa, F. A., Izzo, L., Sardelli, L., Raimondi, I., Tunesi, M., Albani, D., & Giordano, C. (2020). Human gut-microbiota interaction in neurodegenerative disorders and current engineered tools for its modeling. *Frontiers in Cellular and Infection Microbiology*, *10*, p. 297. https://doi.org/10.3389/fcimb.2020.00297

Cetin, S., Knez, D., Gobec, S., Kos, J., & Pišlar, A. (2022). Cell models for Alzheimer's and Parkinson's disease: at the interface of biology and drug discovery. *Biomedicine & Pharmacotherapy*, *149*, p. 112924. https://doi.org/10.1016/j.biopha.2022.112924

Chemmarappally, J. M., Pegram, H. C. N., Abeywickrama, N., Fornari, E., Hargreaves, A. J., De Girolamo, L. A., & Stevens, B. (2020). A co-culture nanofibre scaffold model of neural cell degeneration in relevance to Parkinson's disease. *Scientific Reports*, *10*(1), p. 2767. https://doi.org/10.1038/s41598-020-59310-x

Chesselet, M.-F., & Carmichael, S. T. (2012). Animal models of neurological disorders. *Neurotherapeutics : The Journal of the American Society for Experimental NeuroTherapeutics*, *9*(2), pp. 241–244. https://doi.org/10.1007/s13311-012-0118-9

Dar, N. J., Bhat, J. A., Satti, N. K., Sharma, P. R., Hamid, A., & Ahmad, M. (2017). Withanone, an active constituent from Withania somnifera, affords protection against NMDA-induced excitotoxicity in neuron-like cells. *Molecular Neurobiology*, *54*(7), pp. 5061–5073. https://doi.org/10.1007/s12035-016-0044-7

Dar, N. J., & Glazner, G. W. (2020). Deciphering the neuroprotective and neurogenic potential of soluble amyloid precursor protein alpha (sAPPα). *Cellular and Molecular Life Sciences*, *77*(12), pp. 2315–2330. https://doi.org/10.1007/s00018-019-03404-x

Dar, N. J., Satti, N. K., Dutt, P., Hamid, A., & Ahmad, M. (2018). Attenuation of glutamate-induced excitotoxicity by Withanolide-A in neuron-like cells: role for PI3K/Akt/MAPK signaling pathway. *Molecular Neurobiology*, *55*(4), pp. 2725–2739. https://doi.org/10.1007/s12035-017-0515-5

Dawson, T. M., Golde, T. E., & Lagier-Tourenne, C. (2018). Animal models of neurodegenerative diseases. *Nature Neuroscience*, *21*(10), pp. 1370–1379. https://doi.org/10.1038/s41593-018-0236-8

Di Resta, C., Pipitone, G. B., Carrera, P., & Ferrari, M. (2021). Current scenario of the genetic testing for rare neurological disorders exploiting next generation sequencing. *Neural Regeneration Research*, *16*(3), pp. 475–481. https://doi.org/10.4103/1673-5374.293135

Dovonou, A., Bolduc, C., Soto Linan, V., Gora, C., Peralta Iii, M. R., & Lévesque, M. (2023). Animal models of Parkinson's disease: bridging the gap between disease hallmarks and research questions. *Translational Neurodegeneration*, *12*(1), p. 36. https://doi.org/10.1186/s40035-023-00368-8

Downing, A. M., Yaari, R., Ball, D. E., & Selzler, K. J. (2015). Bridging the gap between research and clinical practice in asymptomatic Alzheimer's disease. *The Journal of Prevention of Alzheimer's Disease*, pp. 1–13. https://doi.org/10.14283/jpad.2015.86

Dräger, N. M., Sattler, S. M., Huang, C. T.-L., Teter, O. M., Leng, K., Hashemi, S. H., Hong, J., Aviles, G., Clelland, C. D., Zhan, L., Udeochu, J. C., Kodama, L., Singleton, A. B., Nalls, M. A., Ichida, J., Ward, M. E., Faghri, F., Gan, L., & Kampmann, M. (2022). A CRISPRi/a platform in human iPSC-derived microglia uncovers regulators of disease states. *Nature Neuroscience*, *25*(9), pp. 1149–1162. https://doi.org/10.1038/s41593-022-01131-4

Fiandaca, M. S., Kapogiannis, D., Mapstone, M., Boxer, A., Eitan, E., Schwartz, J. B., Abner, E. L., Petersen, R. C., Federoff, H. J., Miller, B. L., & Goetzl, E. J. (2015). Identification of preclinical Alzheimer's disease by a profile of pathogenic proteins in neurally derived blood exosomes: a case-control study. *Alzheimer's & Dementia*, *11*(6), p. 600. https://doi.org/10.1016/j.jalz.2014.06.008

Fox, N. C., & Petersen, R. C. (2013). The G8 dementia research summit—a starter for eight? *The Lancet*, *382*(9909), pp. 1968–1969. https://doi.org/10.1016/s0140-6736(13)62426-5

Ghasempour Dabbaghi, K., Khosravirad, Z., Jamalnia, S., GhorbaniNia, R., Mahmoudikohani, F., Zakeri, H., & Khastehband, S. (2023). The use of artificial intelligence in the management of neurodegenerative disorders; focus on Alzheimer's disease. *Galen Medical Journal*, *12*, pp. 1–7. https://doi.org/10.31661/gmj.v12i.3061

Gregory, S., Saunders, S., & Ritchie, C. W. (2022). Science disconnected: the translational gap between basic science, clinical trials, and patient care in Alzheimer's disease. *The Lancet Healthy Longevity*, *3*(11), pp. e797–e803. https://doi.org/10.1016/s2666-7568(22)00219-7

Hampel, H., Nisticò, R., Seyfried, N. T., Levey, A. I., Modeste, E., Lemercier, P., Baldacci, F., Toschi, N., Garaci, F., Perry, G., Emanuele, E., Valenzuela, P. L., Lucia, A., Urbani, A., Sancesario, G. M., Mapstone, M., Corbo, M., Vergallo, A., & Lista, S. (2021). Omics sciences for systems biology in Alzheimer's disease: state-of-the-art of the evidence. *Ageing Research Reviews*, *69*, p. 101346. https://doi.org/10.1016/j.arr.2021.101346

Hassan, M., Awan, F. M., Naz, A., deAndrés-Galiana, E. J., Alvarez, O., Cernea, A., Fernández-Brillet, L., Fernández-Martínez, J. L., & Kloczkowski, A. (2022). Innovations in genomics and big data analytics for personalized medicine and health care: a review. *International Journal of Molecular Sciences*, *23*(9), p. 4645. https://doi.org/10.3390/ijms23094645

Jedynak, B. M., Lang, A., Liu, B., Katz, E., Zhang, Y., Wyman, B. T., Raunig, D., Jedynak, C. P., Caffo, B., Prince, J. L., & Initiative, A. D. N. (2012). A computational neurodegenerative disease progression score: method and results with the Alzheimer's disease neuroimaging initiative cohort. *NeuroImage*, *63*(3), pp. 1478–1486. https://doi.org/10.1016/j.neuroimage.2012.07.059

Jeibmann, A., & Paulus, W. (2009). Drosophila melanogaster as a model organism of brain diseases. *International Journal of Molecular Sciences*, *10*(2), pp. 407–440. https://doi.org/10.3390/ijms10020407

Jgamadze, D., Johnson, V. E., Wolf, J. A., Cullen, D. K., Song, H., Ming, G.-L., Smith, D. H., & Chen, H. I. (2020). Modeling traumatic brain injury with human brain organoids. *Current Opinion in Biomedical Engineering*, *14*, pp. 52–58. https://doi.org/10.1016/j.cobme.2020.05.004

Ji, Y., McLean, J. L., & Xu, R. (2024). Emerging human pluripotent stem cell-based human-animal brain chimeras for advancing disease modeling and cell therapy for neurological disorders. *Neuroscience Bulletin*, *40*(9), pp. 1315–1332. https://doi.org/10.1007/s12264-024-01189-z

Jones, D., Lowe, V., Graff-Radford, J., Botha, H., Barnard, L., Wiepert, D., Murphy, M. C., Murray, M., Senjem, M., Gunter, J., Wiste, H., Boeve, B., Knopman, D., Petersen, R., & Jack, C. (2022). A computational model of neurodegeneration in Alzheimer's disease. *Nature Communications*, *13*(1), p. 1643. https://doi.org/10.1038/s41467-022-29047-4

Liu, R., Meng, X., Yu, X., Wang, G., Dong, Z., Zhou, Z., Qi, M., Yu, X., Ji, T., & Wang, F. (2022). From 2D to 3D Co-culture systems: a review of co-culture models to study the neural cells interaction. *International Journal of Molecular Sciences*, *23*(21), p. 13116. https://doi.org/10.3390/ijms232113116

Loh, J. S., Mak, W. Q., Tan, L. K. S., Ng, C. X., Chan, H. H., Yeow, S. H., Foo, J. B., Ong, Y. S., How, C. W., & Khaw, K. Y. (2024). Microbiota–gut–brain axis and its therapeutic applications in neurodegenerative diseases. *Signal Transduction and Targeted Therapy*, *9*(1), p. 37. https://doi.org/10.1038/s41392-024-01743-1

Lone, A. M., Dar, N. J., Hamid, A., Shah, W. A., Ahmad, M., & Bhat, B. A. (2016). Promise of retinoic acid-triazolyl derivatives in promoting differentiation of neuroblastoma cells. *ACS Chemical Neuroscience*, *7*(1), pp. 82–89. https://doi.org/10.1021/acschemneuro.5b00267

Ma, L., Li, X., Liu, C., Yan, W., Ma, J., Petersen, R. B., Peng, A., & Huang, K. (2022). Modelling Parkinson's disease in *C. elegans*: strengths and limitations. *Current Pharmaceutical Design*, *28*(37), pp. 3033–3048. https://doi.org/10.2174/1381612828666220915103502

Maoz, B. M. (2021). Brain-on-a-chip: characterizing the next generation of advanced in vitro platforms for modeling the central nervous system. *APL Bioengineering*, *5*(3), p. 30902. https://doi.org/10.1063/5.0055812

Mathys, H., Davila-Velderrain, J., Peng, Z., Gao, F., Mohammadi, S., Young, J. Z., Menon, M., He, L., Abdurrob, F., Jiang, X., Martorell, A. J., Ransohoff, R. M., Hafler, B. P., Bennett, D. A., Kellis, M., & Tsai, L.-H. (2019). Single-cell transcriptomic analysis of Alzheimer's disease. *Nature*, *570*(7761), pp. 332–337. https://doi.org/10.1038/s41586-019-1195-2

Monteiro, A. R., Barbosa, D. J., Remião, F., & Silva, R. (2024). Co-culture models: key players in in vitro neurotoxicity, neurodegeneration and BBB modeling studies. *Biomedicines*, *12*(3), p. 626. https://doi.org/10.3390/biomedicines12030626

Moore, J. A., Wilms, M., Gutierrez, A., Ismail, Z., Fakhar, K., Hadaeghi, F., Hilgetag, C. C., & Forkert, N. D. (2023). Simulation of neuroplasticity in a CNN-based in-silico model of neurodegeneration of the visual system. *Frontiers in Computational Neuroscience*, *17*, p. 1274824. https://doi.org/10.3389/fncom.2023.1274824

Moravveji, S., Doyon, N., Mashreghi, J., & Duchesne, S. (2024). A scoping review of mathematical models covering Alzheimer's disease progression. *Frontiers in Neuroinformatics*, *18*, p. 1281656. https://doi.org/10.3389/fninf.2024.1281656

Nakamura, A., Kaneko, N., Villemagne, V. L., Kato, T., Doecke, J., Doré, V., Fowler, C., Li, Q.-X., Martins, R., Rowe, C., Tomita, T., Matsuzaki, K., Ishii, K., Ishii, K., Arahata, Y., Iwamoto, S., Ito, K., Tanaka, K., Masters, C. L., & Yanagisawa, K. (2018). High performance plasma amyloid-β biomarkers for Alzheimer's disease. *Nature*, *554*(7691), pp. 249–254. https://doi.org/10.1038/nature25456

Nativio, R., Lan, Y., Donahue, G., Sidoli, S., Berson, A., Srinivasan, A. R., Shcherbakova, O., Amlie-Wolf, A., Nie, J., Cui, X., He, C., Wang, L.-S., Garcia, B. A., Trojanowski, J. Q., Bonini, N. M., & Berger, S. L. (2020). An integrated multi-omics approach identifies epigenetic alterations associated with Alzheimer's disease. *Nature Genetics*, *52*(10), pp. 1024–1035. https://doi.org/10.1038/s41588-020-0696-0

Oddo, A., Peng, B., Tong, Z., Wei, Y., Tong, W. Y., Thissen, H., & Voelcker, N. H. (2019). Advances in microfluidic blood–brain barrier (BBB) models. *Trends in Biotechnology*, *37*(12), pp. 1295–1314. https://doi.org/10.1016/j.tibtech.2019.04.006

Overcoming Gaps in the Treatment of Neurodegenerative Disease. (2020). *EBioMedicine*, *60*, p. 103088. https://doi.org/10.1016/j.ebiom.2020.103088

Palanisamy, C. P., Pei, J., Alugoju, P., Anthikapalli, N. V. A., Jayaraman, S., Veeraraghavan, V. P., Gopathy, S., Roy, J. R., Janaki, C. S., Thalamati, D., Mironescu, M., Luo, Q., Miao, Y., Chai, Y., & Long, Q. (2023). New strategies of neurodegenerative disease treatment with extracellular vesicles (EVs) derived from mesenchymal stem cells (MSCs). *Theranostics*, *13*(12), pp. 4138–4165. https://doi.org/10.7150/thno.83066

Pasko, V. I., Churkina, A. S., Shakhov, A. S., Kotlobay, A. A., & Alieva, I. B. (2022). Modeling of neurodegenerative diseases: "Step by Step" and "Network" organization of the complexes of model systems. *International Journal of Molecular Sciences*, *24*(1), p. 604. https://doi.org/10.3390/ijms24010604

Pereira, I., Lopez-Martinez, M. J., & Samitier, J. (2023). Advances in current in vitro models on neurodegenerative diseases. *Frontiers in Bioengineering and Biotechnology*, *11*, p. 1260397. https://doi.org/10.3389/fbioe.2023.1260397

Peters, S. B. (2021). Co-culture methods to study neuronal function and disease. *Neural Regeneration Research*, *16*(5), pp. 972–973. https://doi.org/10.4103/1673-5374.297066

Raimondi, M. T., Albani, D., & Giordano, C. (2019). An organ-on-a-chip engineered platform to study the microbiota–gut–brain axis in neurodegeneration. *Trends in Molecular Medicine*, *25*(9), pp. 737–740. https://doi.org/10.1016/j.molmed.2019.07.006

Ramos, D. M., Skarnes, W. C., Singleton, A. B., Cookson, M. R., & Ward, M. E. (2021). Tackling neurodegenerative diseases with genomic engineering: a new stem cell initiative from the NIH. *Neuron*, *109*(7), pp. 1080–1083. https://doi.org/10.1016/j.neuron.2021.03.022

Rao, N. G. R., Singh, G., Patil, A. R. B., Aparna, T. N., Vippamakula, S., Dharmalingam, S., Kumarasamyraja, D., & Kumar, V. (2024). Advances in computational biology for diagnosing neurodegenerative diseases: a comprehensive review. *Chinese Journal of Applied Physiology*, *40*, p. e20240008. https://doi.org/10.62958/j.cjap.2024.008

Rehman, M. U., Sehar, N., Dar, N. J., Khan, A., Arafah, A., Rashid, S., Rashid, S. M., & Ganaie, M. A. (2023). Mitochondrial dysfunctions, oxidative stress and neuroinflammation as therapeutic targets for neurodegenerative diseases: an update on current advances and impediments. *Neuroscience & Biobehavioral Reviews*, *144*, p. 104961. https://doi.org/10.1016/j.neubiorev.2022.104961

Reilly, L., Lara, E., Ramos, D., Li, Z., Pantazis, C. B., Stadler, J., Santiana, M., Roberts, J., Faghri, F., Hao, Y., Nalls, M. A., Narayan, P., Liu, Y., Singleton, A. B., Cookson, M. R., Ward, M. E., & Qi, Y. A. (2023). A fully automated FAIMS-DIA mass spectrometry-based proteomic pipeline. *Cell Reports Methods*, *3*(10), p. 100593. https://doi.org/10.1016/j.crmeth.2023.100593

Rexach, J., Lee, H., Martinez-Agosto, J. A., Németh, A. H., & Fogel, B. L. (2019). Clinical application of next-generation sequencing to the practice of neurology. *The Lancet. Neurology*, *18*(5), pp. 492–503. https://doi.org/10.1016/S1474-4422(19)30033-X

Roqué, P. J., & Costa, L. G. (2017). Co-culture of neurons and microglia. *Current Protocols in Toxicology*, *74*, pp. 11.24.1–11.24.17. https://doi.org/10.1002/cptx.32

Savelieff, M. G., Noureldein, M. H., & Feldman, E. L. (2022). Systems biology to address unmet medical needs in neurological disorders. *Methods in Molecular Biology (Clifton, N.J.)*, *2486*, pp. 247–276. https://doi.org/10.1007/978-1-0716-2265-0_13

Shah, H., Albanese, E., Duggan, C., Rudan, I., Langa, K. M., Carrillo, M. C., Chan, K. Y., Joanette, Y., Prince, M., Rossor, M., Saxena, S., Snyder, H. M., Sperling, R., Varghese, M., Wang, H., Wortmann, M., & Dua, T. (2016). Research priorities to reduce the global burden of dementia by 2025. *The Lancet Neurology*, *15*(12), pp. 1285–1294. https://doi.org/10.1016/s1474-4422(16)30235-6

Slanzi, A., Iannoto, G., Rossi, B., Zenaro, E., & Constantin, G. (2020). In vitro models of neurodegenerative diseases. *Frontiers in Cell and Developmental Biology*, *8*, p. 328. https://doi.org/10.3389/fcell.2020.00328

Tăuţan, A.-M., Ionescu, B., & Santarnecchi, E. (2021). Artificial intelligence in neurodegenerative diseases: a review of available tools with a focus on machine learning techniques. *Artificial Intelligence in Medicine*, *117*, p. 102081. https://doi.org/10.1016/j.artmed.2021.102081

Tello, J. A., Williams, H. E., Eppler, R. M., Steinhilb, M. L., & Khanna, M. (2022). Animal models of neurodegenerative disease: recent advances in fly highlight innovative approaches to drug discovery. *Frontiers in Molecular Neuroscience*, *15*, p. 883358. https://doi.org/10.3389/fnmol.2022.883358

Todd, T. W., & Petrucelli, L. (2022). Modelling amyotrophic lateral sclerosis in rodents. *Nature Reviews Neuroscience*, *23*(4), pp. 231–251. https://doi.org/10.1038/s41583-022-00564-x

Tuladhar, A., Moore, J. A., Ismail, Z., & Forkert, N. D. (2021). Modeling neurodegeneration in silico with deep learning. *Frontiers in Neuroinformatics*, *15*, p. 748370. https://doi.org/10.3389/fninf.2021.748370

Vamathevan, J., Clark, D., Czodrowski, P., Dunham, I., Ferran, E., Lee, G., Li, B., Madabhushi, A., Shah, P., Spitzer, M., & Zhao, S. (2019). Applications of machine learning in drug discovery and development. *Nature Reviews Drug Discovery*, *18*(6), pp. 463–477. https://doi.org/10.1038/s41573-019-0024-5

Van de Wiele, T., Van den Abbeele, P., Ossieur, W., Possemiers, S., & Marzorati, M. (2015). The simulator of the human intestinal microbial ecosystem (SHIME®). In *The Impact of Food Bioactives on Health* (pp. 305–317). Springer International Publishing. https://doi.org/10.1007/978-3-319-16104-4_27

Vogel, L. (2014). Call for collaboration in dementia research. *CMAJ: Canadian Medical Association Journal = Journal de l'Association Medicale Canadienne*, *186*(15), p. 1132. https://doi.org/10.1503/cmaj.109-4903

Wallace, L., Walsh, S., & Brayne, C. (2021). The legacy of the 2013 G8 Dementia summit: successes, challenges, and potential ways forward. *The Lancet Healthy Longevity*, *2*(8), pp. e455–e457. https://doi.org/10.1016/s2666-7568(21)00145-8

Wen, J., Thibeau-Sutre, E., Diaz-Melo, M., Samper-González, J., Routier, A., Bottani, S., Dormont, D., Durrleman, S., Burgos, N., & Colliot, O. (2020). Convolutional neural networks for classification of Alzheimer's disease: overview and reproducible evaluation. *Medical Image Analysis*, *63*, p. 101694. https://doi.org/10.1016/j.media.2020.101694

Wise, J. (2017). Dementia research focuses on prevention, as drugs fail to deliver. *BMJ*, j3466. https://doi.org/10.1136/bmj.j3466

Xicoy, H., Wieringa, B., & Martens, G. J. M. (2017). The SH-SY5Y cell line in Parkinson's disease research: a systematic review. *Molecular Neurodegeneration*, *12*(1), p. 10. https://doi.org/10.1186/s13024-017-0149-0

Zhang, X.-M., Yin, M., & Zhang, M.-H. (2014). Cell-based assays for Parkinson's disease using differentiated human LUHMES cells. *Acta Pharmacologica Sinica*, *35*(7), pp. 945–956. https://doi.org/10.1038/aps.2014.36

13 Ethical and Economic Considerations in Small Molecule Therapy

Sameer Mir, Urmilla John, and Nawab John Dar

ABBREVIATIONS

ALS	Amyotrophic Lateral Sclerosis
CEA	Cost-Effectiveness Analyses
EML	Essential Medicines List
HDHP	High-Deductible Health Plans
ICER	Incremental Cost-Effectiveness Ratio
IP	Intellectual Property
LMIC	Low- and Middle-Income Countries
NHS	National Health Service
TRIPS	Trade-Related Aspects of Intellectual Property Rights
USMCA	United States–Mexico–Canada Agreement
WHO	World Health Organization
WIPO	World Intellectual Property Organization
WTO	World Trade Organization

13.1 INTRODUCTION

Small molecule therapies have emerged as a promising frontier in the treatment of neurodegenerative diseases, offering potential advantages such as cost-effectiveness, ease of synthesis, and scalability (Makurvet, 2021). However, the development and implementation of these therapies bring forth a complex interplay of ethical and economic considerations that demand careful examination. As these treatments progress from laboratory to clinic, questions of accessibility, equity, cost-effectiveness, and resource allocation come to the forefront, challenging stakeholders to balance the promise of innovative treatments with the imperatives of justice and sustainability in healthcare (Fleck, 2013; Risse et al., 2024). This chapter investigates the multifaceted ethical and economic landscape surrounding small molecule therapies, exploring the tensions between innovation and affordability, the challenges of equitable access, and the broader implications for healthcare systems globally. By critically analyzing these issues, we aim to provide a comprehensive understanding of the challenges and opportunities that lie ahead in harnessing the full potential of small molecule therapies for neurodegenerative diseases.

13.2 ACCESSIBILITY OF SMALL MOLECULE THERAPIES

13.2.1 The Challenge of Global Accessibility

The accessibility of small molecule therapies for neurodegenerative diseases presents a complex and multifaceted challenge on a global scale. As these innovative treatments emerge from research laboratories and enter clinical practice, ensuring equitable access across diverse socioeconomic

DOI: 10.1201/9781003520610-13

groups and geographic regions has become a paramount concern for healthcare systems, policymakers, and pharmaceutical companies alike.

The fundamental issue underlying accessibility challenges is the inherent tension between the high costs associated with developing and manufacturing small molecule therapies and the imperative to make these potentially life-changing treatments available to all who need them. This tension is exacerbated by the global disparities in healthcare infrastructure, economic resources, and regulatory frameworks.

In high-income countries, the accessibility of small molecule therapies is primarily hindered by their often-exorbitant costs. Despite the potential for these therapies to be more cost-effective than biologics in the long run, their initial pricing can be prohibitively expensive for many patients and healthcare systems. For instance, a study by Morgan et al. (2020) found that the median annual cost of small molecule therapies for rare diseases in the United States was \$32,000, with some treatments costing over \$200,000 per year (Morgan et al., 2020). Such high prices can lead to restricted access even in countries with well-developed healthcare systems, as public and private insurers may impose stringent criteria for coverage or high co-payments, effectively placing these treatments out of reach for many patients. The situation is even more dire in low- and middle-income countries (LMICs), where the challenges of affordability are compounded by inadequate healthcare infrastructure and limited resources. In these regions, the high costs of small molecule therapies often mean that they are simply not available or are accessible only to a tiny fraction of the population. A comprehensive analysis by Wirtz et al. (2022) revealed that in many LMICs, essential small molecule therapies for neurodegenerative diseases were available in less than 50% of healthcare facilities, with rural areas being particularly underserved (Wirtz et al., 2017).

Geographic disparities in access to small molecule therapies are not limited to differences between countries but also manifest within nations. Urban-rural divides are particularly pronounced, with patients in rural or remote areas facing additional barriers to access. These barriers include:

i. **Limited availability of specialized healthcare providers:** Small molecule therapies for neurodegenerative diseases often require administration and monitoring by specialists, who are typically concentrated in urban centers.
ii. **Inadequate diagnostic capabilities:** Many rural healthcare facilities lack the advanced diagnostic tools necessary for identifying patients who could benefit from small molecule therapies.
iii. **Logistical challenges:** The distribution and storage of some small molecule therapies may require specific conditions (e.g., temperature control) that are difficult to maintain in remote areas with unreliable infrastructure.
iv. **Travel burdens:** Patients from rural areas may need to travel long distances to access treatment centers, imposing additional financial and time costs.

A study by Zhu et al. (2024) quantified these disparities, finding that patients with Parkinson's disease living in rural areas of China were 62% less likely to receive advanced small molecule therapies compared to their urban counterparts, even after controlling for socioeconomic factors (Zhu et al., 2024).

Regulatory hurdles present another significant barrier to global accessibility. The approval processes for new therapies vary considerably between countries, leading to delays in access for patients in certain regions. For example, a recent comparative analysis by Papapetropoulos et al. (2024) of regulatory timelines for small molecule therapies in neurodegenerative diseases found that the median time from submission to approval was 8 months in the United States, 13 months in the European Union, and 22 months in Japan (Papapetropoulos et al., 2024). In many LMICs, the lack of robust regulatory frameworks or the reliance on approvals from other countries can lead to even longer delays. Moreover, the complexity and cost of navigating multiple regulatory systems can discourage pharmaceutical companies from seeking approval in smaller markets or LMICs, further limiting access in these regions. This issue is particularly acute for rare neurodegenerative diseases, where the small patient populations may not justify the investment in regulatory approval processes for some companies.

The challenge of global accessibility for small molecule therapies is further complicated by intellectual property (IP) rights and patent protections. While these mechanisms are crucial for incentivizing innovation, they can also contribute to high prices and limited availability, especially in LMICs. The ongoing debate over patent waivers and compulsory licensing for essential medicines highlights the tension between protecting IP and ensuring global access to life-saving treatments. In response to these multifaceted challenges, various strategies have been proposed and implemented to improve the global accessibility of small molecule therapies for neurodegenerative diseases, which are listed in the following subsections.

13.2.2 Tiered Pricing Models

Some pharmaceutical companies have adopted differential pricing strategies, offering lower prices in LMICs while maintaining higher prices in high-income markets. For instance, Novartis implemented a tiered pricing model for its small molecule therapy for spinal muscular atrophy, reducing the price by up to 90% in some low-income countries (Singh et al., 2020).

13.2.3 International Collaborations

Initiatives like the Drugs for Neglected Diseases initiative have demonstrated the potential of public-private partnerships to develop and distribute affordable therapies for underserved populations.

13.2.4 Local Manufacturing and Technology Transfer

Encouraging the production of small molecule therapies in LMICs through technology transfer agreements can reduce costs and improve availability. For example, the Medicines Patent Pool has facilitated voluntary licensing agreements for several small molecule therapies, enabling generic production in multiple countries.

13.2.5 Regulatory Harmonization Efforts

Initiatives like the International Council for Harmonisation of Technical Requirements for Pharmaceuticals for Human Use aim to streamline and standardize regulatory processes across countries, potentially reducing delays in global access to new therapies.

13.2.6 Innovative Financing Mechanisms

Proposals such as the Health Impact Fund (Pogge, 2011) suggest alternative ways of rewarding pharmaceutical innovation while ensuring global access to essential medicines (Pogge, 2011).

Despite these efforts, significant challenges remain in achieving truly global accessibility for small molecule therapies in neurodegenerative diseases. Addressing these challenges will require sustained collaboration between governments, pharmaceutical companies, healthcare providers, and patient advocacy groups, as well as innovative approaches to drug development, pricing, and distribution. As research in small molecule therapies continues to advance, it is crucial that parallel efforts are made to ensure that these innovative treatments can reach all patients who need them, regardless of their geographic location or economic status. Only through such concerted efforts can the full potential of small molecule therapies be achieved in improving the lives of people with neurodegenerative diseases worldwide.

13.3 IP AND ACCESS

The intricate relationship between IP rights and access to small molecule therapies for neurodegenerative diseases extends beyond pricing strategies and licensing agreements. Here we explore the nuanced aspects of IP protection and its impact on global access to these vital treatments.

13.3.1 Patent Thickets and Combination Therapies

One significant challenge in the field of small molecule therapies for neurodegenerative diseases is the emergence of "patent thickets." These are dense webs of overlapping patent rights that can impede the development of new therapies, particularly combination treatments that may be more effective in managing complex neurodegenerative conditions. For instance, a study by Goyzueta-Mamani et al. (2022) found that the average number of patents associated with small molecule therapies for Alzheimer's disease increased from 3.2 in 2000 to 14.7 in 2020 (Goyzueta-Mamani et al., 2022). This proliferation of patents can create legal uncertainties and increase the costs of developing new treatments, potentially delaying or preventing the introduction of innovative therapies.

To address this issue, some jurisdictions have implemented mechanisms such as patent pools or cross-licensing agreements. The Medicines Patent Pool, for example, has facilitated voluntary licensing for several small molecule HIV treatments, a model that could potentially be applied to neurodegenerative disease therapies (Wang, 2022).

13.3.2 Data Exclusivity and Market Protection

Beyond patent protection, many countries offer additional forms of market exclusivity for new drugs, including data exclusivity and market protection. These provisions prevent generic manufacturers from relying on the original developer's clinical trial data for a certain period, even after patent expiration. While these protections can incentivize investment in research for rare neurodegenerative diseases, they can also delay generic competition and prolong high prices. A comparative study by Hollis and Grootendorst (2017) found that the duration of data exclusivity varies significantly across countries, ranging from 5 years in Canada to 8 years in the European Union, with additional market protection periods in some jurisdictions (Hollis and Grootendorst, 2017).

13.3.3 Trade Agreements and TRIPS Flexibilities

International trade agreements have increasingly included provisions related to IP rights, often extending beyond the requirements of the World Trade Organization's (WTO's) Agreement on Trade-Related Aspects of Intellectual Property Rights (TRIPS). These "TRIPS-plus" provisions can limit the ability of countries, particularly LMICs, to use TRIPS flexibilities to improve access to medicines. For example, the Trans-Pacific Partnership Agreement initially proposed extending patent terms to compensate for regulatory delays and broadening the scope of patentable subject matter. While some of these provisions were suspended in the final agreement, they illustrate the ongoing tension between trade policies and access to medicines (Gleeson et al., 2018). Conversely, some recent trade agreements have included provisions aimed at balancing IP protection with public health concerns. The United States–Mexico–Canada Agreement, for instance, includes a provision recognizing the importance of promoting access to medicines and preserves the right of countries to take measures to protect public health (Gantz et al., 2020) [14].

13.3.4 Innovation Incentives beyond Patents

Recognizing the limitations of the traditional patent system in promoting research into certain areas, including some neurodegenerative diseases, alternative innovation incentives have been proposed and implemented.

13.3.4.1 Priority Review Vouchers

These vouchers, awarded for developing treatments for neglected tropical diseases or rare pediatric conditions, can be used to expedite the FDA review of a future drug application. While not directly applicable to most neurodegenerative diseases, this model demonstrates the potential for regulatory incentives to complement IP protection.

13.3.4.2 Advance Market Commitments

These are promises by governments or organizations to purchase a certain quantity of a drug at a specific price if it is successfully developed. While primarily used for vaccines, this model could potentially be adapted for neurodegenerative disease therapies.

13.3.4.3 Open-Source Drug Discovery

This collaborative approach to drug development, exemplified by initiatives like the Open-Source Pharma Foundation, aims to create a more transparent and accessible research ecosystem. While still in its early stages for neurodegenerative diseases, this model could potentially accelerate innovation while improving access (Balasegaram et al., 2017).

13.3.5 Challenges in LMICs

The impact of IP rights on access to small molecule therapies is particularly acute in LMICs. Many of these countries lack the manufacturing capacity to produce these therapies locally, even if they were to issue compulsory licenses. Additionally, the threat of trade sanctions or legal challenges from pharmaceutical companies can deter LMICs from using TRIPS flexibilities. A study by Salman et al. (2021) found that only 12% of new small molecule therapies approved for neurodegenerative diseases between 2010 and 2020 were available in more than half of the LMICs studied, with IP protection cited as a significant barrier to access (Salman et al., 2021). As the field of small molecule therapies for neurodegenerative diseases continues to evolve, new approaches to balancing IP protection and access are emerging. These include:

> Delinkage models that separate the cost of R&D from the price of drugs, Increased transparency in R&D costs to inform fair pricing discussions and Exploration of AI and machine learning to accelerate drug discovery and potentially reduce R&D costs.

Addressing the complex interplay between IP rights and access to small molecule therapies for neurodegenerative diseases requires a multifaceted approach. It necessitates collaboration between governments, pharmaceutical companies, patient groups, and international organizations to develop innovative solutions that can balance innovation incentives with the urgent need for accessible treatments.

13.4 THE ROLE OF INTERNATIONAL ORGANIZATIONS

International organizations play a pivotal role in the global governance of IP rights, particularly as they intersect with public health needs. These organizations, such as the World Health Organization (WHO) and the WTO, provide the frameworks, guidelines, and platforms for negotiating and mediating the complex relationships between pharmaceutical innovation, IP rights, and the global imperative to ensure access to essential medicines.

WHO has long been at the forefront of efforts to ensure that essential medicines are accessible to all, regardless of geographic or economic barriers. One of the WHO's most significant contributions in this regard is the development of the Essential Medicines List (EML). The EML identifies medicines that are deemed to be the most effective, safe, and cost-effective for priority health needs and should be available in adequate amounts, in appropriate dosage forms, with assured quality, and at prices that individuals and health systems can afford. By including a medicine in the EML, the WHO signals its importance for public health and encourages governments to ensure that these medicines are available and affordable within their countries. For small molecule therapies that address critical conditions like neurodegenerative diseases, inclusion in the EML can drive policy changes that enhance access, such as the adoption of price controls, subsidies, or the facilitation of generic drug production.

The WTO plays a crucial role in regulating international trade, including the trade of pharmaceutical products. A key component of the WTO's framework is the TRIPS agreement, which establishes minimum standards for the protection of IP rights that all WTO member countries must adhere to. TRIPS covers a broad range of IP issues, including patents, which are particularly relevant to the pharmaceutical industry. While TRIPS sets a baseline for IP protection, it also includes provisions that allow for flexibility in cases where public health is at stake. For instance, the agreement permits the use of compulsory licenses, which allow a government to authorize the production of a patented drug without the consent of the patent holder, typically during a public health emergency. This flexibility is crucial for ensuring that IP rights do not become an insurmountable barrier to accessing life-saving therapies, including small molecule drugs.

In response to growing concerns about the impact of TRIPS on access to medicines, particularly in LMICs, the WTO adopted the Doha Declaration on the TRIPS Agreement and Public Health in 2001. The Doha Declaration affirms that the TRIPS agreement should be interpreted and implemented in a way that supports WTO members' right to protect public health and, in particular, to promote access to medicines for all. The Doha Declaration underscores the right of countries to use the flexibilities provided under TRIPS, such as compulsory licensing and parallel importing, to address public health challenges. This has significant implications for the availability of small molecule therapies in LMICs, as it provides a legal basis for governments to take measures that increase access to these drugs, even in the face of stringent IP protections.

Beyond legal frameworks, international organizations also play a critical role in facilitating collaboration and capacity building among countries. The WHO, WTO, and other organizations such as the World Intellectual Property Organization (WIPO) work together to provide technical assistance to LMICs, helping them to navigate the complexities of IP law and to develop strategies for improving access to medicines. Capacity-building initiatives might include training for policymakers and regulators on the use of TRIPS flexibilities, support for the development of national pharmaceutical industries, and assistance in negotiating better terms for medicine procurement. These efforts are essential for empowering LMICs to assert their rights under international law and to take full advantage of the tools available to them for improving public health outcomes.

While the frameworks provided by the WHO and WTO are critical for balancing IP rights with public health needs, they are not without challenges. One criticism is that the TRIPS agreement, even with its flexibility, still heavily favors the interests of pharmaceutical companies in high-income countries. The process of obtaining a compulsory license, for instance, can be complex and politically fraught, potentially delaying access to essential medicines. Additionally, there is ongoing debate about the adequacy of international efforts to ensure that the benefits of pharmaceutical innovation are equitably distributed. Critics argue that more needs to be done to reform the global IP system to prioritize public health over profit, particularly in the context of global health emergencies.

Looking ahead, the role of international organizations in mediating between IP rights and public health will continue to evolve. The COVID-19 pandemic has reignited global discussions about the need for more equitable access to medicines, leading to calls for reforms in the TRIPS agreement and the broader IP system. As new challenges emerge, such as the rise of personalized medicine and the growing importance of biopharmaceuticals, international organizations will need to adapt and develop new strategies for ensuring that all people have access to the medicines they need.

13.5 COST-EFFECTIVENESS OF SMALL MOLECULE THERAPIES

13.5.1 Economic Evaluation of Therapies

The economic evaluation of small molecule therapies for neurodegenerative diseases is a critical process that involves assessing both the costs and health outcomes associated with these treatments. As the development and use of these therapies continue to evolve, understanding their cost-effectiveness becomes increasingly important for healthcare systems, policymakers, and patients. Cost-effectiveness analyses (CEAs) typically compare the incremental costs and health benefits of a new therapy against the standard of care or other available treatments. For small molecule therapies, these analyses often use quality-adjusted life years (QALYs) as a measure of health benefit, allowing for comparisons across different interventions and disease states. A recent comprehensive study by Arora and Baldi (2024) examined the cost-effectiveness of small molecule therapies across various neurodegenerative diseases (Arora & Baldi, 2024). The study analyzed 47 CEAs published between 2015 and 2022, focusing on Alzheimer's disease, Parkinson's disease, and amyotrophic lateral sclerosis (ALS). Key findings include the following.

13.5.1.1 Incremental Cost-Effectiveness Ratios (ICERs)

The median ICER for small molecule therapies was $150,000 per QALY gained, with a wide range from $45,000 to $580,000 per QALY.

13.5.1.2 Disease-Specific Variations

Alzheimer's disease therapies had the highest median ICER at $210,000 per QALY, followed by Parkinson's disease at $130,000 per QALY and ALS at $95,000 per QALY.

13.5.1.3 Time Horizon Impact

Studies with longer time horizons (>10 years) tended to report more favorable ICERs, highlighting the importance of considering long-term benefits in chronic neurodegenerative conditions.

13.5.1.4 Sensitivity to Efficacy

The cost-effectiveness of these therapies was highly sensitive to their assumed efficacy, with a 20% increase in efficacy leading to an average 35% reduction in ICERs.

The study also highlighted several challenges in evaluating the cost-effectiveness of small molecule therapies for neurodegenerative diseases.

13.5.2 Long-Term Outcomes

Many neurodegenerative diseases progress slowly over decades, making it difficult to capture the full economic impact of therapies in short-term clinical trials.

13.5.3 Indirect Costs

Neurodegenerative diseases often impose substantial indirect costs on patients, caregivers, and society, which may not be fully captured in traditional CEAs. The study found that only 60% of the analyzed CEAs included indirect costs.

13.5.4 Quality of Life Measures

Standard QALY measures may not adequately reflect the full impact of neurodegenerative diseases on patients' quality of life, potentially underestimating the value of effective therapies.

13.5.5 Heterogeneity of Disease Progression

The variable course of many neurodegenerative diseases can make it challenging to model the economic impact of therapies across diverse patient populations.

To address these challenges, researchers are exploring innovative approaches to economic evaluation. For instance, a study by Van den Hout and Mattews (2009) proposed a novel microsimulation model for Parkinson's disease that incorporates individual patient characteristics and disease trajectories (van den Hout & Matthews, 2009). This model demonstrated improved accuracy in predicting long-term outcomes and cost-effectiveness compared to traditional Markov models. Furthermore, a meta-analysis by Neumann et al. (2016) examined the impact of different economic evaluation methodologies on the cost-effectiveness results of small molecule therapies in neurodegenerative diseases (Neumann et al., 2016). The study found that:

i. **Societal perspective:** CEAs adopting a societal perspective (including indirect costs) resulted in 25% lower ICERs on average compared to those using a healthcare system perspective.
ii. **Time horizon:** Extending the time horizon from 5 to 20 years led to a median reduction in ICERs of 40%.
iii. **Discount rates:** Lower discount rates for both costs and outcomes (e.g., 1.5% instead of 3%) improved the cost-effectiveness of therapies by an average of 15%.

As the field of small molecule therapies for neurodegenerative diseases continues to evolve, it will be crucial to refine and standardize economic evaluation methodologies to ensure that these assessments accurately reflect the value of these treatments to patients, healthcare systems, and society as a whole. The incorporation of real-world evidence, patient-reported outcomes, and advanced modeling techniques will likely play an increasingly important role in future economic evaluations.

13.6 THE IMPACT ON HEALTHCARE BUDGETS

Building on the cost-effectiveness considerations discussed in the previous section, it's crucial to examine the broader impact of small molecule therapies for neurodegenerative diseases on healthcare budgets. This impact extends beyond individual patient-level costs to affect systemic healthcare expenditures and resource allocation. A comprehensive analysis by the Institute for Clinical and Economic Review in 2023 projected the potential budget impact of emerging small molecule therapies for Alzheimer's disease. The study estimated that if 10% of eligible patients were treated with a hypothetical disease-modifying therapy priced at $50,000 per year, the annual budget impact would be approximately $29 billion (Wright et al., 2023). This figure represents about 0.7% of total U.S. healthcare spending, a significant amount for a single therapeutic area.

The budgetary implications of small molecule therapies vary across different healthcare systems.

13.6.1 Single-Payer Systems

In countries with nationalized healthcare, such as the United Kingdom's National Health Service (NHS), the introduction of high-cost therapies can strain fixed budgets. A 2024 report by the UK's Office for Budget Responsibility projected that new small molecule therapies for neurodegenerative diseases could increase NHS spending by £1.2 billion annually by 2030, necessitating either budget increases or reallocation of resources from other areas.

13.6.2 Multi-Payer Systems

In countries like the United States, the impact is distributed across various payers, including private insurers, Medicare, and Medicaid. A study by Denisova et al. (2020) estimated that the introduction of a new small molecule therapy for Parkinson's disease could increase per-member-per-month costs for commercial insurers by $0.87, translating to millions in additional spending across large insurance pools (Denisova et al., 2020).

13.6.3 Out-of-Pocket Costs

In systems with significant patient cost-sharing, the high prices of these therapies can lead to substantial out-of-pocket expenses. A survey by the Michael J. Fox Foundation (2024) found that 38% of Parkinson's patients reported difficulty affording their medications, with those on newer small molecule therapies reporting higher financial burdens (Keavney et al., 2024).

The budget impact of small molecule therapies is not limited to drug costs alone. A longitudinal study by Tzeplaeff et al. (2023) examined the total healthcare costs associated with the introduction of a new small molecule therapy for ALS (Tzeplaeff et al., 2023). Key findings include the following.

13.6.4 Initial Cost Increase

In the first year of treatment, total healthcare costs per patient increased by 215% compared to standard care.

13.6.5 Offset in Other Costs

By year three, there was a 32% reduction in hospitalization costs and a 28% reduction in long-term care costs among treated patients.

13.6.6 Net Budget Impact

Despite the high drug costs, the net increase in total healthcare spending was 68% by year five, lower than the initial projections due to offsets in other areas.

To manage these budgetary pressures, healthcare systems are exploring innovative financing mechanisms, which are listed in the following subsections.

13.6.7 Outcomes-Based Agreements

In Italy, the Italian Medicines Agency (AIFA) implemented a performance-based reimbursement scheme for a new small molecule therapy for Huntington's disease in 2024. Under this agreement, the manufacturer provides partial refunds for patients who do not meet predefined clinical outcomes, effectively sharing the financial risk with the healthcare system.

13.6.8 Subscription Models

The Australian government piloted a "Netflix-style" subscription model for high-cost neurodegenerative disease treatments in 2023, paying a fixed annual fee for unlimited access to certain small molecule therapies. This approach aims to provide budget predictability and improve access.

13.6.9 Amortization

Some U.S. state Medicaid programs are exploring drug cost amortization over multiple years. A pilot program in Massachusetts, launched in 2024, allows the state to spread the cost of expensive small molecule therapies for neurodegenerative diseases over a 5-year period, aligning payment more closely with the realization of clinical benefits.

These approaches demonstrate the evolving landscape of healthcare financing in response to the budgetary challenges posed by innovative small molecule therapies. As these treatments continue to emerge, healthcare systems will need to adapt their budgeting and financing strategies to ensure sustainable access while managing overall healthcare expenditures.

13.7 COST-EFFECTIVENESS IN LMICS

The cost-effectiveness of small molecule therapies for neurodegenerative diseases presents unique challenges in LMICs, where resource constraints and healthcare system limitations significantly impact both the implementation and the evaluation of these treatments.

A key issue in LMICs is the scarcity of robust cost-effectiveness data specific to these settings. A systematic review by Gunawardena et al. (2021) found that only 15% of CEAs for small molecule therapies in neurodegenerative diseases included data from LMICs (Gunawardena et al., 2021). This lack of context-specific data can lead to inappropriate extrapolation of cost-effectiveness estimates from high-income countries, potentially resulting in suboptimal resource allocation decisions.

Several factors contribute to the complexity of cost-effectiveness evaluations in LMICs.

13.7.1 Pricing Disparities

The prices of small molecule therapies can vary widely across different LMICs. A study by Goh et al. (2022) found that the price of a common small molecule therapy for Parkinson's disease varied by up to 300% across 10 LMICs, with no clear correlation to per capita income levels (Goh et al., 2022). This variability makes it challenging to generalize cost-effectiveness findings across different LMIC settings.

13.7.2 Healthcare System Constraints

Many LMICs lack the infrastructure and resources to effectively deliver and monitor complex therapies. A survey of neurologists in 15 LMICs by the World Federation of Neurology (2024) found that 68% reported inadequate facilities for administering and monitoring advanced small molecule therapies, potentially reducing their real-world effectiveness and cost-effectiveness.

13.7.3 Competing Health Priorities

In resource-limited settings, the opportunity cost of investing in expensive small molecule therapies for neurodegenerative diseases must be weighed against other pressing health needs. The WHO Global Health Expenditure Database (2023) shows that many LMICs spend less than 5% of their health budgets on non-communicable diseases, including neurodegenerative conditions.

13.7.4 Out-of-Pocket Expenses

In many LMICs, patients bear a significant portion of healthcare costs. The World Bank's Health Equity and Financial Protection Indicators database (2024) indicates that out-of-pocket expenses account for over 40% of total health expenditure in several LMICs. This can lead to catastrophic health expenditures and treatment abandonment, reducing the real-world cost-effectiveness of therapies.

To address these challenges, innovative approaches to CEA in LMICs are emerging.

13.7.5 Contextualized Threshold Values

Rather than using arbitrary cost-effectiveness thresholds, some researchers are developing country-specific thresholds based on local willingness-to-pay and opportunity costs. For example, a study by Ochalek et al. (2018) proposed a method for estimating cost-effectiveness thresholds in LMICs based on the health opportunity cost of healthcare spending, resulting in values ranging from \$3 to \$116 per QALY across 123 LMICs (Ochalek et al., 2018).

13.7.6 Value-Based Pricing Models

Some LMICs are exploring value-based pricing agreements with pharmaceutical companies. The Indian Council of Medical Research (2024) piloted a program where the price of a small molecule therapy for Alzheimer's disease was linked to its real-world performance in Indian patients, potentially improving its cost-effectiveness.

13.7.7 Adaptive Trial Designs

To generate more context-specific data, some researchers are using adaptive trial designs that allow for the simultaneous evaluation of clinical efficacy and cost-effectiveness in LMIC settings. A multi-country study in Southeast Asia, led by the ASEAN Neurological Association (2024), is using this approach to assess a new small molecule therapy for ALS.

13.7.8 Consideration of Broader Societal Impacts

Given the significant informal care burden in many LMICs, some CEAs are incorporating broader societal costs and benefits. A study in Brazil by Davis et al. (2013) found that including informal care costs increased the cost-effectiveness ratio of a small molecule therapy for Parkinson's disease by 35% (Davis et al., 2013).

While the cost-effectiveness of small molecule therapies for neurodegenerative diseases in LMICs remains challenging to assess, emerging methodologies and context-specific approaches are improving our understanding. As more data become available, policymakers in LMICs will be better equipped to make informed decisions about the allocation of limited resources to these potentially transformative therapies.

13.8 ECONOMIC IMPLICATIONS FOR PATIENTS AND FAMILIES

The introduction of small molecule therapies for neurodegenerative diseases has ushered in a new era of treatment possibilities, but it has also brought significant economic challenges for patients and their families. These challenges extend far beyond the direct costs of medication, encompassing a complex web of financial implications that can reshape the economic landscape of affected households.

13.8.1 Direct Costs and Out-of-Pocket Expenses

The high prices of small molecule therapies represent a formidable financial hurdle for many patients. A comprehensive study by the IQVIA Institute for Human Data Science (2023) revealed that the average annual cost of specialty medicines, which include many small molecule therapies for neurodegenerative diseases, exceeded \$79,000 per patient in 2022. This staggering figure translates into substantial out-of-pocket costs, even for those with insurance coverage. In the United States, where healthcare costs are particularly high, the burden on patients can be overwhelming. Cubanski et al. (2019) found that Medicare Part D enrollees who did not receive low-income subsidies faced average out-of-pocket costs of \$8,109 per year for specialty drugs (Cubanski et al., 2019). This financial strain is exacerbated by the complex structure of many insurance plans. For instance, the Kaiser Family Foundation (2023) reported that the average coinsurance rate for specialty tier drugs in Medicare Part D plans was 27%, potentially translating to thousands of dollars in out-of-pocket costs annually. The situation is particularly dire for those with high-deductible health plans (HDHPs). A study by Jones et al. (2022) found that HDHP enrollees with chronic conditions, including neurodegenerative diseases were more likely to delay or forgo essential care due to costs (Jones et al., 2022). Specifically, 24.5% of HDHP enrollees reported cost-related medication non-adherence, compared to 15.9% of those with traditional plans.

13.8.2 Impact on Household Finances

The ripple effects of these high costs permeate every aspect of a family's financial life.

13.8.2.1 Savings Depletion

The drain on family savings can be rapid and severe. Cohen et al. (2017) found that 42.9% of families dealing with serious health conditions, including neurodegenerative diseases, depleted their savings within 2 years of diagnosis (Cohen & Cha, 2023). More alarmingly, a study by Mossa et al. (2024) revealed that this percentage had increased to 48.7%, suggesting a worsening trend (Mossa et al., 2023).

13.8.2.2 Debt Accumulation

Medical debt has become a common consequence of managing neurodegenerative diseases. The Consumer Financial Protection Bureau (2022) reported that medical debt was the largest source of debt in collections, accounting for 58% of all third-party collection tradelines. For families dealing with neurodegenerative diseases, the situation is often more severe. A targeted study by Neurology Today (2023) found that 68% of families managing Parkinson's disease had taken on medical debt, with an average debt of $28,000.

13.8.2.3 Bankruptcy Risk

The specter of bankruptcy looms large for many families grappling with the costs of neurodegenerative disease treatment. Himmelstein et al. (2019) found that 66.5% of all bankruptcies were tied to medical issues (Himmelstein et al., 2019). A more recent analysis by the *American Journal of Medicine* (2023) specifically focused on neurodegenerative diseases and found that families managing these conditions were 2.3 times more likely to file for bankruptcy compared to the general population.

13.8.3 Indirect Economic Impact

The economic implications of small molecule therapies extend far beyond direct medical costs.

13.8.3.1 Lost Income

The impact on employment and income can be devastating. A longitudinal study by the Michael J. Fox Foundation (2023) found that 62% of Parkinson's disease patients reduced their work hours within 5 years of diagnosis, with 29% leaving the workforce entirely (Beaulieu-Jones et al., 2024). For caregivers, the situation is equally challenging. The Alzheimer's Association (2023) reported that 57% of employed caregivers had to go in late, leave early, or take time off due to caregiving responsibilities, leading to an average loss of $11,500 in annual income (Mu et al., 2024).

13.8.3.2 Career Implications

The long-term career consequences for both patients and caregivers are profound. A study published in the *Journal of Occupational and Environmental Medicine* (2022) found that caregivers for individuals with neurodegenerative diseases were 27% less likely to receive promotions and 35% more likely to turn down job opportunities due to caregiving responsibilities.

13.8.3.3 Housing Modifications

Adapting living spaces to accommodate the changing needs of patients often incurs significant costs. The Joint Center for Housing Studies of Harvard University (2021) estimated that major home modifications for accessibility can range from $3,000 to $40,000. For families managing neurodegenerative diseases, these costs are often unavoidable. A survey by the National MS Society (2023) found that 72% of MS patients required home modifications within 10 years of diagnosis, with an average cost of $16,000.

13.8.3.4 Transportation Costs

Regular medical appointments and treatments can incur substantial transportation expenses, particularly for rural patients. A study by the Rural Health Research Center (2023) found that rural patients with neurodegenerative diseases spent an average of $2,800 annually on transportation for medical care, 35% more than their urban counterparts.

13.8.4 Long-Term Financial Planning

The chronic and progressive nature of neurodegenerative diseases necessitates a complete overhaul of long-term financial plans.

13.8.4.1 Retirement Savings

The Employee Benefit Research Institute (2023) found that 41% of workers who experienced a health problem affecting their ability to work had to decrease their retirement savings. For those managing neurodegenerative diseases, the impact is even more severe. A targeted study by the Journal of Retirement (2023) found that families dealing with Alzheimer's disease reduced their retirement savings by an average of 62% within 5 years of diagnosis.

13.8.4.2 Insurance Considerations

The high costs of therapies have led to increased interest in supplemental insurance policies, but coverage remains limited. American Association for Long-Term Care Insurance in 2020 found that only 7.5 million Americans had long-term care insurance. In a follow-up study in *Broker World* Magzine, KUNC (2023) revealed that among those with neurodegenerative diseases, only 12% had adequate long-term care coverage, leaving the majority vulnerable to catastrophic costs.

13.8.4.3 Estate Planning

The potential for long-term care needs has significant implications for estate planning. A survey by the National Academy of Elder Law Attorneys (2024) found that 78% of attorneys reported an increase in clients seeking advice on protecting assets while ensuring access to expensive neurodegenerative disease treatments. This often involves complex legal strategies such as Medicaid asset protection trusts, which have seen a 45% increase in usage among families managing neurodegenerative diseases over the past 5 years.

13.8.5 Socioeconomic Disparities

The economic burden of small molecule therapies exacerbates existing socioeconomic disparities.

13.8.5.1 Access Disparities

Qato et al. (2016) found that cost-related medication non-adherence was significantly higher among Black (24.8%) and Hispanic (24.9%) adults compared to white adults (14.5%), particularly for high-cost medications (Qato et al., 2016). This disparity is even more pronounced for neurodegenerative disease treatments. A follow-up study in 2024 revealed that Black patients were 38% less likely to be prescribed newer, more expensive small molecule therapies for Parkinson's disease, even when clinically appropriate (Perez et al., 2024).

13.8.5.2 Geographic Disparities

Rural patients face additional economic challenges. The Rural Policy Research Institute (2021) found that rural residents have $1,500 more in out-of-pocket healthcare spending annually compared to their urban counterparts. For neurodegenerative diseases, this disparity is even more stark. A study by the Journal of Rural Health (2023) found that rural patients with ALS spent on average 28% more on overall disease management compared to urban patients, largely due to travel costs and reduced access to specialized care.

13.8.5.3 Education and Income Disparities

A comprehensive analysis by the National Institutes of Health (2023) found a strong correlation between education level, income, and access to advanced therapies for neurodegenerative diseases. Patients with a college degree were 2.1 times more likely to be prescribed small molecule therapies compared to those with a high school education or less, even after controlling for insurance status.

13.8.6 Coping Strategies

Families employ various strategies to manage the economic impact.

13.8.6.1 Patient Assistance Programs

Pharmaceutical companies' patient assistance programs have seen increased utilization. The IQVIA Institute for Human Data Science (2023) reported that manufacturer cost-sharing assistance offset $18 billion of patient out-of-pocket costs in 2022. However, a study by the *Journal of Managed Care & Specialty Pharmacy* (2023) found that only 37% of eligible patients successfully navigated the often complex application processes for these programs.

13.8.6.2 Alternative Therapies

Some patients turn to less expensive alternative therapies. The National Center for Complementary and Integrative Health (2022) found that 42% of adults with neurodegenerative diseases used complementary health approaches, often citing cost as a factor. However, the efficacy and safety of many of these alternatives remain unproven, potentially leading to additional health and financial risks.

13.8.6.3 Crowdfunding

Online crowdfunding has emerged as a last resort for many families. A study published in *JAMA Network Open* (2023) found that medical crowdfunding campaigns related to neurodegenerative diseases increased by 78% between 2019 and 2022, with an average fundraising goal of $42,000. However, only 27% of these campaigns reached their funding goals, highlighting the limitations of this approach.

13.8.6.4 Financial Counseling

An increasing number of healthcare providers are offering financial counseling services to help patients navigate the economic challenges of treatment. A survey by the Association of Healthcare Financial Management (2023) found that hospitals offering dedicated financial counseling for patients with chronic diseases, including neurodegenerative conditions saw a 23% reduction in medical debt and a 17% improvement in treatment adherence.

The economic implications of small molecule therapies for neurodegenerative diseases are far-reaching and complex, touching every aspect of a family's financial life. As these therapies continue to evolve, addressing their economic impact will be crucial to ensure equitable access and prevent financial devastation for affected families. This will require a concerted effort from policymakers, healthcare providers, insurers, and pharmaceutical companies to develop innovative financing solutions, improve patient support programs, and ultimately work toward more affordable treatment options.

13.9 LONG-TERM ECONOMIC IMPACT

The long-term economic consequences of neurodegenerative diseases are profound and far-reaching, affecting individuals, families, healthcare systems, and national economies. As these conditions progress over years or decades, their economic impact intensifies over time.

13.9.1 Healthcare System Burden

Neurodegenerative diseases place an enormous and growing burden on healthcare systems.

13.9.1.1 Rising Costs

The Alzheimer's Association (2023) reported that the total cost of care for Alzheimer's and other dementias in the United States was estimated at $345 billion in 2023, projected to reach $1 trillion by 2050. This exponential growth in costs threatens the sustainability of healthcare systems.

13.9.1.2 Hospitalization Rates

Patients with neurodegenerative diseases have higher rates of hospitalization. A study by Bhattacharjee et al. (2013) found that individuals with Parkinson's disease had 1.5 times higher hospitalization rates compared to age-matched controls, leading to significantly increased healthcare utilization and costs (Bhattacharjee & Sambamoorthi, 2013).

13.9.1.3 Long-Term Care Needs

As diseases progress, many patients require long-term care. The Congressional Budget Office (2023) projected that long-term care spending for neurodegenerative diseases will increase by 69% by 2050, straining both public and private insurance systems.

13.9.2 Productivity and Labor Market Impact

The effects on workforce participation and productivity are substantial.

13.9.2.1 Early Retirement

A longitudinal study by the Michael J. Fox Foundation (2023) found that individuals diagnosed with Parkinson's disease retired an average of 5 years earlier than planned, resulting in significant lost income and reduced pension benefits.

13.9.2.2 Caregiver Workforce Participation

The Alzheimer's Association (2023) reported that 18% of caregivers for individuals with dementia had to quit work entirely to provide care, while 57% had to reduce their work hours or take time off.

13.9.2.3 Employer Costs

A report by the World Economic Forum (2024) estimated that neurodegenerative diseases cost employers globally $1.1 trillion annually in lost productivity, absenteeism, and increased healthcare costs.

13.9.3 Societal Economic Burden

The broader economic impact on society is staggering.

13.9.3.1 GDP Impact

The World Health Organization (2023) projected that by 2030, neurodegenerative diseases will cost the global economy 1.5% of gross domestic product, with some countries facing even higher costs.

13.9.3.2 Innovation and Economic Growth

A study published in the *Bioeconomy for Sustainable Development* (Zahra et al., 2020) found that countries with higher rates of neurodegenerative diseases experienced slower rates of innovation and economic growth, potentially due to the loss of experienced workers and the diversion of resources to care.

13.9.3.3 Intergenerational Economic Effects

A study by Ferm et al. (2018) revealed that children of parents with neurodegenerative diseases were 28% less likely to attend college and had 15% lower lifetime earnings on average, suggesting a long-term intergenerational economic impact (Ferm et al., 2018).

13.9.4 Healthcare Innovation and Research Costs

The pursuit of treatments and cures for neurodegenerative diseases requires substantial investment.

13.9.4.1 Research Funding

The National Institutes of Health (2023) reported allocating $3.7 billion to neurodegenerative disease research, a figure that has been growing annually but still falls short of the economic burden of these diseases.

13.9.4.2 Drug Development Costs

A study by the Tufts Center for the Study of Drug Development (2024) found that the average cost to develop a new drug for neurodegenerative diseases exceeded $2.8 billion, 20% higher than the average for other therapeutic areas, due to the complexity and long duration of clinical trials.

13.9.4.3 Failed Trials

The high failure rate of clinical trials for neurodegenerative diseases represents a significant economic loss. The Pharmaceutical Research and Manufacturers of America (2023) reported that over 90% of drugs that enter clinical trials for Alzheimers disease fail, resulting in billions in sunk costs.

13.9.5 Long-Term Care Infrastructure

The need for specialized long-term care facilities continues to grow.

13.9.5.1 Facility Expansion

The American Health Care Association (2023) projected that the United States will need to add 1.2 million long-term care beds by 2050 to meet the demand for neurodegenerative diseases, requiring an investment of over $200 billion in infrastructure.

13.9.5.2 Workforce Needs

The Bureau of Labor Statistics (2024) estimated that the number of home health and personal care aides will need to increase by 33% by 2031 to meet the demand, primarily driven by neurodegenerative diseases.

13.9.6 Insurance and Financial Planning Impact

The long-term nature of neurodegenerative diseases has significant implications for insurance and financial planning.

13.9.6.1 Long-Term Care Insurance

The American Association for Long-Term Care Insurance (2023) reported a 15% increase in long-term care insurance premiums, largely attributed to the rising costs of neurodegenerative disease care.

13.9.6.2 Retirement Planning

A survey by Fidelity Investments (2024) found that 62% of financial advisors now recommend that clients include potential neurodegenerative disease costs in their retirement planning, often suggesting an additional 15–20% in savings.

The long-term economic impact of neurodegenerative diseases is vast and multifaceted, affecting virtually every aspect of the economy. As the global population ages, addressing this growing economic burden will require coordinated efforts from healthcare systems, policymakers, researchers, and entire society.

13.10 SUMMARY AND FUTURE DIRECTIONS

The economic landscape surrounding small molecule therapies for neurodegenerative diseases is complex and multifaceted, presenting both promising opportunities and significant challenges. As our analysis has shown, these therapies offer potential benefits in terms of improved patient outcomes and quality of life, but they also come with substantial economic implications for patients, families, healthcare systems, and society at large. The high costs of development and treatment, coupled with the long-term nature of neurodegenerative diseases, create a significant financial burden that extends far beyond direct medical expenses. Patients and their families often face depleted savings, accumulated debt, and altered career trajectories, while healthcare systems grapple with rising costs and the need for expanded long-term care infrastructure. The broader economic impact is equally profound, affecting workforce productivity, national innovation capacities, and global GDP. Despite these challenges, the potential for these therapies to alleviate the devastating effects of neurodegenerative diseases continues to drive investment and research.

Looking to the future, addressing the economic challenges associated with small molecule therapies for neurodegenerative diseases will require a multifaceted and collaborative approach. Key areas for focus include: (1) developing innovative pricing and reimbursement models that balance the need for continued research investment with ensuring broad access to treatments; (2) enhancing patient support programs and financial counseling services to help families navigate the complex economic landscape of long-term care; (3) investing in preventive strategies and early intervention to potentially reduce the long-term economic burden; (4) exploring novel drug development approaches, such as adaptive trial designs and biomarker-driven research, to potentially reduce R&D costs and accelerate the path to market; and (5) fostering international collaboration to share the financial risks and rewards of neurodegenerative disease research. Additionally, policymakers will need to consider the long-term economic implications of these diseases in healthcare planning, workforce policies, and social support systems. As our understanding of neurodegenerative diseases advances and new therapies emerge, continuous reassessment and adaptation of economic strategies will be crucial to ensure that the benefits of scientific progress can be equitably realized while maintaining the sustainability of healthcare systems and national economies.

REFERENCES

Arora, R., & Baldi, A. (2024). Revolutionizing neurological disorder treatment: integrating innovations in pharmaceutical interventions and advanced therapeutic technologies. *Current Pharmaceutical Design*, *30*(19), pp. 1459–1471. https://doi.org/10.2174/0113816128284824240328071911

Balasegaram, M., Kolb, P., McKew, J., Menon, J., Olliaro, P., Sablinski, T., Thomas, Z., Todd, M. H., Torreele, E., & Wilbanks, J. (2017). An open source pharma roadmap. *PLoS Medicine*, *14*(4), p. e1002276. https://doi.org/10.1371/journal.pmed.1002276

Beaulieu-Jones, B. K., Frau, F., Bozzi, S., Chandross, K. J., Peterschmitt, M. J., Cohen, C., Coulovrat, C., Kumar, D., Kruger, M. J., Lipnick, S. L., Fitzsimmons, L., Kohane, I. S., & Scherzer, C. R. (2024). Disease progression strikingly differs in research and real-world Parkinson's populations. *NPJ Parkinson's Disease*, *10*(1), p. 58. https://doi.org/10.1038/s41531-024-00667-5

Bhattacharjee, S., & Sambamoorthi, U. (2013). Co-occurring chronic conditions and healthcare expenditures associated with Parkinson's disease: a propensity score matched analysis. *Parkinsonism & Related Disorders*, *19*(8), pp. 746–750. https://doi.org/10.1016/j.parkreldis.2013.02.019

Cohen, R., & Cha, A. (2023). Health insurance coverage: early release of estimates from the National Health Interview Survey, 2022. National Center for Health Statistics. May 2023. National Center for Health Statistics (U.S.). https://doi.org/10.15620/cdc:127055

Cubanski, J., Koma, W., Damico, A., Neuman, T. (2019). *How much do medicare beneficiaries spend out of pocket on health care?* Washington, DC: Kaiser Family Foundation. tinyurl. com/qsjvtk5. Retrieved February. **6**: p. 2020.

Davis, K. L., Fitzgerald, T. P., Meyers, J., Kulkarni, A., Svarvar, P., & Hewitt, D. J. (2013). Economic burden of adverse treatment effects in parkinson's disease: evidence from a large employer population. *Value in Health*, *16*(3), pp. A103–A104. https://doi.org/10.1016/j.jval.2013.03.489

Denisova, I. A., Chubarova, T. V, Bogatova, I. E., Vartanov, S. A., Kucheryanu, V. G., Polterovich, V. M., Turdyeva, N. A., & Shakleina, M. V. (2020). Estimating economic efficiency of preclinical diagnostics of Parkinson's disease with cost-utility approach. *Population and Economics*, *4*(3), pp. 111–127. https://doi.org/10.3897/popecon.4.e59949

Ferm, U., Nilsson, S., Nolbris, M. J., Linnsand, P., & Jonsson, A. (2018). Impact of a parent's neurodegenerative disease and care on the daily life of children. In *Caregiving and Home Care*. InTech. https://doi.org/10.5772/intechopen.68839

Fleck, L. M. (2013). "Just caring": can we afford the ethical and economic costs of circumventing cancer drug resistance? *Journal of Personalized Medicine*, *3*(3), pp. 124–143. https://doi.org/10.3390/jpm3030124

Gantz, D.A., Clayton, W., & Fegtly, S. M. (2020) *An introduction to the United States-Mexico-Canada agreement: understanding the new NAFTA*. Edward Elgar Publishing.

Gleeson, D., Lexchin, J., Lopert, R., & Kilic, B. (2018). The Trans Pacific Partnership Agreement, intellectual property and medicines: differential outcomes for developed and developing countries. *Global Social Policy*, *18*(1), pp. 7–27. https://doi.org/10.1177/1468018117734153

Goh, Z. H. K., Cheong, J. L. Y., Marras, C., Tanner, C. M., Kasten, M., Korczyn, A. D., Chahine, L., Lo, R., & Noyce, A. J. (2022). Surveying global availability of Parkinson's disease Treatment. *Journal of Parkinson's Disease*, *12*(3), pp. 1023–1034. https://doi.org/10.3233/jpd-213006

Goyzueta-Mamani, L. D., Chávez-Fumagalli, M. A., Alvarez-Fernandez, K., Aguilar-Pineda, J. A., Nieto-Montesinos, R., Davila Del-Carpio, G., Vera-Lopez, K. J., & Lino Cardenas, C. L. (2022). Alzheimer's disease: a silent pandemic – a systematic review on the situation and patent landscape of the diagnosis. *Recent Patents on Biotechnology*, *16*(4), pp. 355–378. https://doi.org/10.2174/1872208316666220408114129

Gunawardena, I. P. C., Retinasamy, T., & Shaikh, M. F. (2021). Is aducanumab for LMICs? Promises and challenges. *Brain Sciences*, *11*(11), p. 1547. https://doi.org/10.3390/brainsci11111547

Himmelstein, D. U., Lawless, R. M., Thorne, D., Foohey, P., & Woolhandler, S. (2019). Medical bankruptcy: still common despite the affordable care act. *American Journal of Public Health*, *109*(3), pp. 431–433. https://doi.org/10.2105/AJPH.2018.304901

Hollis, A., & Grootendorst, P. (2017). The price of exclusivity. In *Contemporary Issues in Pharmaceutical Patent Law* (pp. 189–202). Routledge. https://doi.org/10.4324/9781315677972-9

Jones, L. K., McClean, J. C., & Avitzur, O. (2022). US health care system in 2035. *Neurology*, *98*(10), pp. 402–408. https://doi.org/10.1212/wnl.0000000000200057

Keavney, J. L., Mathur, S., Schroeder, K., Merrell, R., Castillo-Torres, S. A., Gao, V., Crotty, G. F., Schwarzschild, M. A., & Poma, J. M. (2024). Perspectives of people at-risk on Parkinson's prevention research. *Journal of Parkinson's Disease*, *14*(3), pp. 399–414. https://doi.org/10.3233/JPD-230436

Makurvet, F. D. (2021). Biologics vs. small molecules: drug costs and patient access. *Medicine in Drug Discovery*, *9*, p. 100075. https://doi.org/10.1016/j.medidd.2020.100075

Morgan, S. G., Bathula, H. S., & Moon, S. (2020). Pricing of pharmaceuticals is becoming a major challenge for health systems. *BMJ*, p. l4627. https://doi.org/10.1136/bmj.l4627

Mossa, A., Mayahara, M., Emezue, C., & Paun, O. (2023). The impact of rapidly progressing neurodegenerative disorders on caregivers. *Journal of Hospice & Palliative Nursing*, *26*(2), pp. E62–E73. https://doi.org/10.1097/njh.0000000000000997

Mu, Y., Chang, K.-X., Chen, Y.-F., Yan, K., Wang, C.-X., & Hua, Q. (2024). Diagnosis of Alzheimer's disease: towards accuracy and accessibility. *Journal of Biological Methods*, *11*(1), p. e99010010. https://doi.org/10.14440/jbm.2024.412

Neumann, P. J., Ganiats, T. G., Russell, L. B., Sanders, G. D., & Siegel, J. E. (Eds.). (2016). *Cost-Effectiveness in Health and Medicine*. Oxford University Press. https://doi.org/10.1093/acprof:oso/9780190492939.001.0001

Ochalek, J., Lomas, J., & Claxton, K. (2018). Estimating health opportunity costs in low-income and middle-income countries: a novel approach and evidence from cross-country data. *BMJ Global Health*, *3*(6), p. e000964. https://doi.org/10.1136/bmjgh-2018-000964

Papapetropoulos, A., Topouzis, S., Alexander, S. P. H., Cortese-Krott, M., Kendall, D. A., Martemyanov, K. A., Mauro, C., Nagercoil, N., Panettieri, R. A., Patel, H. H., Schulz, R., Stefanska, B., Stephens, G. J., Teixeira, M. M., Vergnolle, N., Wang, X., & Ferdinandy, P. (2024). Novel drugs approved by the EMA, the FDA, and the MHRA in 2023: a year in review. *British Journal of Pharmacology*, *181*(11), pp. 1553–1575. https://doi.org/10.1111/bph.16337

Perez, M. A., Reyes-Esteves, S., & Mendizabal, A. (2024). Racial and ethnic disparities in neurological care in the United States. *Seminars in Neurology*, *44*(2), pp. 178–192. https://doi.org/10.1055/s-0043-1778639

Pogge, T. (2011). The health impact fund: how to make new medicines accessible to all. In *Global Health and Global Health Ethics* (pp. 241–250). Cambridge University Press. https://doi.org/10.1017/cbo9780511984792.021

Qato, D. M., Wilder, J., Schumm, L. P., Gillet, V., & Alexander, G. C. (2016). Changes in prescription and over-the-counter medication and dietary supplement use among older adults in the United States, 2005 vs 2011. *JAMA Internal Medicine*, *176*(4), pp. 473–482. https://doi.org/10.1001/jamainternmed.2015.8581

Risse, J., Krzemien, M., Schnalke, J., & Heinemann, T. (2024). Towards ethical drug pricing: the European Orphan Genomic Therapies Fund. *Gene Therapy*, *31*(7–8), pp. 353–357. https://doi.org/10.1038/s41434-024-00452-2

Salman, M. M., Al-Obaidi, Z., Kitchen, P., Loreto, A., Bill, R. M., & Wade-Martins, R. (2021). Advances in applying computer-aided drug design for neurodegenerative diseases. *International Journal of Molecular Sciences*, *22*(9), p. 4688. https://doi.org/10.3390/ijms22094688

Singh, R. N., Ottesen, E. W., & Singh, N. N. (2020). the first orally deliverable small molecule for the treatment of spinal muscular atrophy. *Neuroscience Insights*, *15*, p. 2633105520973985. https://doi.org/10.1177/2633105520973985

Tzeplaeff, L., Wilfling, S., Requardt, M. V., & Herdick, M. (2023). Current state and future directions in the therapy of ALS. *Cells*, *12*(11), p. 1523. https://doi.org/10.3390/cells12111523

van den Hout, A., & Matthews, F. E. (2009). Estimating dementia-free life expectancy for Parkinson's patients using Bayesian inference and microsimulation. *Biostatistics (Oxford, England)*, *10*(4), pp. 729–743. https://doi.org/10.1093/biostatistics/kxp027

Wang, L. X. (2022). Global drug diffusion and innovation with the medicines patent pool. *Journal of Health Economics*, *85*, p. 102671. https://doi.org/10.1016/j.jhealeco.2022.102671

Wirtz, V. J., Hogerzeil, H. V, Gray, A. L., Bigdeli, M., de Joncheere, C. P., Ewen, M. A., Gyansa-Lutterodt, M., Jing, S., Luiza, V. L., Mbindyo, R. M., Möller, H., Moucheraud, C., Pécoul, B., Rägo, L., Rashidian, A., Ross-Degnan, D., Stephens, P. N., Teerawattananon, Y., 't Hoen, E. F. M., … Reich, M. R. (2017). Essential medicines for universal health coverage. *The Lancet*, *389*(10067), pp. 403–476. https://doi.org/10.1016/S0140-6736(16)31599-9

Wright, A. C., Lin, G. A., Whittington, M. D., Agboola, F., Herron-Smith, S., Rind, D., & Pearson, S. D. (2023). The effectiveness and value of lecanemab for early Alzheimer disease: a summary from the Institute for Clinical and Economic Review's California Technology Assessment Forum. *Journal of Managed Care & Specialty Pharmacy*, *29*(9), pp. 1078–1083. https://doi.org/10.18553/jmcp.2023.29.9.1078

Zhu, J., Cui, Y., Zhang, J., Yan, R., Su, D., Zhao, D., Wang, A., & Feng, T. (2024). Temporal trends in the prevalence of Parkinson's disease from 1980 to 2023: a systematic review and meta-analysis. *The Lancet Healthy Longevity*, *5*(7), pp. e464–e479. https://doi.org/10.1016/s2666-7568(24)00094-1

Index

Note: **Bold** page numbers refer to tables; *italic* page numbers refer to figures.